Principles of
Fire Protection Chemistry

Principles of Fire Protection Chemistry

by Richard L. Tuve

470 Atlantic Avenue, Boston, MA. 02210

About the Author

Richard Larsen Tuve, D.Sc., has thirty-two years of experience with the U.S. Naval Research Laboratory where he was technical consultant for fire engineering matters and founder and director of the Navy's Fire Research Laboratory. In 1963 his discovery of the water-film-forming capacities of fluorocarbon surface-active agents on hydrocarbon brought about a threefold efficiency in fuel fire extinguishment by foam. He also initiated the research that proved the efficacy of potassium dry chemical. Dr. Tuve is a member of many prominent chemical, research, and scientific associations. As a result of his outstanding achievements in the fire protection field, Dr. Tuve, a long-time member of the NFPA Foam Committee, is a recipient of the National Fire Protection Association's Technical Committee Service Award; the NFPA award is only one of many prestigious awards he has received. Dr. Tuve is presently affiliated with the Applied Physics Laboratory of Johns Hopkins University where he is serving as consultant for fire research problems on a program sponsored by the National Science Foundation.

NFPA No. TXT-2
Standard Book No. 87765-080-2
Library of Congress No. 76-26781
Printed in U.S.A.

Dedication

The material in this book along with whatever usefulness it might attain no longer belongs to the author, and is hereby dedicated principally to the thoughtful and curious student and reader. It is also dedicated to my wife and helpmate, Maxine, whose continual faith, patience, literary acumen, and typing skill helped make this book possible.

Contents

Preface

In today's complicated civilization, individuals who are devoting their careers to the fire service profession must possess knowledges never before required. In order to better understand the complexities of the new technologies and products involved in fire fighting, they must continually further their knowledge of specialized areas such as chemistry. The need for this increased awareness is due to the fact that the steadily increasing development of complicated material hazards and their fire and toxic characteristics has far outdistanced the methods involved in older fire fighting skills such as the simple cooling attack of water on hot or burning surfaces. Therefore, the main objective of *Principles of Fire Protection Chemistry* is to help provide the fire science student with a knowledge of the chemistry of materials and their physical properties as these subjects relate to the fire services.

The teaching approach used in *Principles of Fire Protection Chemistry* is an inductive approach — an approach that utilizes explanation and justification. Thus, the student need not possess an extensive background in chemistry in order to fully grasp the general principles of chemistry presented herein.

The text, subject matter, and arrangement of *Principles of Fire Protection Chemistry* have been especially designed for use in community colleges, universities, and other schools that offer fire science courses and programs, as well as for use in fire service personnel training situations. Because of this, all material has been prepared to reflect the many concerns of fire service specialists and safety officials involved in fire protection, fire prevention, and fire fighting.

In portions of this text, a kinship exists between stated material and the National Fire Protection Association's *Fire Protection Handbook*, 14th Edition. Because the *Fire Protection Handbook* contains objective information concerning fire-related subjects — such information having been contributed to by experienced authorities in specialized areas of expertise — reference to it will greatly enhance the study of the principles of chemistry contained herein.

The goal of the author during the writing of *Principles of Fire Protection Chemistry* was the effort to promote the union and companionship of science, technology, and the scientific method with methods of fire protection of all types, and with the needs of the fire service.

Richard L. Tuve

Acknowledgments

The author is highly indebted to his many professional associates over a span of almost 50 years of thinking, doing, and being educated in the fields of chemistry and fire protection. The contributions of each of these associates, no matter how large or small, have helped determine the directions of the author's zeal and enthusiasm in both professions.

Grateful thanks are tendered to several fire science instructors of the author's acquaintance who reviewed portions of the manuscript and offered constructive criticisms as to content and presentation. Special thanks are due to Dr. Lawrence W. Hunter of the Johns Hopkins University's Applied Physics Laboratory for his careful review of several technical chapters. Gratitude is also due to Keith Tower, Editor, and Chester I. Babcock, Director, Editorial Division, of the NFPA for their careful editing and encouraging reviews of the text during its preparation. Many commercial and publication sources of text material are deservedly thanked for their permission to include text and appendix material in this book.

Introduction

An old-time fire fighter once remarked, "You can't fight fire with books." However, without books and the knowledge gained from them, we could never hope to successfully combat the new and different types of fires we had never seen or experienced. Accordingly, this text has been planned as an aid to help further the technical and chemical knowledge of the specialist in the field of fire science by presenting the principles of fire protection chemistry.

In the past, the words "chemistry" and "chemicals" may have conjured visions of strange and dangerous compounds or powders that reacted violently. Chemists were often pictured as white-coated individuals who presided over mysterious glass beakers and tubes containing bubbling, varicolored mixtures of uncertain — but always dangerous — components. Chemistry itself, with its numbers and complicated formulas, may have made the science of chemistry seem incomprehensible to the nonscientific person. Today, after many generations of education and understanding, there is an increasing awareness that all substances are chemicals of one kind or another, and that these chemicals are not all deadly, dangerous, or mysterious. Simultaneously, there is a growing awareness that chemistry is a study of the properties of all of the materials of the world, as well as of how these materials react under physical or chemical change.

The teaching thrust of this text is based on the fact that the study of chemistry as it applies to the protection of materials and persons from unwanted or hostile fire requires both a basic and an applied type of learning. By introducing a basic study of the facts concerning the composition of the world around us and the forces existing in it, an understanding of how materials are formed and how they react is acquired. By applying this basic information to particular cases involving fire, the acquired knowledge of chemistry is expanded to incorporate specific examples of fire in materials. Such applied learning helps broaden the student's basic knowledge and familiarity with the many ways in which chemistry is involved in the study of the science of fire.

From a descriptive viewpoint, the chemistry of fire is concerned with the chemistry of materials ("chemicals") and their reactions when exposed to heat, high temperatures, and ignition sources while they are surrounded by or contacted by oxidizing atmospheres such as air, oxygen, oxygen-yielding substances, or even other reactants that promote heat evolution. When substances are acted on by both heat and oxygen or air, they change in their chemical constituents or composition. They become new chemical compounds with different characteristics and properties. Obviously, all materials do not undergo the "fire" or combustion reaction in the same manner. Descriptions of such reactions are an integral part of this text.

The text begins with a presentation of the basic structure of atoms and molecules — the "building blocks" of the universe. How these "blocks" are put together, or "built," and how chemistry looks at these structures and their reactivity, one with another, is introduced. Basic interacting forces in the material world, such as heat and energy, are dealt with; both chemistry and fire are highly involved with these forces and their production. The many standard methods for describing and referring to substances and the relationship of this terminology to chemical reactions, some of which are fire reactions, are included. Also, this text deals with the purely physical forces of heat flow and how the presence or absence of vapor or other characteristics that occur in our world are related to the effects of fire and combustion, and *vice versa*. Individual cases of the different classes of materials and the properties of each that help make them either more or less capable of ignition and combustion, and why this is so, are presented.

In this text, as in any scientific study, numbers constitute the basic form for describing the dimensions or limits of the area or conditions under which discussion of the subject is more meaningful. Numbers and units of measurements are the fundamental language for indicating extent or degree of physical (or chemical) size. In the study of the fire sciences, numbers and units of measurement are needed to provide a description of the extent of the fire situation or the relative conditions under which operations must be conducted. Fire science is deeply involved with the use of numbers to describe water pressures, fire sizes, amounts of fuel, time intervals, and temperatures produced. Without numbers and units of measurement, all descriptions of conditions under which observations are made become meaningless and without dimension or value. Similarly, the study of the chemistry of fire must use numbers and descriptive units of measurement to denote relative conditions under which reactions occur. Therefore, just as the fire science student must understand all of the chemicals of our daily lives in order to know how each can burn or be prevented from burning, so must that fire science student also understand the units of measurement needed to communicate relevant facts regarding size, volume, capacity, and temperature. These numbers and units of measurement are the fundamental language for communicating the extent or the degree of physical or chemical size. Thus, in this text, all units of measurement of the customary type (English: foot/pound/second) are followed by numbers and units in parentheses. The latter figures are given in SI (Système International) units of measurement. The SI system is a relatively new system for an international language of units. It is similar to the metric system of units, and is explained more fully in Chapter 2, "The Language of Science."

Because it is impossible to completely explain and clarify all of the facets of all of the technical problems involved in any scientific or technical subject, a completely comprehensive study of such subjects can never be totally contained within the pages of any single text. This is particularly so in the field of fire science in which the diversity of materials and the extreme variability of fire situations can give rise to questions that require assistance from more than one source. For this reason, the student should become familiar with many other sources of information concerning the reactions of materials in hazardous or fire situations. References of many types should be consulted before data and information are finally assembled and conclusions made. It has been rightfully stated that, "There is no better method for the construction of knowledge than by its careful extraction from authoritative sources." Therefore, each chapter in this text ends with a selected list of reference materials given in conjunction with the subject matter of the accompanying chapter. Some of these references constitute only an enlargement of the subject covered in the chapter; others give detailed treatment of the specifics contained in the chapter.

The final chapters of the text contain useful information concerning the ways we can describe the components involved in a fire, and the chemistry and physics involved when fires propagate and are extinguished. The text culminates with a presentation of the mechanisms and tools by which different extinguishing agents accomplish their tasks.

1

The Basic "Stuff" of Our World

MATTER AND ENERGY

To accurately describe our physical world, it is necessary to describe its basic constituents — or the "stuff" of which it is made. To describe its basic constituents in simple terms, only two categories are needed: these two categories are "matter" and "energy." Everything that is tangible can be classified under these two categories; that is, everything that can be sensed in some way.

There are a great variety of substances around us such as water, air, wood, stone, and living organisms of all kinds. These substances belong to the broad class called matter. The two basic characteristics possessed by all forms of matter are: (1) that it occupies space, and (2) that it has weight. Matter can always be measured in some dimension of size, volume, and weight or mass, and can be described according to its properties.

Although the properties of gases, liquids, and solids are different from each other, they are all matter. The properties of matter can be experienced by our senses of sight, smell, taste, or touch. Chemistry's major focus of the study of matter centers on the changes in matter's properties.

Energy is not as easily defined as matter because our senses can only describe what energy does. Energy can be defined as the capacity or ability to do work; however, we can only "see" work as the manifestation of energy. For example, although we cannot "see" the energy stored in a stretched-out rubber band, we can see, and perhaps hear and feel, the work produced by the stored energy when the stretched-out rubber band is released. It is important to remember that energy is the basic force that brings about both chemical changes and physical changes in matter. There are many forms of energy, some of which will be taken up in more detail later in this chapter.

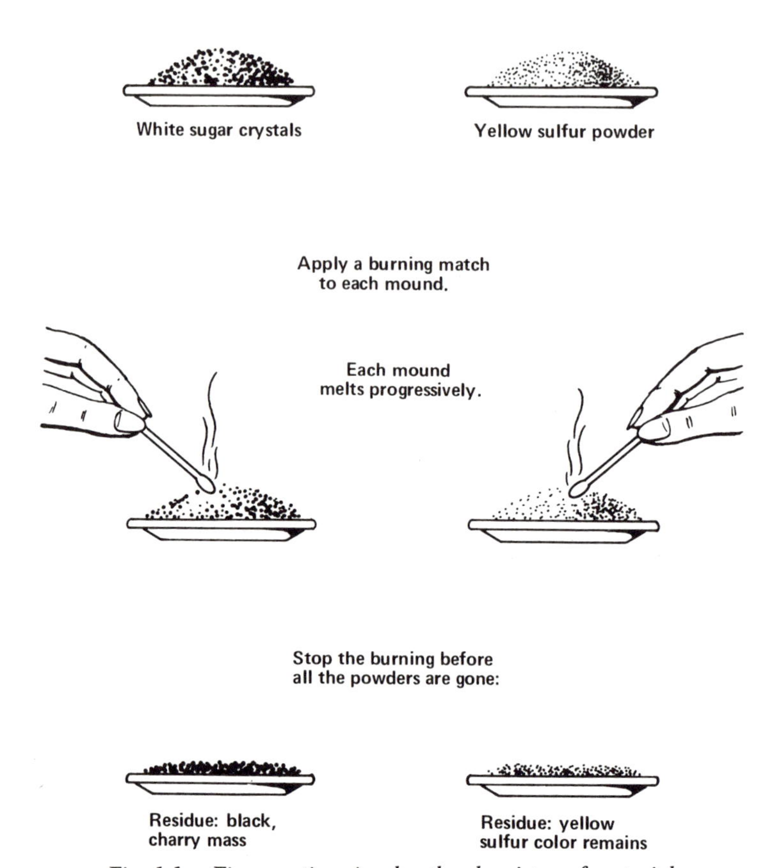

Fig. 1.1. Fire reactions involve the chemistry of materials.

CHEMICAL AND PHYSICAL CHANGES

Matter can undergo two distinct kinds of changes: chemical changes and physical changes. Chemical changes are the kind of deep-seated, almost irreversible changes that take place in a substance when the substance's properties and characteristics completely change to new ones. For example, the process of burning or combustion is a chemical process or a chemical change. When charcoal, wood, or paper burns, it unites with oxygen in the air and changes from its original solid form into a colorless, odorless gas named carbon dioxide. Any ash that is left does not resemble the original solid in shape, texture, or form. Similarly, when iron rusts, it combines with oxygen and becomes a reddish, flaky solid that differs from the original silvery gray iron metal in both form and properties. The iron has undergone a chemical change; it cannot be restored to its original metallic form without undergoing extensive processes.

Physical changes are simpler changes, and are usually connected with what the physicist calls changes in state. When sugar is dissolved in water, a syrupy liquid solution is formed. The white, crystalline sugar is changed from a solid state to a solution that, although it looks like water, tastes quite different. When water is boiled, it changes from a liquid state to a vapor or gaseous state that cannot be distinguished from the gases in ordinary air unless it is slightly cooled to produce steamy fog. Unlike chemical changes, physical changes can be easily reversed by simple physical processes such as cooling or evaporating. (See Fig. 1.2.)

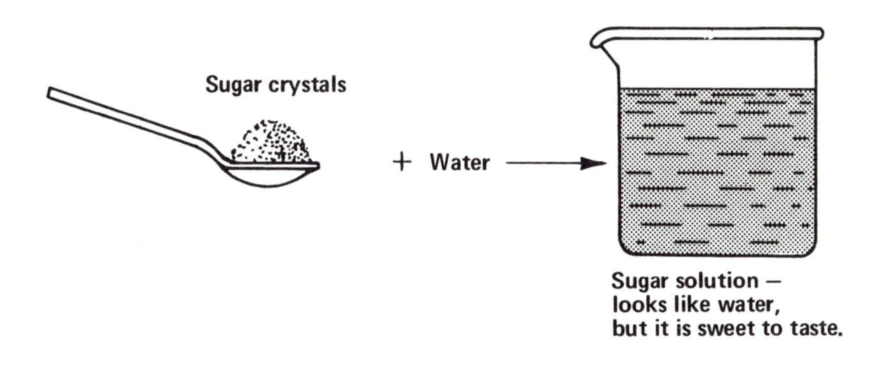

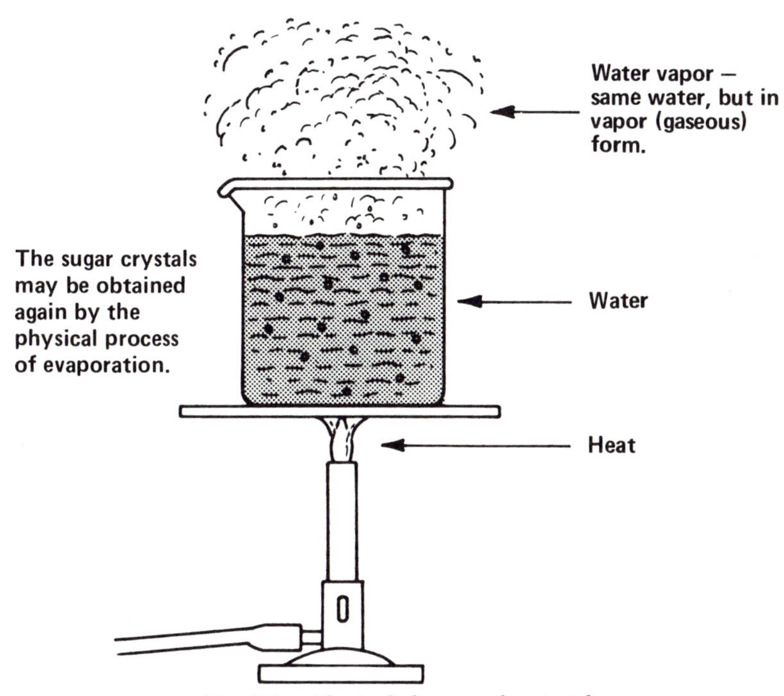

Fig. 1.2. Physical changes of materials.

ATOMS

All matter is composed of elements of one kind or another. Some elements in matter are combined with each other (such as carbon dioxide gas, which is composed of the element carbon and the element oxygen). Other matter may consist of only one element (such as metallic iron, which consists of the element iron only). All elements are composed of atoms. Atoms are the smallest particles that together constitute the matter of an element or substance. Each atom of an element is spherical in shape and is extremely small: even with today's most powerful magnification methods, the atom cannot be seen. To better comprehend the size of an atom, imagine a piece of matter so small that if it was placed in a straight line with other similarly sized pieces of matter, about four million pieces would be required to reach across the head of a common pin.

It was originally thought that the atom was indivisible and was the smallest particle that existed. Today we know that the atom is really a system of particles consisting of a dense core or nucleus around which other particles are continually moving in orbits or circular motions in well-defined paths. Similar to this, but on a much larger scale, is the solar system with the regular and constant movements of the earth and the other planets around the sun.

THE NUCLEUS AND PLANETARY ELECTRONS

The weight or mass of an atom (which varies in every element) is almost totally concentrated in the atom's nucleus. This nucleus (or core) consists of protons and neutrons. Protons are particles that carry a positive electrical charge. An equal number of neutrons, which are electrically neutral particles with no electrical charge, are also contained within the nucleus. Outside the nucleus are different levels in which other particles, called electrons, are rotating in orbits similar to the planetary form. These electrons carry a negative electrical charge. The energy levels or orbits that these particles exist in are called shells. The number of shells varies from one element to another.

The properties of elements and their inclination to react chemically is determined by the number of electrons in the outermost shell of an element. The innermost shell of all elements — the orbit closest to the nucleus — always contains a maximum of two electrons (except in hydrogen, which only contains one). Proceeding outward, each shell of an element contains more than the minimum number. If the outer shell contains an odd number of electrons, or if there are less than eight electrons, the element will have a tendency to react with other elements. Figure 1.3 illustrates the reactivity of seven elements, from the nonreactive, rather light element known as neon, to more reactive elements such as aluminum and calcium.

In fire situations there are many instances that support the statement that the number of the electrons in the outer shell of an element determines the element's activity toward combination with another element. For example, a carbon (such as coal, charcoal, wood, etc.) has only four electrons in the outer shell,

An atom of neon gas

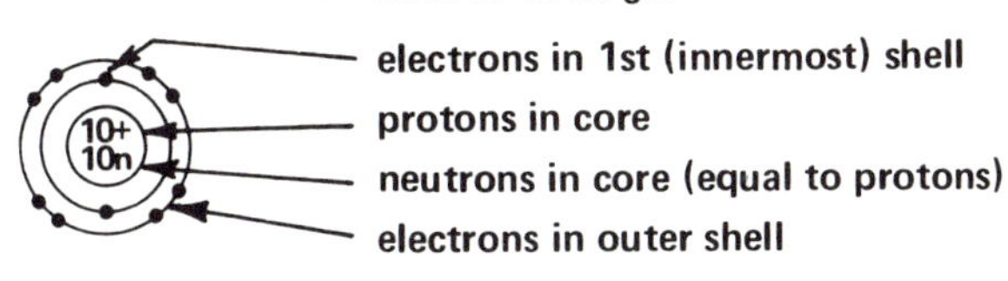

A nonreactive element,
equal number of 8 electrons
in outer shell.

An atom of aluminum metal

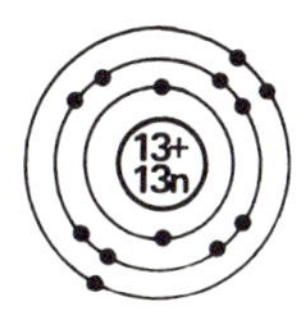

A reactive element,
unequal number of electrons
in outer shell.

An atom of calcium metal

A reactive element,
electrons in outer shell
off-balance and less
than 8.

An atom of carbon nonmetal

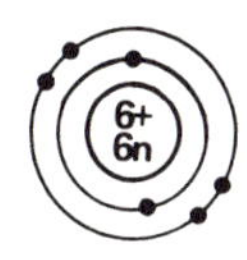

A reactive element,
less than 8 electrons
in outer shell.

An atom of oxygen gas

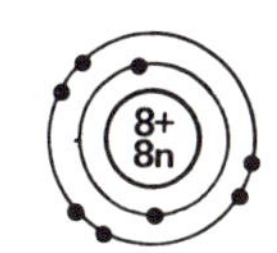

A reactive element,
less than 8 electrons
in outer shell.

An atom of magnesium metal

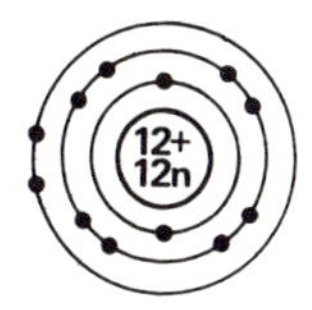

A reactive element,
electrons in outer shell
off balance and less
than 8.

An atom of hydrogen gas

A reactive element,
unequal number of electrons
in outer shell and less than 8.

Fig. 1.3. The nucleus and planetary electron structure of several elements.

and oxygen has only six electrons in the outer shell. As previously stated, eight electrons are needed to make an element nonreactive; therefore, oxygen is ready to react. If the two elements are brought together, a reaction takes place easily when a small input of energy in the form of an ignition source (activation energy) is added to start the reaction.

A similar example of this type of reaction can be seen with magnesium metal and oxygen. In this instance, the element magnesium has only two electrons in the outer shell and is, therefore, ready to react with oxygen at any time. Hydrogen, with its one electron in the outer shell, is also quite reactive, combining with oxygen if even a small spark is present to start the reaction. When these two elements combine, water is formed.

NUCLEAR REACTIONS AND ISOTOPES

All of the chemical reactions that have been discussed are associated with the electrons contained in the outer shell of each of the elements. Another way of saying this is that energy transfer (chemical reactivity) is easily accomplished by means of a gain or a loss of the amount of electrons in the outer shell. As the number of electrons in the inner shells of an atom are changed, or "lifted," to the next outer shell, the task requires more and more energy. When the distribution of protons and neutrons or their number within the dense core or nucleus of an atom is changed, a large amount of energy is needed. When certain elements such as uranium are bombarded by high-energy neutrons, a splitting of the nucleus takes place resulting in the formation of new elements. As this nuclear reaction occurs, large amounts of energy are given off; radioactive particles result from these high-energy reactions.

When certain elements are acted upon by particles such as neutrons, which can be emitted during the nuclear reaction of uranium, they become transmuted (changed or altered in form) into other elements that may not have the normal balance of protons and neutrons in their nucleus. Some elements occur in nature with a mixture of atoms having different atomic weights; these unbalanced elements are called isotopes. Although isotopes are chemically similar to their parent element, they contain different numbers of neutrons in their nucleus. As a result, they are heavier or lighter than they should be. The radioactive isotopes are of most importance because they are used as radiotracers for studying many chemical and biological phenomena, and for medical purposes. Because radioactive isotopes emit radioactive particles, they require special handling precautions.

ATOMIC WEIGHTS

All atoms have a weight or mass; since they constitute the matter of our world, they must possess weight and occupy a space. To understand the chemistry of materials, the student needs to know both the relative weights of atoms and the amount of atoms of one element that combine with an amount of atoms of another element to result in a new compound. Without these facts, it is impossible to describe what took place in any chemical reaction and how much was involved in the reaction. (See Table 1.1.)

Table 1.1. Table of Relative Atomic Weights

(Based on the Atomic Mass of $C^{12} = 12$)

NAME	SYMBOL	ATOMIC NUMBER	ATOMIC WEIGHT	NAME	SYMBOL	ATOMIC NUMBER	ATOMIC WEIGHT
Actinium	Ac	89	....	Krypton	Kr	36	83.80
Aluminum	Al	13	26.9815	Lanthanum	La	57	138.91
Americium	Am	95	....	Lawrencium	Lw	103	....
Antimony	Sb	51	121.75	Lead	Pb	82	207.19
Argon	Ar	18	39.948	Lithium	Li	3	6.939
Arsenic	As	33	74.9216	Lutetium	Lu	71	174.97
Astatine	At	85	....	Magnesium	Mg	12	24.312
Barium	Ba	56	137.34	Manganese	Mn	25	54.9380
Berkelium	Bk	97	....	Mendelevium	Md	101	....
Beryllium	Be	4	9.0122	Mercury	Hg	80	200.59
Bismuth	Bi	83	208.980	Molybdenum	Mo	42	95.94
Boron	B	5	10.811	Neodymium	Nd	60	144.24
Bromine	Br	35	79.909	Neon	Ne	10	20.183
Cadmium	Cd	48	112.40	Neptunium	Np	93	....
Calcium	Ca	20	40.08	Nickel	Ni	28	58.71
Californium	Cf	98	....	Niobium	Nb	41	92.906
Carbon	C	6	12.01115	Nitrogen	N	7	14.0067
Cerium	Ce	58	140.12	Nobelium	No	102	....
Cesium	Cs	55	132.905	Osmium	Os	76	190.2
Chlorine	Cl	17	35.453	Oxygen	O	8	15.9994
Chromium	Cr	24	51.996	Palladium	Pd	46	106.4
Cobalt	Co	27	58.9332	Phosphorus	P	15	30.9738
Copper	Cu	29	63.54	Platinum	Pt	78	195.09
Curium	Cm	96	....	Plutonium	Pu	94	....
Dysprosium	Dy	66	162.50	Polonium	Po	84	....
Einsteinium	Es	99	....	Potassium	K	19	39.102
Erbium	Er	68	167.26	Praseodymium	Pr	59	140.907
Europium	Eu	63	151.96	Promethium	Pm	61	....
Fermium	Fm	100	....	Protactinium	Pa	91	....
Fluorine	F	9	18.9984	Radium	Ra	88	....
Francium	Fr	87	....	Radon	Rn	86	....
Gadolinium	Gd	64	157.25	Rhenium	Re	75	186.2
Gallium	Ga	31	69.72	Rhodium	Rh	45	102.905
Germanium	Ge	32	72.59	Rubidium	Rb	37	85.47
Gold	Au	79	196.967	Ruthenium	Ru	44	101.07
Hafnium	Hf	72	178.49	Samarium	Sm	62	150.35
Helium	He	2	4.0026	Scandium	Sc	21	44.956
Holmium	Ho	67	164.930	Selenium	Se	34	78.96
Hydrogen	H	1	1.00797	Silicon	Si	14	28.086
Indium	In	49	114.82	Silver	Ag	47	107.870
Iodine	I	53	126.9044	Sodium	Na	11	22.9898
Iridium	Ir	77	192.2	Strontium	Sr	38	87.62
Iron	Fe	26	55.847	Sulfur	S	16	32.064

(Continued)

Table 1.1. Table of Relative Atomic Weights (Continued)

NAME	SYMBOL	ATOMIC NUMBER	ATOMIC WEIGHT	NAME	SYMBOL	ATOMIC NUMBER	ATOMIC WEIGHT
Tantalum	Ta	73	180.948	Tungsten	W	74	183.85
Technetium	Tc	43	....	Uranium	U	92	238.03
Tellurium	Te	52	127.60	Vanadium	V	23	50.942
Terbium	Tb	65	158.924	Xenon	Xe	54	131.30
Thallium	Tl	81	204.37	Ytterbium	Yb	70	173.04
Thorium	Th	90	232.038	Yttrium	Y	39	88.905
Thulium	Tm	69	168.934	Zinc	Zn	30	65.37
Tin	Sn	50	118.69	Zirconium	Zr	40	91.22
Titanium	Ti	22	47.90				

The lengthy processes by which atomic weights have been accurately determined have involved a large amount of experimental chemistry covering a period of almost a century. For purposes of this textbook, they will not be detailed herein. However, it is important to know that our knowledge of the weights of atoms of all the elements was arrived at by: (1) carefully measuring the amounts of elements that reacted with each other to form new substances, and (2) analyzing certain substances for the amount of each element that they contained.

Thus, our knowledge of atomic weights is purely relative. We know the weights of the atoms only by comparing the atoms with each other. At first, the element hydrogen was used as a basis for comparison with an atomic weight of 1.00; then oxygen was internationally agreed upon as a standard with an atomic weight of 16.00. In 1961 it was decided that carbon, with an atomic weight of 12.00, would be the standard. Besides these elements, water, with its components of simple atoms of hydrogen and an atom of oxygen, played an important role in the laborious process of determining atomic weights.

ATOMIC NUMBERS AND THE PERIODIC TABLE

As scientists continued to work with the relative weights of the atoms, their structures, and their properties, they began to notice certain relationships that were important. When the elements are arranged in an order of increasing atomic weight and are then assigned whole numbers in that order, beginning with hydrogen as No. 1 and helium as No. 2, it can be proven by experiment that we now are denoting the number of positive charges (protons) on the nucleus of an atom of that element. Since all elements contain an equal number of neutrons in the nucleus (except for their isotopes), these are also equal to that number. Since an atom must have an equal number of planetary electrons in orbits around the nucleus as it does protons, the number of the element is also the number of planetary electrons in the structure of an atom of that element. This is called the atomic number of the element, and is useful in chemical studies. (See Table 1.1.)

Another outcome of these studies of elements concerned the similarity with which they reacted, both chemically and physically. Scientists began to arrange the elements in groups according to their properties, and the Periodic Table is the final result of these studies. (See Table 1.2.)

MOLECULES: DIATOMIC AND MONATOMIC ELEMENTS

As we progress from the small subdivision of matter, the atom, we come to the next larger size of matter, molecules, which are groups of two or more atoms. This classification of substances is a large one, as it includes all of the materials in which one element has reacted with another element to form a compound. Each small subdivision of such a compound is referred to as a molecule, not an atom. A molecule includes those elements that are gaseous at ordinary temperatures, and are reactive. These are not compounds; however, it has been shown that hydrogen, nitrogen, oxygen, fluorine, chlorine, and bromine are molecules consisting of two atoms of the element. We call these diatomic molecules or elements. All the elements that are solids at ordinary temperatures are found to react as single atoms, called monatomic elements.

All chemical combinations of elements or compounds such as sugar, salt, cotton, plastics, etc., are molecules that contain combined atoms of the elements. Molecules of sugar contain atoms of carbon, hydrogen, and oxygen; salt consists of molecules that are composed of atoms of the element sodium and the element chlorine. Cotton and plastics contain a mixture of molecules of different compounds and, as will be seen later in this text, are more difficult to define.

Compounds are formed in definite proportions from the elements; their molecular weights (of each molecule) may be determined by adding the atomic weights of the elements or atoms of each element comprising the element. Definite identification of the compound must be made before this can be done. For example, ordinary bicarbonate of soda (sodium bicarbonate) contains a certain definite ratio of sodium atoms to hydrogen atoms to carbon atoms to oxygen atoms. The formula for this compound is: $NaHCO_3$. This means that one atom each of sodium, hydrogen, and carbon, and three atoms of oxygen were used to form this compound. Once this information is determined, the molecular weight of sodium bicarbonate can be computed.

$$\begin{aligned} Na &= 23. \text{ (atomic weight)} \\ H &= 1. \\ C &= 12. \\ 3 \text{ O's at 16. each} &= \underline{48.} \\ \text{Total } & 84. = \text{the molecular weight of } NaHCO_3 \end{aligned}$$

THE KINETIC-MOLECULAR THEORY OF STATE

The modern concept views molecules as particles that are in constant motion; however, the scope of this motion is so small that it cannot be seen. Figure 1.4

Table 1.2. Periodic Table of the Elements

I	II											III	IV	V	VI	VII	VIII
1 **H** Hydrogen												Metalloids and Nonmetals					2 **He** Helium
3 **Li** Lithium	4 **Be** Beryllium					METALS						5 **B** Boron	6 **C** Carbon	7 **N** Nitrogen	8 **O** Oxygen	9 **F** Fluorine	10 **Ne** Neon
11 **Na** Sodium	12 **Mg** Magnesium					Transition Metals						13 **Al** Aluminum	14 **Si** Silicon	15 **P** Phosphorus	16 **S** Sulfur	17 **Cl** Chlorine	18 **Ar** Argon
19 **K** Potassium	20 **Ca** Calcium	21 **Sc** Scandium	22 **Ti** Titanium	23 **V** Vanadium	24 **Cr** Chromium	25 **Mn** Manganese	26 **Fe** Iron	27 **Co** Cobalt	28 **Ni** Nickel	29 **Cu** Copper	30 **Zn** Zinc	31 **Ga** Gallium	32 **Ge** Germanium	33 **As** Arsenic	34 **Se** Selenium	35 **Br** Bromine	36 **Kr** Krypton
37 **Rb** Rubidium	38 **Sr** Strontium	39 **Y** Yttrium	40 **Zr** Zirconium	41 **Nb** Niobium	42 **Mo** Molybdenum	43 **Tc** Technetium	44 **Ru** Ruthenium	45 **Rh** Rhodium	46 **Pd** Palladium	47 **Ag** Silver	48 **Cd** Cadmium	49 **In** Indium	50 **Sn** Tin	51 **Sb** Antimony	52 **Te** Tellurium	53 **I** Iodine	54 **Xe** Xenon
55 **Cs** Cesium	56 **Ba** Barium	57 **La** Lanthanum	72 **Hf** Hafnium	73 **Ta** Tantalum	74 **W** Tungsten	75 **Re** Rhenium	76 **Os** Osmium	77 **Ir** Iridium	78 **Pt** Platinum	79 **Au** Gold	80 **Hg** Mercury	81 **Tl** Thallium	82 **Pb** Lead	83 **Bi** Bismuth	84 **Po** Polonium	85 **At** Astatine	86 **Rn** Radon
87 **Fr** Francium	88 **Ra** Radium	89 **Ac** Actinium															
Lanthanides (Rare Earth Metals)				58 **Ce** Cerium	59 **Pr** Praseodymium	60 **Nd** Neodymium	61 **Pm** Promethium	62 **Sm** Samarium	63 **Eu** Europium	64 **Gd** Gadolinium	65 **Tb** Terbium	66 **Dy** Dysprosium	67 **Ho** Holmium	68 **Er** Erbium	69 **Tm** Thulium	70 **Yb** Ytterbium	71 **Lu** Lutetium
Actinides				90 **Th** Thorium	91 **Pa** Protoactinium	92 **U** Uranium	93 **Np** Neptunium	94 **Pu** Plutonium	95 **Am** Americium	96 **Cm** Curium	97 **Bk** Berkelium	98 **Cf** Californium	99 **Es** Einsteinium	100 **Fm** Fermium	101 **Md** Mendelevium	102 **No** Nobelium	103 **Lw** Lawrencium

Molecules of a solid. Closely packed in a crystalline pattern. Vibrating back and forth and up and down.

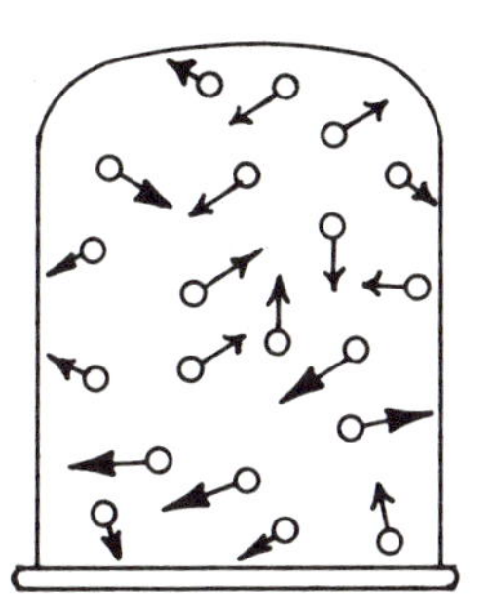

Molecules of a gas or vapor. Relatively far apart, moving randomly. Colliding with each other and the walls of the vessel.

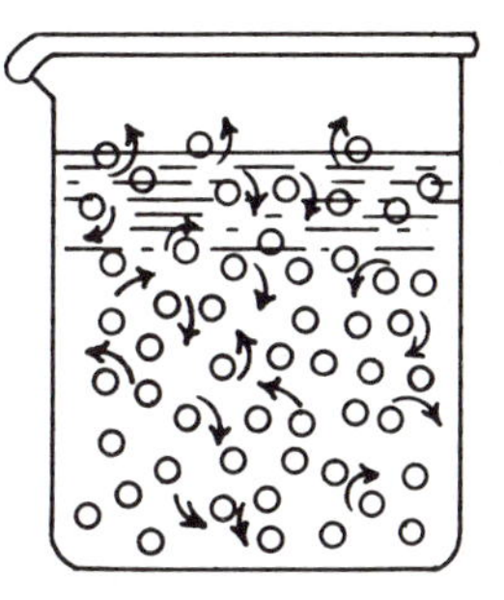

Molecules of a liquid. Almost as closely packed as in a solid. Gliding over one another in a random manner and escaping from the surface (vapor pressure).

In each state -- solid, liquid, and gas -- as the temperature increases, the kinetic energy of the molecules increases and they may change from a solid to a liquid to a gaseous state.

Fig. 1.4. The kinetic-molecular theory of state.

shows a hypothetical picture of the kinetic-molecular theory of motion when molecules of a gas like hydrogen are imprisoned in a vessel.

CHEMICAL SYMBOLS

The elements that have been discussed in this chapter have a variety of names that came from many sources over the last 200 or so years. To write out these names each time a particular element, molecule, or compound is referred to would be tedious. Therefore, chemists have devised a system of letter symbols for each element. The Periodic Table of the Elements and their atomic numbers, symbols, and weights are given in Tables 1.1 and 1.2. The symbols in these tables have been agreed upon internationally, and their use in equations and formulas of compounds constitutes a sort of universal language.

Some symbols seem to bear no similarity to the name of the element. This has happened because of the early usage of another name for the element; the symbol of the earlier name was retained so that early writings about that element could be more easily translated or used. For example, K, the symbol for potassium, originated from the early Latin name for potassium, Kalium. Similarly, Na, for sodium, originated from the Latin name for sodium, Natrium.

Chemical formulas are always written with whole numbers before the symbols to indicate numbers of molecules, and with subscripts following the constituent elements to indicate numbers of atoms. Thus, 2 $KClO_3$ (potassium chlorate) means two molecules of potassium chlorate, which consists of one atom of potassium, one atom of chlorine, and three atoms of oxygen.

To read a symbol aloud, $Ba(ClO_3)_2$ (barium chlorate) would be: "B-A-C-L-O-three, taken twice," which means that the two elements chlorine and oxygen (a radical) react with barium as though they were one element, and that it takes two of this combination of chlorine with three atoms of oxygen in order to react with one atom of barium.

CHEMICAL EQUATIONS

A chemical equation is merely a type of convenient "shorthand" to show what, and how many, atoms or molecules reacted together to result in a new or different compound or compounds. In some cases, the gas or the heat that might have been evolved from the reaction is shown. For example, the complete burning of charcoal could be stated as:

Charcoal burned in the oxygen of air gives
(or yields) carbon dioxide gas.

With a chemical equation, this same thing would be simply stated as:

$C + O_2 \rightarrow CO_2 \uparrow$. (The final arrow indicates that the product was gaseous and escaped.)

Another example is the burning of hydrogen gas:

$$2\,H_2 + O_2 \rightarrow 2\,H_2O \text{ (vapor)}.$$

This equation tells us that it takes two molecules of hydrogen gas (diatomic) to react (burn) with one molecule of oxygen (diatomic), and two molecules of water vapor are produced. It also tells us that 4 weight units (atomic weight of H = 1) of hydrogen will react with 32 weight units (atomic weight of O = 16) of oxygen to produce 36 weight units of water (2 by 2 + 2 by 16), either in vapor (steam) or liquid form. It also tells us that for these two diatomic gases, one volume of oxygen requires two volumes of hydrogen to form two volumes of gaseous (vapor) water or steam.

All chemical reactions involving chemical changes where new compounds with new properties are formed can be written in the chemical equation form. These may be very simple:

$$\underset{\text{(sulfur)}}{S} + \underset{\text{(oxygen)}}{O_2} \rightarrow \underset{\text{(sulfur dioxide)}}{SO_2 \uparrow}\ .$$

Or they may be very complex:

$$\underset{\text{(sodium bromide)}}{2\,NaBr} + \underset{\text{(sulfuric acid)}}{2\,H_2SO_4} + \underset{\text{(manganese dioxide)}}{MnO_2} \rightarrow$$

$$\underset{\text{(manganese sulfate)}}{MnSO_4} + \underset{\text{(sodium sulfate)}}{Na_2SO_4} + \underset{\text{(water)}}{H_2O} + \underset{\text{(bromine)}}{Br_2}\ .$$

In all of the preceding equations, there are exactly as many atoms of each element on the right-hand side (which is the reacted side) of the reaction mark (the arrow, →) as there are on the left-hand side of the equation. This leads to the statement of an important natural law.

THE LAW OF CONSERVATION OF MASS

Chemical reactions often result in changes in the state of matter. In the preceding simple reaction (in which sulfur reacts or burns in the oxygen of air) the sulfur changes from a yellow crystalline solid to a colorless gas, sulfur dioxide. Notice, however, that the atomic weight or mass of the sulfur has not changed, but has merely been transformed into the gaseous compound containing oxygen. In every chemical reaction and in the written equations that describe these reactions, the mass or amount of each element is the same in both the unreacted and reacted, or final, form. In all ordinary changes in matter, the total mass is neither increased nor decreased; this law is known as the Law of the Conservation of Mass. The law states that, in all the chemical reactions that have ever occurred, matter is never lost or created by chemical changes or in any other of the reactions between elements.

The language of this law has been changed since the discovery of nuclear reactions and nuclear energy. The word "ordinary" has been inserted before "changes in matter" because of the developments in nuclear science since 1938. In certain nuclear reactions a change in mass does occur, but this is not an ordinary change such as a chemical change.

ENERGY AND WORK

Energy is the other basic constituent of our world that governs our existence and that can be sensed by the changes it brings about in the matter of our world. As we have said earlier, energy is only defined by a description of what it does, yet it is used and observed everywhere. Energy has the capacity to do work or to produce change of some sort, usually in conjunction with matter. When heat is supplied to water, it expands and finally boils; its steam causes increases in pressure that can be used in work. When energy is stored in a battery, it can be used to turn an electric motor or to light an incandescent bulb.

FORMS OF ENERGY

Energy can exist in many forms. Familiar forms of energy are: heat energy, electrical energy, mechanical energy, light energy, nuclear energy, and chemical energy. These forms of energy exist in two states: kinetic and potential. Kinetic energy is the energy of moving bodies when their motion is arrested in some way, like a flower pot falling on the head of an unsuspecting passerby. Potential energy is energy that a body of matter has stored up, awaiting release in some form or another.

An example of the kinetic and potential states of energy can be seen in the fluorescent ceiling lights in a room. This form of energy would be electrical energy. When the lights are turned off, the form of the electrical energy would still exist; however, it would be in a "stored up" capacity — a potential energy state. By flicking the light on, the stored capacity would be released thereby converting the potential electrical energy to a kinetic state.

All matter exerts some form of energy, or even absorbs some form of energy, when it changes from one form into another. In burning fires, wood or paper is converted to carbon dioxide gas and ash in a process that gives off much heat energy and even radiant light energy. When water is applied to fires, the water changes to steam and the process absorbs heat energy in this change of form. The steam produced can also exert energy in the form of steam pressure if it is contained or imprisoned in some way.

THE TRANSFORMATION OF ENERGY

Energy seems to undergo continual transformation from one form to another. The example of an ordinary electrical power station generating electricity and distributing it to homes and factories is typical of this never-ending energy change. In these installations, the potential energy in coal or oil is changed to heat energy by burning, which causes water to change to steam or pressure energy that is used as kinetic energy to cause blades of an electric generating turbine to rotate with mechanical energy. This mechanical energy then becomes electrical energy and is conducted by transmission lines to the point where it is utilized. At this point, the electrical energy is again transformed into many kinds of energy: light, heat, mechanical, nuclear, hydraulic (a well-water pumping system in a farm home), cooling (absorbtion of heat energy), and other forms of work-energy. In all these transformations of energy a definite amount of energy of one kind always produces an equivalent amount of energy of another kind. There are certain losses of efficiency in these transformations due to our poor methods of harnessing all available energy when such transformations are made. For instance, the Mazda incandescent light bulb loses some radiant or light energy in the form of heat when it is connected to a source of electrical energy. In precise experiments, however, it is possible to calculate and to determine equivalent amounts of mechanical energy from potential electrical energy or heat energy when it is transformed to kinetic energy (in the proper terms of each, of course).

THE LAW OF CONSERVATION OF ENERGY

These considerations about the transformation of energy were studied in detail by scientists. More than a century ago, a definite statement or law concerning energy was formulated. This "Law of Conservation of Energy" stated: energy in all ordinary transformations is neither created nor destroyed, but is merely changed in form. This means that energy can neither be irretrievably lost, nor can we produce new energy that has not existed before in some form. Electrical energy is not newly created when coal is burned in a boiler to produce steam from water and the steam pressure generates electricity from turbines: it is transformed. Similarly, the mass of the coal is not destroyed; rather, it is transformed into the products of combustion: carbon dioxide and ash. Its energy has been converted to heat, which then does useful work in generating electricity, a form of energy that existed before in the form of a lump of unburned black coal and in the oxygen from the air.

Another interesting example of the laws of conservation of mass and energy is the photographer's flash bulb. Before being flashed, the bulb is weighed on a very sensitive balance or scale. It is then used in a camera by the application of a slight amount of electric current, which is called activation energy. This energy causes a large output of light energy in a short time. If the used bulb is again weighed, it will still weigh the same as it did before it was flashed. The metal foil in the bulb, composed of magnesium metal, has chemically combined with the oxygen gas surrounding it to form a white deposit of magnesium oxide, which has the same weight as the original metal and the uncombined oxygen gas. This simple experiment proves that: (1) no mass is lost by this chemical reaction, and (2) the light energy

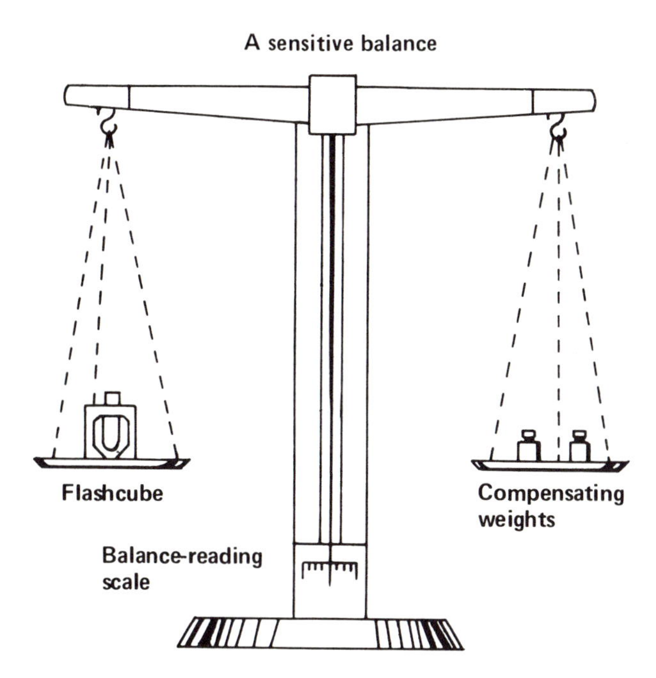

When the photoflash bulb is flashed, its weight is the same before and after the flash although it has undergone a chemical reaction that produces light.

Fig. 1.5. The conservation of mass and energy.

resulting from the activation energy does useful work in the formation of a picture. The light energy started from the flash initiates certain chemical changes on the film in the camera that could not be done without the light energy. Thus, the light energy was not lost, but was transformed and did work.

ACTIVITIES

1. (a) Describe the composition of the "matter" that is contained in a helium-filled, sightseeing blimp.
   (b) Why does the blimp rise with all this "matter" aboard?
2. Describe what happens to the energy of a clock (electrical, spring, and falling weight types) when running, and when stopped.
3. Name the form of energy demonstrated by each of the following: (a) a waterfall, (b) a flashlight battery, (c) a firecracker or pyrotechnic, (d) an automobile engine, and (e) an electric motor.
4. In the city or town where you live, the electrical energy supplied to your home is derived from two sources in the plant where the electricity is generated: one is a steam electric generator, and the other is an oil-fueled turbo-electric generator. Explain the transformations of energy undergone by the primary sources of fuel, coal, and fuel oil before useful electricity is put into the cables to your home.
5. Write an explanation for the chemical reaction of hydrochloric acid (HCl) with ordinary lye, which is sodium hydroxide (NaOH), to produce salt and water.
6. What is the most useful isotope of uranium? Why?
7. In view of the law of conservation of mass, how can you explain the fact that fine steel wool burns in air and the resulting residue weighs more than the original piece of steel wool.
8. Give examples, other than those presented in this chapter, of each of the following: (a) a physical change, (b) a chemical change, (c) a unit of matter sensed by smelling, and (d) a system where we find flowing electrons instead of outer shell electrons, as in the structure of atoms.
9. Based on what you have learned in this chapter, write an explanation describing the makeup of the Periodic Table of the Elements.
10. In your own words, describe the makeup of an atom.

SUGGESTED READINGS

Bennett, C. E., *College Physics*, Barnes and Noble, New York, 1960.

Brandwein, P. F., Stollberg, R., Burnett, R. W., *Matter, Its Forms and Changes*, Harcourt, Brace and World, New York, 1968.

Casey, J. F., *Fire Chief's Handbook*, Reuben Donnelly, New York, 1967.

Elements of Combustion and Extinction (Book 1), Manual of Firemanship, Her Majesty's Stationery Office, London, 1974.

Fire Protection Handbook, 14th Edition, National Fire Protection Association, Boston, 1976.

Hudiburg, E. (Ed.), *Fundamental Principles of Science Applied to the Fire Service*, Department of Fire Protection, College of Engineering, Oklahoma State University, Oklahoma.

Gamow, G., *Matter, Earth and Sky*, Prentice-Hall, Englewood Cliffs, New Jersey, 1963.

Lewis, J. R., *College Chemistry*, Barnes and Noble, New York, 1969.

Mack, E., Garrett, A. B., Haskins, J. F., Verhoek, F. H., *Textbook of Chemistry*, Ginn, Lexington, Massachusetts, 1956.

Perros, T. P., *Chemistry*, American Book, New York, 1967.

Resnick, R. and Halliday, D., *Physics*, John Wiley and Sons, New York, 1966.

Taffel, A., *Visualized Physics*, Oxford Book, New York, 1954.

2

The Language of Science

THE PROBLEM OF COMMUNICATION

The study of the physical science of chemistry requires a familiarity with many words and terms not ordinarily used in daily conversation. Over the years, the development of the words and terms used to express facts about the physical world evolved simultaneously with the need for investigators and scientists to communicate these facts to one another. In today's technical and industrial world, a high degree of familiarity with technical terms is necessary in order for people from all walks of life to communicate with and understand each other.

Because the professional fire officer's daily work with combustible materials is basically concerned with chemical reactions, a fire officer's vocabulary must include the terminology that accompanies the science of chemistry. For example, when the "first in" fire officer enters a factory that manufactures plastics, it is necessary for that fire officer to instantly know that a fire in a container marked "TRIETHYL-ALUMINUM—Net Weight 3-KILO," is a serious fire requiring special treatment with a dry chemical extinguishing agent only, and that water must not be used. Also, it is important for the fire officer to understand that because the 3 kilograms of material in the container equals only 6.6 pounds, the fuel cannot burn very long even if extinguishment cannot be secured. Recognition of the chemical name of the material and its relationship to its chemical properties of combustion and reaction immediately become important.

One of the fundamental problems connected with the communication of technical facts is that of standardizing the units by which measurements are made so that every person understands accurately what is meant by a given or stated dimension. How far is a mile? How heavy is a pound? How long is a metre? In early times every language of every country developed its own terminology for describing necessary terms denoting dimensions such as length, weight, and force.

Until approximately 1790 it was almost impossible to describe or to communicate to people of differing languages the accurate length of a piece of string or a volume measure of milk or water. Confusion among technologists who wanted to share accurate information concerning their discoveries about the physical world and about chemistry reached a cataclysmic state, and in 1791 the French National Academy of Science began the development of the international language of measurement. The outcome of this effort was the metric system of measurement (also known as the "cgs," centimetre-gram-second system). This method of measurement became the language of the physical sciences and, because it was accepted as a national measurement standard by almost all of the European countries, it also became a sort of European standard. The United Kingdom of Great Britain and the United States of America remained with the English system of measurement (a medieval system also known as the "fps," foot-pound-second system). A great advantage of the metric system is its constant relationship of values of the common units (except for the time units) by factors of tens or decimals, thus facilitating arithmetical calculations. The English units require variable multiplicands: 16 ounces to the pound, 32 ounces liquid to one quart liquid measurement, etc.

It is important to note that the English spelling of *meter* and *liter*, which have been customarily used in the metric system in the U.S. (and in Britain) have been changed to *metre* and *litre* (pronounced the same as before) in the SI system. This change (from metric) occurred when the first translation into English was made of the officially agreed-upon SI document, "Le Système International d'Unites." It represents a worldwide uniform spelling of these units.

THE INTERNATIONAL SYSTEM OF UNITS

Difficulties were encountered with the use of the English units of measurement because one must commit to memory the correct relationship of each term of measurement within each class of length, weight, etc. (*i.e.*, 12 inches = one foot, 3 feet = one yard, 1,760 yards = one mile), and with the problems of converting the metric measurements of the language of science (and of almost all European countries) to English units. These difficulties finally brought about the 1963 adoption of a new system of measurements called "Système International" (SI) by the National Bureau of Standards of the U.S. Department of Commerce.

The SI system was agreed upon by international conference and is much the same as the metric system, with the addition of some new names for measurement of new physical values. It facilitates calculations because of its constant factor of 10 relationship between larger and smaller units of each class of weight, volume, length, etc., measurement. (See Table 2.1, "Multiple and Submultiple Terminology.") As with the metric and English systems, the SI system utilizes the present time units of measurement (60 seconds = 1 minute, 60 minutes = 1 hour).

Another important change to note involves the measurement of temperature. The SI system bases all temperature measurements on the *absolute zero* scale of degrees. This is called *degrees Kelvin*. Because degrees Kelvin is a difficult system to use, *degrees Celsius*, which is based on the freezing and boiling points of pure water, is more often used. As will be seen later in this chapter, this system is quite different from our present Fahrenheit scale of degrees of temperature.

Predictably, the enormous task of learning to use a new system of measurement will require many years — and perhaps generations — of workers; however, in addition to the use of English units in technical reports and instructional manuals, many organized areas of technicians (such as the U.S. Armed Forces) have already begun to require the use of the SI system. In the United Kingdom where a changeover to SI units is beginning, instructive articles have appeared in

Table 2.1. SI Units: Multiple and Submultiple Terminology

QUANTITY BEING MEASURED	UNIT OF MEASUREMENT				
"Length"	Metre (39.37 inches U.S.)	10 units dekametre*	100 units hectometre	1000 units kilometre	1,000,000 units megametre
		1/10 unit decimetre	1/100 unit centimetre	1/1000 unit millimetre	1/1,000,000 unit micrometre
"Weight"	Gram (0.035 ounces avd.)	10 units dekagram*	100 units hectogram	1000 units kilogram	1,000,000 units megagram
		1/10 unit decigram	1/100 unit centigram	1/1000 unit milligram	1/1,000,000 unit microgram
"Volume"	Litre (33.81 fluid ounces U.S.)	10 units dekalitre*	100 units hectolitre	1000 units kilolitre	1,000,000 units megalitre
		1/10 unit decilitre	1/100 unit centilitre	1/1000 unit millilitre	1/1,000,000 unit microlitre

* Not recommended, but sometimes used.

journals devoted to fire protection. In England where the use of degrees Celsius for temperature measurement has become a national standard, the Fahrenheit scale is fast disappearing.

The international changeover to the SI system of measurement is of importance to the U.S. Fire Service. Since this text is about the science of fire protection chemistry, and since the language of the sciences uses the metric system (which is almost identical to the SI system), many of the units herein used are given in the metric system; such units are immediately followed by the equivalent SI (metric) units in parentheses. Similarly, where the English units are most familiar (especially in regard to engineering units of measurement), they are used in this text; such units are immediately followed by the equivalent SI value and unit in parentheses.

The important methods of conversions of the two systems — English and SI (metric) — will be found later in this chapter in a section devoted especially to this subject.

Table 2.2. A Comparison of Approximate Ranges of Usage of English and SI Units for the Measurement of Mass (Weight, Volume, and Size)

	ENGLISH SYSTEM		SI SYSTEM
		compares with usage as	
Weight	Tons (long)		Megagrams
	Tons (short)		Metric Tons
	Hundred weight (Brit.)		Kilograms
	Stones (Brit.)		Kilograms
	Pounds		Grams
	Ounces (avdp.)		
	Grains		Milligrams
(decreasing)	Carats		
Volume	Hogsheads (U.S.)		Kilolitres
	Barrels		
	Gallons		Litres
	Quarts (fluid)		Litres
	Pints (fluid)		
	Gills		
	Ounces (fluid)		Millilitres
	Drams		
(decreasing)	Minims		
Size	Miles		Kilometres
	Yards		
(by spatial	Feet		Metres
units)	Inches		Centimetres
	Mils		Millimetres
(may be square or cubic) (decreasing)			

UNITS OF MEASUREMENT FOR MATTER OR MASS

In Chapter 1 of this text, matter was defined as something that possesses two basic characteristics: (1) weight, and (2) volume (or size). The measurement of these two characteristics should require that only two different units be used; however, the process has been complicated by naming various magnitudes of weight and volume in different ways. The English system of units has a great variety of units to express the measurements of weight and volume. The SI system (and the metric) have comparatively few units, using numbers to express magnitudes. Table 2.2 presents a comparison of the approximate ranges of measurement between English and SI units.

UNITS OF MEASUREMENT FOR ENERGY OR WORK

As previously stated, the problem of outlining the terms by which we describe the capacity to do work, which is the manifestation or end product of energy, is complicated by the many forms in which work appears.

There are several forms of energy that are used in, and are important to, the science of fire protection. In this text, consideration is limited to the following of these forms: heat energy, electrical energy, mechanical energy, pressure energy, and force energy. While many other forms of energy occur during energy transformation (such as light energy, etc.), such forms are beyond the scope of this book. Perhaps the most easily understood presentation for these units of energy and their equivalents would be in tabular form, as shown in Table 2.3.

FIRE SERVICE QUANTITY ELEMENTS

Quantity elements of importance to the fire service, together with the conversion of these elements to the SI system, are shown in Table 2.4. (See Appendix I, "Conversion Factors," for a more comprehensive listing of conversion factors for units, and Appendix II, "Temperature Conversion," for conversion of Celsius to Fahrenheit.)

SOME METHODS FOR THE CONVERSION OF UNITS

Although some units of the SI system have been in use as the metric system for many decades by the fire services of other countries, there are considerable problems involved in converting to SI units the ordinary units of measurement that have been used by the fire service of the United States for over 100 years. However, the changeover is imminent. Currently there are only certain units of importance in the day-by-day routine of the fire service, and the conversion factors for these quantity elements should be mastered by technicians and specialists in fire protection. There are simple and easy relationships for the two systems of units that can help enable us to begin "thinking in two languages" — English and SI.

Table 2.3. A Comparison of English and SI (Metric) Units for the Measurement of Various Forms of Energy

FORM OF ENERGY (WORK)	ENGLISH SYSTEM UNITS	SI SYSTEM UNITS
Heat	Calorie	Joule
	Btu (British thermal unit)	Joule
Electrical	Watt	Watt
Mechanical	Horsepower	Watt (in terms of electricity)
	Foot pounds	Joule (in terms of heat)
Pressure (per unit area)	Pounds per square inch	Newton per square meter or Pascal
Force	Pounds	Newton per metre

Temperature Conversion

As previously stated, the scales of measurement of temperature are based on the boiling and freezing points of pure water. In 1736 a German physicist, Gabriel D. Fahrenheit, decided that there should be an arbitrary spread of 180 degrees between the freezing point of water and its boiling point. In 1744 a Swedish astronomer, Celsius, decided that we should work in systems of tens with temperature, just as we do with other Arabic numerals. Thus, Celsius divided the thermometer into 100 points between the freezing and boiling points of water. Since both systems were viable, many have pondered why we should have two temperature scales. With the Fahrenheit scale's smaller degree breakdown and a zero below the freezing point, meteorologists and other "close degree followers" found this scale to be more precise. (The Dominion of Canada and the United States use Fahrenheit.) Celsius is used for a greater mathematical simplicity, as 0° becomes freezing and 100° becomes boiling.

In the early part of the twentieth century more was learned about the state of change in gas volumes with temperature change; it was also learned that the point at which there occurs a complete lack of change is at minus 273° Celsius (or Centigrade). Minus 273° Celsius is called absolute zero, a temperature at which all gases become liquids. This was named the Kelvin scale of temperature, starting with zero degrees K at minus 273° Celsius. The representation of these three systems [Fahrenheit, Celsius (or Centigrade), and Kelvin] is shown in Figure 2.1.

The process of converting these temperature scales can be simplified if it is kept in mind that: (1) as a rough approximation, °F are *about* 2 times °C; (2) °F usually have 32 degrees more than °C (see Fig. 2.1); (3) going from °C to °F is in the normal alphabetical direction (a, b, c, d, e, f) and requires a multiplier *larger* than

Table 2.4. Fire Service Quantity Elements and Their Conversion Factors to SI (Metric) Equivalent Units

QUANTITY ELEMENT	ENGLISH UNIT	SI UNIT AND SYMBOL	CONVERSION FACTOR: ENGLISH TO SI MULTIPLY BY:
Time	second	second, s	(same)
Temperature	degree Fahrenheit	degree Celsius, °C	(°F − 32) × (0.55)
Temperature	degree Celsius	degree Kelvin, °K	°C + 273 (algebraic)
Length	foot	meter, m	0.305
Length	inches	centimetres, cm	2.54
Length	miles	kilometres, km	1.61
Area	square feet	square metre, m^2	0.09
Weight (mass)	pound (avdp.)	kilogram, kg	0.453
Weight (mass)	ounces	grams, g	28.3
Volume	gallon (U.S.)	cubic metre, m^3	0.0037
Volume	gallon (U.S.)	litre, ℓ	3.78
Volume	quart	litre, ℓ	0.95
Volume	cubic feet	cubic metres, m^3	0.028
Volume Flow	gallons per minute	litre per second, ℓ/s	0.06
Pressure	pounds per square inch	Newton per square metre, N/m^2 (or Pascal, Pa)	6894.8
Pressure*	atmospheres	bar	1.0 (approx.)
Pressure (Atmospheric)	atmospheres	Newtons per square metre, N/m^2	101,325.
Pressure (Head)	inch (water)	Newton/metre, N/m^2	249.
Pressure (Head)	inch (Hg)	kilo Newton/metre^2, kN/m^2	3.39
Power	horsepower	watts, w	745.
Density	pounds per gallon	grams per centimetre^3	0.12
Density	pounds per cubic foot	kilograms/metre^3	16.0
Velocity	miles per hour	metres per minute	26.82
Velocity	miles per hour	kilometres per hour, km/hr	1.61
Thermal Conductivity	Btu/ft/hr/°F	Joule/metre/sec. Kelvin, J/m. s. K	1.73
Energy (work, quantity of heat)	calories	Joule, J	4.185
Radio Frequency	cycles per second	Hertz, Hz	1.00
Electric Current	ampere	ampere, A	(same)
Electromotive Force (potential difference)	volt	volt, v	(same)
Electric Resistance	ohm	ohm, Ω	(same)
Electric Capacitance	farad	farad, F	(same)
Specific Heat	calorie/gram	Joule/kilogram, J/kg	4.19
Luminous Flux	candle power	lumen, lm	12.57

** The practical problems concerned with expressing hydraulic pressures in SI units of Newton/m² or Pascal or kilo-Pascal units can easily be seen. Fire protection*

authorities are now considering the use of "bar" as a unit of pressure. A "bar" equals 1.013 atmospheres. A small error of 1.3% is introduced if one "bar" equals one atmosphere. Since one atmosphere equals 14.7 psi, the conversion problem is eased considerably in everyday practice by the use of the "bar."

Note: The modern electronic calculator makes it quite easy to reverse *the conversion process if it is desired to convert* SI values *to* English. *This is done as follows: Find the reciprocal value of the above applicable conversion factor; that is, set the fraction 1 ÷ X (the factor) into the device. The resulting answer is the conversion factor for SI values to English. Or if your calculator has a 1/X key, use it by setting in the above applicable English to SI factor and use the 1/X key to obtain the conversion factor for SI values to English.*

(Of course the above conversions may be accomplished by division *with the applicable above conversion factor.)*

Examples: (a) Find the velocity (speed) in kilometres per hour of an engine truck going 40 mph.

$$40 \times 1.61 = 64.4 \text{ km/hr.}$$

(b) Find the velocity in mph of an engine truck going 64.4 km/hr.

$$1 \div 1.61 \times 64.4 = 40 \text{ mph.}$$
$$(\text{or, } 64.4 \div 1.61 = 40 \text{ mph.})$$

1.0; and (4) going from °F to °C is in a *counteralphabetical* direction (a, b, c, d, e, f) and requires a multiplier *smaller* than 1.0.

This can be explained in formula fashion, as in the following examples:

Converting from °C to °F: (alphabetically) $\frac{9}{5}$ (or 1.8) × °C + 32 = °F.

(Since all °F are 1.8 or $\frac{9}{5}$°C, multiply by this first and then add 32 to get °F.)

Converting from °F to °C: (counteralphabetically) °F − 32 × $\frac{5}{9}$ (or 0.55) = °C.

(Since all °F are higher than °C by 32 degrees, subtract this first and then multiply by $\frac{5}{9}$ or 0.55 to get °F.)

The conversion of the Celsius scale to the Kelvin scale is comparatively simple and involves only the algebraic addition of 273, as in the following examples:

Converting from °C to °K:

$$\begin{array}{r} +\ 10°C \\ +273 \\ \hline +283°K \end{array}$$

$$\begin{array}{r} -\ 10°C \\ +273 \\ \hline +263°K \end{array}$$

Length Conversions

The problem of thinking in two systems or languages for the measurement of lengths, distances, or sizes requires that some approximations be used to accomplish comparisons between the English and the SI systems of measurement. For example, a yardstick is 3 feet long; however, if a yardstick were made to equal 1 metre (a metrestick), it would be just a little longer (1.09 yards, or almost 40 inches). Thus, it can be approximated that 1 metre equals a little over 3 feet, or a little more than 1 yard. When converting smaller dimensions, it can be approximated that 1 inch equals 25 millimetres, or 2.5 centimetres; thus, each time there is a multiple of 25 millimetres, each millimetre will equal 1 inch (*i.e.*, a 75-millimetre diameter equals a 3-inch diameter). When converting miles to kilometres or vice versa, the process becomes more complicated. Because a kilometre is about six-tenths of a mile, distances in kilometres are almost double the distances in miles; to convert miles to kilometres, it is necessary to multiply by 1.6 (approximately $1\frac{1}{2}$ or $\frac{3}{2}$). Similarly, miles per hour equal 1.6 kilometres per hour, or kilometres per hour equal half that in miles per hour (0.625 times kmph).

In summary, the following *approximations* of length or distance can be used for making comparisons between the English and SI units of measurement:

1 metre = a little over 3 feet = a little over 1 yard
1 inch = about 25 millimetres = 2.5 centimetres
1 foot = about 30 centimetres (300 millimetres) (or 25 times 12)
1 kilometre = about 0.6 mile or a little over $\frac{1}{2}$ mile
1 mile = a little more than $1\frac{1}{2}$ kilometres (1.6) (about $\frac{3}{2}$)
1 mile per hour = a little more than $1\frac{1}{2}$ kilometres per hour (1.6)
1 kilometre per hour = about $\frac{1}{2}$ mile per hour (0.625) (about $\frac{3}{5}$)

Size or Area Conversions

Area or size can be visualized more easily when approximations of the equalities of the two units (English and SI) are made. For example, small areas are measured in SI in square centimetres. One square inch equals very close to 6.5 square centimetres. One square metre equals a little over 10 square feet (10.7 square feet); thus, for approximately every 10 square feet, an equivalence of a little more than one square metre can be estimated. An acre contains a little less than $\frac{1}{2}$ (0.4) hectare (the SI unit for land areas), or 4,000 square metres. It takes about 2.5 acres to equal one hectare. It takes 250 acres to equal a square kilometre.

Also, a square kilometre equals about one-third of a square mile (0.386 square miles). Thus, it takes almost 3 square kilometres to equal 1 square mile. To convert from square miles to square kilometres, it is necessary to multiply by 2.59, which is the reciprocal of 0.386 (1/0.386).

In summary, the following approximations of area can be used for making comparisons between the English and SI units of measurement:

1 square inch = about 6.5 square centimetres
1 square metre = about 10 square feet
1 square kilometre = about 250 acres
1 square kilometre = one-third of a square mile
About 3 square kilometres = one square mile

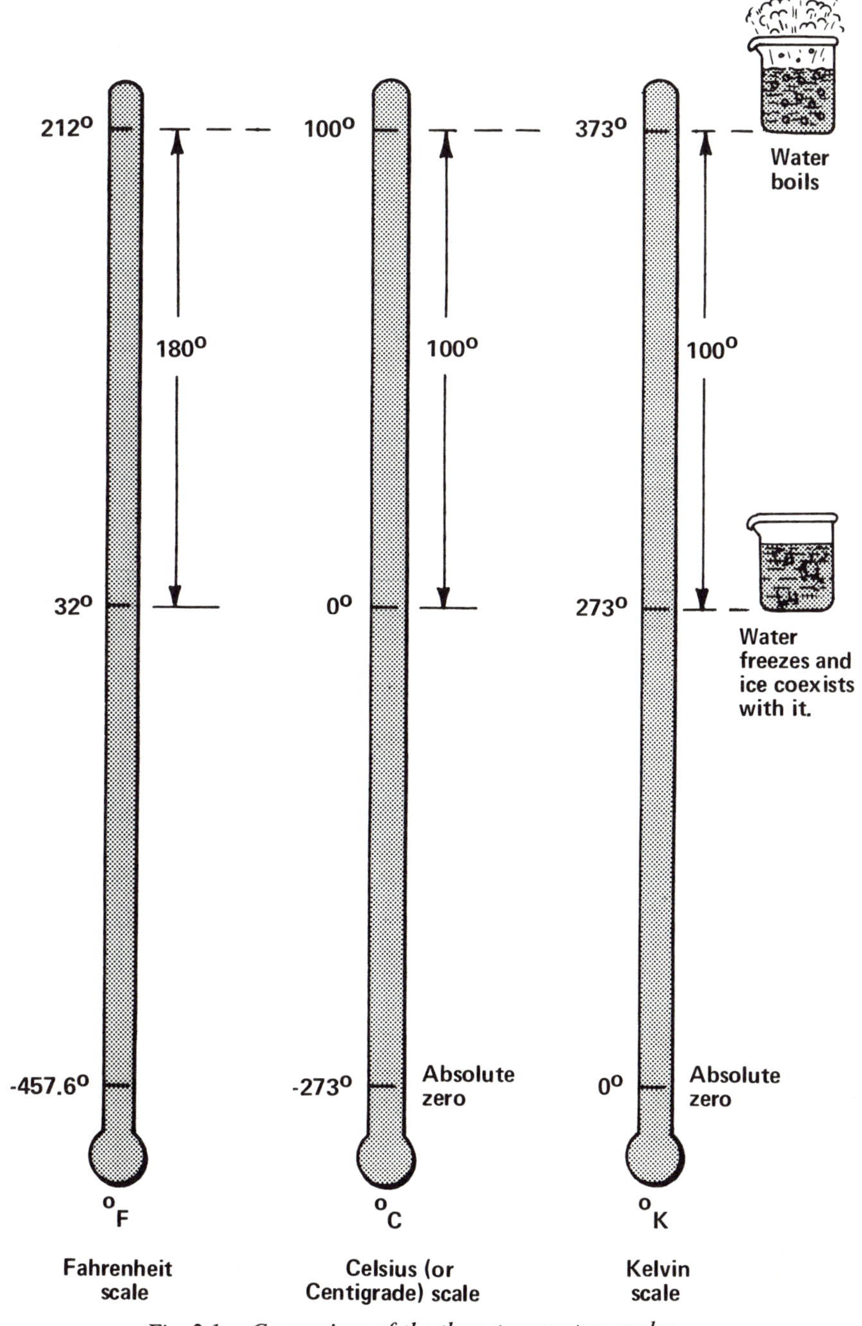

Fig. 2.1. Comparison of the three temperature scales.

Volume or Capacity Conversions

In order to convert liquid volumes or bulk capacity or equivalencies from the familiar U.S. gallon and quart designations, it is necessary to think in SI litres, cubic metres, and cubic millilitres.

A U.S. gallon equals a little less than 4 litres (actually, 3.8 litres) and each one of the 4 quarts that make up a gallon equals a little less than a litre (actually, 1 quart = 0.9 litre). Conversely, a litre equals a little more than a quart (1 litre = 1.06 quarts).

In volume terms, one cubic metre = 1,000 litres (a kilolitre), which equals about 260 gallons (264.2 gallons), or about 35 cubic feet (35.32 ft^3). A cubic foot equals $28\frac{1}{3}$ litres (almost 30 litres).

Since a quart is a little larger than a litre and 1,000 millilitres = 1 litre, a quart equals a little less than 1,000 millilitres (actually, 946 millilitres). A pint equals one-half of this, or 473 millilitres.

In summary, the following approximations of volume and capacity can be used for making comparisons between the English and SI units of measurement:

1 gallon = a little less than 4 litres (3.8 litres)
1 quart = a little less than 1 litre (0.9 litre)
1 litre = a little more than 1 quart (1.06 quarts)
1 cubic metre = 1 kilolitre = a little more than 250 gallons (264.2 gallons)
1 kilolitre = about 35 cubic feet (35.3 cubic feet)
1 quart = a little less than 1,000 millilitres (946 millilitres)
1 cubic foot = almost 30 litres (28.3 litres)

Weight Conversions

To use the SI system of measurement, it is necessary to convert the English denominations of ounces, pounds, and tons into grams, kilograms, and metric tons or megagrams (1,000 kilograms). This is not difficult if precision is not required. For example, a U.S. pound equals a little less than one-half a kilogram (actually, 0.45 kilogram), and there are a little over 2 kilograms in a pound (actually 2.2 kilograms = 1 pound). Therefore, to arrive at the number of pounds in a container that is labeled in kilos or kilograms, it is merely necessary to multiply by 2 and add a little more.

The conversion of ounces to grams or kilograms becomes more involved because there are 16 ounces in a pound, as compared to 1,000 grams in a kilogram. When converting ounces to grams, it should be kept in mind that because there are about 30 grams (actually, 28.3) in an ounce, it is merely necessary to multiply by 30. When converting from grams to ounces, the same factor of 3 is involved, but moved one decimal place to the left; that is, the number of grams are *divided* by three and the decimal is moved one place to the left; *i.e.*, 900 grams ÷ 3 = 300 = 30.0 ounces.

The quick and approximate conversion of pounds to grams is governed by the fact that there are almost 500 grams in 1 pound (actually, 453.4 grams). An approximate equivalency in grams can be arrived at by multiplying the number of pounds by 500. The conversion of grams to pounds can be approximated by *dividing* the number of grams by 5 and moving the decimal two places to the left, as in the following example:

1,850 grams ÷ 5 = 370 = 3.7 pounds. (The correct figure is 4.07 pounds.)

Large weight figures that are given in English (or U.S.) tons are stated in metric tons or kilograms in the SI system. Our U.S. ton (a short ton, which is 2,000 pounds) equals nine-tenths of a metric ton, or 907 kilograms. Thus, it takes 1.1 tons to equal a metric ton (which equals 1,000 kilograms, or one megagram), or 2,204.6 pounds to equal a metric ton.

In summary, the equivalencies of weight in the English and SI systems are as follows:

1 pound = a little less than one-half kilogram (0.45 kilogram)
1 kilogram = a little more than two pounds (2.2 pounds)
1 ounce = almost 30 grams (28.3 grams)
1 ton U.S. (2,000 pounds) = 0.90 metric ton or 907 kilograms
1 metric ton = 1.1 U.S. tons, or 2,200 pounds

Flow Rate Conversions

Members of the fire service require a knowledge of fluid-flow terminology in order to appropriately describe water-usage rates and nozzle capacities. In flow rate conversions it is necessary to use SI units of litres per second or minute instead of gallons per second or minute, or the English unit of cubic feet per minute.

A flow rate of 1 gallon per minute is equivalent to a little less than 4 litres per minute (actually, 3.78 litres/minute), and a rate of 1 litre per minute is a little less than $\frac{1}{3}$ of a gallon per minute (actually, 0.26 gallon/minute). When working in cubic feet per minute of flow, it is necessary to keep in mind that 1 cubic foot per minute is equivalent to about $7\frac{1}{2}$ gallons per minute (actually, 7.48 gallons/minute), or 1 gallon per minute equals 0.13 cubic feet per minute. Then, 1 cubic foot per minute equals almost 30 litres per minute (actually, 28.3 litres/minute).

In summary:

1 gallon per minute = 3.78 litres per minute (almost 4)
1 litre per minute = 0.26 gallon per minute (almost $\frac{1}{3}$)

Pressure Conversions

In many countries, pressure has been customarily measured in atmospheres (1 atmosphere = 14.7 pounds per square inch). In other countries, the term kilograms per square centimetre is used to express pressure (1 kilogram per square centimetre = 14.2 pounds per square inch). For many years the latter has been extensively used in the metric system. Currently in the United Kingdom there is a movement to utilize "metre-head" as a unit of pressure measurement (1 metre-head = 0.1 atmosphere = 1.47 psi). The SI unit for pressure is Pascal (Pa) or kiloPascal (kPa). In terms of force (newton) over a unit area, where one could use N/m^2 (newtons per square metre), it has been decided to use the name "Pascal," which is equal to a force of one N/m^2. The multiple term of "kiloPascal" is preferred. One psi equals 6.9 kiloPascal. Thus it can be seen that a gage that reads 75 psi would read 520 kPa.

Recently it has been found that the unit "bar" is the most convenient for fire service use. The bar is equal to 1.013 atmospheres. Also, one psi is equal to 0.07 bar. In daily use one could use the equivalence of one atmosphere equals one bar and an error of only 1.3 percent would be introduced. (See Appendix 1, "Conversion Tables.")

ACTIVITIES

1. (a) Show, by calculation, why minus 40°F = minus 40°C.
   (b) Would minus 10°F = minus 10°C? Why or why not?
2. You would like to purchase a piece of portable fire fighting equipment from Germany. The equipment's weight is 79 kilograms.
   (a) What would be the equipment's weight in pounds?
   (b) How many fire fighters would be needed to transport this apparatus to the fire ground?
3. You are a paramedic and have been called to the scene of a highway accident. Before performing any type of emergency medical treatment, you have been instructed to determine the victim's body temperature.
   (a) What is the normal temperature of the human body in degrees Celsius?
   (b) What would be the Celsius temperature of a victim with a 99° Fahrenheit temperature?
   (c) The temperature of the pavement the victims are lying on is 150° Celsius. What would the same temperature be in Fahrenheit?
4. Convert the following equipment for ladder trucks and elevating platforms into SI terminology:
   (a) Two pick-head axes, 6 pounds.
   (b) Two crowbars, 50-inch minimum.
   (c) Two 6-foot pike poles.
   (d) Two 8-foot pike poles.
   (e) One 125-foot rope ($\frac{5}{8}$-inch diameter).
   (f) One 125-foot rope ($\frac{3}{4}$-inch diameter).
5. (a) What was your maximum speed of travel in kilometres per hour the last time you looked at the speedometer of your automobile?
   (b) If a highway sign reads "Roadside rest area 4 miles," how many kilometres would this be?
6. As the officer in charge of rescue operations at a high-rise hospital fire, you must determine how many fire fighters should be used to transport patients down an aerial ladder.
   (a) How many fire fighters would be needed to help lift a 250-kilogram patient down the ladder?
   (b) In kilograms and grams, how much would a 145$\frac{1}{2}$-pound patient weigh? Could a fire fighter weighing 99 kilograms easily transport this patient to safety?

7. A Swedish fire protection specialist who speaks English visits your community's fire department to learn more about elevating platforms. How would you convert the following information to SI terminology for the benefit of your visitor?
   (a) The telescopic type platform is able to carry a 700-pound payload.
   (b) The 150-foot elevating platform is used extensively in fighting high-rise building fires.
   (c) The 85-foot elevating platform has a platform weight capacity of 750 pounds.
8. The equivalent values for a pressure of one atmosphere (absolute) are: 14.7 psia; 29.72 inches of mercury; 33.90 feet of water; 0.0 psig; 760 mm mercury; 101,325 newtons per square metre; and 101.3 kilo N/m^2. During a pump drafting operation a pump suction gage reads 9.97 inches of mercury.
   (a) How many psia does this equal?
   (b) How many newtons per square metre?
   (c) How many atmospheres?
9. The pressure gage on an operating fire pump reads 98.5 psig.
   (a) What designation and number of units would this be in SI terminology?
   (b) What pressure and flow rate would be stated for a nozzle giving 60 gpm at 100 psig?
10. Write a brief statement justifying why you do or do not feel that a professional fire officer's daily work should include the terminology that accompanies the science of chemistry. Defend the reasoning in your statement by citing at least two specific examples in which the usefulness or lack of usefulness of such terminology would be of major importance.

SUGGESTED READINGS

"Abbreviated Guide for the Use of the SI and Conversion Tables," *Chemical Engineering Progress*, Vol. 67, No. 5, May 1971.

Branley, F. M., *Think Metric*, T. Y. Crowell, New York, 1972.

"Brief History of Measurement Systems," Department of Commerce, National Bureau of Standards, Washington, D.C., October 1972.

Buffington, A. V., *Meters, Liters and Grams*, Random House, New York, 1974.

Chisholm, L. J., "Units of Weight and Measure," Department of Commerce, National Bureau of Standards, Washington, D.C., 1967.

DeGaeta, P. F., "Inching Up to Use of Metric System in U.S.," *Fire Engineering*, Vol. 128, No. 3, March 1975.

Gilbert, F. T., *Thinking Metric*, John Wiley and Sons, New York, 1973.

Gray, A., "The Fire Service and Metrication," *Fire*, July–August, 1969.

Hopkins, R. A., "The International (SI) Metric System and How It Works," Polymetric Services, Tarzana, California.

Jaeger, T. W., and Winquist, C. J., "Better Learn Your Metric System Terms!," *Fire Command!*, Vol. 39, No. 10, October 1972.

"Metrication," *Fire Command!*, Vol. 42, Nos. 8–12, August–December 1975.

The Chemical World and Some of Its Reactions

THE STATES OF MATTER

One of the purposes of this textbook is to help the student understand the "building blocks" of the material universe by looking at the surrounding world in a basic manner. To help achieve this, this chapter presents some of the ways that matter and materials of all kinds can exist and make an impression on our three "matter-detecting" senses of sight, smell, and touch. Also included is how these impressions may be described in a uniform, scientific manner.

Solids, Liquids, and Gases

All of the matter that makes up our material world exists in one of the three physical states: (1) solid, (2) liquid, and (3) gas (or vapor). When we speak of ice and of iron, we are speaking of substances that are solids. When we speak of water and of mercury, we are speaking of liquids. And, when we speak of steam and of air, we are speaking of gases. Steam is a gaseous vapor of water, and air is the colorless, odorless gas we live in and breathe. We usually cannot see vapors except in rare cases such as the brown vapors of the element bromine, or the greenish color of chlorine gas. However, we can observe what effects are caused by gases when they blow dust and smoke around, or when they make rubber balloons expand. These three basic classifications of matter — solid, liquid, and gas — are an integral part of the basic vocabulary of technology and science, and can usually be defined by our senses of sight, smell, and touch.

Temperature Dependencies

In the preceding examples of solids, liquids, and gases, the various forms of water were given for each state. This was done to help illustrate the fact that temperature is important to the description of a substance. At ordinary temperatures (when the human body is comfortable), water is a liquid. Ordinary temperatures or room temperatures are in the area of 70°F (21.1°C). Water is a solid if the temperature reference is at the freezing point of water, 32°F (0°C), or lower. Water at higher temperatures, such as 212°F (100°C) or above, is always a vapor or gas: water cannot exist as a liquid or a solid (ice) at these high temperatures. Thus, it can be said that the description of any substance must be related in some way to the temperature surrounding it. Standard usage assumes that ordinary room temperatures are meant unless otherwise denoted.

Pressure is also a governing factor in determining the accuracy of the state in which we find a substance. The influence of pressure is somewhat more complicated and less obvious than the effects of temperature, and is more generally seen in the case of gases than in other states of matter. The influence of pressure is discussed in greater detail later in this text. For present purposes, descriptions of the state of matter assume an ordinary atmospheric pressure of 0.0 pounds per square inch gage, or 14.7 pounds per square inch absolute, or one atmosphere, absolute (1.01 newtons per square metre) (one bar).

COMPOUNDS AND MIXTURES

To better understand the characteristics of the matter that constitutes the material universe, an understanding of the union or nonunion of the elements is necessary. The elements that are found in the earth and in its atmosphere may unite completely with other elements, or they may refuse to unite or join up with other elements and may merely mix with them on a temporary basis. The process of the uniting of the earth's basic elements is called compound formation. When two or more elements become united, the resultant chemical reaction is the loss of original character (or chemical-physical properties) for the original reacting partners, and a new substance is formed with entirely new properties or characteristics. If two or more elements do *not* join together, or if conditions for such a union are *not* correct, the elements merely *mix* with each other; no new compound is formed, and the elements may be separated again.

Therefore, it can be concluded that the characteristics of the mixture of several elements are similar to the characteristics of each component element making up the mixture.

The elements that unite to form common, everyday water are the two gases, hydrogen and oxygen. These two gases will unite and form a new compound if their temperatures are raised to a point that gives them sufficient energy to encourage their mutual reaction. This temperature point is at the "burning" temperature of hydrogen gas in a surrounding atmosphere of oxygen gas.

When this temperature is reached (it may be started with a match flame or a spark), the result of the union of the gases is water which, at such an elevated temperature, is a vapor (see Fig. 3.1). The vapor can then be cooled by some process such as contact with a cold surface on which liquid water drops will form again and again.

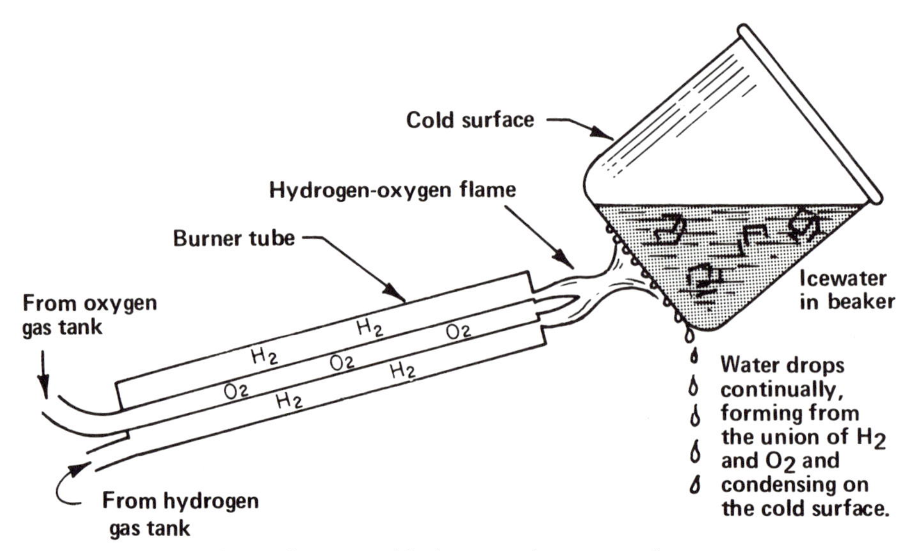

Fig. 3.1. The union of hydrogen and oxygen to form water.

Another example of the union of substances to form new compounds is the burning of an ordinary wax candle. Burning candles serve the useful purpose of giving off light, and the burning process results in the consummation of the wax and the paraffin that formed them. In the burning process, a new compound is formed that is of particular use to the fire services. This compound is carbon dioxide gas, the well-known fire extinguishing agent.

Wax and paraffin are compounds that contain a high proportion of carbon in combination with hydrogen and other elements. When a candle wick is lit, some wax is melted in the area of the burning wick. This liquid wax travels upward in the cotton strands of the wick and burns or oxidizes when it contacts the flame. The carbon of the wax unites with oxygen, and water vapor is also formed. Since carbon dioxide will not support further combustion, its extinguishing effect can easily be demonstrated by collecting some of this newly formed gas from the burning wax. The burning candle experiment shown in Figure 3.2 illustrates the effect of carbon dioxide on fire and flame.

It is difficult to restore, to their original waxen state, the new compounds (carbon dioxide gas and water vapor) formed by the burning wax. It is also difficult to break down or separate the carbon dioxide into its constituents, carbon and oxygen, or the water into its original elements, oxygen and hydrogen. The carbon dioxide and water are true and stable new compounds. However, these two compounds exist as a gas and vapor *mixture* in the area above the flame. In order to separate these two compounds, it is merely necessary to hold a cold surface above the flame. The cold surface causes the water vapor to change to water droplets, and the carbon dioxide to disappear into the air.

As previously discussed, compounds result from the chemical unions of materials (such as wax and the oxygen of air in a hot flame). Mixtures of materials, however, are simply nonunited collections of two or more materials. When two substances do not unite in any kind of reaction with one another, they remain unchanged in their character as solids, liquids, or gases. For example, when a small amount of ordinary cooking oil is placed in a glass that is partially filled with water, the oil will "bead" on the surface of the water. No amount of stirring or shaking will cause the two substances to combine.

Mixtures of two liquids take on a blend of the properties of each of the components. This can be illustrated by adding a teaspoonful of ordinary tap water to a teaspoonful of glycerin, a heavy viscous liquid. The resulting liquid is more viscous than the ordinary water, but less viscous than the original glycerin. Also, at a much lower temperature the mixture would freeze and become a solid. If the mixture was poured into an open saucer and exposed to air for a few hours, the part of the liquid that was water would evaporate into the air; the glycerin would be left in its original state.

Similarly, when milk of magnesia is mixed with water, a milky, white liquid results. The resulting liquid is more watery and thinner than the original thick and heavy milk of magnesia, but is not as thin and transparent as the original plain water. If this milky liquid is left undisturbed for a few hours, the milky, white

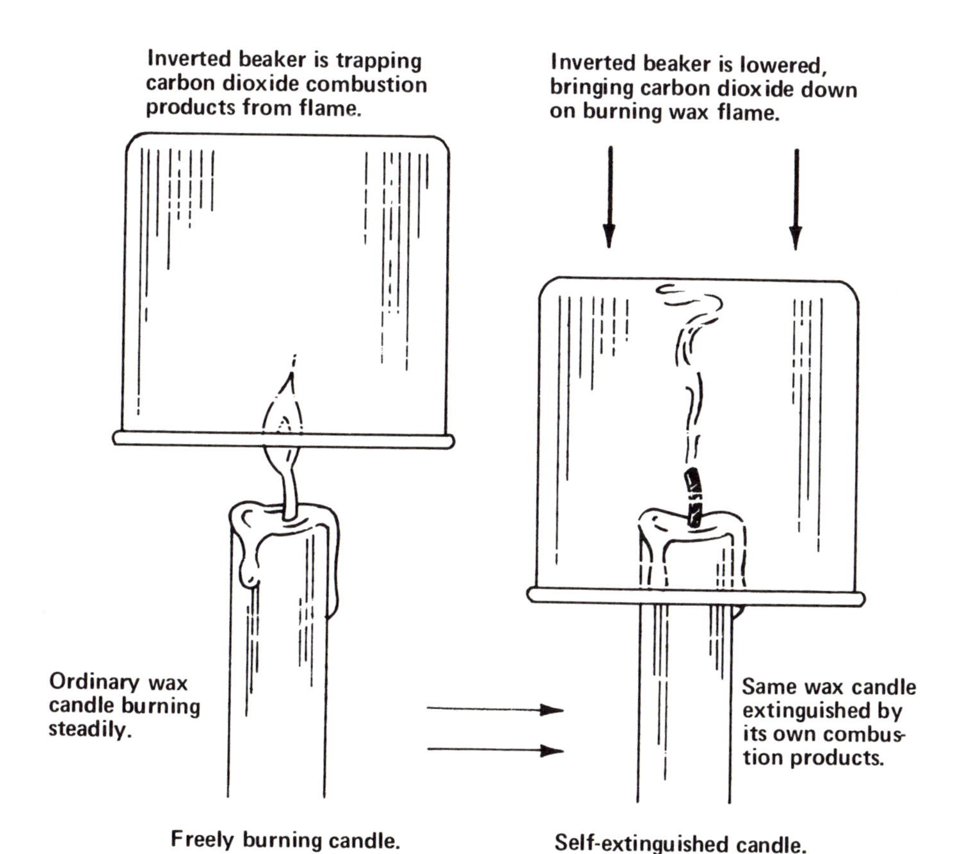

Fig. 3.2. The union of candlewax with oxygen.

particles of magnesia will settle to the bottom of the container, and the plain water can be poured (or decanted) from the top of the container. This process can be repeated until only the original, undiluted milk of magnesia remains.

The process of decanting water from a liquid mixture, and the process of water evaporating from a glycerin and water mixture are known as physical processes. Physical processes employ forces of nature that do not change the properties of the original material. Chemical processes result in changes in the original material so that the material possesses entirely new properties (as with candle wax, which becomes a gas and water vapor when burned or chemically united with oxygen).

SOLUTIONS AND SOLVENTS

In our world of matter — in addition to solids, liquids, gases, and the many compounds and mixtures of each — there is another important classification. This classification is known as solutions. Basically, any perfect mixture of one substance with another can be called a solution (such as common salt crystals mixed with water to form a salt solution, or ordinary wax dissolved in gasoline to form a wax solution). The properties of the solution become a blend of the properties of each of the components that make up the solution. In the preceding examples, the water and the gasoline liquids are called solvents. The salt solution is a salty tasting, watery liquid from which the water solvent can be separated by simple evaporation or drying; the wax solution becomes a slippery, waxy liquid from which the gasoline may be evaporated or vaporized to leave a thin, surface coating of wax.

Solutions are commonly thought of as liquids. However, in the chemical world some solids are also considered to be solutions. For example, consider some of the important metal alloys such as brass, coin silver, and stainless steel. Ordinary brass is a mixture of 67 percent copper, and 33 percent zinc. Both copper and zinc are solid, metallic elements. When melted together at high temperatures, the resulting solution or mixture (or alloy, as it is more usually called) becomes — when cooled to ordinary temperatures — brass, with characteristics that are a blend of the two original, solid metals. The dilution of the bright, chocolate-brown color of copper with the grayish, bright-metal color of zinc gives brass its yellowish coloring. Brass alloy is more easily worked into various shapes and into metal items that are more serviceable and more durable than are soft copper items or items that are made from the harder, more brittle, zinc metal.

In a similar manner, the alloy (or solution, or mixture) of 90 percent silver and 10 percent copper gives a new metallic mixture with more useful and durable properties than pure silver. When 79.8 percent iron, 16 percent chromium, 2 percent nickel, 1 percent manganese, 1 percent silicon, and 0.2 percent carbon are combined at high temperatures, a stainless steel alloy is produced. This stainless steel alloy has useful and important properties that cannot be obtained with any of the single constituent metals that make up the composition of the mixture.

Another type of solution of importance to the chemical world involves gas in a liquid solvent. Perhaps the most common example of such a solution is ordinary soda or "seltzer" water (the basis of soft drinks). When carbon dioxide gas bubbles up through water (carbonation), the slowly dissolving gas gives the water different properties and an entirely new taste. When gas is charged through the water at a higher than ordinary atmospheric pressure, even more carbon dioxide is dissolved in the water and, when the pressure is released, the surplus gas "pops" out of the solution as tiny bubbles. When a soft drink is tasted, these gas bubbles give a pleasurable sensation in the mouth.

Carbon dioxide solution in water is often referred to as carbonic acid, as though a new compound had been formed by physically mixing a gas and a liquid. The water solution of carbon dioxide is a weak acid, and in some cases reacts in a manner similar to other acids. It forms carbonates. However, in this case the strength of the atomic bonds that bind the atoms together is very weak and can be easily broken. The carbon dioxide can be removed from water solutions by merely heating it. Other gases dissolve in water quite easily too. Air dissolves in water when freshly boiled water is exposed to air. The process of aeration (in which the air is more intimately mixed with water and caused to circulate in the liquid) will more quickly aerate or dissolve air in water.

Table 3.1 is a description of the many possible kinds of solutions. In Table 3.1 the word "suspensions" occurs. This is because some solutions are time dependent or temporary in nature. Such solutions are not true molecular mixtures in that the material in the solvent (smoke particles in gaseous air, or water droplets in air that appear as "fogs") is large in size and is subject to gravitational forces. Suspensions will settle out their suspended material when it is in a liquid or gaseous solvent and the mixture separates in this manner. In solids, a gas may become imprisoned and cannot separate. Foam rubber or foamed plastics exhibit this property. "Occluded" water is another special case. Certain crystals form in nature under ordinary conditions with "water of crystallization" attached to the crystal or molecular structure. Epsom salts are a compound of this type. When ordinary epsom salts crystallize from their water solution, they are obtained as a dry, free-flowing crystal

Table 3.1. Various Formations of Solutions

		OCCURRENCE	TYPES	EXAMPLES
Solids in	Solids:	Common	Alloys	Bronze, steel, etc.
	Liquids:	Very common	Liquid solutions	Salt water, lime water
	Gases:	Common	"Suspensions"	Smokes, etc.
Liquids in	Solids:	Rare	Various salts or minerals	"Occluded" water
	Liquids:	Very common	Solutions, tinctures, etc.	Perfumes, etc.
	Gases	Common	"Suspensions"	Fogs, mists, etc.
Gases in	Solids:	Rare	"Imprisoned suspensions"	Light minerals, plastic foams
	Liquids:	Common	Dissolved gases	Soda water, beer, etc.
	Gases:	Common	Gas mixtures	Air, helium-oxygen

powder. These crystals (of magnesium sulfate) contain 50 percent water; however, since the salts are in "dry" powder form, the water cannot be detected by ordinary means. The powder may be heated to drive off the water of crystallization. The very dry crystals then are spoken of as "anhydrous" material.

With solutions, the most important physical force that can be brought to bear is the application or extraction of heat. When heat is applied to solids and liquids, their universal action is to promote the solubility of other substances in them: the only exception is the dissolving of *gases* in liquids and in some solids. Here the opposite is true. The temperatures must be *lowered* in order to promote the solubility of gases in such materials.

Common examples of hot water dissolving soluble substances (like sugar or soap powders) quicker and more completely than does cold water can be found in many everyday situations. Alloys of metals cannot be made until their high temperature melting points are reached and they mix and dissolve into each other as liquids. An example of solubility is the auto engine that is stalled due to cold weather. At ordinary temperatures, gasoline fuel will contain a small amount of water dissolved in it. As temperatures are lowered, this water separates from the gasoline solution in the form of tiny, suspended droplets. If temperatures are lowered below the freezing point of water, the droplets freeze into crystals that collect and block the fuel line into the engine. The addition of small amounts of isopropyl alcohol to such fuels will act as a more powerful solvent for the water, and the resulting gasoline-alcohol-water mixture (or solution) will not freeze in very cold weather.

The solubility of gases in liquids can be demonstrated by again using the example of carbon dioxide in water to make soft drinks. The colder the liquid, the more carbon dioxide gas the liquid will hold; and, as a soft drink warms up, the bubbles of gas form more rapidly and are released from the container. The phenomenon of bubbles forming on a glass of cold drinking water that has been left in a warm room is still another example of gas (air dissolved in the cold water) not remaining in the dissolved state as the temperature of the water increases.

In certain large refrigeration or industrial ice manufacturing plants, ammonia gas is used as a refrigerant. This gas is very soluble in cold water. Cold water sprays are suitable for dealing with the emergency escape of this gas in the event of a fire in such an installation. However, care must be taken to ensure that the cold water solution of ammonia gas does not become heated from the hot fire or from heated nearby surfaces. The diminishing solubility of gas in heated water would be quickly evident.

When considering solutions, it is important to note that in most solids, gases are much less soluble than in liquids. However, exceptions to this are the special cases of imprisoned gases suspended in solids like foam rubber, or in porous substances like certain charcoals or other similar absorbent materials. When a liquid in which gas has been dissolved is cooled to below its freezing point so that the liquid is solid and crystalline, any dissolved gas will be released from the material. This accounts for the fact that a capped bottle of soft drink or beer may generate enough carbon dioxide gas pressure within the bottle to explode the bottle if the liquid is completely frozen.

The solubility of gases in liquids and in some solids is also influenced by pressure increase. If gases are caused to permeate liquids at elevated pressures above atmospheric pressure, more gas will be dissolved in the substance. The release of this pressure will sometimes result in an almost explosive loss of gas from the solution. One need only to open a bottle of champagne to discover this fact.

OXIDATION REACTIONS

Of major importance to professional fire officers are the chemical reactions that take place with oxygen — principally with the oxygen in the air we breathe. These are oxidation reactions. Fire is intimately concerned with the process of oxidation, usually a very rapid oxidation. New compounds are formed in this chemical process, and an understanding of fire must be based on a knowledge of these reactions.

We live at the bottom of a sea of air. Air is our atmosphere, and is the only atmosphere in the universe that supports life processes. This air is a constant ratio mixture of approximately twenty-one percent by volume oxygen gas and seventy-nine percent by volume nitrogen gas. There are other gases present too. They are called the "rare" gases, examples of which are xenon, krypton, neon, and argon. However, these gases do not figure in the process of oxidation or fire because they are present in such small amounts (about 1 percent), and are almost totally inert to chemical reactions of any kind.

It is both interesting and fortunate that this twenty-one to seventy-nine percent ratio remains constant. The life processes of plant and animal life on our earth, and the natural cycles of the world's weather phenomena operate in such a way that air does not vary in a significant manner from its ratio of nitrogen to oxygen. Even when large catastrophic fires occur in certain areas and oxygen is consumed rapidly in the wholesale burning process, new air rushes into the area and the original ratio is quickly established again.

Air is not the only source of oxygen for the burning or fire process. There are many substances that contain oxygen and are capable of oxidizing other substances in the rapid manner that is associated with burning. All pyrotechnics contain oxidizers such as nitrates in their composition. These materials react and burn without the aid of the oxygen in air. In most cases involving the burning of pyrotechnic compositions, they have been intentionally provided with more oxygen-carrying material than could be obtained from the oxygen-nitrogen ratio in air; thus, they burn more vigorously. The continued burning of these compositions cannot be easily halted during the oxidation process other than by removing the oxidizer from the mixture: this effort is accomplished by mechanically breaking up the solid mass in its container and isolating its components from further reaction.

The fire process of oxidation can be compared to the life process of the human body. We breathe air and use the oxygen in it to oxidize the combustible substances we eat: fats, sugars, and other materials. Although this is an oversimplified description of a much more complicated process, it illustrates how our useful energy is obtained and how our constant body heat is developed in an oxidative manner.

We even exhale carbon dioxide in much the same manner that a flame produces carbon dioxide when carbonaceous materials burn.

Processes of Combustion

From a basic chemical viewpoint, the dissection of the oxidation process during the burning of a fire is a complicated task. For most purposes, the explanation of the types of combustion may be separated into two simple categories.

Type 1: Direct oxidation of combustible gases, liquids, or solids that need not undergo decomposition or change by pyrolysis* in order to oxidize or burn.

Type 2: Sequential oxidation of combustible liquids or solids that must pyrolyze in order to oxidize or burn.

With Type 1 combustion or burning, all that is necessary to promote a fire is an ignition point; or, more exactly, the substance must be raised to its ignition temperature in order to oxidize rapidly with the evolution of heat and light. For example, when a burning match is applied to a flow of natural gas (which is used universally as a source of heat for domestic and industrial uses), the gas immediately bursts into flame. This is because the match flame has raised the gas to its ignition temperature within the vicinity of the burning match, and the natural gas is influenced by this heat to combine with oxygen in the same gaseous envelope. The flame of burning gas propagates (at a finite and measurable rate) in admixture with air, and continued combustion ensues. The chemical equation for this combustion would be:

$$\underset{\text{Methane}}{CH_4} + \underset{\text{Oxygen}}{2O_2} \rightarrow \underset{\substack{\text{Carbon}\\\text{dioxide}}}{CO_2} + \underset{\substack{\text{Water}\\\text{(vapor)}}}{2H_2O}.$$

Similarly, the burning of ordinary sulfur is a Type 1 combustion. When a burning match or source of heat is applied to solid sulfur or sulfur powder, the material melts first. Then it vaporizes to produce a mixture of sulfur vapor and air around the burning match flame. This mixture reaches its ignition point and unites with oxygen in the air. The heat from this reaction melts more sulfur, thus vaporizing more of the solid, and the process continues. A fire can then be said to be established. The chemical equation for this combustion would be:

$$\underset{\text{Sulfur}}{S} + \underset{\text{Oxygen}}{O_2} \rightarrow \underset{\substack{\text{Sulfur dioxide}\\\text{(gas)}}}{SO_2}.$$

The burning of paper or wood is an example of Type 2 combustion or burning. Paper or wood are complicated chemical materials containing carbon, hydrogen, and oxygen united into a mixture of organic molecules. In order to burn, these

* Chemical decomposition or other chemical change brought about by the action of heat, regardless of the temperature involved.

molecules must break down, or pyrolyze, under the influence of heat to produce simple, combustible gases and solids that will unite with the oxygen in air. When a burning match is applied to a piece of paper, the carbon, hydrogen, and oxygen compounds begin to decompose and break down into different compounds or into their constituent elements. Some pyrolysis products of paper will be combustible gases and will ignite in the burning match flame. Another product of decomposition will be carbon, which will ignite in the flame envelope. Almost all pyrolysis products of paper or wood will burn and generate heat, which then feeds back to continue the combustion process.

There are many carbon-containing solids (except carbon or charcoal themselves) that undergo the Type 2 combustion process. In the real world of combustible substances, almost all solids and liquids having in their composition carbon or hydrogen (or both) will burn by pyrolysis and sequential oxidation as described by the Type 2 process of combustion. A brief comparison of common combustible materials classified according to their types of burning processes is presented in Table 3.2.

Oxidizing Agents

When oxygen unites with a substance in the burning process or the "fire reaction," it is called an *oxidizing agent.* There are many oxidizing agents other than the gaseous oxygen contained in air. Some oxidizing agents, although not composed of oxygen, can cause reactions that give off heat (but not necessarily flames) that can be considered as *fire.*

It is difficult to define oxidizing agents as a class of substances without going into considerable detail about atomic structure and bond strengths. For purposes of this text, oxidizing agents are considered on the basis of what is contained in their composition.

In general, the oxidizing agents that may be concerned in fire reactions contain the following elements:

Oxygen (peroxides).
Oxygen and nitrogen (nitrates and nitrites).
Oxygen and chlorine, bromine, etc. (chlorates, etc.).
Oxygen and sulfur (persulfates).
Fluorine, chlorine, bromine (gaseous elements, etc.).

There are many other types of oxidizing agents of lesser importance to the problems of general fire protection. A compilation by sources within the chemical field is presented in NFPA 491M, *Manual of Hazardous Chemical Reactions.*

REDUCTION REACTIONS

In the previous section of this chapter, the addition of oxygen to substances by the process of oxidation was discussed. In the world of chemical reactions, the opposite to this — the subtraction of oxygen from substances that contain oxygen

Table 3.2. Brief Comparison of Common Combustible Materials Classified According to Their Types of Burning Processes

SUBSTANCES THAT BURN BY THE TYPE 1 PROCESS OF COMBUSTION	SUBSTANCES THAT BURN BY THE TYPE 2 PROCESS OF COMBUSTION	
Methane (natural gas), CH_4	Wood	"Celotex" insulation
Acetylene, C_2H_2	Paper	Rubber
Sulfur, S	Paints (oil)	Dry leaves
Gasoline vapor, C_8H_{18}	Cooking oils	Bread
Alcohol, C_2H_5OH	Plastics	Roofing tar
Charcoal, C	Fabrics	Leather

— is of importance. This process is called reduction. Because of the excess of oxygen that usually exists during unwanted fire and combustion processes, the reduction process is not often encountered in fire fighting situations.

The oxygen reduction reaction takes place only when there is a deficiency of oxygen in the combustion process that generates a *reducing atmosphere* in the fire. When these conditions occur, reducing agents or compounds capable of extracting oxygen or reacting rapidly with oxygen are produced in the fire zones. The presence of these reducing agents in the reducing atmosphere of a fire of this type is of extreme importance to the fire officer.

When combustion of ordinary materials takes place in a closed space and oxygen is hindered from entering the space, the combustion process does not continue at its normal rate and the fire is said to be "starved" for oxygen. When this occurs, the normal burning of carbonaceous materials can no longer continue to produce the completed combustion product of carbon dioxide according to the following complete combustion reaction:

$$C + O_2 \rightarrow CO_2.$$

Instead, a reducing agent — carbon monoxide gas — is produced under this restricted burning process:

$$2C + O_2 \rightarrow 2CO.$$

The presence of carbon monoxide in the fire zone results in a reducing atmosphere that seeks oxygen in any form. If a piece of old tarnished red or oxidized copper were heated by the fire in this reducing atmosphere, the carbon monoxide would react with the reddish copper oxide on the old metal and a bright shiny surface would result according to the following reaction:

$$\underset{\text{copper oxide}}{\overset{\text{Red}}{CuO}} + CO \rightarrow CO_2 + \underset{\text{copper}}{\overset{\text{Bright}}{Cu}}.$$

Copper oxide is said to be *reduced* in this reaction.

Additionally, the generation of the highly combustible and very poisonous carbon monoxide gas in a reducing atmosphere poses an explosion threat to the fire fighter if ventilation of the space suddenly occurs. The breathing of such atmospheres without self-contained oxygen breathing equipment is highly dangerous.

Other reducing atmospheres can be produced in fires where oxygen starvation has occurred. Under some conditions of combustion, dense and finely divided carbonaceous particle smoke or suspensions may be generated. These smokes are reducing atmospheres capable of rapid combination with oxygen when the space is ventilated.

There is another reaction, a reaction that produces a reducing and highly flammable atmosphere, that occurs in fire fighting operations under certain conditions. This may take place in an area that may *not* be starved for oxygen. If the two materials — carbon and water vapor (steam) — are present, and both are heated to a high temperature by the fire, the following reaction may take place:

$$C + H_2O \rightarrow CO + H_2.$$

This reaction is known as the "water gas" reaction, and may take place at any time during an intensely hot fire. Both the hydrogen gas and the carbon monoxide are flammable reducing agents.

Reducing Agents

There are many types of reducing agents in the chemical world. Most of them are never encountered in the field of fire protection. The only ones that concern us are carbon, hydrogen, and carbon monoxide, all of which are highly flammable. It is well to remember that almost all flammable gases or vapors can develop a reducing atmosphere if oxygen starvation takes place in a fire involving these materials.

EXOTHERMIC AND ENDOTHERMIC REACTIONS

The science of chemistry, like all other sciences, deals with negative effects as well as positive effects in relation to all characteristics possessed by matter of all kinds. When we consider the combustion reactions of materials and the chemistry of these reactions, it would seem that all chemical reactions produce heat in varying amounts and rates. This is not so. Substances may undergo reactions and chemical changes in either one of two ways: (1) they may give off heat in the course of the reaction, or (2) they may absorb heat or energy during the reaction. The ordinary oxidation or combustion of substances gives off heat or energy, and is called an exothermic reaction. The burning of charcoal or methane is an exothermic reaction. The opposite type of reaction, in which substances absorb heat and require a continuous input of heat or energy to sustain the reaction, is called an endothermic reaction. When water is decomposed by having an electric current change it into its hydrogen and oxygen constituents, it absorbs energy and the process is called an endothermic reaction. Similarly, mercuric oxide can be decomposed into its components, mercury metal and oxygen gas, by the continuous application of heat. This is an endothermic reaction. Many chemical reactions of the reduction type, as mentioned earlier, are endothermic in character.

ACTIVITIES

1. Write detailed explanations for each of the following:
   (a) What happens when a hospital patient being given oxygen decides to smoke a cigarette?
   (b) What happens if a glass bottle containing a soft drink is placed in a freezer for an appreciable amount of time?
   (c) Why is it that blowing on a candle flame extinguishes the fire instead of making the burning more intense? (Compare this with the opening of a window during a fire.)
2. In warm temperatures, water droplets form on the outside of cold-drink glasses. Explain why water droplets don't also form on the outside of warm beverage glasses in cold temperatures.
3. Explain what you think takes place when a water slurry of fresh cement or concrete becomes a dry, solid, stony material after it has "set".
4. Iron rusts quickly (oxidizes) when it is exposed to the rain. However, if iron is submerged in *pure* water in a closed system (like a marine boiler), no rusting takes place. How can you account for this?
5. If a small amount of iron filings are mixed with a small amount of powdered sulfur, a gray-yellow powder is produced. If this is a mixture, not a compound, the iron filings can be retrieved unchanged.
   (a) How can this be accomplished?
   (b) If we expose the same powder to the heat from the flame of a propane gas burner, it would suddenly start to glow and become black. What has happened? Can we now retrieve the iron filings in some way? What has been produced, if anything?
6. Which of the following are Type 1 processes of combustion, and which are Type 2? Why?
   (a) Ignition of a matchstick.
   (b) Burning leaves.
   (c) A burning candle.
   (d) Explosion of an LPG tank's contents.
   (e) Ignition of rugs.
   (f) Ignition of photoflash bulbs.
7. With your classmates, discuss accurate definitions for the terms mixtures, solutions, and solvents. Then give ten examples of each.
8. Explain why you think each of the following terms can or cannot be used to describe ordinary oil-based paints.
   (a) A solution.
   (b) A suspension.
   (c) A mixture.

(d) A compound.
(e) A mixture of compounds.
(f) A liquid.
(g) A solvent.
(h) A substance capable of oxidation.
(i) A combustible.

9. Alcohol and gasoline are both excellent fuels. Both alcohol and gasoline undergo Type 1 processes of combustion. However, alcohol burns with a nonluminous flame, and gasoline has to have air forcibly supplied to it in order to burn without a luminous flame containing burning carbon particles. Explain why this is so.

SUGGESTED READINGS

Brandwein, P. F., Stollberg, R., Burnett, R. W., *Matter, Its Forms and Changes*, Harcourt, Brace and World, New York, 1968.

Clark, G. L. and Hawley, G. G., *The Encyclopedia of Chemistry*, 2nd Edition, Van Nostrand-Reinhold, New York, 1966.

Fire Protection Handbook, 14th Edition, National Fire Protection Association, Boston, 1976.

Johnson, R. H. and Grunwald, E., *Atoms, Molecules and Chemical Change*, 2nd Edition, Prentice-Hall, Englewood Cliffs, New Jersey.

Mack, E., Garrett, A. B., Haskins, J. F., Verhoek, F. H., *Textbook of Chemistry*, Ginn, Lexington, Mass., 1956.

Maher, J. T., *Essentials of Fire Science*, Grid, Columbus, Ohio, 1973.

Perros, T. P., *Chemistry*, American Book, New York, 1967.

Sorum, C. H., *Fundamentals of General Chemistry*, Prentice-Hall, Englewood Cliffs, New Jersey, 1963.

The Van Nostrand Scientific Encyclopedia, 3rd Edition, D. Van Nostrand, New York, 1959.

Timm, John A., *General Chemistry*, 4th Edition, McGraw-Hill, New York, 1966.

The Relationship of Fire and the Physical World

THE PHYSICAL CHARACTERISTICS OF MATERIALS

The principal emphasis in the previous chapters of this text has been on the chemical characteristics of the material composition of our physical world, and on how chemical reactions completely change the properties of the components of substances. This chapter presents the effects of changes in the *physical* characteristics of materials when acted on by physical forces such as those caused by fire — *i.e.*, heat.

The physical properties or characteristics of a compound or a substance are those properties that change only when acted on by external forces of energy (such as heat energy), and in which no change in the composition of the substance occurs, nor does the substance combine with its surroundings.

As explained in Chapter 3, "The Chemical World and Some of Its Reactions," substances exist in one of three physical states: (1) solid, (2) liquid, and (3) gas. When these purely physical states of matter are changed (as from a solid state to a liquid state — such as from ice to water), the material undergoes a physical change. However, the material's chemical properties do *not* change. As will be explained, there are many physical properties of substances.

THE GENERAL FIRE CHARACTERISTICS OF SOLID MATERIAL

Very few of the statements that can be made about the combustion characteristics of solid materials (as a large, main class) can be consistently applied to the many types of solids that unite with oxygen in fires. That is, each type of combustible solid substance has its own peculiar reactions when exposed to heat, sources of ignition, and fire conditions.

The temperature and heat input of solid materials are important factors in fire situations. For example, as soon as elemental (or yellow) phosphorus contacts air at room temperature, it unites with oxygen to produce a low, luminous flame. To prevent this burning reaction, it must be stored under water. Magnesium metal in the form of a ribbon or wire can be burned if held over a burning match so that its temperature is raised to about 1,100°F (593.3°C). This, of course, could not be done if the metal was in the form of a thick bar or rod.

As explained in Chapter 3, "The Chemical World and Some of Its Reactions," fire and combustion may proceed in many ways in solids. The most frequently used combustible solids (such as wood, paper products, fabrics of all kinds, and plastics) contain complicated compounds of carbon and hydrogen, with other elements such as oxygen, sulfur, nitrogen, etc. These materials must undergo a decomposition and pyrolysis from heat before ignition can occur.

It is important to note that the flames that occur during the combustion of all solids are always the result of the rapid oxidation of the gases and vapors that are produced during the pyrolysis process. When pyrolysis can be retarded by some means (cooling with water or removing heat or oxygen), combustible gases and vapors will stop evolving and combustion will cease.

LIQUIDS: THEIR VAPOR PRESSURE, BOILING POINT, AND VAPOR DENSITY

All substances in the form of liquids (and even many solids) possess a type of molecular motion that results in the escape of molecules from their surface in the form of vapor when they are not confined. When water is left in an open container at room temperature, its molecules evaporate. When water is confined in a partially full container that is *closed*, the molecules of water will continue to escape from the surface; however, because they cannot escape from the closed container, some of the molecules will return to the water. Within a short time an equilibrium will be set up between the numbers of molecules escaping from the surface and those returning to the surface of the water. When this equilibrium occurs, a certain pressure will be exerted in the empty space above the liquid in the closed container. This is called the vapor pressure of the liquid water. In accordance with the molecular kinetic theory of gases (see page 9), the numbers of molecules — and thus the magnitude of the pressure exerted on the vessel — rises with rising temperature. The molecules become more agitated as heat is added. Vapor pressures vary widely among solid and liquid compounds.

There is a continuous atmospheric pressure of 14.7 pounds per square inch absolute (1.01 newtons per square metre, or 1 bar) continuously being brought to bear on the surface of a liquid contained in an *open* container. As previously stated, a rise in temperature produces a rise in the vapor pressure above the liquid and, thus, a greater rate of escape of molecules (or vapor). If the temperature of a liquid (such as water in an open container) continues to rise due to the application of heat, large bubbles will begin to rise from the liquid and burst at its surface at a rapid rate. A thermometer immersed in the rapidly bubbling and vaporizing water would read 212°F (100°C); the thermometer would stay at this temperature as long as any liquid water remained in the container, and as long as heat continued to be applied to the container. Also, if the *upward* pressure of the vapor above the bubbling surface of the water was measured, it would equal the *downward* normal atmospheric pressure pressing on the liquid in the open container. This, then, is called the boiling point of the liquid water. At this point the vapor pressure of the water equals the atmospheric pressure pressing upon it; as long as heat is supplied to it, the liquid boils in the attempt to release its molecules to the vapor state (see Fig. 4.1).

The boiling points of different types of liquids vary widely. They are an important physical characteristic both of liquids and of the many solids that melt to become liquids and then boil at a certain characteristic temperature.

Atmospheric pressure has considerable bearing on boiling points — the temperature at which vaporization takes place at a fast rate. When pressures on the liquid surface are high, such as in a closed marine high pressure power plant system, water boils at much higher temperatures than normal; when atmospheric pressures are low, such as at the top of Mount Everest, water boils at much lower temperatures. (For this reason, it takes 8 minutes to boil an egg on Mount Everest, while it takes 3 minutes to boil an egg elsewhere.)

Vapor density is a physical property of major importance to members of the fire service. Because the vapor density varies with the total weight of all the atoms in a molecule of the vapor of a substance, if the chemical composition of the substance comprising the vapor is known, then the weight or density of its vapor when compared to air can be determined as in the following relationship:

$$\text{Vapor density} = \frac{\text{Molecular weight of the substance being vaporized}}{\text{29 (The molecular weight of air)}}.$$

From this formula it can be seen that any vapor from a substance with a molecular weight of 29 will give a vapor density of 1.0. A substance with a higher molecular weight gives a vapor density over 1.0, and lower will be less than 1.0.

IGNITION TEMPERATURE

All combustible substances do not "catch fire" or begin to burn at the same or at any minimum temperature. The point at which they *ignite* is characteristic of each substance, and is determined by the composition and properties of the particular substance. As stated earlier, burning and combustion are chemical reactions. In order for these chemical reactions to take place, the molecules that comprise a combustible substance must be brought up to a certain temperature by the addition of heat energy so that the substance's molecules are ready to combine with the oxygen molecules in the air. This temperature is called the ignition temperature. At this temperature, the combustion reaction continues without any external input of heat since the substance gives off heat by its own combustion and the burning becomes self-sustaining.

(1) Water ready to be heated

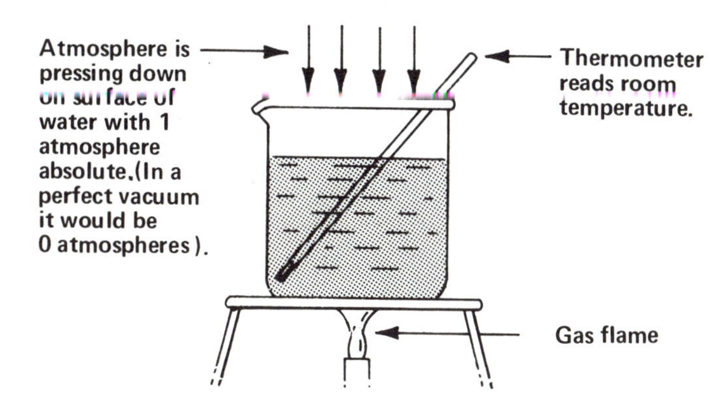

(2) During heating process

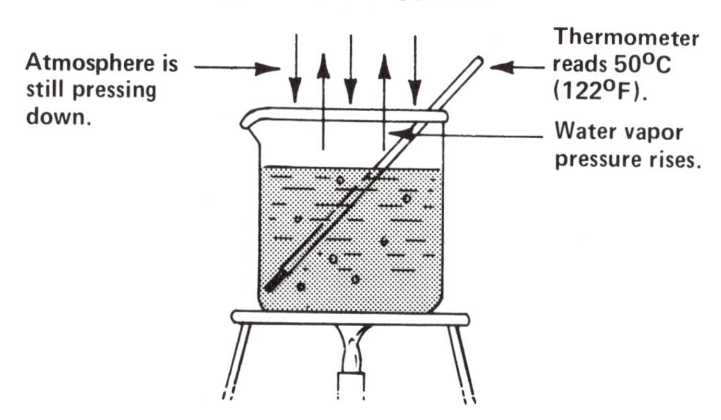

(3) At the boiling point of water

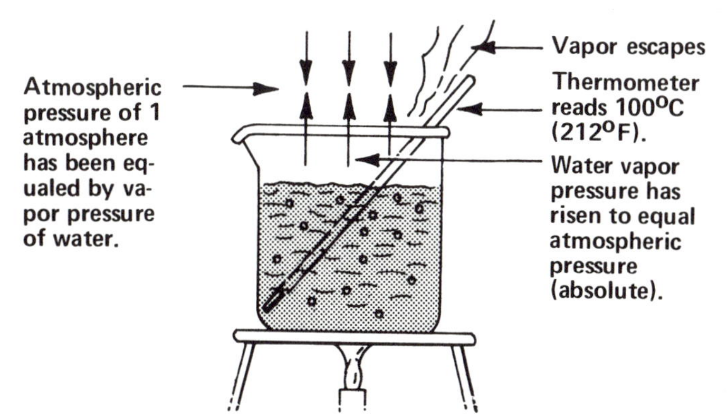

Fig. 4.1. How water boils.

The ignition temperatures of combustible materials are neither accurate nor dependable values because of the varied conditions of differing fire situations. Although test methods have been used to add heat to substances and to measure the temperatures at which substances begin to burn, the fire science student should keep in mind that these tests are only approximations. While a substance may burn at a certain temperature in a particular fire situation, if the situation is changed even slightly the ignition temperature will also change. In actual fires, the conditions surrounding combustible materials may vary in many ways; for this reason, ignition temperatures of materials are only rough approximations made from synthetic tests.

Flash Point

There are a few accurate and dependable measured tendencies of combustion that are of use when considering the fire characteristics of combustible and flammable liquids. One of the measured tendencies of combustion that is both accurate and dependable is the flash point of the liquid.

The flash point of a liquid is determined by slowly raising the liquid's temperature (raising its vapor pressure) until it gives off flammable vapors in the right amount of mix with air, and a flammable mixture of vapor and oxygen is formed. This flammable vapor mixture is easily detected by passing a source of ignition (a flame) through it, thereby flashing the mixture. Momentarily, a flame will occur in the mixture. The temperature at which this occurs is the flash point of the liquid. (The flame does not continue to burn at the flash point when the source of ignition is removed: it only flashes.) The flash point is a distinctive characteristic of each flammable or combustible liquid, or of mixtures of them.

Because it is an indicator of the degree of safety of a material, the flash point of a liquid is one of the most important fire characteristics of substances. At its flash point, a liquid continuously produces flammable vapors at the right rate and amount (volume) to give a flammable and even explosive atmosphere if a source of ignition should be brought into the mixture. Flammable liquids (like gasoline) with a flash point of minus 45°F (minus 42.9°C) continually give off vapors that can burn or explode (depending on the confinement of the mixture) at ordinary temperatures. However, fuel oil (such as that used in home-heating furnaces) with a flash point of 130°F (54.4°C) does *not* give off vapor that can burn until heated above its flash point. Such temperatures are seldom encountered in ordinary living conditions.

> The question of how to fire an ordinary oil furnace in a cold condition is often asked. The atomization of fuel oil by the nozzle in the cold furnace produces many small droplets, each of which is surrounded by a vapor pressure of oil at low temperatures. When a high voltage spark from the igniter courses through this flammable vapor, some of it is in the right admixture with air to ignite. This produces heat. In turn, the heat raises the temperature of other droplets, thereby increasing their vapor pressure. Thus, the process continues at a rapid rate until a complete jet of burning droplets issues from the nozzle.

Fire Point

When flammable and combustible liquids continue to be heated above their flash points, they reach a temperature at which their output of flammable vapors is at a balanced rate with air so that their vapors continue to burn even after the source of ignition has been removed. This is known as the fire point of a substance. The fire point of liquids is always a few degrees above the flash point temperature.

This characteristic property is not widely used, and is usually considered to be of minor importance to members of the fire services. The primary fire danger of materials is that point in time and temperature when potential fire conditions first begin to exist; from that point on they become increasingly more serious. In the case of flammable and combustible liquids, that first point of danger begins at the flash point. Temperatures of liquids must be reduced below this to halt vapor emission. The fire point designates only a degree of hazard, and lowering temperatures below this point does not remove the hazard.

Autogenous Ignition Temperature (AIT)

Earlier in this chapter, the ignition temperatures of substances were discussed. It was pointed out that the ignition temperatures of various materials are not reliable and cannot be depended upon because of the varied conditions of differing fire situations. There exists another type of ignition temperature that must also be considered as an approximation. This ignition temperature characteristic is known as the autogenous ignition temperature, or AIT. It is most frequently used when referring to combustible or flammable liquids.

When the temperature of a liquid is quickly raised above its flash point, its fire point, and perhaps even above its boiling point, it reaches a temperature at which it burns without requiring a source of ignition. This temperature, the autogenous ignition temperature, is the temperature at which ignition of a liquid or its vapor occurs automatically (without a source of ignition such as a match). Each combustible or flammable liquid demonstrates different autogenous ignition temperatures. The determination of these temperatures may be different, depending upon the test methods used.

Because of the variable ways in which fire occurs in liquids, the test values for autogenous ignition temperatures must be regarded as approximations in the practical world of fire protection.

Spontaneous Combustion

It has been established that: (1) combustion and fire are chemical reactions involving oxidation, (2) ignition marks the beginning of burning and combustion, and (3) substances will burn at certain higher temperatures, even without a source of ignition. Another type of "automatic" ignition is known as spontaneous combustion. Spontaneous combustion, as explained in the following paragraph, occurs partly because of chemical reactions and partly because of physical effects on materials.

Slow combustion (in which substances take up oxygen very slowly at comparatively low temperatures) can best be contrasted with *rapid* combustion and fire by using as an example thc simple and slow oxidation of linseed oil. Linseed oil, the essential ingredient of varnishes and oil-based house paint of the "drying oil" type, contains certain organic chemical compounds called esters. Esters are "unsaturated" and have an affinity for combining with oxygen. When these compounds oxidize in air, a certain amount of heat is liberated and the "unsaturated esters" become more fully oxidized and chemically "satisfied." The more fully oxidized compounds are solids and act as protective films. (This is the process by which oil-based paints are able to provide a solid and decorative paint film on wood or metal surfaces.) If the liquid "unsaturated esters" in linseed oil are confined within a layer of thickness of material that insulates and restricts the escape of heat from the oxidation reaction of the esters and yet air has access to the oil, heat generation will continue until it has reached the autogenous ignition temperature of the oil; at this point, rapid combustion will ensue. This is spontaneous combustion. Spontaneous combustion needs no external heating to take place because the insulating effect of the confining material surrounding the oxidizing material holds the heat in, and the temperature and reaction rate of the oil increases rapidly to the self-ignition point.

CLASSIFICATION OF COMBUSTIBLE AND FLAMMABLE LIQUIDS

A discussion of the subjects of flammability, flash points, and other characteristics of liquids pertaining to fire would be incomplete without some information concerning present standards of relative fire hazard classification of flammable and combustible liquids. There are two such standards: (1) the Underwriters Laboratories, and (2) the NFPA National Fire Codes. (See also Appendix III of this text.)

Underwriters Laboratories Classification

This older system of classifying flammable and combustible liquids uses common liquids as reference points with arbitrary "degrees of danger" for each type of liquid, employing a spread of ten numbers with each class, with 100 being the most hazardous (and having the lowest flash point).

Liquid	Relative Flammability Hazard
Ethyl Ether Class	100
Gasoline Class	90 to 100
Ethyl Alcohol Class	60 to 70
Kerosine Class	30 to 40
Paraffin Oil Class	10 to 20

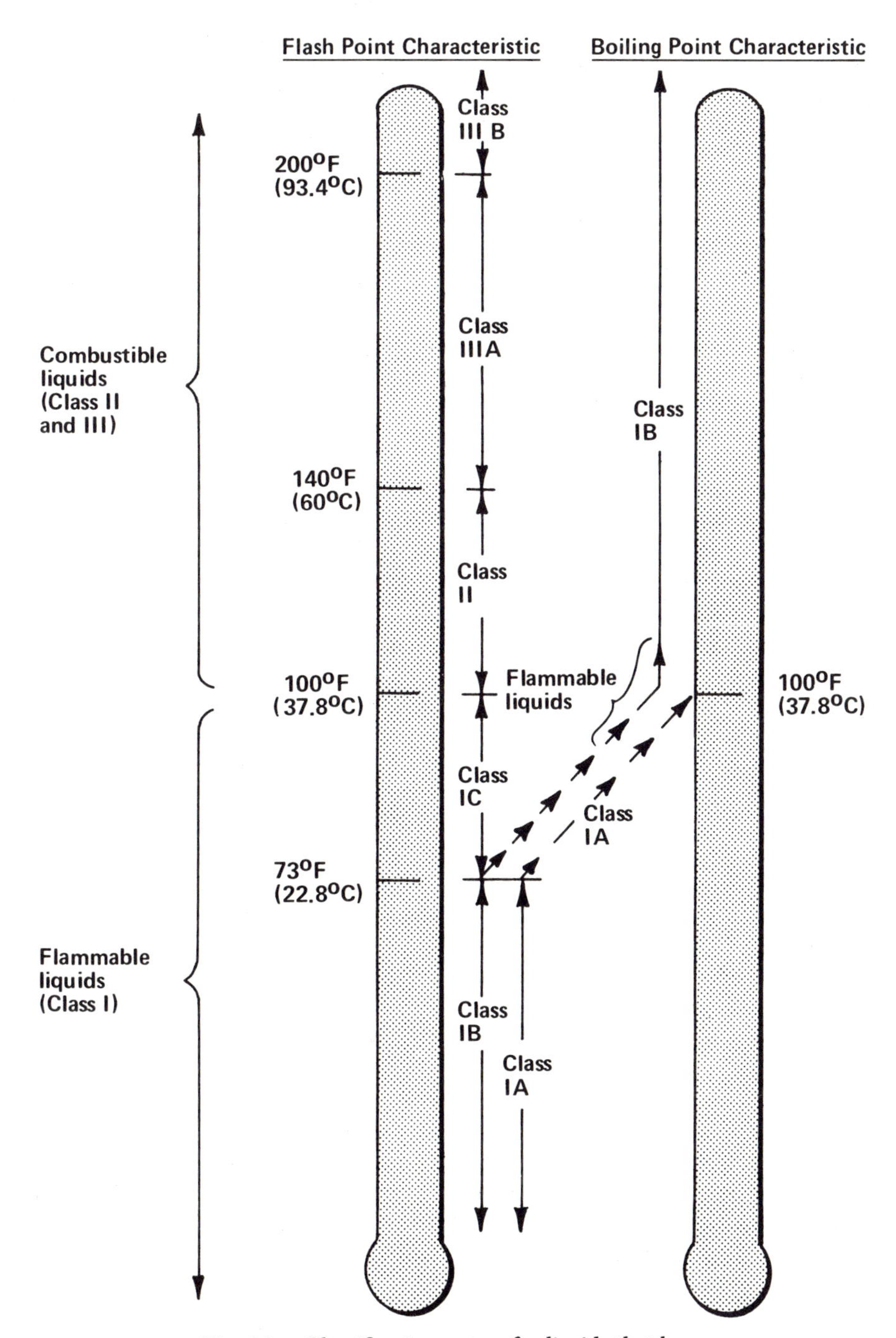

Fig. 4.2. Classification system for liquids that burn.

National Fire Codes Classification

The most useful system for indicating the degree of hazard or fire danger of liquids capable of being ignited and their classifications and labeling is the system developed by the National Fire Protection Association's technical committees on flammable liquids. The NFPA system is reported in detail in NFPA 321, *Standard on Basic Classification of Flammable and Combustible Liquids.* The following definitions are from NFPA 321.

> Liquids are those fluids having a vapor pressure not exceeding 40 psi absolute (2068.6 mm mercury) (3.76 bar) at 100°F (37.8°C) (approximately 25 psi gage) (1.74 bar). (Materials that are solid at 100°F are not included in this classification.)
>
> **Flammable Liquids** are fluids having a flashpoint below 100°F (37.8°C) and having a vapor pressure not exceeding 40 pounds per square inch absolute (3.76 bar) absolute at 100°F (37.8°C). These are Class I liquids. They are subdivided as follows:
>
> *Class IA* liquids are those having flashpoints below 73°F (22.8°C) and a boiling point below 100°F (37.8°C).
>
> *Class IB* liquids are those having flashpoints below 73°F (22.8°C) and having a boiling point at or above 100°F (37.8°C).
>
> *Class IC* liquids are those having flashpoints at or above 73°F (22.8°C) and below 100°F (37.8°C).
>
> **Combustible Liquids** are those liquids that have flashpoints at or above 100°F (37.8°C). They may be subdivided as follows:
>
> *Class II* liquids include those having flashpoints at or above 100°F (37.8°C) and below 140°F (60°C).
>
> *Class IIIA* liquids include those having flashpoints at or above 140°C (60°C) and below 200°F (93.4°C).
>
> *Class IIIB* liquids include those having flashpoints at or above 200°F (93.4°C).

A graphic thermometric representation of this classification system is depicted in Figure 4.2.

FLAMMABLE GASES AND VAPORS

The fire characteristics of flammable gases and vapors are closely related to the special physical (and chemical) characteristics of this type of matter. All gases and vapors actually consist of molecules in free and random motion. These molecules are capable of mixing with other gaseous molecules at rapid rates, and the mixture formed is uniform throughout its mass. Mixtures of air with its oxygen content and flammable gases usually burn at very rapid rates of flame propagation called explosive rates of burning.

There are a large number of gases and vapors that are capable of uniting with the oxygen of air. The gases that do not burn are relatively few and are called the inert gases. Examples of inert gases are: nitrogen, helium, carbon dioxide, steam, carbon tetrachloride vapor, etc.

There are oxidizing gases other than oxygen that can propagate flame when mixed with combustible gases. Chlorine gas is capable of reacting much like oxygen under certain conditions, as is nitrous oxide gas, a compressed gas used as an anesthetic.

Flammable gases, vapors, and even suspensions of very small particles of combustible solids (such as smokes and dusts, which may be considered as having the characteristics of a gas) are the basic compositions of flames. All solids and liquids

Table 4.1. Limits of Flammability of Some Ordinary Gases and Vapors

GAS OR FLAMMABLE LIQUID	LIMITS OF FLAMMABILITY (PERCENT IN AIR, BY VOLUME)	
	LOWER	UPPER
Acetone (Solvent)	2.6	12.8
Acetylene	2.5	81.0
Alcohol (Liquor)	4.3	19.0
Benzene	1.3	7.1
Butane (LPG)	1.9	8.5
Gasoline (Motor)	1.4	7.6
Hydrogen	4.0	75.0
Jet Fuel	0.6	3.7
Kerosine	0.7	5.0
Methyl Ethyl Ketone (Solvent)	1.8	10.0
Natural Gas	3.8	17.0
Propane (LPG)	2.2	9.5
Sewage Gas	6.0	17.0

must undergo processes of either decomposition or vaporization before combustion can take place. These processes produce gases or vapors that become flames when suitably ignited. Thus, the production of flammable gases *must first occur* before a flame can be produced in a fire.

FLAMMABLE LIMITS OF GASES AND VAPORS

One of the most characteristic properties of flammable gases and vapors is their tendency to burn or to explode when present in air in different amounts. However, no ignition will result when only a small amount of flammable gas (such as natural gas) is released into a large volume of air (such as a roomful) and the mixture is probed throughout with a spark or a match. This is because the mixture is too diluted to burn; because it is too "lean," it is nonflammable. Similarly, no ignition will result when a much larger amount of flammable gas is released into a roomful of air (flammable gas equal to 20 percent of the volume of the room) and the mixture is probed throughout with a spark or a match. This is because the mixture contains an excess of gas in comparison with the oxygen that is present: because it is too "rich," it is nonflammable.

All combustible gases, vapors, and smoke or dust suspensions possess definite limits of concentrations in air below which, or above which, they will not propagate or sustain flame. These limits are called the lower limit (lean) and the upper limit (rich) of flammability. They may also be called the lower and upper limits of combustibility, or the explosive limits of a gas or vapor.

The limits of flammability of a combustible gas or vapor are directly related to the chemistry of the combustion of each compound making up the gas or vapor. Thus, each combustible substance has its individual limits, as shown in Table 4.1.

The values for the flammable ranges of the materials included in Table 4.1 are based on gases and air that have been well-mixed. In the real-life experiences of fire protection personnel, such mixtures might never occur. Sometimes when flammable gases or vapors arise from a source they are not well-mixed, and certain areas in the mixture might contain the correct concentration of gas and air; an ignition source could ignite such an area, and flames could result. Although such flames would probably be temporary, a full-scale burn or explosion could ensue if the gas or vapor evolution were to continue.

There are three other important factors that influence the flammable ranges of gases and vapors. These are temperature, pressure, and oxygen content in the atmosphere. Although in actual practice these factors may not be present, under certain conditions their effect can be seen.

At high ambient or existing temperatures (such as one would find in an oven or infrared tunnel for "baking" paint coatings), the lower flammable limit decreases and smaller amounts of flammable gas become quickly ignitable. Similarly, if one were to decrease the pressure on a given volume of gas and air mixture, the flammable limits would decrease.

When an increase in oxygen content occurs (as in a hyperbaric chamber), the upper flammability limit of gases and vapors increases drastically so that much wider flammability limits are produced for any flammable gas or vapor in this kind of oxygen-enriched atmosphere. Thus, a much greater fire hazard exists when a combustible liquid or a gas in present in air that contains high oxygen concentrations. This is also true of combustible solids once they become ignited and flammable gases are produced to react with the higher oxygen content of such an atmosphere.

ACTIVITIES

1. Kerosine has a flash point of about 145°F (62.8°C) and will not flash or sustain burning at ordinary temperatures. However, when kerosine is allowed to come to the top of a cotton wick, it will continue to burn at room temperatures. Explain why this happens. (*Hint:* Lamp wicks are slowly consumed and have to be "turned up" periodically to sustain flame.)
2. Explain why lubricating oil cannot be responsible for spontaneous combustion under any circumstances.

3. Write an explanation for each of the following:
   (a) Why does a covered pan of liquid boil quicker than an uncovered pan?
   (b) Why is it a good idea to keep a wet towel or a pan of water in an extremely dry room?
   (c) What is the principle of a room humidifier?
4. Filling station gasoline pumps have signs that read: NO SMOKING. However, very few fires occur during gasoline pumping operations. What might be the reasons for this?
5. Make a list of at least ten examples of physical changes in characteristics of materials that you have brought about in the past 24-hours by ordinary daily life. Then, with a group of your classmates, discuss exactly how you brought about some of these changes.
6. As a fire protection specialist, you have been assigned the task of explaining "flash point," "ignition temperature," and "fire point" to a class of fire science students. How would you explain each term, and how might you demonstrate an example of each?
7. Calcium hypochlorite is a white powder used for disinfecting swimming pools and other areas because it gives off chlorine very easily. Why is it dangerous to store calcium hypochlorite near kerosine containers that might leak or have some liquid on their surfaces?
8. Ordinary fingernail polish consists of a solution of types of lacquer and coloring combined into a solvent that evaporates and leaves an attractive coating on the fingernails.
   (a) Is this a physical or a chemical change?
   (b) Why is it easy to remove, while varnish or paint has to be sandpapered or burned from a coated surface?
9. Solid, metallic lead has a definite melting point and a much higher boiling point. Magnesium metal also has these properties. What is the difference between these two metals as their temperatures are raised in air to above their boiling points?
10. Arc welding of certain metals is sometimes done with a constant flow of argon gas in the hot metal zone. What useful action is accomplished here?

SUGGESTED READINGS

Coward, H. F. and Jones, G. W., "Limits of Flammability of Gases and Vapors," Department of the Interior, Bureau of Mines, Washington, D.C., 1952.

Fire Protection Handbook, 14th Edition, National Fire Protection Association, Boston, 1976.

Lapedes, D. N., *McGraw-Hill Encyclopedia of Environmental Science*, McGraw-Hill, New York.

Semat, H. and Baumel, *Fundamentals of Physics*, 5th Edition, Holt, Rinehart and Winston, New York, 1974.

Van Name, F. W., *Elementary Physics*, Prentice-Hall, Englewood Cliffs, New Jersey, 1966.

5

Heat and Its Effects

HEAT EVOLUTION FROM FIRES

One of the basic characteristics of a fire is the emission of heat. All combustion reactions, or oxidation reactions, are exothermic (exo = out, thermo = heat), meaning they give off heat. The rate and the extent to which this heat is given off is highly variable and depends upon many factors. In the case of most fires, it is difficult — if not impossible — to identify the fire's rate and extent. Because of this, these factors can only be determined in a general way.

When a fire fighter is faced with a fire situation, an almost instantaneous decision must be made concerning the relative importance of at least three principal governing factors that control heat evolution. These factors are determined by asking the following questions:

1. What is burning?
2. How hot is it now?
3. What is ready to burn, or likely to begin burning?

The facts needed to answer the first question are related to the chemistry of combustion of all combustible substances. The most plentiful and most frequently encountered substances that burn readily are those containing carbon and hydrogen in varying proportions. These two elements form most of our common combustible solids, liquids, and gases, and give off large amounts of heat during their oxidation. These so-called organic substances are most commonly found in nature where they have been formed by biological processes involving organisms. Wood and paper, or cellulosic materials such as cotton, are examples of solid carbon-hydrogen combustibles that also contain oxygen in their molecules. Petroleum hydrocarbons such as gasoline and kerosine are typical liquid carbon-hydrogen flammables, and natural gas or methane is our most common gaseous carbon-hydrogen compound. With the addition of some other elements such as nitrogen, oxygen, and even sulfur, chemists have learned how to make large numbers of useful, but highly flammable, substances from these two elements.

As a general rule, the amounts of heat given off during the combustion of these organic substances is nearly the same on a weight basis, depending, of course, on their carbon-hydrogen content. However, the rate at which the heat is given off depends on how closely the substances approach the gaseous state. Since the oxygen in air is a gas, it can react rapidly only with another gas (that is, rapid ignition and combustion in air is governed by the state of subdivision of a combustible). Flammable liquids vaporize when they are heated; wood shavings burn at a fast rate by rapidly breaking down into flammable gases and vapors, while solid lumber is slower to pyrolyze and burn. (There are many other combustible substances that give off heat and that are encountered in fires; some of these substances are presented as special hazards later in this text.)

The second question, "How hot is it now?," concerns the state of the heat of a fire and involves temperature. Temperature is only a measure of the extent to which a substance has been heated. Heat and temperature should not be confused with each other; heat is a form of energy, while the degree to which heat has been exerted on a substance results in raising its temperature. Thus, temperature is a measure of the intensity of heating of a material.

When a fire fighter senses the temperature of a fire situation by sight, touch, or by the amount and type of smoke or gas odor smelled, such information must be translated into the "extent of heat" that has been evolved by the fire. High temperatures mean that ordinary combustibles not already burning are rapidly reaching their ignition points; in order to slow up the evolution of heat, the prompt preventative efforts of cooling are necessary. Ordinary carbon-hydrogen substances produce temperatures during burning of 1,100° to 1,800°F (593° to 982°C), whereas their ignition points may be only 350° to 1,000°F (177° to 538°C).

The third question, "What is ready to burn, or likely to begin burning?," involves the need to recognize the physical and chemical properties of combustibles. With this recognition, immediate action can be taken to prevent further evolution of heat by their ignition. In the case of ordinary combustibles (except for gaseous combustibles such as methane), the cooling action of water in some form — perhaps in the form of foam for liquid flammables — is the most effective deterrent for heat evolution following ignition and during burning by pyrolysis or vaporization. (The special ignition or burning prevention needs of various hazardous substances are identified in detail later in this text.)

The fire prevention action of cooling with water is the most economical and effective method of keeping combustibles from burning. This is also the case when combustibles are "ready to burn."

THE MEASUREMENT OF HEAT

As in other forms of energy, it is impossible to measure heat in terms of its own character. Heat can only be measured by what it does, and by measuring the work it accomplishes. In this situation, work means raising the kinetic energy level of the molecules of a substance to increase their vibrational speed. (See Chapter 1, "The Basic 'Stuff' of Our World," the section titled "Kinetic-Molecular Theory of State.")

As this energy transfer occurs, the temperature of a substance increases; conversely, if one impinges water in a liquid form at a temperature lower than its boiling point onto a hot surface, the heat will be removed from the hot surface to the water. The hot surface loses heat, thereby decreasing its temperature; the water rises in temperature, thereby gaining heat energy.

Heat is described in several different ways, all of which bear a definite relationship to each other. The most familiar unit of heat is the British Thermal Unit (Btu). The Btu was standardized many years ago, and is defined as follows:

> One Btu is the amount of heat required to raise the temperature of one pound of water one degree Fahrenheit (when the measurement is performed at 60°F).

Another familiar unit of heat is the calorie. In many ways, this is easier to use than other heat units because it bears a decimal relationship to its weight denominator and to water (as indicated by the following standardized definition):

> One calorie is the amount of heat required to raise the temperature of one gram (1/254 of a pound) of water one degree Centigrade, or Celsius (when the measurement is performed at 15°C [59°F]).

The equivalency between a Btu and a calorie is:

One Btu = 252 calories

The SI unit for heat is the Joule (pronounced jowel). It has a standardized equality to our older units of heat, shown as follows:

One Btu = 1,055 Joules

One calorie = 4.18 Joules

It is important to note that all exothermic burning of combustible materials gives off energy in the form of heat that can be measured and described in the preceding units. For example, if one pound of propane gas (such as is contained in an LPG tank) were burned, it would generate 21,646 Btu. (Or, 21,646 by 252 = 5,454,792 calories; or, 21,646 by 1,055 = 22,836,530 Joules.) If this amount of heat were completely transferred to 1,000 pounds (453,592.0 grams, or 454 kilograms) of water (this is equal to 19.8 gallons), the temperature of that amount of water would rise from 60°F to 81.6°F (or from 15°C to 27.6°C).

The preceding example of heat evolution by a fuel, its transfer to water, and the resultant measurement of the rise in temperature of the 1,000 pounds of water is a calorimetric (calor = heat, metric = measurement) measurement of the heat of combustion of propane gas.

HOW SUBSTANCES DIFFER IN "HEAT" RELATIONSHIPS

All substances that are combustible or that are capable of oxidation in a burning reaction do not give off the same amount of heat during their oxidation. Thus, it can be said that substances differ in their "heat" relationships.

Heat of Combustion

When standard amounts of materials of different kinds in the form of the weight (not the volume) of a substance are burned under perfect conditions of combustion, varying outputs of heat are evolved when measured by a calorimeter. These outputs of heat are called the heat of combustion of the substance. Table 5.1 shows how the heats of combustion of substances differ, depending upon the composition of the substance.

Table 5.1. The Heats of Combustion of Different Substances

SUBSTANCE	HEAT OF COMBUSTION (Btu PER POUND)
Alcohol (ethyl)	12,800
Aluminum (metal)	13,300
Benzene	18,028
Carbon	13,480
Charcoal	12,920
Coal (anthracite)	13,000
Coke (petroleum)	15,800
Gasoline	20,100
Heptane	20,657
Lubricating oil	20,400
Newspaper (and paper)	7,883
Paraffin wax	20,100
Phosphorus	10,580
Rags (cotton)	7,165
Wood bark (pine)	9,496
Wood (oak)	7,180

In combating actual fires, there are many variable conditions of combustion that influence the output of heat from burning materials. Substances with comparatively low heats of combustion may yield extremely high heat exposures during their burning. The factors that govern this are: (1) the amount of area of solid combustibles exposed to heat and oxygen (*i.e.*, its state of subdivision), (2) the area of free spread of the liquid (in the case of liquid flammable materials), (3) the ability of a flammable liquid to give off vapor pressure, and (4) the heat conductivity of a solid (metals *vs.* wood), which can influence the amount of heat given off when materials are burning.

Specific Heat

In the preceding example of heat emission by burning propane gas, its absorption, and the resulting temperature elevation, water was used as the heat absorber. If another liquid had been used (such as alcohol or lubricating oil), the temperature rise would have been different because all substances vary in their ability to respond to heat intake when their temperature elevations are measured after exposure to a given amount of heat.

This varying ability to take up energy in the form of heat is called the specific heat of substances. Earlier it was stated that the standardized definition of a calorie is the amount of heat that raises the temperature of one gram of water one degree Centigrade. This, then, becomes the *specific heat* of water — *1.000* calorie per gram per degree. Table 5.2 gives the specific heats of some other situations as compared to this value for water.

Table 5.2. The Specific Heats of Various Substances

SUBSTANCE	SPECIFIC HEAT (IN CALORIES PER GRAM PER DEGREE C)
Water	1.000
Acetone	0.528
Air	0.240
Aluminum metal	0.217
Butane (in LPG)	0.549
Carbon (charcoal)	0.165
Carbon tetrachloride	0.201
Copper metal	0.091
Glass	0.161
Gold	0.031
Iron metal	0.113
Lubricating oil	0.51
Mercury	0.033
Paraffin wax	0.70
Tin metal	0.054
Wood	0.42

From the values given in Table 5.2, it can be seen that one reason why water is our most effective cooling agent is that it possesses the capability of taking up more calories or heat units (per weight of water) during its rise in temperature. Because of this, every common substance other than water is a less efficient heat absorber or cooling agent than water.

Heat of Fusion

When substances change in state (*i.e.*: when solids melt or when liquids become gaseous vapors) they require heat energy to accomplish this change. The amount of heat needed to melt the solid form of a substance and thus transform it into a liquid is called the heat of fusion of the substance. The heat of fusion of a substance is measured in terms of the number of heat units per weight of substance needed to transform the substance from a solid to a liquid at the melting point. Table 5.3 presents the variability of amounts of heat of fusion for some different substances.

Table 5.3 shows the large amount of heat absorbed when water changes from a solid form (ice) to a liquid form. This accounts for the utilization of mixtures of ice and water to cool materials such as milk and cold drinks. For example, a glass of ice cubes stays cool longer than a glass containing solid alcohol cubes.

Table 5.3. The Heats of Fusion of Various Substances

SUBSTANCE	HEAT OF FUSION (IN CALORIES PER GRAM AT THE MELTING POINT)
Water (ice)	79.7
Acetone	23.4
Benzene	30.1
Carbon dioxide (solid to gas)	45.30
Carbon tetrachloride	4.16
Ethyl alcohol	24.9
Lead metal	5.42
Mercury	2.77
Paraffin wax	35.0
Tin metal	14.0

Heat of Vaporization

The most important heat-removing property of substances concerns the change of state of various liquids when they become gaseous vapors. Large amounts of heat are needed to accomplish this change, which is called the heat of vaporization. As with other forms of heat, this heat absorption varies with the chemical character of the substance. Table 5.4 illustrates the wide range of heat amounts needed to cause substances to change from liquids to vapors.

Table 5.4. The Heats of Vaporization of Various Substances

SUBSTANCE	HEAT OF VAPORIZATION (IN CALORIES PER GRAM AT THE BOILING POINT)
Water	539.6
Acetone	124.5
Alcohol (ethyl)	204.0
Benzene	94.3
Carbon tetrachloride	46.4
n-Octane (gasoline)	70.9
Lead (liquid)	222.6
Nitrogen (liquid)	47.8
Propane (liquid, LPG)	98.0

The relative values of the heat of vaporization of the substances listed in Table 5.4 illustrate the superiority of water (in comparison to the other substances in the Table) in its ability to take up heat while going from a liquid to a vapor or gaseous form. Added to this ability is the fact that a gram (or a pound, or a kilogram, or a ton) of water (at 211.9°F [99.9°C]) can absorb more than eleven times as much heat as can equal amounts of super-cold liquid nitrogen at a tremendously cold temperature of minus 320.6°F (minus 195.7°C). The high heat of vaporization of water is perhaps the major reason for its popularity as a fire extinguishing agent.

Thermal Conductivity of Materials

In any consideration of the effects of the heat of fires, there is another important property of the materials that are involved in the propagation of flames. This property has to do with the materials' rate of heating, or ability to conduct heat.

When substances are heated by some source, heat is conducted in all directions away from the point of heating. However, all materials vary in their rates of heat conduction. Because of this, heat that is added to materials to reach ignition points (temperatures that cause pyrolytic breakdown in the case of combustible solids like wood or paper) will also vary.

When one speaks of heat conduction or the relative rate of travel of heat through a substance (solid, liquid, or gas), the function of heat insulation comes to mind as a closely associated phenomenon. In reality, the mechanism of complete insulation from heat conduction, or flow, is impossible from a practical viewpoint. Heat insulation is actually a relative term denoting a lesser rate of travel of heat possessed by a substance when compared to other materials. Certain heat insulation materials are spoken of as "lagging" materials. This is a more accurate term, since heat insulators merely slow down the rate of heat conduction through a system. They cause it to "lag" in its rate of travel. Materials used as heat "insulators" do not really stop the flow of heat.

The thermal conductivity of materials is measured in standard terms of amount of heat conducted per unit thickness of material per unit degree of temperature. The relative figures obtained are spoken of as the "k factor" of the material, or its thermal conductivity coefficient.

Table 5.5. Thermal Conductivity of Materials

MATERIAL	RATE OF HEAT CONDUCTIVITY (k) (IN CALORIES PER SEC PER CM PER °C) (*NOTE:* MULTIPLY ALL VALUES BY 1/1,000)
Aluminum (metal)	500.0
Brick (common)	1.7
Charcoal	0.21
Concrete	4.1
Copper (metal)	910.0
Corkboard	0.1
Fiberboard ("celotex" type)	0.14
Glass	2.3
Iron (metal)	150.0
Marble	6.2
Mineral wool (blanket)	0.1
Paper	0.3
Plaster	1.7
Plastics (solid)	0.45
Vermiculite	0.14
Wood (oak)	0.41
Wood (white pine)	0.29
Wool (loose clothing)	0.8

The values of "k" for different materials are obtained by methods of measurement and computation that take into account the density, thickness, and area of the materials. Table 5.5 gives the "k factors" for various substances. These values are directly comparable with each other.

Table 5.5 shows the great difference that exists between metals and other materials of construction. A fire that impinges heat on metal framing or on metal supports within other types of construction material such as concrete or plaster will more quickly transmit (or conduct) the effects of heat to its surroundings, and thus increase the scope of the fire much faster. The low values shown for heat conductivity of plaster and mineral wool blanket insulation also account for the fact that fires in rooms with modern ceiling insulation construction rarely penetrate overhead attic spaces unless structural ceiling collapse occurs, or unless the fire is of long duration.

Thermal Expansion

There is another heat effect that arises from fires which, because of its disaster-causing characteristics during the progress of a fire, must not be overlooked. This effect involves the way in which materials expand during their process of heating. The expansion of materials of construction causes structural failures in buildings, which is a leading factor in the collapse of walls and supports. Many fire service personnel lose their lives from this cause.

When materials of construction such as iron, brick, concrete, and wood are heated by flame exposure, they expand differently in size or length. The effect this has on a structure is that it ruptures the holding or supporting surface of one type of material to another. Strength and adherence is suddenly lost, and the material separates from its bond to another surface or joint. Collapse can then occur.

The evaluation of the comparative values of expansion has been standardized so that figures are obtained that show the relatively small (in millionths of an inch or millionths of a centimetre) increases in length that a certain material an inch or a centimetre in length undergoes when heated with just a one-degree Centigrade temperature rise. (See Table 5.6.) The values in Table 5.6 are comparable, one to another.

The Table 5.6 values for the relative properties of materials in regard to their expansion under heat attack do not tell the complete story when it comes to real-life fires. For example, Table 5.6 indicates that brick, iron, and concrete construction expand in almost equal amounts when exposed to heat. From this, it could be expected that a wall composed of brick, iron, and concrete should remain integrally sound during a fire, and should not separate and possibly collapse. Actually, the opposite action usually occurs during a fire involving such construction: the iron separates from any brick and concrete, and any brick and concrete connection breaks loose with subsequent loss of strength and adherence.

The reason for this is indicated in Table 5.5, which shows a large difference in *rate* of heat conductivity between iron and brick or concrete, and a more than double rate of heat conductivity between concrete and brick. The iron does its expanding very rapidly, losing contact with the concrete and brick in the process.

Table 5.6. Thermal Expansion of Materials

MATERIAL	COEFFICIENT OF LINEAR THERMAL EXPANSION PER DEGREE C (*NOTE:* MULTIPLY ALL VALUES BY 1/100,000 FOR CM)
Aluminum	25.0
Brass	18.7
Brick	9.5
Concrete	10.0 to 14.0
Copper	16.6
Glass	9.2
Iron	11.5
Plaster	4.0 to 7.0
Wood (oak, lengthwise with grain)	4.92

Such expansion is followed by a more rapid heating of the concrete than the brick; in the process the concrete destroys any masonry bond with the brick, and the wall loses its strength.

Another interesting example of how substances differ in their "heat" relationships concerns the use of large wooden trusses (usually laminated and glued for strength purposes) used for roof supports in large, open-area occupancies such as supermarkets, high-ceilinged churches, and athletic gymnasiums. Since wood is combustible, it would seem that for firesafety purposes, steel trusses should be used in such constructions so that fire exposure from the large, open volume below would lessen fire damage to the trusses.

In Table 5.6 it can be seen that in their linear dimension, steel members expand a little more than twice what a wooden member would expand when exposed to the same heat attack. Also, a steel member conducts heat 360 times more rapidly than does a wooden member. Thus, it could be assumed that the following sequence of events might occur during fire exposure to wooden and to steel roof support truss construction:

> The *wooden* truss heats up very slowly due to its low rate of heat conductivity. Surface burning takes place from the heat and flame exposure. As charring continues, the rate of heat absorption slows down due to the lower rate of conductivity of charred depth (one-half the rate of wood). Lateral expansion of the entire wooden span to increase its length is so slow that outward thrust against sidewall masonry supports would not occur until very late stages in the fire.
>
> In comparison, the *iron* truss heats up rapidly and quickly conducts its heat to all parts of the truss construction. Lateral expansion takes place, with the result that masonry-bearing supports on the walls are forced outward, thus losing integrity and causing collapse. Another detrimental effect of heat on iron supports can take place if fire attack is not halted; this effect concerns the fact that iron loses its ability to withstand compression loads when it reaches temperatures of 1,500°F (815°C).

HEAT TRANSFER

When energy is transformed into heat, and when this energy can be measured to see what effect it has, or the useful work it can do, we can then make use of one or more of three basic methods by which heat travels from one point to another. Heat travel, the method by which heat travels from one point to another, is more scientifically called the transfer method of heat energy.

The *transfer* of heat is quite different, however, from the *transformation* of heat. When heat is transferred, it is *transported* from place to place without being changed into another form of energy. When heat is transferred from an electrical hot plate to a tea kettle with water in it, the water boils due to heat absorption. The steam that issues from the spout of the kettle becomes a transformation product of heat and is another form of energy — steam pressure energy. Steam pressure energy is the energy used to drive locomotives and heavy equipment.

There are three basic methods by which the heat from the combustion of materials is transferred: conduction, radiation, and convection.

Conduction

Heat energy flows in all directions from where it is generated; it flows unimpeded through all the matter that constitutes our physical world, be it solid, liquid, or gas. The flow of heat cannot be stopped by a solid substance in the same manner in which the flow of water can be halted by an obstruction in a pipe. Because all matter consists of atoms and molecules that are in constant motion and that can transport heat from one point to another, conduction of heat can occur by this process. However, substances differ in their ability to accomplish this conduction. Gaseous substances like air conduct heat very slowly; when air is trapped within small solid cavities (as in some solid plastic foams), the air acts as a heat insulator. Such so-called heat insulators merely act as deterrents to slow down the conduction of heat, not to completely stop it. Another name for this is the "lagging" of the flow of heat of conduction by means of a heat insulator.

Heat transfer by conduction has been responsible for many fires. The most familiar examples of such fires are the chimney and the fireplace hearth-support fires, where wood contacts the brick or masonry construction of flues, chimneys, or the poured-concrete hearth foundations of open fireplaces. In the construction of chimneys or flues, it is often convenient to support wooden joists near or in the brick or masonry fabrication that constitutes the avenue for hot gas discharge from flames below. If the travel of hot gases is intermittent in such a flue, the masonry does not accumulate much heat; however, if there is a constant discharge of hot gases in such a flue, the masonry gradually becomes hot. Any wood in contact with the masonry will absorb heat and, after a period of time, will reach its self-ignition temperature and begin to burn. This may take days or even weeks of constant heat flow. The same mechanism occurs in hearth constructions where wooden forms might have been left in close contact with a poured concrete base construction for the fireplace hearth. Fires of these types are very difficult to find and to extinguish.

Another familiar type of fire that evolves from the conduction of heat is the ordinary kitchen fire in which a hot electric-range element contacts a pan filled with fat. If heat input from the element continues heating the fat by conduction, the temperature of the fat continues to rise until it reaches its A.I.T. (auto-ignition-temperature) and begins burning.

Radiation

Radiative heat transfer is when heat is transferred from one area to another without being in direct contact with the area, and without there being any circulation of hot gases to help "bathe" the area with heat. Heat energy occurs in exactly the same form as light; that is, it may be in the form of *rays* of heat. These rays travel through space (which may be considered a gaseous form with varying amounts of gas, down to the vacuum of outer space) as units of electro-magnetic-energy. Heat in the form of rays is the infrared portion of the entire radiation scale of energy known as the electro-magnetic-spectrum of radiation.

Radiation energy travels in straight lines. Thus, heat absorbed from a pinpoint source would be much less than the same amount of heat absorbed from a large, radiating surface, providing the body also had an appreciable surface area (see Fig. 5.1).

Infrared, or heat radiation, has no easily measurable effect until the rays contact a substance and the energy of the rays is transferred into heat energy. The incandescent light bulb is an example of this. When a hand is held near the lighted bulb, the infrared portion of the light rays can be felt by a rise in the temperature of the exposed side of the hand. The same sensation is found in bright sunlight. It is the infrared heat rays in sunlight that accomplish the heating of any area upon which sunlight falls.

Another example of radiation is the long tubular fluorescent light fixture. In this device the light radiation that is emitted has a high proportion of visible light and a low proportion of infrared rays. The "hand exposure test" on a light source of this type results in only a slight heating of the skin, as these fixtures are

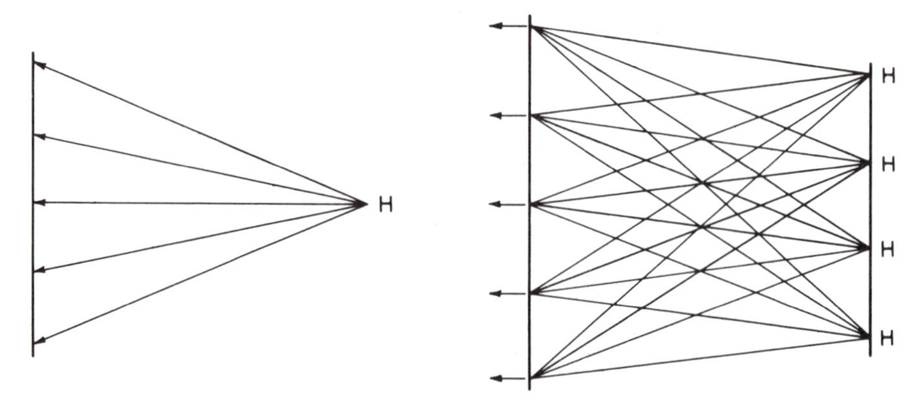

Fig. 5.1. A comparison of heat absorption by surfaces of a similar area from a pinpoint source (left) and a large radiating surface (right). (From *Fire Protection Handbook*, NFPA.)

a "cooler" light than the incandescent electric bulb. As a result, they are more efficient, since it is the light rays, not the heat rays, that are desired.

Although infrared heat rays are invisible, they are similar in action to light rays. They pass through any visibly transparent medium such as air, glass, water, and transparent plastics with only a small absorption of energy that only slightly heats the transparent substance. Heat rays may be reflected like light rays, and may be concentrated by lenses like light rays. The familiar sun ray "burning" glass uses this property.

The degree to which heat rays are absorbed is influenced by the color and the degree of opacity of the substances upon which they fall. Black or dark-colored fabrics or solid materials readily absorb heat rays or infrared rays; as a result, they become warm or hot to the touch. For this reason, white or light-colored clothing is cooler and more comfortable to wear in the summertime.

Heat radiation may be reflected from surfaces in much the same way that light rays are reflected. The polished, mirror-like surfaces of metals can be employed to direct heat away from an object in line to receive it. This principle is used in the aluminum-coated fabrics that make up the exposure suits worn by fire fighters when entering flaming areas (such as a crashed aircraft from which the rescue of trapped personnel is necessary). The radiation of heat from flames is reflected away by the shiny surfaces of the suit; however, if the metallic-coated surface becomes dirty or black from soot, it readily begins to absorb radiated heat.

Heat exposure of combustible substances by radiation is a danger during fires in which flames of large size and volume are produced, and where radiation from combustible substances falls on nearby materials that can become heated to the point of ignition. Protective cooling of radiation heated materials with water must be initiated quickly in such exposure situations. It should be realized that in such flame exposures, the water-film bathing of an exposed surface does not act to halt the heat radiation received by the surface; rather, the heat from the surface is transferred by conduction to the water and is cooled by this action.

There is one interesting action in which water can be used to block the radiation from flames. Earlier in this chapter it was stated that because of its transparency, water does not stop infrared rays. However, if water is broken up into very small droplets such as those produced by a fog nozzle or a fine-spray nozzle or head, the many tiny droplets of water in the discharge act as a nontransparent whitish screen. Heat rays falling on the milk-like large volume of droplets are reflected in all directions, and *attenuation* of the infrared rays takes place. (Attenuation means a "loss of strength" or a "thinning out" has occurred.) In this case the amount of radiation is decreased by the simple process of reflecting it away from its original straight-line path. This mechanism is used extensively by fire fighters assigned to hoseline nozzles when advancing on a burning fire. It is also an important feature of dry chemical nozzle discharges, as will be explained in a later chapter.

Convection

The process of heat transfer by the convection of flames and hot gases involves the earth's gravity and the relationship of heated, mobile substances to cool, mobile substances. (Mobile, used here, means movable.) Mobile molecules change position by gravitational means, without mechanical assistance. Gases, vapors, mists, and liquids are typical examples of substances capable of such action.

Convective heat transfer is perhaps the most important and most frequently observed mechanism of heat transfer in the propagation of fires involving combustible solids. To a large extent, fire "spread" or propagation within enclosures exhibits forms of heat transfer by convection.

When a gas, vapor, or liquid is heated by some means, the kinetic energy of its molecules is increased and the sphere of vibration of these molecules enlarges and expands so that less molecules fill larger spaces. This means that a volume of heated gas will be *lighter* than the same volume of the same cold gas. (See the kinetic-molecular theory of state, Chapter 1, "The Basic 'Stuff' of our World.") Since there are fewer particles possessing weight in this heated volume of gas than in the cold volume, there is less gravitational force exerted on it, allowing it to rise to a new level above the cold gas. At this higher level, the heated volume of gas may give up some of its heat, transfer it, and thus progressively raise the temperature of the area or its surroundings. This is known as convective heat transfer. Simply stated, hot gases, vapors, and liquids rise by convection (a function of lesser gravitational attraction) to exert transfer of heat.

The common example of heat transfer by convection is the wood- or coal-burning stove used to heat rooms in houses that do not have central, forced-air heating plants. A layer of paper wads is placed inside such stoves, followed by sticks of thin wood upon which even larger-diameter sticks or pieces of wood are placed. Often, a thin layer of coal lumps is placed on top of the paper and wood. Such stoves usually have a large outlet or flue that extends upward for some distance, terminating outside the house at an elevated point.

When a match or other flaming object is placed adjacent to the lowest level of the layer of paper wads, pyrolysis of the paper begins and ignition of the gases of decomposition of the paper quickly takes place. The flaming, burning gases travel upward in all directions, convection of the hot gases begins, and as successive portions of the easily decomposable paper are heated, progressive flaming occurs. The hot burning gases travel upward, bathing the smaller sticks of wood. The hot gases initiate the pyrolysis of the wooden sticks, which give off combustible gases and then ignite. These hot, burning gases convect upward to bathe the larger pieces of wood. The process continues by convection of hot gases until the entire fuel bed has received heat transfer in amounts sufficient to sustain combustion. As the hot gases rise, they also create a motion in the large pipe or flue at the top of the stove due to the fact that they are less dense (or lighter) than the cold gases surrounding the stove. A sustaining motion of these gases upward in the flue then becomes the natural *draft* of the stove. This draft will continue, with varying strength dependent on the temperature of hot gases coming from the burning fuel and their volume, until there is a small difference in temperature between the inside of the stove and the air at the top of the flue.

Heat transfer by convection occurs in fires in buildings and dwellings with the movement of hot gases in a manner similar to that presented in the preceding example of the stove. Even in a slanting lateral direction, the travel of hot gases continues as long as any fuel remains to create heat by combustion.

Heat transfer by convection operates in a multitude of everyday instances. The heating of homes and offices relies upon the fact that warmed air must be supplied at the floor level of a room so that it will travel upward, causing a natural convection circulation in the room without the help of fans or mechanical devices to mix the air. For the same reasons, cooled air must be supplied to the upper levels of rooms, since it is more dense (heavier) than existing warm air; also, cooled air will promote mixing and cooling by its natural descent into the room. In this case, convection of the warm air is still upward in direction; however, the warm air is said to be displaced by the cooler air traveling in a downward direction due to its higher density.

Perhaps the most dramatic example of the ability of warmed or hot air to rise is the hot-air balloon. (The hot-air balloon is merely a light envelope containing hot air only about 20 degrees higher in temperature than the air surrounding the envelope.) However, the hot-air balloon has sufficient lifting power to take several persons and the balloon's equipment aloft for fairly long periods of time. A very small burner flame on the balloon can sustain flight for a much longer duration if it continues to supply heat to the captive hot air in the envelope.

SOME COMMON SOURCES OF HEAT OF IMPORTANCE TO FIRE PROTECTION

Unwanted or accidental fires are caused by many circumstances, any one of which might be singled out as the source from which runaway combustion begins. It is lengthy and difficult to catalog all such possible situations; such cataloging can usually be found in studies of the science and art of arson investigation. However, it is important that certain basic categories of energy sources should be recognized as primary origins by which heat is developed — heat which, if it is not controlled in some way, can result in fires of a disastrous nature.

There are three common energy classifications that statistics tell us are our most frequent sources of heat for the generation of fires. These are: (1) chemical energy, (2) electrical energy, and (3) mechanical energy. In considering these sources, the subject of *fuel* must also be dealt with, since initiation of combustion from any energy source requires that fuel and air for oxidation be suitable for continuation of the combustion process.

Chemical Sources of Heat

This source of energy is often responsible for the initiation or ignition of further chemical reactions, which, as stated earlier, are all actual chemical combustion reactions in the burning process of a fire.

The safety match, with its special striking surface and simple design, is a source of chemical energy and heat. All too often the chemical reactions occurring in the head of the safety match have been responsible for the initiation of accidental fires. Friction, in combination with a chemical reaction, ignites the match. The head of the match is composed of antimony trisulfide mixed with potassium dichromate and glue to cause the mixture to adhere to the stiff paper or wooden stick of the match. This paper or stick has been pretreated with ammonium phosphate to prevent glow after the flame has been extinguished and part of the length of the paper or stick has been dipped in molten paraffin to facilitate burning. Antimony trisulfide (reducing agent) and potassium dichromate (oxidizing agent) mixtures require higher temperatures to ignite than do the mixtures found on ordinary "strike anywhere" matches. The striking surface for safety matches consists of a mixture of antimony trisulfide, red phosphorus (which is a hard to oxidize, amorphous form of white or yellow phosphorus) and powdered glass to increase friction. Glue is also used here to hold the mixture on the paper or match box side. When the match head is drawn across the striking surface, some of the red phosphorus ignites by friction, and a very small point of high temperature is produced in contact with the match head, which starts the reaction between the antimony trisulfide and the potassium dichromate. Flaming ensues, followed by ignition of the matchstick by conduction of heat. (The ignition of the red phosphorus on the striking surface of the match box is interesting to observe in completely dark conditions; in such conditions, a trail of luminescence can be seen during the striking operation.)

Temperatures may increase to the self-ignition temperature and rapid burning may ensue in a slow chemical reaction that takes place when substances capable of oxidizing (like linseed oil) generate heat, and if confinement of the heat produced takes place. This spontaneous ignition by chemical reaction is explained earlier in Chapter 4, "The Relationship of Fire and the Physical World."

There are many so-called "chemical" substances that are capable of oxidation in air to produce heat, or that are dangerous when they are stored or when they come in contact with other substances. The most important encyclopedia for reference in this matter is the *NFPA Fire Protection Guide on Hazardous Materials*, which lists these materials, their hazards, and their reactivities. (See also Appendix III of this text.)

Electrical Sources of Heat

Because it is used in many ways, electrical energy is a common cause of unwanted fires. In home or industrial oil burners, electrical sparks are used to start and to sustain the ignition of the heating oil. In the kitchen electrical resistance heating elements or other forms of electric heating energy are needed to supply cooking temperatures. In certain living or working areas, electric energy is used to provide comfortable temperatures by means of resistance heaters. In our various appliances electricity is used in many ways, each of which is a potential fire hazard to nearby combustible materials. Unless electrical energy is used in an efficient manner, it will almost always produce heat as an unwanted by-product. Another factor that must continually be remembered is that the fire hazards of electrical energy are almost always concealed. In efforts to beautify and hide electrical wires and equipment, they are often installed in walls and partitions or enclosures so that malfunctions or dangerous conditions that can initiate a fire are difficult to detect. In addition, rather than installing additional wall outlets when they are needed, the tendency is to overload existing ones.

Fig. 5.2. Triple lightning bolt lighting up Milwaukee's Water Tower Park during a thunderstorm. (Associated Press.)

There are five forms of electrical heating energy that must be guarded against from the viewpoint of fire initiation. These five forms of electrical heating energy are resistance, arcing, sparking, static, and lightning.

Resistance heating results from overloading electrical conductors or from allowing electrical equipment that produces heat during its useful function to be situated near combustible materials. The electrical wiring, or switch, or connection that heats up during transmission of electricity because it is carrying too large a load (20 amperes at 110 volts instead of 15 amperes), is dangerous to nearby combustibles. The large incandescent bulb in a small enclosed lamp fixture (200 watt bulbs in sockets and lamps listed for 100 watt bulbs) is a hazard to furnishings and to combustible materials in its vicinity.

Arcing heating takes place when electrical energy is caused to discharge across an air gap due to a break in electrical continuity in the connection to an appliance, or in switches and fuse blocks. When a good electrical connection is not made in a switch or fuse block or in some appliance where a screw is loose or a metal-to-metal contact is poorly secured, arcing may take place and heat is quickly developed at the joint. This heat may be sufficient enough to cause the ignition and burning of insulation or nearby combustible materials. Unfortunately, arcing of electrical contacts is usually not observed until heat is already developed and ignition has already occurred.

Sparking heating is different from arcing heating in that it may occur only once or it may be the result of a high-voltage discharge with a low energy output. Sparks of this type are not capable of large, sustained heating effects, but if the sparking occurs in a flammable atmosphere of gas and air, or fuel vapor and air, ignition of the fuel may ensue. In an oil burner, the high voltage spark discharging in the vicinity of the atomized particles of liquid fuel oil constitutes a continuous source of sufficient heat to ignite the oil.

Static sparking heating is similar to sparking heating except that it occurs under many different conditions of movement of liquids and of finely divided or pulverized solids. Its detrimental effects are concerned with the fact that static sparks

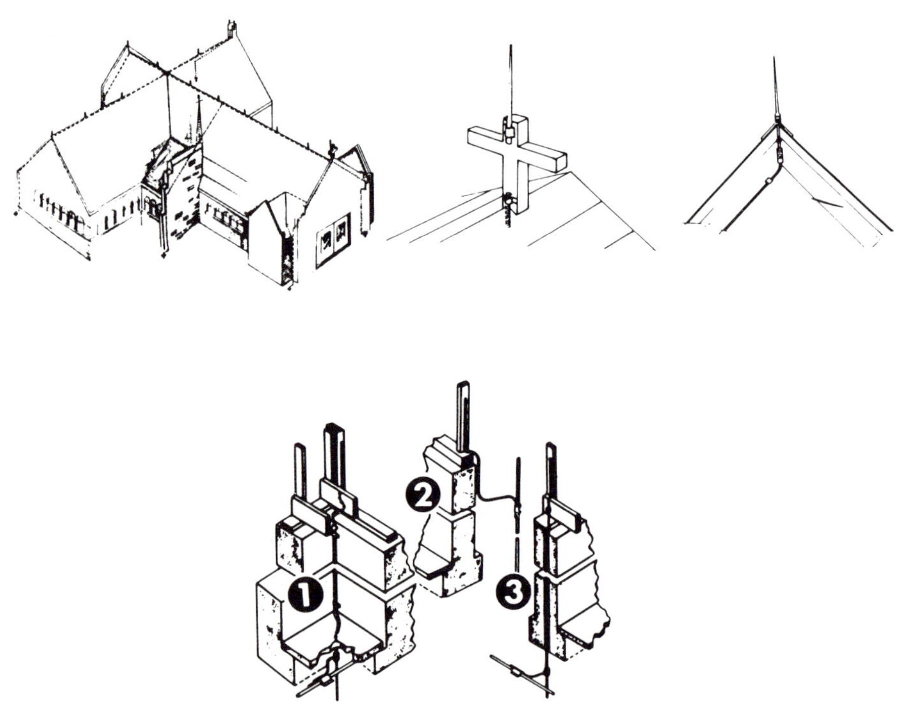

Fig. 5.3. Illustration of a modern lightning protection system for a church (top left); some details and installation pointers are also shown. Top center illustrates that terminal points may be custom-designed to fit into the architecture. The rod may even be a part of the cross. The top right figure shows air terminals pushed up through a hole cut in the roof permitting connectors to be concealed in the roof framing spaces. The figures at the bottom show: No. 1, how a conductor might be carried down along the wall to a grounding rod beneath a floor slab; No. 2, how the down conductor might be brought outside to a grade-level trench and then out two feet to the ground rod; No. 3, the grounding conductors coursed down inside the wall.

are hot enough to ignite flammable gas, vapor, or dust-air mixtures, thus causing a fire. Static sparks are low in energy and, as a result, do not heat up the areas in which they occur.

Lightning heating is a powerful source of electrical heating energy, even though it occurs infrequently. When a lightning discharge contacts wood, masonry, or other construction materials, its heating effects cause immediate charring and destruction of wood by instantaneously vaporizing the occluded moisture in the thickness of the wood. This heating "explodes" masonry by the same mechanism and, because of its high energy, ignition of combustible materials is easily accomplished during its travel over surfaces to reach the ground. The safe dissipation of the energy of a lightning discharge can only be ensured by providing metallic electrical conducting paths (large cables and wire conductors) for its travel to ground. The most widely used form of conducting path is the metal lightning rod. Designed to attract the electrical energy from a lightning discharge or stroke, the rod thus diverts the electrical energy away from nearby poor and flammable conductors such as roof shingles and wooden sheathing by safely conducting it to the ground by means of cables.

Mechanical Sources of Heat

The most frequently occurring cause of fire (or source of heat) generated by purely mechanical means involves situations in which friction and continuous rubbing or mechanical stress causes heat build-up within a combustible material. For example, when an automotive truck or trailer is driven long distances with a partially or noninflated tire, flexing and road stress cause heat to build up in the tire until it reaches its self-ignition point and begins flaming. This can become a serious type of fire, capable of igniting nearby combustibles. The same heat build-up can occur in any installation of rapidly moving kinetic or mechanical forces transforming their mechanical energy into heat energy by means of friction. The degree of hazard of fuels in the vicinity of such friction heat production governs the likelihood of ignition and subsequent flame propagation. Obviously, it is more dangerous to have a fuel oil leak in the vicinity of a rubbing oil burner fan on nearby wood than if the oil were not present to facilitate ignition.

ACTIVITIES

1. (a) Table 5.6 in this chapter, titled "Thermal Expansion of Materials," indicates that brick, iron, and concrete construction expand in almost equal amounts when exposed to heat. Thus, it could be expected that a wall constructed of brick, iron, and concrete would remain structurally sound during a fire, and would not separate and possibly collapse. However, the opposite usually occurs during a fire involving such construction: the iron separates from any brick and concrete, and any brick and concrete connection breaks loose with subsequent loss of strength and adherence. Explain why this is so in terms of rates of heat conductivity. (Refer also

to Table 5.5, "Thermal Conductivity of Materials," to formulate your answer.)
   (b) Describe, by presenting another example, how substances differ in their "heat" relationships.
2. (a) From a standpoint of heat transfer, why is an ordinary home fireplace an inefficient room heating device? What means can be supplied to make the device more efficient? Why would they help?
   (b) Describe the heat sources and the heat transfer methods involved in portable, electric hot water room heaters. Even though they are more expensive and no more efficient than electric "glowing element" heaters, why do you think they might be apparently more effective?
3. In the early days of civilization, habitable shelters that were built with thick, light-colored stone or thick, white-washed walls that were constructed from wood were more comfortable in summer than in winter. Explain the technical reasons for this.
4. If you were to thrust two metal rods (one made of copper and the other made of aluminum, but both of the same diameter) into a hot furnace (like a coal or gas-fired burner furnace), which one would you have to let go of first because of its uncomfortable warmth?
   (a) Why would one heat faster than the other?
   (b) What is this type of heating called?
5. (a) With a group of your classmates, construct a labeled diagram that shows conduction, radiation, and convection flow of heat from a burning chair near a window curtain in a room that is filled with wooden furniture and pine paneling.
   (b) Display the diagram to the entire class, eliciting comments on how the diagram might be improved.
6. While on a camping trip in the woods, you discover that the striking surface of your safety matches has become wet and useless, but the matches are dry.
   (a) What different ways could you devise to start the chemical reaction in the match head?
   (b) What principles of ignition would you be using in each case mentioned in (a)?
7. The hot or cold contents of sealed, vacuum-insulated bottles ("Thermos") will equal the temperature of their surroundings if left undisturbed for a period of time.
   (a) Why is there only one *principal* method of heat transfer operating in this situation?
   (b) What is the method?
8. Why is it that at ordinary room temperatures, objects that are made of metal and glass feel colder to the touch than do objects that are made of paper, wood, or foamed plastics?
9. When a loaded electrical circuit is switched off, a hot spark usually occurs between the switch contacts during their motion. Switch boxes or enclosures designed for use in hazardous atmospheres are not completely gas tight. Why are they built so thick and strong?

10. Fires involving burning magnesium metal are very difficult to extinguish. Water and carbon tetrachloride are some of the substances that cannot be used here because of their rapid reaction with the burning magnesium. From purely physical aspects, what noncombustible substances could be used to diminish this type of fire if they were in granular form? Why?

SUGGESTED READINGS

Casey, J. F., *The Fire Chief's Handbook*, 3rd Edition, R. H. Donnelly, New York, 1967.

Fire Protection Handbook, 14th Edition, National Fire Protection Association, Boston, 1976.

Glazebrook, R., *A Dictionary of Applied Physics* (Vol. 1), *Mechanics, Engineering and Heat*, MacMillan, New York, 1950.

Hudiberg, E., *Fundamental Principles of Science Applied to the Fire Service*, Department of Fire Protection, College of Engineering, Oklahoma State University, Oklahoma.

Semat, H., *Fundamentals of Physics*, 4th Edition, Rinehart, New York, 1966.

Shortley and Williams, *Principles of College Physics*, 2nd Edition, Prentice-Hall, Englewood Cliffs, New Jersey, 1967.

Van Name, F. W., *Elementary Physics*, Prentice-Hall, Englewood Cliffs, New Jersey, 1966.

Properties of Solids Important to the Fire Sciences

COMMON COMBUSTIBLE SOLIDS

Many of the principal problems in fire protection arise from the chemical and physical properties of combustible solids. When ordinary wooden construction and fabrication burns, noxious smokes and gases are given off; thus, a surface cooling is required before combustion can be halted. If the construction is coated with paint or varnish, a different type of fire control problem is constituted. Polymer and plastic fabrications burn in various ways, sometimes giving off highly toxic gases. They may also melt during the burning process and may become flowing liquid fires. Paper and cellulosic solids may be involved in deep-seated fires that are difficult to uncover and extinguish. Fabric compositions also burn in entirely different ways, depending upon their chemical constituents and their physical or geometric state. Even certain metallic fabrications may be heated to temperatures that cause them to oxidize readily and burn.

The only general statement that can be made about the combustion mechanism of common combustible solids is that almost all of them must be heated at their surface by some external means, such as a flame or an impinging spark, until that surface reaches a temperature where combustible vapors or gases are given off and are subsequently ignited. These vapors may arise because of a chemical decomposition of the solid under the flame attack, or because of a vaporization of the solid combustible substance that will evolve combustible vapors and gases.

The process of heating with subsequent burning of solids may occur even in the absence of initiation by a flame or a spark. If a solid is steadily heated by contacting (or being irradiated by) a surface having an elevated temperature sufficiently high to cause decomposition or vaporization of the combustible solid, the exposed surface of the solid may reach its spontaneous ignition temperature. (This is identical to its autogenous ignition temperature, or its self-ignition temperature.) Flaming combustion can proceed without initiation by an external flame or spark.

An example of the flaming type of combustion of solids concerns the glowing and relatively slow combustion that can be observed when charcoal, anthracite coal and coke, and compressed cotton (such as a mattress or upholstery padding) burn slowly in air after their surfaces have become ignited. In this type of combustion, very little flaming or visible smoke is produced. However, the burning of the solid can continue either as a glowing red color at the surface of the solid that is exposed to air, or deep inside pile until all combustible material is consumed.

This form of combustion is almost always found where a solid combustible of low heat conductivity has been exposed only for a short time to temperatures sufficient for its surface to reach its ignition temperature or its spontaneous ignition temperature. When the external heat source is removed, the glowing type of burning without flaming may continue on the surface of the solid or, in the case of a cotton mattress or a cotton bale, the glowing combustion may penetrate deeply into the core of the bulk material.

In such cases, the absence of luminous flames in glowing combustion of this sort may be due to either one or both of two characteristics: (1) the solid combustible (such as charcoal) is almost pure carbon and does not contain organic compounds that break down under the influence of heat to produce flammable gases that burn with luminous flames, or (2) the slow, glowing combustion on the surface of the combustible (such as a cotton pad) releases its flammable gases of decomposition at such a low rate and with very little heat convection so that they burn slowly or gently at the glowing surface, or some gases may escape without luminous ignition in the relatively small heated zone.

An important consideration to keep in mind concerning the combustion characteristics of solids is the fact that the ease of ignition and the rate of burning of all combustible solids is governed by their physical shape or their geometric configuration. More accurately, it can be said that the state of subdivision of solids *governs* these burning characteristics. When a match is applied to wood shavings or wood excelsior (also called "wood wool," similar to "steel wool"), ignition will occur almost immediately; if the match is applied to a two-inch cube of wood, no ignition will take place even after the match is burned to the end. Similarly, when newspaper material is shredded into ribbons or when it is separated into a single page and crinkled into a light ball, it will ignite by touching a match to it and will burn readily. If the same paper in an unseparated or bulk form is exposed to an igniting flame, it resists flame propagation and ignites very slowly.

An analysis of this situation reveals two factors that control the speed of ignition and the flame propagation in solids: (1) their heat conductivity, and (2) the extent to which each combustible surface is surrounded by air or oxygen so that burning can proceed. In the case of the solid block of wood, heat conduction or penetration into the block is very slow because of the block's low rate of conduction (that is, wood — or paper — is a good heat insulator) and whatever particles of the solid block are heated to their ignition point are not surrounded and fully exposed to oxygen so that union or burning does not take place. In the case of the wood shavings or the paper surfaces, temperature rise takes place rapidly because there is a very small mass of wood to be heated by the match; also, because each wood or paper surface is surrounded by air, burning takes place quickly.

Wood, Paper, and Cotton (Cellulose)

Ordinary woods of construction, the many types of paper that constitute our packaging material and printing or writing surfaces, and the cotton that is fabricated into clothing and fabrics of all types are combustible solids of vegetable or plant origin. They are all forms of cellulose that contain, in nature, various elements other than the carbon, hydrogen, and oxygen of the pure cellulose molecule. Wood and its close derivative, paper, contain small amounts of organic compounds of nitrogen and potassium. In addition, its carbon, hydrogen, and oxygen content are present as other organic compounds in the wood cell, such as lignin and resin. Cotton is the purest form of cellulose in nature. It contains more than 90 percent cellulose, which can be treated to obtain the pure compound having an empirical formula of $C_6H_{10}O_5$. Its structure is complicated and contains multiple amounts of these elements in varying bond relationships.

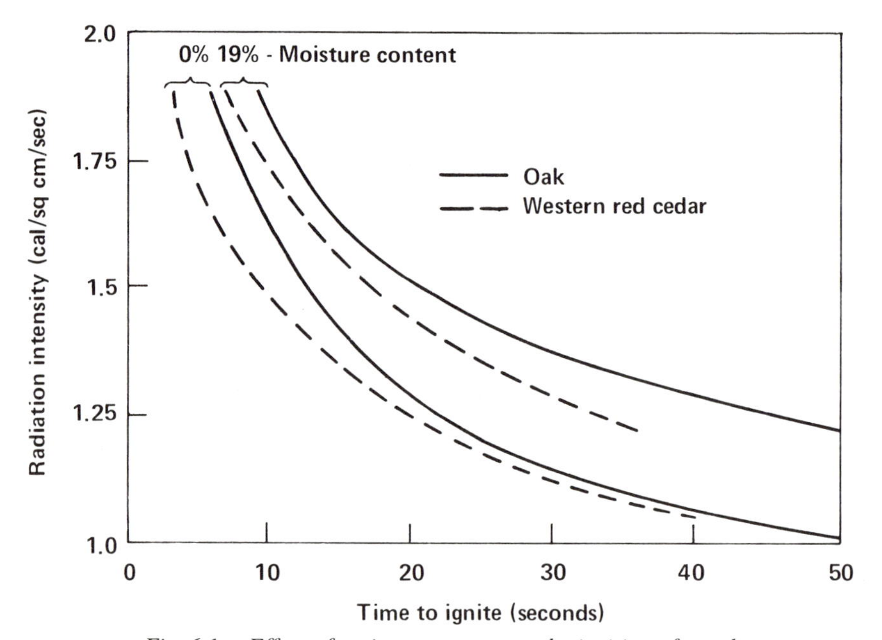

Fig. 6.1. Effect of moisture content on the ignition of wood.

The physical properties of these naturally occurring cellulosic combustible solids are of great importance to fire protection. Wood, as used in ordinary construction, is composed of many thousands of cells per volume. In their natural state, these wood cells formerly contained sap and aqueous solutions; in their dried form, they present open cells that contain surfaces and interstices for the absorption of water vapor. This has the effect of varying the moisture content of wooden construction, depending on its surroundings. The moisture content of wood has a great influence on its tendency toward ignition with a given heat exposure. Fires in wooden buildings constructed in humid, tropical surroundings spread more slowly than do fires in similar constructions situated in dry, cold regions. Figure

6.1 shows a plot of test results of the effect of two moisture contents of two woods. Figure 6.1 also illustrates the effect of moisture content on the ignition-time susceptibility of wood.

The low heat conductivity of wood is another important physical property of this combustible material, and is of considerable importance in some fire situations. If one considers the situation of a moderate heat or flame exposure of considerable thicknesses of wood, charring immediately takes place from the exposed surface inward. The char formation possesses an even lower heat conductivity than does wood (about one-half; see Table 5.5, Chapter 5, "Heat and Its Effects"), and the rate of heat penetration into the wood is progressively lowered by this mechanism. In the case of heavy timber construction or where large, laminated timber trusses are employed, the structural integrity of a building may be preserved for longer lengths of time under fire exposure than if the building had been constructed with steel framing of a type for similar stresses. The higher heat conductivity of the metal would be more quickly vulnerable to strength loss due to rapid penetration by heat attack.

It is difficult to separate and assess the variables involved in the penetration of wood by high temperatures so that ignition and combustion take place. As stated earlier, if they are to burn, wood masses must be heated to a point where combustible gases are evolved at the wood's surface. Gases of this type require ignition sources of temperatures in the range of 800°F (427°C) to 1,000°F (538°C) before combustion can occur. (Most open or "naked" flames of almost any type are in the region of 2,000°F [1,093°C].) As a consequence, it is relatively simple to describe the ignition point of wood as being in the vicinity of 350°F (178°C) to 390°F (200°C) if an open flame is present to act as an ignition point for the gases evolved from wood at these temperature ranges. (*N.B.:* The density, state of subdivision, amount of pitch or resin content, state of dehydration, and rate of heating of the wood, influence the ignition temperature of the wood in some proportional manner.)

Concerning the spontaneous ignition temperature of wood, the influence of the variables involved is even more indeterminate than where an igniting flame for the evolved gases is present. Considering the fact that wood slowly changes its chemical composition under sustained heat attack lesser in amount than is necessary to cause ignition of the combustible gases evolved, the end point of such an exposure over long periods of time is the formation of extremely porous charcoal. Charcoal is a material that is capable of absorbing gases and vapors. If these gaseous materials are combustible and are capable of further slow breakdown with the evolution of heat within the charcoal, a temperature may be reached that is capable of spontaneous ignition of the charcoal and of any wood in contact with it.

Many tests have been conducted for the purpose of determining the spontaneous ignition point of wood. A wide variation of results has been obtained from 421°F (216°C) to 507°F (264°C). The general conclusion has been drawn that wood or wood products must be prevented from decomposition if self-ignition is not to occur. It is generally agreed that wood must not be exposed to temperatures higher than 150°F (62.8°C) for long periods of time if risk of spontaneous ignition is to be eliminated.

The combustion of wood gives rise to varying amounts of gases, vapors, and smokes of different compositions and concentrations, depending upon the conditions in which the combustion takes place, and upon the type of wood fuel. To the fire fighter, the most important of these products of combustion are carbon monoxide, carbon dioxide, toxic or irritating organic compounds, and acids, all of which are present in varying composition in the smokes and vapors evolved from the burning wood.

The concentration of all of these substances is, of course, dependent upon the amount of ventilation during their period of combustion. The toxic gases that are most affected by ventilation of the fire are carbon monoxide and carbon dioxide. They are evolved in copious amounts when the air or oxygen supply to the fire is diminished because of "flashover" conditions. "Flashover" occurs in a space or compartment when combustible substances therein have been heated to their ignition point by a fire in the space. Almost simultaneous combustion of the materials occurs under these conditions, and in seconds temperatures rise dramatically to five times their earlier values. When flashover occurs, a sharp diminution of oxygen takes place, carbon monoxide concentrations rise far above lethal levels, and carbon dioxide increases markedly. If flashover occurs in a space or room that is open so that some ventilation can take place, the oxygen concentration is lowered by lesser amounts and the carbon dioxide may not rise above lethal amounts. However, the carbon monoxide may still evolve in concentrations much above the lethal concentration, which is usually quoted as "0.05 percent, which can be survived for one hour."[1]

The combustion of wood under conditions of an insufficient supply of oxygen produces copious amounts of smoke and organic vapors that induce coughing and lacrimation (tearing of the eyes) upon exposure. Thus, the fire fighter must assume that dangerous amounts of carbon monoxide are being produced simultaneously under these conditions, and that closed breathing apparatus should immediately be employed.

The fire and combustion properties of paper, which is a very close cellulosic derivative of wood, differ from wood only to a minor degree because of the slightly different physical characteristics of paper materials when they are fabricated for diverse purposes. In general, paper is produced in the form of relatively thin sheets or rolls and may be fabricated into light, thin boxes (cardboard) or, in the case of poorer grades of paper, may be mixed with other cellulosic material and compressed into building boards or tiles.

Paper and constructions formed from paper are subject to the same mechanisms of burning as wood, except that the formation of charcoal at the surface of paper that has been exposed to a heat source presents a less conductive material than the charcoal formed on wood surfaces. This is due to the fact that the contact of paper charcoal to the unburned surface beneath it is not close, and heat conduction is of a low order. In the case of piled sheets of paper, the rate of penetration of heat into the interior of the pile is slowed by virtue of the slight air gap between each sheet. Similarly, the combustion of compressed paper and cellulose raw materials

[1] Butler, C. P., "Measurements of the Dynamics of Structural Fires," OCD Work Unit 2561A, SRI Project PYU-8150, Stanford Research Institute, California, Aug. 1970.

used as building wall insulation or as blocks for ceiling construction is quite slow due to the low density of the material; such low density affects its rate of conductivity of heat.

The ignition temperatures of paper and paper products are approximately the same or a little lower than the ignition temperatures of wood because of paper's

Table 6.1. Self-Ignition Temperature of Solids *

MATERIAL	TYPE OF SPECIMEN	°F
Woods and Fibrous Materials		
Short-leaf pine	Shavings	442
Long-leaf pine	Shavings	446
Douglas fir	Shavings	500
Spruce	Shavings	502
White pine	Shavings	507
Paper, newsprint	Cuts	446
Paper, filter	Cuts	450
Cotton, absorbent	Roll	511
Cotton, batting	Roll	446
Cotton, sheeting	Roll	464
Woolen blanket	Roll	401
Viscose rayon (parachute)	Roll	536
Nylon (parachute)	Roll	887
Silk (parachute)	Roll	1058
Wood fiberboards	Piece	421 to 444
Cane fiberboard	Piece	464
Synthetic rubber		
GR-S (R-60) black	Coagulum	590
GR-S (R-60) black	Buffings	374
GR-S, black	Coagulum	563
GR-S, black	Buffings	320
GR-S, Indulin	Crumb	824
Metals		
Aluminum paint flakes	Fine powder	959
Tin	Fine powder	842
Tin	Coarse powder	1094
Magnesium	Fine powder	883
Magnesium	Coarse powder	950
Magnesium ribbon	Cuts	1004
Magnesium, cast	Piece	1144
Magnesium-Al-Zn-Mn alloys (Mg 89 percent or more)	Piece	860 to 1256
Zinc	Fine powder	1202
Miscellaneous		
Nitrocellulose film	Roll	279
Matches (strike anywhere)	Heads	325
Carbon soot	Dust	366
Crude pine gum	Powder	581
Shellac	Scales	810
Paint film, oxidized linseed oil-varnish	Powder	864

* National Bureau of Standards, 1947.

tendency to deform under heat. The spontaneous ignition temperature of paper is also similar to that of wood. (For comparison, see Table 6.1, "Self-Ignition Temperature of Solids.")

When referring to temperature, we are not expressing the amount of heat that is being brought to bear in order to bring substances to the required temperature of ignition. In the case of paper, careful measurements made at the Borehamwood (England) Fire Research Station[2] showed that the minimum irradiance of heat impinged on corrugated paper wrapping carton stock was about one-third that required for spontaneous combustion of soft wood. This was evidently due to the tendency of paper to glow in spots under heat attack; these glowing spots acted as ignition points. Wood does not exhibit this effect under heat attack.

The smokes and gases evolved during the combustion of paper and paper products of different types are similar to those from the burning of wood, and identical precautions need to be taken in questionable atmospheres surrounding these fires.

The combustion properties, ignition temperature, and the smokes and gases evolved from the burning of cotton and other materials fabricated from this form of cellulose are similar to those of wood in all respects but one. The one difference concerns cotton's "glowing combustion" characteristic. When cotton materials are ignited by an open flame or by the glowing end of a cigarette, some flaming may temporarily occur at the ignited surface of the cotton. This is followed by a slow, glowing type of burning that continues without change, progressively consuming the combustible fibers in all directions.

The slow combustion of cotton can easily be verified if it is kept in mind that all cotton materials are composed of fibers surrounded by small columns of air and, when cotton is even slightly compressed, the heat conductivity of the nearby fibers is raised to a point where transmission of heat from fiber to fiber is promoted. External heat loss from the glowing combustion mass is prevented by the insulating properties of the unburned cotton surrounding it. The glowing combustion of many solids, including carbon and charcoal, will continue at lowered oxygen concentrations in air down to about sixteen percent.

Keeping in mind the property of continued combustion of cotton in an atmosphere of insufficient oxygen, it is easy to see why large amounts of carbon monoxide can be generated from this type of slow burning. Glowing charcoal fires and cotton mattress fires evolve dangerous amounts of carbon monoxide. If ventilation of the gas from such fires in closed areas does not take place, a gaseous atmosphere will form that will endanger life safety. In the case of mattress fires, including cotton bedclothing, sleeping individuals who are close to the source of carbon monoxide will inhale toxic quantities of gas prior to any type of warning.

Many of the combustion gases and smokes from the burning of all types of cellulose possess distinctive odors, some of which cause eye irritation. The detection of these effects indicates the presence of toxic gases; when a fire fighter enters a poorly ventilated area, breathing apparatus should be quickly employed.

The amounts of heat developed by burning cellulosic materials are not great when compared to other combustible solids, as indicated in Table 6.2. However,

[2] Wraight, H., "The Ignition of Corrugated Fibreboard (cardboard) by Thermal Radiation," Fire Research Note 1002, Fire Research Station, Borehamwood, Hertfordshire, England, Feb. 1974.

they do pose serious problems to fire protection specialists because of their abundance and their wide usage in many different forms.

Table 6.2. Heats of Combustion of Cellulosic Solids and Comparative Materials *

SUBSTANCE	CALORIFIC VALUE	
	Btu PER LB	KILOJOULES/KILOGRAM (kj/kg)
Oak wood	7,180	16,701
Pine wood	8,080	18,794
Combed cotton	7,160	16,654
Corrugated paper (carton material)	5,970	13,885
Wrapping paper	7,106	16,528
Anthracite coal	12,520 to 13,830	29,121 to 32,168
Coke	11,690 to 12,810	27,191 to 29,796
Asphalt	17,160	39,914
Wool rags	8,876	20,646
Silk rags	8,391	19,517

* From *Fire Protection Handbook*, 13th Edition, National Fire Protection Association, Boston, 1969.

Wool, Silk, and "Animal" Fabrics.

In addition to combustible solids from plant sources, our daily existence is dependent upon a number of materials of animal origin. The chemical structure of all of these materials is even more complex than the chemical structure of the cellulosic material. In addition to containing the elements carbon, hydrogen, and oxygen, they contain nitrogen bonded to hydrogen in the form of an amino group, which is:

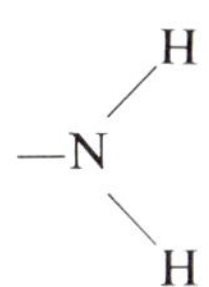

In order to greatly simplify the constitution of all combustible solids of animal origin, it can be said that they all contain proteinaceous matter of the keratin type. Keratin is found in human hair and fingernails, and in animal horns and skins.

From the standpoint of combustion hazards, fabrics and materials of animal origin are less susceptible to ignition and flame propagation than are the cellulose compounds or most of the synthetic fabrics. The decomposition of protein-type nitrogen compounds under flame attack yields many gases and vapors that do not burn. Most of these are unpleasant to smell, and some are toxic.

It is of interest to compare the characteristics of combustion of wool padding or thick woolen fabrics with the characteristics of combustion of materials of cotton composition. With wool (in which each fiber is surrounded by air, as it is with similar cotton fabrics), an igniting flame will scorch and char the area of wool it contacts, and burning will not continue if the flame is removed. However, with materials of cotton composition (as illustrated in the earlier example of a cotton pad), a deepening, glowing area of burning will continue after the source of ignition has been removed.

Considerable fire retardant characteristics are demonstrated by woolen compositions. Over the years, this was an advantage of woolen rugs and brushed woolen fabrics; this advantage was not realized until the recent advent of "fabric" rugs and "fluffy" sweaters made from synthetic fibers. Where the woolen fabric only scorches and burns reluctantly without vigorous flame propagation, some synthetic floor-covering materials and brushed, fluffy, synthetic fabrics burn rapidly.

The determination of the ignition temperature of wool is not of practical value because of: (1) the variability of its decomposition products which, because of such variability, will not sustain combustion, and (2) the many forms in which it may be tested. However, the spontaneous ignition temperature (self-ignition) of wool is not too different from other substances when compared to those listed in Table 6.1, "Self-Ignition Temperatures of Solids." Table 6.1 was compiled by the National Bureau of Standards in 1947, employing tests in which sample materials were slowly heated in a current of air until they burst into flame.

Silk is another material with fire hazard characteristics similar to those of wool. Formerly an important fabric of interest to the fire technologist, silk has largely been replaced by the synthetic fabrics. It is a combustible material from animal sources. The chemical structure of silk is similar to that of wool in that it is protein in character; its complicated molecule contains large amounts of nitrogen. Silk also contains fibroin, which is a type of protein closely related to the amino acids. Similarly to wool, silk decomposes under heat attack; also, its flame propagation properties are similar to those of wool. At its point of ignition silk burns with reluctance, demonstrating scorching and charring without vigorous flame propagation. It is of interest to note from Table 6.1 that the spontaneous ignition temperature of silk is approximately twice that of other naturally occurring combustible solids.

Furs and leather are also combustible solids of animal origin containing proteinaceous compounds. Like wool and silk, furs and leather resist combustion and decompose into scorched and charred material without vigorous flame propagation. Even though fur clothing is in a form that will burn readily, it will not flash in upward flame propagation.

COMBUSTIBLE SYNTHETIC POLYMER SOLIDS & PLASTICS

In the twentieth century it was learned that many organic chemicals could be made to react with themselves or with other organic materials to produce new substances with qualities capable of helping to enrich our daily lives. Thus, members of our society have become surrounded by inexpensively manufactured forms of synthetically produced solids of every description — synthetically fabricated items that meet almost every human need. The name "plastics" has become a commonly used term for the synthetic polymer that results from the chemical combination of small molecules into much larger ones by the *polymerization* process (the union of monomers to form polymers). The product of this process is a substance of high molecular weight, usually shaped into its usable form by flow in a forming mold of some configuration.

Polymers, or plastics, may be compounded with plasticizers, fillers, and stabilizers. They may be formed into stronger, useful items in combination with other materials such as glass cloth or metal rods. Polymers of differing structure may also be mixed to give desirable properties to the finished product.

Plasticizers are organic compounds, usually liquids, that confer some flexibility to the finished plastic film or piece. Fillers in plastics usually consist of powdered solids such as cotton flock, clay, or wood flour so that the finished item possesses greater strength or is easier to fabricate. Stabilizers are usually organic compounds added to plastics to retard their tendency to yellow with age, to become brittle, or to oxidize on their surfaces.

The term resins has often been used incorrectly as a name for plastics and polymers. In nature, there are many semi-solid substances that are exuded from plants or trees such as the familiar resin from pine trees. These are resinous materials and should not be confused with the modern synthetic organic polymers and plastics.

Since all polymers and plastics are organic compounds capable of decomposition under heat attack, they are classified as combustible solids. The processes of combustion that they go through are highly different, due to the widely varying chemical composition of plastics and the forms in which they are fabricated. Generalizations of combustion behavior are not practical when considering the fire hazard properties of plastics and polymers. Accordingly, these substances will be described individually in the following paragraphs, with emphasis on their characteristics under heat and flame attack. The classification of polymers and plastics used by the plastics industry, wherein they may be classified as *thermoplastic* or *thermosetting* depending upon basic physical characteristics and properties, particularly the behavior toward heat, will be used herein.[3] Thermoplastic polymers soften without chemical change upon moderate heating and become rigid upon cooling. Thermosetting polymers are those that undergo a chemical change when they are first heated. They become rigid and do not soften or lose form upon reheating.

COMMON THERMOPLASTIC POLYMER PLASTICS

Acetal Plastics

When formaldehyde is polymerized to a high molecular weight, it becomes a hard, tough, and very durable plastic which, although unaffected by organic solvents, is reacted upon by strong acids or bases. It is used for piping and fittings

[3] See NFPA 654, *Standard for the Prevention of Dust Explosions in the Plastics Industry*, National Fire Protection Association, Boston, 1975, p. 4.

and automotive gears. Under the trade name "Delrin," it has a large number of applications where its durability is needed. The acetal materials burn slowly, melting with burning drippings. It softens at 320°F (160°C).

The Acrylics

The acrylic plastics are polymers of the methyl methacrylate type. The commercial acrylic products trade named "Plexiglas" and "Lucite" are widely known. Acrylics are used in many applications where clearness and weather resistance are needed. They may also be composed of chlorinated methacrylate or other polymer mixtures. The acrylics soften and lose form at temperatures only mildly hot to the touch, 125°F to 150°F (46°C to 63°C). Under flame exposure, they bubble and burn at their exposed surfaces without a great deal of smoke; while burning, a fruity odor is expelled. Little char results, but softening of the plastic proceeds as heat penetrates it. Carbon monoxide in ordinary amounts is generated during the combustion stages.

Cellulose Acetate

When cellulose (such as cotton) is suitably reacted with organic acids, it can produce esters (an often fragrant compound formed by the reaction between an acid and an alcohol) that may be processed in the form of transparent plastic films or other useful products. Cellulose acetate is commercially available as photographic film, and is used in many variations of composition of plasticizer to acetate ester. In ordinary formulations it burns easily, giving off some smoke during its dripping combustion. A slight vinegar odor can be noticed during burning.

Cellulose Acetate Butyrate

This plastic is another product of the reaction of cellulose with organic acids, both acetic and butyric. It is a plastic of greater durability than the previously described cellulose acetate, and one of its major uses is in automobile steering wheels. It ignites readily and gives off burning drippings and a black smoke with a vinegarish, rancid butter odor. This odor is the result of the mixed vapors of acetic and butyric acids.

Cellulose Triacetate

When cotton is carefully reacted with the anhydrous form of acetic acid, it forms a material that differs in many important respects to cellulose acetate. Cellulose triacetate is used principally in the form of flexible films for photography, tapes, etc. It burns slowly with burning drips, and has less smoke than other cellulose esters. The vapors rising from its combustion have a vinegarish odor.

The Fluorocarbon Plastics

The element fluorine may be caused to react with ethylene (C_2H_4) and polymerized to form polytetrafluoroethylene (TFE) under certain conditions of high pressure and temperature. The resulting waxy, inert plastic material is commercially available as "Teflon." Other fluorine-containing plastics may be synthesized with chlorine and the propylene group (C_3H_6), and may be added to the fluoroethylene polymer to yield other useful properties to the final plastic polymer. Fluorine also can be reacted with the vinyl groups (CH_2CH) and polymerized to form a plastic material. The fluorine-containing polymers are used where chemically resistant, superior electrical-insulating properties and heat-resistant plastic coatings or fabricated materials are needed. Their trade names may begin with "fluoro-," or they may use the following: "FEP," which is polyfluoroethylene-propylene; "CTFE", which is polychlorotrifluoroethylene; "PVF," which is polyvinyl fluoride; or "PVF2," which is polyvinylidene fluoride. All of the fluorine-containing polymers are difficult to ignite or, as in the case of TFE, bubbling of the plastic at temperatures of about 500°F (260°C) and above will exhibit volatilization and decomposition of the material. The gases arising from this decomposition are acidic in smell and are toxic. CTFE will melt at temperatures of 390°F (200°C); however, the drips from CTFE will not continue to burn. Similarly, PVF2 does not continue to burn.

Polyamide Plastics ("Nylon," etc.)

This class of polymers is the only synthetically produced material that contains a slight rearrangement of a polypeptide link in its structure:

$$\begin{array}{cc} H & O \\ | & \| \\ -N- & C- \end{array}$$

This is also found in the animal fibers, wool and silk. Many of the properties found in wool and silk are demonstrated by the nylon polyamides when they are formed into fibers or threads.

There are a number of types of nylon, each of which is formed by polymerizing long-chain organic acids containing the amino group, H_2N—, or reacting other types of long-chain amino compounds with long-chain organic acids under special conditions of pressure and temperature. In general, all types of nylon exhibit similar properties of toughness and resistance to almost all types of decomposition and attack except in high-temperature exposure. The nylons melt readily under flame or heat attack at about 426°F (220°C) to 510°F (265°C). The thread fibers are spun from molten nylon at 572°F (300°C).

The nylons burn slightly above the melting point with a luminous flame without smoke; the burning continues on the surface of any melted falling or running plastic until it cools and solidifies, after which burning ceases. The black residue contains char and decomposition products that are considered self-extinguishing.

The vapor and slight smoke that form during combustion give off a slightly sweet odor.

In recent years a new polyamide plastic fiber has been produced for special applications requiring resistance to melting at temperatures up to 900°F (483°C). These plastics are synthesized from amino compounds that contain the aromatic or benzene ring. The benzene ring has a high proportion of carbon-to-hydrogen in its structure.

"Nomex" is one of these polymers. Nomex retains almost all of the beneficial characteristics of nylon. It chars without melting below its much higher melting point than nylon, thus effectively removing the problem of burning molten drops of plastic. It is also considered to be self-extinguishing when the source of heat attack is removed from it.

Polycarbonates

This is a comparatively small class of polymers of which only a few different types are manufactured. One important plastic of this type is called "Lexan." The principal method of production consists of coupling and polymerizing a mixture of an aliphatic-modified phenolic aromatic compound with carbonic acid. Because of its aromatic benzene type structure, it is self-extinguishing and melts only at high temperatures. The plastic is highly resistant to impact, and is used in instances requiring good retention of dimensional characteristics at relatively high temperatures — up to about 280°F (138°C).

The Polyethylene Plastics (Poly-olefins)

Thermoplastic polymer plastics made by polymerizing ethylene (C_2H_4) are a large and widely applied class. They consist of several differing density polymers, and in commercial applications these may be mixed with other polymeric materials to yield desirable properties to molded articles. Polypropylene polymers are similar to those of polyethylene. Speaking of both polymers, they would both belong to the class of poly-olefins. (Propylene [C_3H_6] is the homologue of ethylene with an additional CH_2 group.) For purposes of this text, these polymers will be considered together.

Polyethylene is made by subjecting ethylene gas to elevated pressures and temperatures in the presence of a catalyst. The polyethylene polymers are waxy and have little strength unless they are of the high density (and molecular weight) types. Polypropylene is much more durable, exhibits good hardness, and is more resistant to chemical attack.

The poly-olefins are relatively cheap and easy to manufacture into everyday articles. Starting with the low-density (low-molecular weight) polyethylene plastics, which soften and melt at low temperatures, the softening points of higher molecular weight types increase to only about 230°F (115°C) for polypropylene types.

When submitted to a source of flame, polyethylene will ignite and will burn slowly at low temperatures of 662°F (350°C), releasing burning drips. Similarly, polypropylene ignites at somewhat higher temperatures of 1,058°F (570°C), burning slowly and releasing burning drips. A somewhat fruity and sweetish odor can be detected in the slight amount of smoke when either polyethylene or polypropylene burns. Carbon monoxide gas is also evolved during their combustion. There are a large number of trade names for the poly-olefins, including "Marles," "Propylex," "Super Aeroflex," and "Super Dylan."

The Polystyrene Polymers and Their Mixtures

Polystyrene plastics may consist of only the one polymer, polystyrene, or they may consist of several polymers, called copolymers, together in the same material.

When benzene, an aromatic compound, is caused to react with ethylene under certain special conditions, it forms an additive compound that contains the vinyl group $HC{=}CH_2$ substituted in the benzene ring. This is the styrene monomer, which also may be called vinyl benzene. When this single compound is polymerized, polystyrene is formed. It is a plastic with properties of good resistance to heat and high impact strength. Many large moldings for a large variety of applications are formed from polystyrene. A rigid plastic foam for package insulation and for heat and cold insulation is sold under the familiar name of "Styrofoam." Polystyrene molded solid plastic material ignites easily at about 650°F (343°C), softening and generating a dense, black smoke that gathers in the air in the form of floating black clusters of carbon. It does not burn rapidly. Styrofoam rapidly collapses upon heat exposure, igniting and burning similarly to the molded solid form.

"ABS" is the usual designation for acrylonitrile-butadiene-styrene copolymer plastic material. It is even tougher than polystyrene, and is used in many molded forms that require strength, durability, and resistance to heat or solvent attack. Plumbing piping and fittings of this plastic are being used more widely each year. Its softening point is much higher than the acrylics, and higher than polystyrene. Its burning properties are similar to those of polystyrene.

The Vinyl Polymers and Copolymers

This plastics family consists of a wide variety of polymers and copolymers, too extensive to be completely listed here. These plastics are fabricated into many useful articles, both flexible and rigid in form, and may be mixed with different plasticizers, fillers, and stabilizers for the special end uses to which they are put.

The vinyl polymers are synthesized by utilizing the vinyl group $HC{=}CH_2$ (mentioned earlier in connection with polystyrene), that can be produced from acetylene, C_2H_2, or ethylene, $CH_2{=}CH_2$, and reacting this with the desired material — chlorine, acetic acid, butraldehyde, or formaldehyde — to yield the desired polymer.

Polyvinyl Chloride (PVC)

PVC is perhaps the most widely used vinyl polymer. It is made by a chlorination process. It can be fabricated in a rigid form or with suitable plasticizers, and is widely employed for wire insulation or in thin, flexible sheeting for various purposes. PVC softens at various temperatures, depending upon the amount of plasticizer used in its formulation, ignites with some difficulty, burns slowly, and may self-extinguish. While burning, PVC produces white smoke and hydrochloric acid fumes. These fumes are quickly detected because of their acidic taste and the sensation they cause. The concentration of hydrochloric acid vapor from burning PVC is highly toxic; thus, closed breathing equipment must be used as soon as this gas is identified.

When acetic acid is reacted with the vinyl group, vinyl acetate is produced. The latter liquid may be polymerized with vinyl chloride to form a copolymer, polyvinylchloride-acetate, PV-AC. PV-AC is used where flexibility of the polymer is expecially desired. Its burning characteristics are similar to PVC in that it may self-extinguish (depending upon its plasticizer content), with evolution of hydrochloric acid vapor in toxic quantities. Plastic foams derived from PVC or PV-AC have burning properties similar to those of the solid plastics, and toxic hydrochloric acid vapor is evolved.

When the vinyl group is caused to react in special ways so that an alcohol group (—OH) unites with it, the resulting vinyl alcohol may be polymerized to yield a *water-soluble* plastic, polyvinyl alcohol (PVA), which is impervious to attack by solvents, and which can be formed into flexible hose to transport such liquids. PVA is also used for packaging materials that are to be added to water in measured amounts. It does not readily ignite, but chars and caramelizes.

Other vinyl polymers include polyvinyl butyral (PVB), which is used as the flexible glass adhesive in the center of automotive safety glass. It burns reluctantly, and the slight amount of smoke evolved has a rancid odor.

COMMON THERMOSETTING POLYMER PLASTICS

There are a number of polymers that are classified as "thermosetting" plastics, although they do not always require heat to place them in the form in which they will be used. In contrast to the "thermoplastic" plastics, these polymers may not change their form upon application of heat. The chemical reaction that takes place when they are used results in a permanent "set," or "formed-in-place," plastic that cannot revert to a liquid again.

In the "thermosetting" class of plastics, there are many types that are used in conjunction with fillers such as fibrous glass, glass fiber cloth, and other reinforcing materials.

Additionally, there are types of plastics that act as adhesives for many fabricated items. These types of plastics are used in minor proportions in comparison to the other solids that are used with them so they do not constitute a plastics fire hazard. They will not be considered in great detail herein, but some examples of these plastics are:

1. Alkyd plastics (sometimes called resins), used in paints and in molded items with glass fiber laminates. They are self-extinguishing.
2. Epoxy plastics, used as adhesives and as a laminating material with various reinforcing materials. They burn slowly.

3. Polyester plastic-forming material, used in conjunction with glass fiber for "forming-in-place" items such as boat hulls. This type of plastic is also used to form films. In general, the polyesters may burn slowly with some charring and black smoke production. Other formulations are self-extinguishing.

4. Silicone plastics and rubbers, formulated with reinforcing fillers for rigid-form items, or produced as a resilient rubber-like material for heat- and cold-resistant gaskets or similar useful items. By test, the silicones are generally nonburning.

The Amino Plastics (Urea-formaldehyde)

The amino plastics are formed by the reaction of organic compounds such as urea or melamine (which is a derivative of cyanamide), both of which contain the amino group, $—NH_2$, with formaldehyde, HCHO. Formed in molds under heat and pressure and with fillers of many types, the solid or laminated thermosetting plastic material is hard and unaffected by heat and solvent attack. Because of its durability, it is used for dinnerware or countertops. "Formica" is a familiar brand name for the laminated plastic. The amino plastics are ignited only with difficulty. They are self-extinguishing.

The Phenolic Plastics

Plastic materials made from the reaction of phenol with formaldehyde are the earliest form of synthetic polymer. A large number of phenolic plastic items that are both durable and heat-resistant are manufactured under the trade name "Bakelite." Almost all phenol-formaldehyde items contain large proportions of fillers. Electrically insulating plugs and utensil handles are usually made of phenolic plastics, although arcs and sparks will readily form paths, or "tracks," across this material if moisture or other conditions promote their formation. Overloaded or overheated electrical parts utilizing the phenolics give rise to the familiar carbolic acid odor, but ignition of the plastic is difficult and slow. Phenolic plastics will self-extinguish if the external heat attack is discontinued.

The Polyurethane Plastics (Isocyanate Polymers, Polyether Polymers)

The discovery of the polymer formation by a reaction of a polyester (made by reacting a dicarboxylic acid, adipic, with glycerine), with isocyanates to form isocyanate groups (—NCO) at the ends of the polymer chains, was the basis for a whole new family of plastic materials. It was found that the isocyanate polymers could be synthesized in a number of different ways, using reactions of isocyanates of varying structure with castor oils and derivatives of sucrose (sugar) and other compounds, giving resulting polymers of highly varying characteristics: rigid, rubbery, flexible film and foam. The methods of obtaining the various polyurethane materials of commerce are as many as there are trade names for them.

From the fire hazard standpoint, the polyurethane plastic foams used for upholstery and decorating purposes are the most important materials in this family of plastics. Polyurethane foams, whether they are rigid, semi-rigid, or flexible, are easy to ignite and, under some conditions, can even be ignited from a lit cigarette.

They burn rapidly, emitting burning drops; a large amount of yellow to black smoke is evolved. In any form, burning polyurethane plastic material quickly gives off large amounts of carbon monoxide and toxic quantities of hydrogen cyanide gas. Hydrogen cyanide gas (HCN) has a characteristic odor that smells like almond extract. It is a deadly poison, requiring only a few minutes exposure to atmospheres containing only 300 parts per million to cause death. (In contrast, carbon monoxide requires approximately five times this concentration.) Self-contained breathing apparatus is, of course, vitally important to the fire fighter when hydrogen cyanide is detected in *any* quantity.

The Rubbers and Neoprene Synthetics

Although natural rubber, chlorinated rubber, butadiene, and its copolymer with styrene (which is GRS rubber), and neoprene are classified as "rubbers," not plastics, they are polymeric in composition and are used in many instances where ordinary plastic polymers are also used. From the fire protection standpoint, it is worthwhile to note that rubber burns readily, giving off copious volumes of black, foul-smelling smoke. In large storage areas containing rubber tires, it may be necessary to employ "wet water" to fully extinguish fires. Chlorinated rubber may evolve smoke containing choking amounts of hydrogen chloride. Neoprene materials are difficult to ignite, but burn similarly to rubber. When neoprene foam rubber is used as an upholstery material, it is much more difficult to ignite and burn than the polyurethane foams.

IMPORTANT FIRE PROPERTIES OF SYNTHETIC POLYMER FIBERS

The sources of materials for today's clothing, curtains, rugs, and draperies are different from the sources for formerly often-used materials such as cotton and woolen fibers from natural sources: this is due to the fact that plastic polymers have become increasingly important in the field of woven textile fibers. It is important that the fire protection specialist has some knowledge of the different types of synthetic fibers, even though their identification is difficult. Table 6.3 classifies the principal textile and rug fibers. In its gas and smoke evolution properties, the burning characteristics of each polymer fiber is similar to the basic solid polymer.

PROPERTIES OF SOME COMBUSTIBLE METALS AND OTHER ELEMENTS

The world of combustible solids is not limited to those that contain carbon and hydrogen. Many of the elemental metals and their alloys and some of the non-metallic elements are fire hazards common to everyday industry and living. The combustion of the elemental metals and nonmetals proceeds differently from the burning of ordinary solids like wood, paper, or plastics. In the metals, the union

Table 6.3. Properties of Some Synthetic Polymer Fibers *

POLYMER CLASS AND COMMON NAME	PROPRIETARY NAME(S)	SOFTENING POINT °F	(°C)	MELTING POINT °F	(°C)	FLAMING CHARAC-TERISTICS	ORDINARY USES
Cellulose Acetate (Acetate)	Esteron, Celacloud	284	(140)	445	(230)	Melts and burns slowly	Clothing, drapes, knits
Cellulose Triacetate	Arnel	437	(225)	572	(300)	Melts and burns slowly	Clothing, drip-dry wear, knits
Viscose Rayon	Enka, Avisco	—	—	decomposes 392	(200)	Burns readily	Clothing, rugs, curtains
Polyacrylonitrile (Acrylic)	Orlon, Acrilan, Creslan, Cantrece, Verel-copolymer, Dynel-copolymer	374	(190)	500	(260)	Burns slowly	Fabrics, knits, carpets, blankets
Polyamide	Nylon (all types)	392	(200)	419 to 482	(215 to 250)	Melts, flames and drip burns	Fabrics, carpets, upholstery
Polyester	Dacron, Kodel, Fortrel	437	(225)	482 to 544	(250 to 290)	Burns slowly	Fabrics, drapes, ropes, belting
Polyethylene (Olefin)	Alathon, Durethene, Dylan	248	(120)	284	(140)	Burns slowly	Fabrics, ropes
Polypropylene (Olefin)	Herculon, Super Dylan, Versatex	293	(145)	320	(160)	Burns slowly to self-extinguish	Outdoor fabrics, ropes, upholstery
Polyurethane (Spandex)	Lycra, Spandelle	374	(190)	482	(250)	(Toxic gases) Burns easily	Elastic garments, knits
Polyvinylchloride (PVC)	Vinyon, Dynel-copolymer	158	(70)	284	(140)	(Toxic gases) Burns slowly to self-extinguish	Felts, blends with other fibers
Polyvinylidene Chloride	Saran	239	(115)	338	(170)	Difficult to ignite, chars	Outdoor fabrics, curtains, carpets, upholstery
Polytetrafluoroethylene	Teflon	437	(225)	decomposes 572	(300)	No ignition	Tapes, elect. insulation

* Extracted from "Handbook of Tables for Applied Engineering Science," 2nd Edition, CRC Press.

with oxygen in the air takes place directly without decomposition or vaporization of the elemental solid. Once initiated, this union (or burning) is more violent and, in the case of the metals, is more difficult to extinguish as it may require special agents. The observation to be pointed out here is that metals have far superior heat conduction properties than the ordinary solids or the nonmetal combustible elements. In the following descriptions of the fire properties of the elemental combustible substances, note that in most cases the degree of hazard of the solid is in proportion to its extent of subdivision, as in the case of ordinary solids such as wood and paper.

Magnesium and Its Alloys

Pure magnesium metal is almost always alloyed with aluminum, manganese, copper, or zinc in small amounts. In large castings or machined parts it is difficult to ignite because of its high heat conductivity and its ability to radiate heat away from the area being heated. Near its melting point (which may vary due to the type of alloy), the surface of the metal becomes blackened due to formation of an oxide form of magnesium. This catalyzes the rapid oxidation reaction, and the familiar dazzling bright flame ensues.

The ignition temperatures of magnesium alloys are about 1,067°F (575°C), with flame temperatures in the region of 1,040°F (560°C). The flame region is very small, not exceeding two to four inches (five to ten centimetres) in height above the burning surface of metal.

Thin chips, turnings, or ribbons of magnesium alloys ignite more rapidly than do thick sections. The finely divided powders can ignite with a flashing brightness, and a white smoke will evolve if they are heated and mixed with air by some process of falling or agitation. The metal foils and wires in certain photographic flashbulbs consist of magnesium enclosed in an envelope of oxygen. Such flashbulbs are ignited by an electrical filament. The brilliant pyrotechnic search flares utilize mixtures of magnesium powder in an oxidizing composition to give steady combustion and light.

The combustion process of magnesium cannot be halted with water except by the employment of large streams directed into the combustion zone. Water disassociates into its components (hydrogen and oxygen) at these temperatures, giving a secondary combustion and recombination of the hydrogen-oxygen in the burning area. The large streams of water cause momentary disruption because of this reaction, and molten, burning magnesium is propelled away in all directions (with attendant dangers); however, the burning area is made smaller, and cooling finally takes place. Water sprays may also be used to cool the nonburning bulk of metal so that the temperature of the burning zone will subsequently be brought below its ignition temperature.

Titanium

The metal titanium is not widely used because of the expense involved in producing it. It is strong, yet lightweight, which makes it desirable for aircraft parts.

When heated, titanium burns somewhat like magnesium. In the finely divided form, it burns with the same sort of sparking that one sees when steel parts are held to a grinding wheel. This tendency to spark is one of the drawbacks to the use of titanium for aircraft parts. In the event that an aircraft is forced to make a landing on a concrete or asphalt runway and its titanium parts make sliding contact with the hard surface, the resulting shower of sparks is capable of igniting any fuel that may be leaking from the aircraft's damaged fuel tanks.

Sodium Metal

Metallic elemental sodium is produced by an electrolytic process and is used in certain chemical manufacturing processes. It is very soft, and can be cut with an ordinary kitchen knife. A freshly exposed surface of metallic sodium immediately discolors due to oxide formation with air. It reacts and melts almost explosively in any contact with water, and must be shipped and kept immersed in kerosine or under a constant atmosphere of nitrogen. Metallic sodium will burn quietly in air at, or slightly below, its very low melting point of 208°F (99°C). While burning, it gives off a small flame with caustic, choking smoke that contains the hydroxide, NaOH. When the metal contacts water in any form, hydrogen gas is evolved and, under many conditions, the metal will ignite, as will the hydrogen. It floats (density = 0.97) on water, reacting vigorously and melting, and may very often ignite. The combustion reaction of sodium can be halted only by blanketing it with nitrogen gas or smothering it with a nonreactive material (such as sodium chloride) that slowly halts oxygen access to the metal.

Potassium Metal

Metallic elemental potassium is a soft, highly reactive metal similar in almost every respect to sodium metal. It reacts rapidly with water, melting and self-igniting with vigorous sputtering; it evolves a white smoke that is caustic and choking when breathed. It is lighter than sodium metal (density = 0.86) and melts at only 146°F (62°C), usually igniting in air near the melting point. Metallic elemental potassium must also be kept immersed in kerosine or in an atmosphere of nitrogen; only by halting potassium metal's access to oxygen by some means can combustion of the metal be stopped. It reacts violently with all extinguishing agents except special dry chemical ones.

Sodium-Potassium Liquid Alloy (NaK)

A mixture of sodium metal with potassium metal will result in an alloy that is in liquid form at ordinary temperatures. This alloy is used as a heat-transfer liquid in certain special applications. It possesses the same fire-hazard properties as the solid sodium and potassium metals, but has the additional problem of being a liquid that flows like mercury. It reacts with almost everything in its path, even to the point of extracting water from solid concrete and causing the surface of the concrete to spall.

Other Combustible Metals

Lithium metal is similar to sodium and potassium metals; however, lithium metal is less reactive, even to water. Lithium metal will ignite and burn at temperatures near its melting points 356°F (180°C); it will also burn in an atmosphere containing nitrogen.

Zirconium metal is capable of ignition when in a powdered form. For this reason it is kept under water during storage, although it can ignite spontaneously when handled in a moist condition in air.

Uranium is a combustible, radioactive metal. It can be ignited by heating the solid form of the metal, which is similar to magnesium. In a finely divided or powdered form, it can be ignited readily. Uranium scrap or turnings are usually confined under oil for this reason.

Thorium and plutonium are also combustible radioactive metals. Their dry powdered forms are handled under helium or argon atmospheres. Because of its radioactivity, plutonium requires radiological contamination prevention procedures during its handling.

THE ELEMENTAL NONMETAL COMBUSTIBLE SOLIDS

Elemental Phosphorus

Elemental phosphorus exists in two combustible forms: (1) yellow phosphorus, which is a waxy, light-yellow, sometimes white solid that must be kept under water at all times to prevent its spontaneous combustion, and (2) red phosphorus, which is a crystalline form of phosphorus that looks like clumped red powder, and which ignites at much higher temperatures than the yellow variety.

Yellow phosphorus is very poisonous, even when in the air in a diluted vapor form. It will ignite in air if exposed long enough to oxidize slowly so that its surface temperature rises to about 104°F (40°C). It burns with a small, glowing flame, and gives off a white, phosphoric-acid smoke that is toxic after long exposure. It has a bitter taste. Burning phosphorus is easily extinguished by covering it with water.

Red phosphorus is not highly poisonous, and requires heating to about 463°F (240°C) before spontaneous ignition occurs. However, both yellow and red phosphorus ignite immediately at room temperatures by flame or by spark. Phosphorus flames are quickly extinguished by water.

Elemental phosphorus is used only in certain chemical manufacturing processes. When used in mixture with powdered glass and antimony trisulfide, the red form becomes the reddish-brown striking surface used on safety match containers. When the colored matchhead is drawn across this striking surface, a small amount of the red phosphorus is converted to the yellow variety, burning with a sudden burst of heat. This ignites the quickly combustible colored matchhead coating. If observed in darkness, the red phosphorus conversion may be seen as a lighted streak on the striking surface.

Elemental Sulfur

Elemental sulfur is a material used in manufacturing situations and, although it is combustible, it is not considered to be a hazardous material. Elemental sulfur ignites easily at its melting point and burns with a very low, blue flame. While burning, although it evolves little smoke, it produces large quantities of choking, toxic, sulfur dioxide gas. The flames from elemental sulfur are easily extinguished with water.

COMBUSTIBLE DUSTS OF COMMON SOLIDS

Earlier in this text it was established that when a combustible solid is greatly subdivided, or its geometric configuration is such that it presents more area per unit volume to its surrounding air and oxygen, it is more easily ignited and its rate of burning is faster. This section deals with the effects of the ignition of combustible solids when they are pulverized into a fine powder and are completely surrounded (or suspended) in air. This condition of flame propagation presents the maximum ratio of solid-surface-area to volume-of-combustible exposed in an oxidizing gas medium.

When a combustible solid is subdivided by a grinding or rubbing process, the minute particles produced may mix into the surrounding air to act chemically and physically as a heavy, flammable vapor or gas. For example, consider the parallel cases of pouring gasoline liquid from a container into an open pan, and pouring a very fine powder (like cornstarch) into a large box. The heavy gasoline vapors can be seen rising up from, and overflowing the sides of, the open pan. Similarly, while being poured, the powdered cornstarch will produce a white cloud in the top of the large box and, as pouring continues, the powder will overflow the top edges and settle on any nearby surface. If one were to touch a match to the gasoline vapor overflowing the pan edges, it would quickly ignite. The white cloud of cornstarch could be ignited at the edge of the large box in exactly the same manner. Of course, the liquid gasoline surface in the pan would be ignited by the flame propagation to it, whereas the bulk cornstarch powder in the box would not ignite. The reason for this is because no powder (or vapor) is being given off in the case of the quiescent cornstarch bulk in the box.

The burning rate and explosion tendency of combustible solid dusts is in direct proportion to the dust particle size. The finer the dust, the more tendency it has to remain in suspension in air and mix evenly just as heavy vapor will mix in air. Similar to the combustible gases and vapors studied in Chapter 4, "The Relationship of Fire and the Physical World," when combustible dusts become mixed with air, they demonstrate the "too rich to burn" or "too lean to burn" characteristics.

Not all combustible solids yield dusts that are explosive. Some dusts merely burn slowly and do not result in explosion damage. However, there are a large number of agricultural products, plastics, and combustible metals that are capable of developing rapid pressure rises and high explosion pressures that will cause severe structural damage in the manufacturing, storage, and handling areas where these materials are processed. Table 6.4 lists some common substances that may develop severe explosion damage.

Table 6.4. Common Combustible Solid Dusts Generating Severe Explosions *

TYPE OF DUST	MAXIMUM EXPLOSION PRESSURE (PSIG/)(BAR)	MAXIMUM RATE OF PRESSURE RISE (PSIG/SEC)(BAR/SEC)
Corn		
(processing)	95 (6.55)	6,000 (413.7)
Cornstarch	115 (7.93)	9,000 (620.5)
Potato starch	97 (6.89)	8,000 (551.6)
Sugar		
(processing)	91 (6.27)	5,000 (344.7)
Wheat starch	105 (7.24)	8,500 (586.0)
Ethyl cellulose plastic		
molding compound	102 (7.03)	6,000 (413.7)
Wood flour filler	110 (7.58)	5,500 (379.2)
Natural resin	87 (6.0)	10,000 (689.5)
Aluminum		
(powder)	100 (6.9)	10,000 (689.5)
Magnesium		
(powder)	94 (6.48)	10,000 (689.5)
Silicon		
(powder)	106 (7.31)	10,000 (689.5)
Titanium		
(powder)	80 (5.52)	10,000 (689.5)
Aluminum Magnesium Alloy		
(powder)	90 (6.20)	10,000 (689.5)

* Extracted from Bureau of Mines Investigations and Reports, No. 5753, RI 5971, RI 6516.

When one considers the fact that brick and concrete construction of processing areas, grain storage elevators, and bulk transportation bins can only withstand less than one pound per square inch (0.07 bar) lateral pressure, the explosion pressures developed by the substances in Table 6.4 up to 115 psi (7.93 bar) can severely damage or collapse a structure.

It must be remembered that the values in Table 6.4 represent the maximum degree of explosivity of these dusts. Lesser severity of explosion, or perhaps even no explosion, might occur if the concentration of combustible dust was below, or greatly above, certain optimum amounts suspended in air.

Just as is the case with combustible gas and vapor concentrations in air, combustible dusts have a lower explosion limit (LEL) below which concentration combustion or explosion cannot occur; likewise, combustible dusts also have an upper explosion limit (UEL) above which combustion or rapid explosion is unlikely to occur.

Obviously, dust explosions take place only when a source of ignition is present in, or at the edge of, a mixture of combustible dust and air. Open flames, electric arcs, suddenly broken electric light bulbs, friction sparks, and even static sparks can initiate dust explosions.

The prevention of accumulations of dust in the air or on horizontal surfaces in combustible solid processing areas is the most effective means for combating the hazard of dust explosions. Other methods of protection are: (1) the use of inert gas in a cavity where dust may be suspended in air, (2) the provision of adequate explosion vents that open to release sudden internal pressures, or (3) the installation of explosion-suppressing devices that sense the onset of an explosion wave and release combustion suppressing vapor throughout the dust cloud within milliseconds.

ACTIVITIES

1. Write your own definition of the term "spontaneous ignition temperature."
2. What one general statement can be made concerning the combustion mechanism of common combustible solids?
3. Two dwellings are built. Both are of similar wood construction, and both are identical in proportion and size. One is situated in Maine, the other in the Bahamas. Both homes catch on fire, but one burns quicker than the other. Which house burned quicker? Why?
4. Explain the difference, from a fire protection standpoint, between the composition of wood, as used in ordinary construction, and the composition of wood in its natural state in the forest.
5. With a group of your classmates, compile a listing of at least ten common "animal" fabrics present in the average home. Then discuss the combustibility properties of each, considering how they would burn in a fire situation.
6. The officer in charge of operations at a plastics fire has informed you and your fellow fire fighters that breathing equipment must be worn while fighting the fire. Based on this fact, determine what different kinds of plastics could be burning.
7. If you were a manufacturer of plastic articles and you decided to market a new "oven-proof" plastic food dish, what class of polymer would you make it from? Which polymer in this class might be best? Explain why.
8. (a) Silk is a material having fire characteristics similar to those of wool. Explain these similarities.
   (b) Since nylon fabrics contain similar nitrogen-carbon-hydrogen linkages, as do wool and silk (which are not serious fire hazardous fabrics), what one characteristic of burning nylon makes it very dangerous for fire fighters' and racing drivers' clothing?
9. What choices should a fire protection specialist make in the purchase of a rug or floor covering? Give detailed reasons for your choices.
10. If you entered a fire area suspected of containing burning plastics, how could you identify the source as polyvinyl chloride other than by breathing the choking fumes from its combustion?

SUGGESTED READINGS

Beall, F. C. and Eickner, H. W., *Thermal Degradation of Wood Components*, Forest Products Laboratory Report 130, Forest Service, U.S. Department of Agriculture.

Bowes, P. C., Edgington, J. A. G., Lynch, R. D., *The Inhalation Toxicity of Polyvinyl Chloride Pyrolysis Products*, Fire Research Note No. 1048, Fire Research Station, England.

Coleman, E. H. and Thomas, C. H., "Products of Combustion of Chlorinated Plastics," *Journal Applied Chemistry*, Vol. 4, No. 379, 1954.

DiPietro, J., Barda, H., Stepnicka, H., "Burning Characteristics of Cotton and Nylon Fabrics," *Textile Chemist and Colorist*, Vol. 3, No. 2, February 1971.

DiPietro, J. and Stepnicka, H., "A Study of Smoke Density and Oxygen Index of Polystyrene, ABS and Polyester Systems," *Journal Fire and Flammability*, Vol. 2, January 1971.

Dust Explosions, Analysis and Control, Factory Insurance Association, Hartford, Connecticut.

Eickner, H. W. and Peters, C. C., "Surface Flammability of Wood Coatings," *NFPA Quarterly*, April 1964.

Einhorn, I. N., *An Overview of the Flammability Characteristics of Materials*, Polymer Conferences Series, University of Utah, June 1971.

Fire Protection Handbook, 14th Edition, National Fire Protection Association, Boston, 1976.

Kishitani, K. and Nakamura, K., "Toxicity of Combustion Products," *Journal Fire and Flammability/Combustion Technology*, Vol. 1, No. 104, 1974.

Journal Fire and Flammability, Vol. 5, October 1974.

Lyons, J. W., *The Chemistry and Uses of Fire Retardants*, Wiley-Interscience, New York, 1970.

Madorsky, S. L., *Thermal Degradation of Polymers*, Reinhold, New York, 1968.

Morris, W. A. and Hopkinson, J. S., *Effects of Decomposition Products of PVC in Fire on Structural Concrete*, Fire Research Note 995, Fire Research Station, England, 1974.

Palmer, K. N., "Dust Explosion Hazards in Pneumatic Transportation," *Fire Prevention Science and Technology*, No. 11, April 1975.

Palmer, K. N. and Butlin, R. N., "Dust Explosibility Tests and their Application," *Powder Technology*, Vol. 6, 1972.

Palmer, K. N., Taylor, W., Paul, K. T., *Fire Hazard of Plastics in Furniture and Furnishings: Characteristics of the Burning*, Building Research Establishment Current Paper No. CP 3/75, Fire Research Station, England, January 1975.

Petajan, J. H., *et al.*, "Extreme Toxicity from Combustion Products of a Fire—Retarded Polyurethane Foam," *Science*, Vol. 187, No. 742, 1975.

Polymeric Materials in Fires, Building Research Establishment Current Paper No. CP 91/74, Fire Research Station, England, October 1974.

Roberts, A. F., "The Behavior of Polyurethane in Fire," *Fire*, March 1972.

Sax, N. I., *Dangerous Properties of Industrial Materials*, 3rd Edition, Reinhold, New York, 1968.

Scheel, L. D., Lane, W. C., Coleman, W. E., "The Toxicity of Polytetrafluoroethylene Pyrolysis Products—Including Carbonyl Fluoride and a Reaction Product—Silicon Tetrafluoride," *American Industrial Hygiene Association Journal*, Vol. 29, No. 1, 1968.

Smith, A. F. and Kuchta, J. M., *Toxic Products from Burning of Fire-Resistant Materials*, Technical Progress Report No. 66, U.S. Department of Interior, Bureau of Mines.

Stark, G. W. V., "Toxic Gases from PVC in Household Fires," *Rubber and Plastics Age*, Vol. 50, No. 4, April 1969.

Tonkin, P. S., "Inerting to Prevent Dust Explosions, *Fire Prevention and Technology*, No. 13, December 1975.

Toxic Atmospheres Associated with Real Fire Situations, National Bureau of Standards Report No. 10-807, U.S. Department of Commerce, February 16, 1972.

Tsuchiva, Y. and Sumi, K., "Thermal Decomposition Products of Polyvinyl Chloride," *Journal Applied Chemistry*, Vol. 17, No. 364, 1967.

Wooley, W. D., "Decomposition Products of PVC for Studies of Fires," *British Polymer Journal*, Vol. 3, July 1971.

7

Common Flammable and Combustible Liquids and Gases

GENERAL CONSIDERATIONS CONCERNING LIQUIDS

The physical state that we call liquid is, in reality, only an interim state for a large number of materials — a state in which the governing temperatures and pressures allow molecules to move freely and to press in all directions. Molecular movement is restricted only by the walls of their containers or by the pressures exerted on their free surfaces. By the simple change of either the pressure or temperature — or both — to which liquids are exposed, they may become solids or gases. This section of this chapter is concerned with liquids that can easily become vapors and gases and that can, in most cases, be ignited. The borderline between these physical states is sometimes indistinct. The following section deals with materials that are in the gaseous state under ordinary conditions, and that can only revert to the liquid state by means of special measures using pressures and lowered temperatures.

In Chapter 6, "Properties of Solids Important to the Fire Sciences," emphasis was placed on the fact that flaming combustion can only take place when flammable vapor or gas is formed at, or above, the surface of a solid. Again, this is the case with liquids; gaseous flammable vapor must be formed by the liquid if fire and flaming are to occur. Because of the greater freedom of movement of molecules in a liquid, escape into air in the form of a vapor capable of burning is made easier than in the case of solids.

CONDITIONS GOVERNING COMBUSTION OF LIQUIDS

It is important to review here some of the physical characteristics of liquids that were considered in Chapter 4, "The Relationship of Fire and the Physical World," that have a direct bearing on the fire properties of liquids that will ignite and burn.

If we were to assign a degree of importance to each physical property of liquids necessary for ignition and burning, the prime quality of liquids would be their vapor pressure. All liquids possess this property, but some exhibit a higher vapor pressure than others. The degree to which their molecules escape (vapor pressure) determines the amount of combustible vapor available for burning at a certain liquid temperature. The function of evaporation of a liquid is directly associated with its vapor pressure; the higher the evaporation rate, the higher is the vapor pressure. The reverse is true concerning the relationship of boiling points of liquids to their vapor pressures; substances with high vapor pressures have low boiling points.

Another important property of a flammable and combustible liquid is its flash point. In fire protection matters, it must be assumed that a source of ignition such as a spark or a flame of some sort might always be present and, depending upon the ambient temperature, the flash point of a burnable liquid spells out the degree of danger involved. It took a number of years before the present classification of flammable and combustible liquids was developed, based almost entirely upon flash point temperatures of liquids. (See Chapter 4, "The Relationship of Fire and the Physical World," Fig. 4.5.) Knowing the temperatures to which a liquid is exposed (whether it is in a drying oven at 150°F [63.0°C] or in a refrigerator at 40°F [2°C]), and given the classification of the liquid, one can quickly assess the danger of development of an explosive or ignitible atmosphere.

From consideration of the flash point of a liquid, one naturally is led to the need for knowledge of the flammable and explosive limits of a liquid. More accurately, this property consists of the flammability characteristics of the *vapor* evolved from a liquid. The determination of these flammable limits is a difficult and time-consuming process. Fortunately, almost all liquids that will burn have been tested for this property, and data are readily available in references and handbooks on this subject. Flammable limits of liquids are given in percentage concentration of vapor at normal temperatures and pressures. Where ambient temperatures are higher than normal, the flammable limit will widen, especially with respect to the upper limit. Seriously abnormal pressures will also somewhat change the values of the limits.

The function of the flammable limits of liquids is recognized when one thinks of the daily operations of automobile service stations. The gasoline fuel pumped from an underground storage tank into the tank of an automobile has a narrow flammable limit — from only 1.4 percent to 7.4 percent. When the station attendant brings the fuel hose to a vehicle tank, any drops of gasoline spilled on the pavement will quickly evaporate and the vapors will be quickly diluted by the influence of wind or motion so that the flammable concentration of 1.4 percent is reached in only a small volume above the spill. As soon as gasoline is pumped into the tank, liquid motion and turbulence cause quick evaporation; the empty space above the level of fuel quickly rises to more than 7.4 percent, or above the flammable limit. Any vapor issuing from the filling opening of the tank becomes diluted below the flammable limit by wind or turbulence around the area. When one realizes the potential for ignition — glowing cigarette sparks, sparks from metal to metal abrasion, etc. — surrounding this fueling action, it becomes obvious that the functions possessed by gasoline vapors "too lean to burn" and "too rich to burn" have prevented many fires during the more than sixty-five-year history of transporting and handling flammable gasoline fuel.

The relationships of the properties of flammable and combustible liquids that have been considered herein are closely related. This is graphically illustrated in Figure 7.1, which shows a combination of values taken from separate determinations of each property. When reading these values, it should be kept in mind that flammable ranges are not a function of temperature or vapor pressure of the substances shown. The flammable range concentrations given are determined at ordinary temperatures and vapor pressures. Flash point temperatures of acetone and ethyl alcohol are also shown. Table 7.1 is presented to further illustrate the wide variations in flammable and explosive ranges of common liquids and some gases. The substances with wide limits of flammability are obviously more hazardous because of the possibility of confrontation with a flammable mixture that can occur over a wide range of circumstances.

There are several less important physical properties of flammable liquids that need to be considered in the study of fire protection for this class of materials; among these is the comparative density of the *vapor* of liquids. As stated in Chapter 4, this value is in proportion to the ratio of the molecular weight of the flammable liquid to the molecular weight of air, which is 29. A great number of

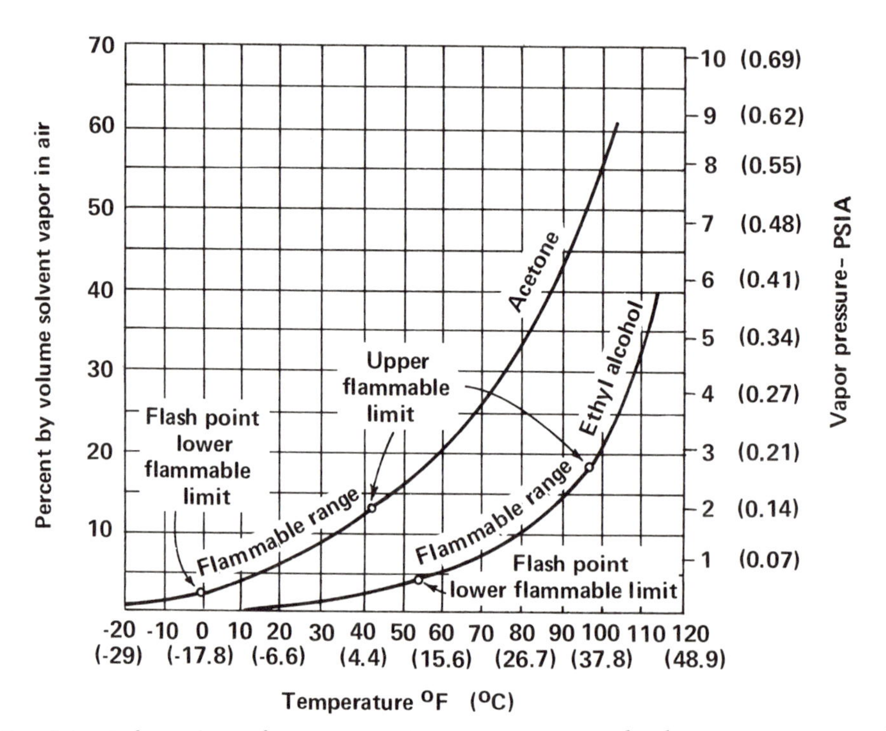

Fig. 7.1. Relationships of temperature to vapor pressure and volume percent vapor of acetone and ethyl alcohol flammable liquids.

Table 7.1. Flash Points and Flammable Limits of Some Common Liquids and Gases

LIQUID (OR GAS AT ORDINARY TEMPS.)	FLASH POINT °F	(°C)	FLAMMABLE LIMITS (PERCENT BY VOLUME)
Acetylene	(Gas)		2.5 to 81.0+
Benzene	12	(−11)	1.3 to 7.1
Ether (ethyl ether)	−49	(−45)	1.9 to 36.0
Fuel oil			
(Domestic, No. 2)	100 (min.)	(38)	None at ordinary temps.
(Heavy, No. 5)	130 (min.)	(54)	None at ordinary temps.
Gasoline (high test)	−36	(−38)	1.4 to 7.4
Hydrogen	(Gas)		4.0 to 75.0
Jet Fuel (A & A-1)	110 to 150	(43 to 65)	None at ordinary temps.
Kerosine			
(Fuel oil, No. 1)	100 (min.)	(38)	0.7 to 5.0
LPG (propane-butane)	(Gas)		1.9 to 9.5
Lacquer solvent			
(butyl acet.)	72	(22)	1.7 to 7.6
Methane (natural gas)	(Gas)		5.0 to 15.0
Methyl alcohol	52	(11)	6.7 to 36.0
Turpentine	95	(35)	0.8–(undetermined)
Varsol (standard solv.)	110	(43)	0.7 to 5.0
Vegetable oil			
(cook, peanut)	540	(282)	(Ignition temp. = 833°F)

flammable liquids have a larger molecular weight than 29; as a result, the density of the vapor of most flammable liquids is heavier than air and will flow downward from an exposed surface in the air near a container or tank of the liquid.

Similarly, the density of a flammable liquid plays a part in its fire protection. Most liquids are lighter than water; hence, they float on water and continue evolving flammable gases. Water is of little use in their fire extinguishment.

Many flammable liquids are soluble in water and mix with it, thus lowering their tendency to burn or to remain ignited. This action is temperature dependent because of the vapor pressures involved.

When flammable liquids consist of a mixture of two or more components of different chemical constitution, the general physical properties of the mixture initially assume the properties of the component having the lowest flash point and boiling point and the highest vapor pressure. As evaporation or heat from combustion of the liquid consumes the component with the lowest boiling point, the characteristics of the mixture change progressively until the mixture finally demonstrates the properties of the component with the highest boiling point. For these reasons, many petroleum fuel mixtures are specified according to their Initial Boiling Point (IBP) and the percentage volume of that component, and on through various temperature region fractional amounts to the Final Boiling Point (FBP) component. Almost all petroleum fuels are mixtures of this type, with some exhibiting narrower boiling point ranges than others.

FIRE CHARACTERISTICS OF BURNING LIQUIDS

As stated, the vapors above flammable and combustible liquids are capable of ignition at the flash point temperature of the liquid. Within a short additional time of exposure to flame, the fire point of the liquid will have been reached by this heat exposure and self-sustained burning will take place. In the case of liquids with flash point temperatures above the ambient temperature, the application of a source of ignition to one spot will initiate burning in that area, and the heat of combustion will travel outward by radiation and conduction until the entire surface of the liquid is burning. Although the speed with which this occurs is dependent upon many factors, the principal factor is the difference between ambient temperatures and the flash point of the liquid. However, the depth rate at which liquids burn may be determined if their burning area remains constant. Rates of burning are inversely influenced by the boiling point of the liquid or its largest fraction if it is a mixture with a wide boiling point range. They are independent of the size of the burning area once equilibrium has been achieved. The burning rates given in Table 7.2 are for liquids in common use and are applicable in bulk storage areas where tanks or pools of fuel are burning.

Table 7.2. Burning Rates of Ordinary Petroleum Fuel Liquids

	BURNING RATE		
LIQUID FUEL TYPE	INCHES DEPTH BURNED PER HOUR	INCHES DEPTH BURNED PER MIN (AV.)	CM. DEPTH BURNED PER MIN (AV.)
Motor gasoline	6 to 12	0.15 = 5/32	.38
Aviation gasoline	10 to 12	0.18 = 3/16	.45
Kerosine (No. 1, FO)	5 to 8	0.10 = 3/32	.25
Diesel oil	6 to 9	0.12 = 1/8	.30
Fuel oil (Heavy, No. 5)	5 to 7	0.10 = 3/32	.25

There is another interesting characteristic about the processes that occur during the burning of liquids stored in tanks, especially those liquid mixtures containing a variety of petroleum derivatives that exhibit a wide boiling range. Crude oils and heavy fuel oils are examples of this wide variation in boiling point fractions. When crude oil and heavy fuel oil burn in a confined area such as a storage tank, some of the heat produced is caused to travel downward into the bulk of unheated oil. This occurs because of differences in the density of the fractions present in the fuel mixture. As the lower density or light fractions of the oil burn at the surface, the heavier fractions with a higher boiling point do not burn, but become hot and sink. When they progress downward, heat transfer takes place and the lesser density fractions rise to become separated by the combustion process, making the heavier fractions again sink. By this means a circulation is set up in the unburned oil near the surface that steadily progresses downward as the fire continues. Figure 7.2 shows how this mechanism takes place.

The zone of heated oil in such a fire is called a *heat wave*. Because of the fact that the oil in this heated volume is much above the boiling point of water, several phenomena may be demonstrated if water present in the tank or water from fire fighting efforts should come in contact with it.

When a tank of crude oil or heavy fuel oil has been burning for a sufficient length of time so that the heat wave progression (at 9 to 15 inches downward travel per hour rate [23 to 38 centimetres], depending on the petroleum fuel type) has reached the bottom of the tank where water and bottom settlings have accumulated, the water will form steam. The steam acts like a piston under pressure, forcing the hot contents of the tank upward and outward. Since the entire contents of the tank are now above the flash point of the oil — at least 450°F (233°C) — this almost explosive expulsion, which is called a *boilover*, assumes the character of a volcano spewing burning lava.

Safe clearance for fire fighting personnel attending such a fire has been said to be at least one-half mile (0.80 kilometres) for a 100-foot (30.5-metre) diameter fuel tank boilover.

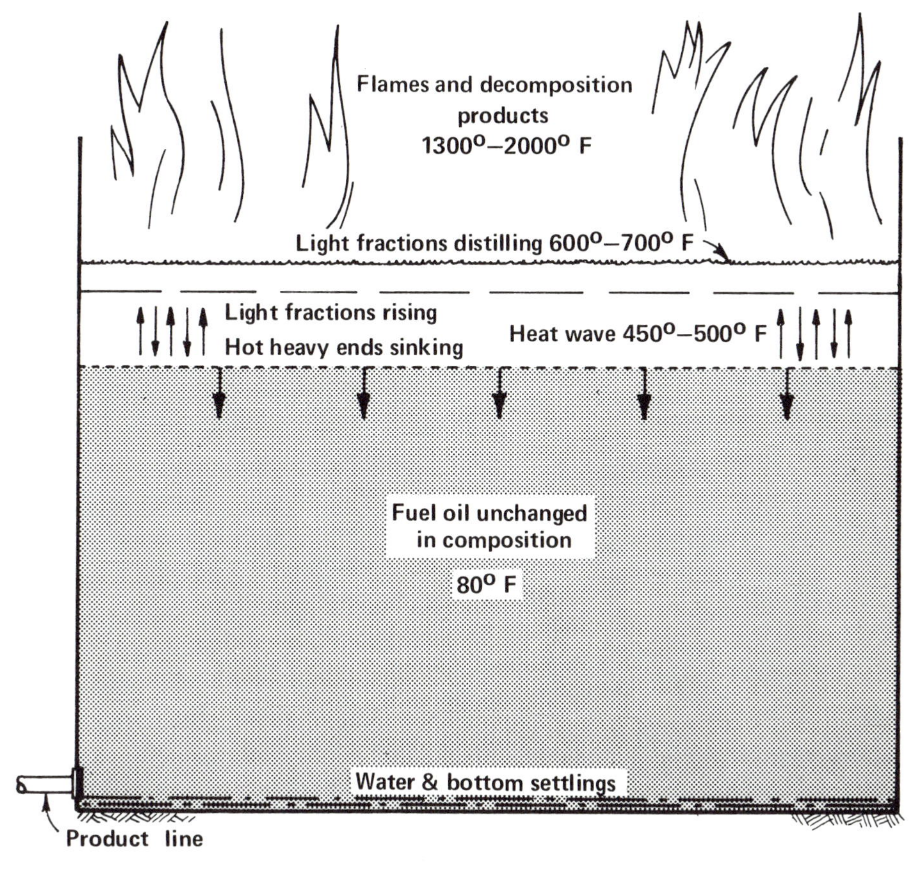

Fig. 7.2. Heat wave propagation in fuel oil tank fires.

A more common reaction to water can be seen during fire fighting efforts on petroleum fires where hot fuel layers are produced at the top of the fuel in the tank. This is the *slopover* of a frothy mixture of oil and steam bubbles which, because it becomes too large in volume for the capacity of the tank, issues from any open area at the tank top. A slopover can easily result during extinguishment efforts on heavier fuels when water or wet foam first contacts the hot oil. An emulsion of large volume quickly forms because of the sudden expansion of the water into steam. The viscosity of the suddenly cooled oil holds the steam in the form of a froth. Slopovers are confined to the area around the wall of the fired tank and can be beneficial to the fire fighting operations because they remove a volume of hot oil from the burning area.

It is important to point out that petroleum mixtures with narrow boiling ranges such as gasolines, kerosine, and fuel oil No. 1, form very thin layers of heated fuel — one-half inch (1.25 centimetres) or less — below the flame zone in a bulk-burning fire, and they do not progress downward to form a heat wave.

Since almost all water-insoluble flammable and combustible liquids are lower in density than water, water is not fully effective as an extinguishing agent for fires in these liquids. Air foams must be used to: (1) cool the burning surface liquid, and (2) to lower the rate of escape of flammable vapor to concentrations below the flammable range.

Water application to water-insoluble flammable liquid "spill" fires such as gasoline will cause dangerous spreading of the burning fuel by causing the lighter density fuel to travel on top of the lower liquid water surface, thus "following" the water over an enlarged area.

Water-soluble flammable and combustible liquids may be diluted with water during their combustion to a point where the evolved flammable vapors are below the flammable range. In the case of a liquid spill fire where the volume to area ratio is small, water dilution for extinguishment is rapid and is easily accomplished, but where large bulk-burning is encountered, special "alcohol" type foams must be used to form a cohesive vapor-restricting layer on the burning surface.

GENERAL PROPERTIES OF GASES

Defining a "Gas"

We live and breathe in an "ocean" of gas that we call air. From a purely physical property standpoint, gases are the vapors of liquids. For example, consider the simple fact that under the proper pressures and temperatures, air can be liquefied. Although such a liquid has a high "vapor pressure," it can be kept in that state. Similarly, all matter that we call gases can be transformed into liquids and some can be transformed into a solid state (*e.g.*, carbon dioxide "gas"). Obviously, it becomes difficult to define gases as compared to vapors unless we take into consideration the physical conditions under which we are dealing with a gas as a form of matter. Doing this, we would define a gas as a material at normal conditions with free and random molecular motion that distributes itself uniformly within the confining walls of any container in which it is held. It does not undergo a change of state (into a liquid) unless its temperature and/or pressure are critically changed from that of ordinary or of normal conditions (NTP).

One important characteristic of a gas is its capability of being compressed in volume when it is acted upon by mechanical forces tending to restrict the paths of the free molecular motion of the gas. Liquids and solids are noncompressible by such mechanical applications of force.

The "Gas Laws"

This leads us to the consideration of an important physical law about gases that is useful in fire protection fields. This physical law is known as Boyle's Law. Boyle's Law states that: the volume of a gas varies inversely with the pressure applied to it. Another way of saying this is that a given volume of gas will be compressed to one-half that volume if its pressure is increased by twice the original pressure, or a factor of two. Conversely, it can also be said that if the pressure on a certain volume of gas was cut in half (decreased by a factor of 0.5), the volume occupied by that gas would be twice its original volume. It is important to remember that this law about gases applies to gases that undergo no temperature change during the compression process. Also, we are dealing with absolute pressures (psia) in any calculations because all gases occupy a certain volume at atmospheric pressure (0 psig = 14.7 psia = 1 atmosphere = 1 bar). In order for a gas to occupy zero volume, it would be necessary to go to zero pressure as our starting point for calculation.

Figure 7.3 is a graphical illustration of Boyle's Law concerning gases using a piston in a cylinder with forces acting on the piston equal to 14.7 psia (1 bar) doubling at each change in the position of the piston. If we were to put this law into a mathematical expression in order to find out what volume we would end up with if we exerted a certain pressure on it, we would write:

$$\underset{\text{(new volume)}}{V_2} = \frac{\overset{\text{(initial pressure)}}{P_1} \times \overset{\text{(initial volume)}}{V_1}}{\underset{\text{(new pressure)}}{P_2}}$$

This equation could be used to calculate the residual number of cubic feet or litres of air or oxygen in a partially used breathing apparatus bottle, knowing its original volume and noting its pressure. (P_1 = pressure reading, P_2 = normal barometric pressure, V_1 = original volume.)

Remember, however, that Boyle's Law relates only to the gases that remain in the gaseous state under pressures normally used in compressed gas containers. When gases such as butane are compressed to high pressures, the gas liquefies in the container and its pressure becomes, in effect, the vapor pressure of the liquid at that temperature.

Another important characteristic of gases concerns their ability to expand dramatically when their temperature is raised. Matter in other states such as liquid and solid also expands when heated, but not to the extent exhibited by gases. A

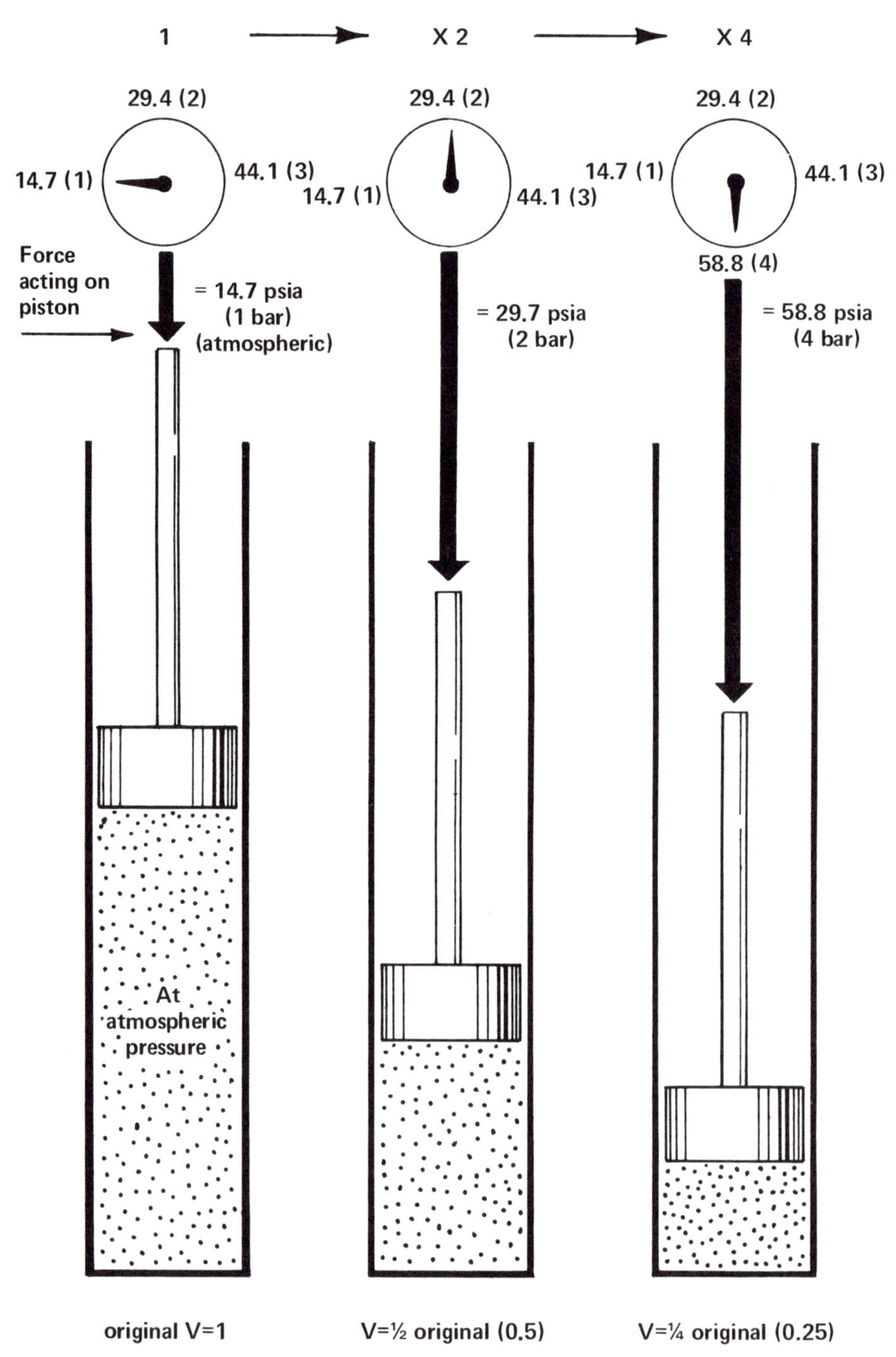

Fig. 7.3. Representation of Boyle's Law. The volume of gas varies inversely with the pressure applied to it.

physical law relative to this was postulated by Jacques Charles in 1787. Charles found that gases expand in direct proportion to the absolute temperatures applied to them. If the absolute temperature (degrees Kelvin or °F absolute = °F + 459) of a gas is doubled, its volume will double. Converting this to a mathematical expression (in order to determine what new volume of gas at a new temperature would be obtained when starting with an initial known volume at an initial known temperature — remembering that degrees Kelvin = °C + 273) produces the following equation:

$$\underset{\text{(new volume)}}{V_2} = \frac{\overset{\text{(new temperature)}}{T_2}}{\underset{\text{(initial temperature)}}{T_1}} \times \overset{\text{(initial volume)}}{V}$$

From this relationship, it is easy to understand why a fire fighter is confronted with a sudden outflow of heated gases when first opening the door to a burning room where temperatures have risen from 293°K (70°F) to at least 811°K (1,000°F).

Note, however, that in the preceding equation concerning temperature and volume changes of gases, nothing has been said about pressures. About twenty years after Charles made his discovery, Joseph Louis Gay-Lussac postulated that when gas temperatures are raised and the volume in which they are confined stays the same, pressures are raised in proportion to the change in absolute temperature of the gas. Converting this to a mathematical expression (in order to show the amount of pressure that a given volume of gas would show when it is confined in a cylinder or gas bottle in which the temperature was raised by a number of degrees Kelvin or degrees F absolute) produces the following equation:

$$\underset{\text{(final pressure)}}{P_2} = \frac{\overset{\text{(final temperature)}}{T_2}}{\underset{\text{(initial temperature)}}{T_1}} \times \overset{\text{(initial pressure)}}{P_1}$$

Again, pressure readings should be in bars (1 bar = 1 atmosphere pressure = 14.7 psia = 0 psig) or in psia.

This relationship is an important one in fire protection technology because it shows the alarmingly high pressures that can exist when cylinders and hemispherical tanks of flammable gases are heated by nearby flames and fires. (This point will be considered in more detail when "BLEVEs" — Boiling Liquid-Expanding Vapor Explosions — are dealt with later in this chapter.)

When dealing with heat input (temperature rise), fire service personnel often need to be concerned with variations in both volume and pressure problems. For this purpose, equations would be used that would take into account *all* of the variables stated herein — volume, temperature, and pressure.

To illustrate this, consider the example of filling a partially used oxygen tank in cold weather. The charging tank and equipment are in an unheated room with a temperature of 32°F (0°C, 273°K). The size of the cavity of the empty tank is known, but we need to know what volume of oxygen *measured at normal temper-*

atures and pressures we have in the tank at a certain pressure in the room when the charging procedure stopped. This could be expressed by the following equation:

$$\underset{\substack{\text{(volume of}\\\text{gas in tank}\\\text{at normal}\\\text{temperature)}}}{V_2} = \frac{\overset{\substack{\text{(normal}\\\text{temperature)}}}{T_2}}{\underset{\substack{\text{(freezing}\\\text{temperature)}}}{T_1}} \times \frac{\overset{\substack{\text{(pressure}\\\text{reading)}}}{P_1}}{\underset{\substack{\text{(atmospheric}\\\text{pressure)}}}{P_2}} \times \overset{\substack{\text{(volume}\\\text{of tank)}}}{V_1}$$

If the same tank should happen to be stored in a room and the tank's temperature was raised abnormally high, the new pressure on the tank could be determined by using the same equation, transposed to find the pressure at this elevated temperature:

$$\underset{\substack{\text{(new}\\\text{pressure)}}}{P_2} = \frac{\overset{\substack{\text{(elevated}\\\text{temperature)}}}{T_2}}{\underset{\substack{\text{(normal}\\\text{temperature)}}}{T_1}} \times \frac{\overset{\substack{\text{(volume at}\\\text{normal}\\\text{temperature)}}}{V_1}}{\underset{\substack{\text{(volume is}\\\text{constant)}}}{V_2}} \times \overset{\substack{\text{(pressure read}\\\text{at normal}\\\text{temperature)}}}{P_1}$$

When using the gas laws, it is important to keep the following facts in mind:

- Always use T_1, P_1, and V_1 for initial conditions. Use T_2, P_2, and V_2 for values of conditions answering, "What you want."
- Use one System of Units for temperatures and pressures (SI = °K, bars; English = °F Absolute (°F + 459) psia (psig + 14.7). Any unit volume can be used if it is consistent.
- Equations may be approximately solved by inspection if fractions showing pressure and temperature are more or less than one.

CLASSIFICATIONS OF GASES

For purposes of this book, we are interested in gases with chemical properties of three types: (1) the flammable gases, which burn with relative ease and which are combustible in an atmosphere of air, (2) the oxidizing gases, which do not burn, but which significantly aid and accelerate the burning of combustible substances, and (3) the inert gases, which have the ability to diminish or to halt combustion of various types.

The gases that we are most likely to encounter are those most often in industrial situations in three different forms: (1) compressed gases are stored in a gaseous form at high pressures up to 3,000 psig (204 bar) in strong containers that contain no liquid phase, (2) cryogenic gases are those that have been transformed into a completely liquid phase and are maintained as low-temperature liquids by containment in vessels with evacuated double walls and with atmospheric vents for the escape of gas from the liquid during storage, and (3) liquefied gases, which are gases with properties enabling them to be relatively easily liquefied and stored under not excessively high pressures in tanks at ordinary temperatures. A liquefied gas container contains both liquid and gas in equilibrium at the storage pressure.

The physical properties of all stable gases that govern whether they can be liquefied and stored at certain pressures are their critical temperature and their critical pressure. The critical temperature for a gas is that temperature above which the gas cannot be liquefied, no matter how great a pressure is exerted upon it. The critical pressure of a gas is that pressure existing at the critical temperature.

Table 7.3 shows the relationships of the critical temperatures and critical pressures of some common gases and their boiling points at ordinary pressures. The upper half of the table consists of gases that cannot be retained in liquid-gas phases in a pressure container at ordinary temperatures. Note the depression of their boiling points at atmospheric pressure from those at their critical temperatures.

Table 7.3. Critical Temperatures and Pressures of Some Common Gases and Their Boiling Points at Atmospheric Pressure

GAS	CRITICAL TEMPERATURE (t_c)		CRITICAL PRESSURE (P_c)		BOILING POINT AT 1 BAR	
	°C	°F	PSIG	BAR	°C	°F
Air	−140.6	−220	546.0	37.2	−195	−320
Hydrogen	−239.9	−400	188.2	12.8	20.4K	−422
Nitrogen	−146.8	−231	492.4	33.5	−195	−320
Oxygen	−118.3	−180	737.0	50.1	−183	−297
Methane	−82.6	−116	667.4	45.4	−434	−259
Carbon dioxide	31.0	88	1070.9	72.8	−78	−109
Chlorine	144.0	291	1118.7	76.1	−34	−30
Butane	152.0	305	554.2	37.7	−0.4	+31
Propane	96.6	206	616.5	41.9	−42	−44

Obviously, according to Gay-Lussac's Law, a gas such as propane (shown in Table 7.3) would show a much lower pressure than 616.5 psig (41.94 bars), and a larger amount of liquid when stored at normal temperatures. Stored at normal temperatures, a gas such as carbon dioxide would show a pressure slightly below it critical pressure of 1,070.9 psig (72.85 bars) in equilibrium with its liquid. A gas such as methane would have no liquid phase in its container, and its pressure would be much higher than 667.4 psig (45.4 bar) at ordinary temperatures.

EXPLOSIVE LIMITS OF GASES

As previously stated, gases are merely vapors; as such, they exhibit the same flame propagation properties. However, gases have flammable and explosive limits that are, in general, somewhat wider than vapors.

One of the most important methods that can be used to diminish and to halt the flammability, combustibility, and explosibility of gases (and of vapors) is by dilution of the gas with an inert gas or a flame-extinguishing vapor. When this is done, two progressive results are demonstrated: (1) the oxygen available for combustion is diminished, as is the amount of combustible gas, and (2) the lowering of the high limit of flammability of the gas is much greater per amount of inert gas introduced than is the case of the lower limit, where the change is minor with increase of inert gas until complete nonflammability is attained.

This mechanism is explained in the curves shown in Figure 7.4a, according to Coward and Jones,[1] where the effect of progressively diluting an atmosphere of hydrogen in air with carbon dioxide and nitrogen is given with respect to the changes in its flammable range. Similarly, in Figure 7.4b, the changes in the flammable range of methane in air when it is diluted with nitrogen are shown. Obviously, hydrogen in air becomes nonflammable when it contains 62 percent carbon dioxide or 75 percent nitrogen; methane-air mixtures are nonflammable when they contain 38 percent nitrogen.

The process of the inert gas flooding of a compartment is based on data such as that shown in Figures 7.4a and 7.4b. It can be seen that the percentage of inert gas required to produce a firesafe atmosphere varies with respect to the nature of the combustible gas, and also the properties of the inerting gas.

COMMON GASES OF INTEREST TO FIRE PROTECTION

Compressed Gases *(see also Appendix III)*

Hydrogen is an odorless elemental gas, light in density and having a high diffusion ratio. It is exceedingly flammable, with a wide flammable range of 4.0 percent to 75 percent by volume. It burns with an almost nonluminous flame, and ignites with extremely low energy sources such as small friction sparks. When mixed in correct stoichiometric* proportions with oxygen (2 volumes H_2 + 1 volume O_2), it explodes with intensity, generating water vapor.

Acetylene is an extremely reactive, flammable gas that cannot be stored in a compressed state alone without a possibility of dissociation into carbon and hydrogen with release of energy. It is stored in cylinders that contain a very porous monolithic mass made of cement, asbestos, diatomaceous earth, and charcoal. The anhydrous filler mass, containing about 80 percent void space, is soaked with acetone. Acetylene gas is pumped into the cylinder and its filler from heavy, small-diameter piping (to withstand decomposition pressure, should it occur at these pressures) until a maximum cylinder pressure of 250 psig (17 bar) is reached. The acetone dissolves 25 times its own volume of acetylene for each 14.7 psig (1 bar) pressure. An ordinary welding type acetylene cylinder contains 5.5 gallons (18.5 litres) of acetone, which is about 43 pounds (19 kilograms) and about 20 pounds (9 kilograms) of acetylene.

Acetylene gas is exceedingly flammable, with a flammable range of 2.5 percent to 81 percent. Under certain conditions it will dissociate at gas concentrations from 81 percent to 100 percent, releasing heat energy in the process.

[1] Coward, H. F. and Jones, G. W., "Limits of Flammability of Gases and Vapors," Department of the Interior, Bureau of Mines, 1952.

* The quantitative relationship of constituents in a chemical entity.

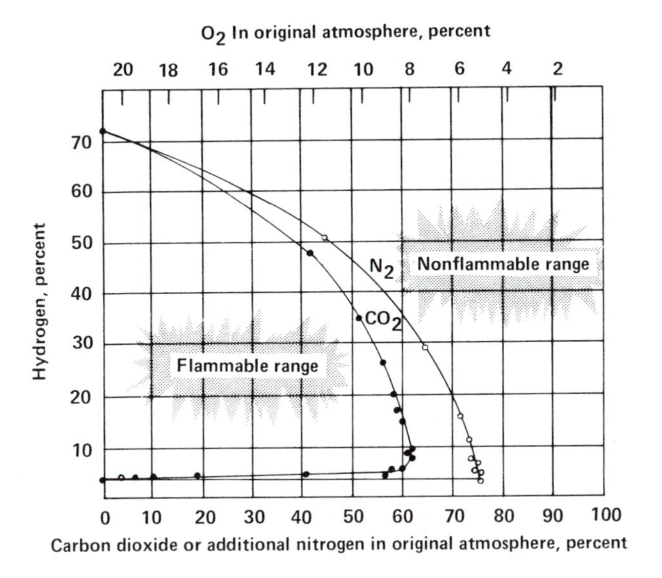

Fig. 7.4a. Influence of temperature on limits of flammability of hydrogen in air (downward propagation of flame).

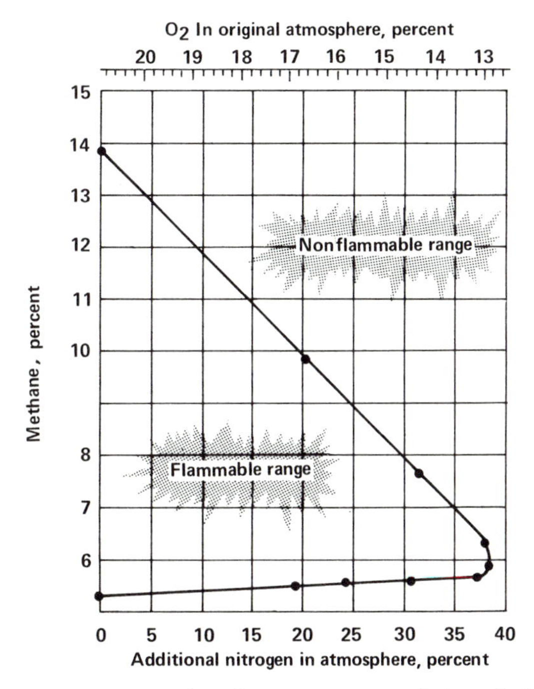

Fig. 7.4b. Limits of flammability of methane in mixtures of air and nitrogen (experiments in large vessels).

Because of the reactivity and unconventional storage method of acetylene gas, all acetylene tanks are provided with fusible plugs that open at about 212°F (100°C); should this occur near an ignition source, a flaming torch of burning gas will issue some distance — 10 to 12 feet (3 to 3.6 metres) — from the opened vent. This need not be extinguished unless it endangers nearby combustibles. After a short time the torch will die down and the cylinder will cool sufficiently to be moved to a safe place where it can continue to vent. Some possibility exists that the flame may propagate back into the cylinder or tank, at which time the tank will heat up and must be cooled with water sprays. Danger of explosion of the tank exists only when it heats to a glowing red color.

Ethylene is transported as a compressed gas at high pressures and is used as a fruit-ripening agent in very low concentrations of 1/10th percent in enclosures. It is highly flammable with a flammable and explosive range of 2.7 percent to 36 percent in air. It reacts vigorously with oxidizing gases and since it has almost the same density as air (air is slightly heavier), it mixes quickly when released into the atmosphere and can form an explosive mixture.

Oxygen is transported as a compressed gas in cylinders at high pressures. It is not flammable, but is capable of supporting and accelerating flame and explosions. It reacts with a large number of substances and is especially reactive with materials of all kinds containing carbon and hydrogen. The gas materially changes the combustibility tendencies of substances to the point that they may burn at explosive rates. In view of the characteristics of oxygen gas for promoting combustion, it should be regarded as a dangerous material even though it is not flammable itself.

Nitrogen gas falls in the class of inert gases that do not react with other substances except under special conditions of temperature and pressure. It is normally transported in both the cryogenic and compressed gas forms. It is not toxic except in its ability to quickly replace oxygen to the extent that there is insufficient oxygen to support life. It can also be used to produce a gaseous atmosphere that will not support combustion.

Liquefied Gases

This classification deals with those gases that are stored under reasonably high pressures — more than 100 psig (6.8 atmospheres) — and contain a liquid portion of gas in equilibrium with a gas portion at normal temperatures. Those liquefied gases that are stored and transported at low temperatures in special containers that are open to the atmosphere will be discussed later in this chapter under the section titled "Cryogenic Gases."

Anhydrous Ammonia is a pungent-smelling gas transported in the liquid-gas phases under pressure. It has a somewhat dangerous flammable and explosive range between 16.0 percent and 25.0 percent in air, and is toxic to human tissue by virtue of its rapid solubility in water to produce caustic ammonium hydroxide ("ammonia" of commerce). It is used as a fertilizer, as a refrigerant, and as a source of hydrogen by decomposition at high temperatures. Its affinity for water leads to the high efficiency of water-fog or spray for disposing of the gas.

Fig. 7.5. The fireball formed in a BLEVE involving LP-Gas railroad tank cars at Crescent City, Illinois, on June 21, 1970. The elevated water tank at lower right is a point of reference in visualizing the tremendous dimensions of the fireball. (Anderson Watseka, Ill.)

Liquefied Petroleum Gases (LPG) are mixtures of propane and butane with much smaller amounts of other hydrocarbon gases such as ethane. They are transported in cylinders in the liquid-gas phases under pressure. Both the gases and liquids in this mixture are highly flammable. In the gaseous form in air they show a flammable and explosive range from 1.55 percent to 9.60 percent. Pressure tanks of LPG are prohibited from storage and from use indoors because of the possibility of liquid phase escape. The liquid produces about 260 volumes of flammable gas per volume of liquid. LPG tanks of all sizes up to railroad transport tanks are subject to a fire phenomenon known as a "BLEVE" (Boiling Liquid-Expanding Vapor Explosion) when they are heated by the combustion of accidentally escaping gas to a point where the confining steel container shell fails by fracture caused by heating. When this happens, the gaseous-phase contents escape and burn in a large fireball, and the liquid phase immediately boils and expands to flammable volumes 260 times the liquid amount present. Obviously, rapid and uniform water cooling of tanks containing LPG must be started when they are exposed to flame. (See Fig. 7.5.)

LPG is a product with a wide usage, ranging from cigarette lighter fuel to large industrial furnace gaseous fuel. In order to detect leaks, the stored gases are slightly odorized with organic sulfur compounds called mercaptans.

Chlorine is an elemental gas that is transported in cylinders in the liquid-gas phases under pressure. It produces a choking sensation in the throat, damages the bronchial tract, and is highly toxic. It is nonflammable, but promotes combustion and explosion upon reaction with hydrogen-containing substances such as acetylene, LPG, and anhydrous ammonia. Water sprays will dissolve chlorine gas and will cause any free liquid phase to boil and dissipate.

Carbon Dioxide is a completely inert gas that is transported in cylinders and containers in the liquid-gas phases under pressure. It is nontoxic, but is a dangerous asphyxiant in quantities higher than five to seven percent in air. It is a heavy gas that will collect in low excavations or openings where air admixture is slow. At temperatures above 87.8°F (30°C), carbon dioxide exists only in the gaseous phase in its container. However, rapid release of pressure of the gas (as in a carbon dioxide extinguisher discharge) will cool the carbon dioxide gas so that it forms particles of solid carbon dioxide (referred to as "snow").

Cryogenic Gases

When any gas is compressed and cooled below its critical temperature and pressure, it assumes a liquid phase. If the cooling process can be continued, the boiling point of the liquid gas at atmospheric pressure can be reached or exceeded, and the liquid can then be contained in a specially built container with heat-insulating walls. Since the liquid gas will have a vapor pressure, the container must be open to the atmosphere to allow the vapor (gas) to escape slowly. Gases that are liquefied and transported in this form are said to be cryogenic gases. It must be emphasized that this name refers to the preceding physical conditions, and not to a type of gas or a chemical property of a group of gases.

With the exception of acetylene, all of the common gases that we dealt with earlier are transported and are used as cryogenic gases. In terms of their chemical and physiological characteristics, their properties as liquids are similar to those found in the gaseous state. However, emphasis must be directed to the fact that when gases are liquefied they represent a high concentration of the gas. When liquid oxygen is mixed with another liquid combustible, or when it is dispersed in a solid combustible (*e.g.*, carbon), the resulting mass will burn at explosive rates. Similarly, liquid chlorine in a mixture with a combustible will react (oxidize) rapidly, sometimes with only the slightest ignition. Chlorine gas and hydrogen gas can even react at explosive rates by ignition from sunlight alone.

From the standpoint of their hazards when released rapidly from their insulated containment, the cryogenic gases possess the dangerous quality of being lower in density at the low temperatures just above their boiling points. A cryogenic gas with a density equal to or less than air at normal temperatures will be much heavier than air when it first vaporizes into gas from its liquid phase; if it is a flammable gas, it will remain at a low point for some distance from the point at which it vaporized from the liquid, thus extending the fire hazard area from any accidental ignition. As the cold gas warms up, it will rise and approach normal density.

Liquefied Natural Gas (LNG) has become an important material of commerce. It is transported in quantity only as a cryogenic gas because of the economy involved. The low gas volume of natural gas transported in high pressure tanks is not economically feasible. (Containment of liquid-gas phases of natural gas in closed tanks is not possible due to its low critical temperature; see methane in Table 7.3.)

Both the liquid phase and the gas vaporized from liquefied natural gas are highly flammable with a flammable and explosive range of the vapor and gas of 5.3 percent to 14.0 percent by volume of gas.

When an amount of the liquid gas vaporizes rapidly (due to spillage or container fracture), the cold gas phase will "fog" with water vapor in the air, rolling low to the ground until heating takes place. As the warming gas decreases in density, it rises and dissipates in air currents. During this process the gas will go through its explosive range; air-gas mixtures in large volumes have been known to cause tree and surface structure destruction when ignition of the envelope took place.

Marine transportation of LNG is steadily increasing in amount. Ocean-going vessels constructed with large, insulated, double-shell tanks are used. It has been shown that fracture or accidental spillage of their liquid gas contents into relatively warm water causes vigorous explosion-like reactions. These are not ignitions, but a result of a reaction caused by inequalities of heat transfer rates to the cold liquid gas in its effort to vaporize.

The use of carefully directed water sprays on LNG liquid spills and on the cold gas vapor is perhaps the only action that can be taken toward dissipation of the gas, thereby lessening the hazards of ignition. High-expansion foam directed onto the surface of a burning pool area of liquid natural gas has shown some promise of lowering the rate of gas evolution by a process of freezing the layer of foam adjacent to the cold liquid. However, the complete halting of the process of vapor evolution is almost impossible.

GENERAL FIRE PROBLEMS OF GASES

The cardinal rule in the control of gas hazards of all types is that the flow of gas must first be halted. Extinguishment of flame before stopping an evolution of a flammable gas may lead to generation of an explosive atmosphere that could result in disastrous effects should reignition occur. In cases of escape of nonflammable gases of the oxidizing or the inert types, ventilation and dilution with air is of primary importance. This would, of course, also be necessary in cases of non-ignition of "free" flammable gases.

The identification of the gas or gases involved in a hazardous situation is also of major importance. Gas analysis need not always require an elaborate chemical analyzer or sampler. There are many means of identification of gas sources such as the type of containment or obvious source; even careful sniffing of the atmosphere when the gas is first encountered can provide important information.

Since there are few common gases that can support life processes, the use of self-contained breathing apparatus must be considered early in the planning of any procedures intended to decrease the hazards involved. Gaseous atmospheres are capable of serious explosions over a wide portion of their flammable range.

Furthermore, most gases have wider total ranges of flammability than do vapors from liquids.

The use of water in spray form is an important tool in many situations involving burning gases or in the escape of oxidizing gases such as chlorine. Where combustibles are in the proximity of burning gas, they must be cooled by judicious use of water. Where tanks of liquefied gases are being heated by gas flames they must be cooled to prevent the occurrence of tank fracture with a resultant "BLEVE," and also to prevent or decrease the release of flammable gas by operation of overpressure safety equipment on the tank. Many oxidizing gases such as chlorine are soluble in water and may be diluted and "washed" from air mixtures by the use of water fog patterns.

If it becomes necessary and advisable to extinguish "torch" flames issuing at high pressures from containers, it is first necessary to safeguard and cool nearby sources of ignition of any type. Dry chemical nozzle discharges at the base of the flame, projected in the same general direction as the jet of gas, will extinguish the burning torch. Dry chemical is also efficient in halting "not too large" burning areas of liquefied flammable gases.

GASES DEVELOPED BY FIRES

It is important to keep in mind that under certain conditions the combustion reaction produces flammable gases that were not originally present. The production of carbon monoxide by a diminished oxygen supply in a combustion reaction involving carbonaceous materials is the most common of the "fire-gas" situations. A closed atmosphere in which ordinary combustibles such as paper or cloth are vigorously burning may produce sufficient carbon monoxide gas over a length of time to result in an explosive combustion of the gas phase when the enclosure is vented, or when the air supply is increased by opening a door.

Similarly, hydrogen gas and carbon monoxide may be rapidly produced by the "water-gas" reaction if steam from a water jet comes in contact with the glowing interior of a vigorously burning carbon material. This reaction was formerly used for the production of ordinary "illuminating" or "city" gas:

$$C + H_2O \text{ (steam)} \rightarrow CO + H_2.$$

Very hot burning coke was used as the heat and carbon source. It has been replaced with natural gas from pipeline sources.

ACTIVITIES

1. A storage tank filled with crude oil is burning, and you are called upon to fight the fire.
   (a) Would you use water to extinguish the fire? Why or why not?
   (b) Is a quarter of a mile a safe distance for personnel to fight a fire in a 100-foot diameter fuel tank?
   (c) What are the benefits of using air foams for this type of fire?

2. Based on what you have learned in this chapter, write your own explanation of the importance of vapor pressure in liquids.
3. Explain why water or wet foam has a beneficial effect on a petroleum fire.
4. A mixture of acetone and benzene is formed. Acetone's flash point is 0°F, while benzene's flash point is 12°F.
   (a) What component's properties would the mixture assume?
   (b) As the mixture is heated, what properties would the mixture assume? Why?
5. Cite an example of what you think might happen if water from fire fighting efforts should come in contact with the zone of heated oil burning in a confined area such as a storage tank.
6. If the initial volume of gas in a room were 2, and if its original temperature was 80°F, what would be the new volume of the gas if the room were burning at a temperature of 210°F? What law have you employed?
7. In a compressed gas container, the original pressure is 44.1 psia (3 bars) and the original volume is 2.
   (a) If the pressure were to be doubled, what would be the new volume of the compressed gas?
   (b) What law would be employed in this computation?
8. In the following descriptions, identify the gas described and state whether it is a compressed, liquefied, or cryogenic gas.
   (a) Transported in cylinders and containers in liquid-gas phases under pressure; dangerous asphyxiant in quantities higher than five to seven percent in air.
   (b) Can be transported in ocean-going vessels; the cold gas phase will "fog" with water vapor in the air.
   (c) Extremely flammable; storage tanks are provided with fusible plugs; will exhibit "torching."
   (d) Nonflammable, but can support and accelerate flame and explosions; reacts with materials containing carbon and hydrogen.
   (e) Inert gas; can be used to produce a gaseous atmosphere that will not support combustion.
   (f) Extremely flammable and explosive; toxic; can be used as a fertilizer.
   (g) Mixtures of propane and butane; can cause BLEVEs.
   (h) Nonflammable; highly toxic; can be dissolved by water sprays.

SUGGESTED READINGS

Atallah, S. and Allan, D. S., "Safe Separation Distances from Liquid Fires," *Fire Technology*, Vol. 7, No. 1, February 1971.

"Carbon Monoxide," Data Sheet 415, National Safety Council, Chicago, 1976.

Coffee, R. D., "Evaluation of Chemical Stability," *Fire Technology*, Vol. 7, No. 1, February 1971.

Conference Proceedings on LNG Importation and Terminal Safety, Committee on Hazardous Materials, National Academy of Sciences, Washington, D.C., June 1972.

Coward, H. F. and Jones, G. W., *Limits of Flammability of Gases and Vapors*, Department of the Interior, Bureau of Mines, 1952.

Encyclopedia of Chemical Technology, 2nd Edition, Interscience, New York, 1963–1969.

Fawcett, H. H. and Wood, W. S., *Safety and Accident Prevention Chemical Operations*, Interscience, New York, 1965.

Fire Protection Handbook, 14th Edition, National Fire Protection Association, Boston, 1976.

Fundamental Principles of Science Applied to the Fire Service, Department of Fire Protection, College of Engineering, Oklahoma State University, Oklahoma.

Handbook of Propane-Butane Gases, 4th Edition, Chilton, Los Angeles, 1962.

Handbook of Tables for Applied Engineering Science, 2nd Edition, CRC Press, Chemical Rubber Company, 1973.

Hilado, C. J., "A Method for Estimating Limits of Flammability," *Journal of Fire and Flammability*, Vol. 6, April 1975.

Kelley, A. L., "Volunteers Prepared When Butane Tank Explodes," *Fire Command!*, Vol. 43, No. 6, June 1976.

"MAPP Industrial Gas," Loss Prevention Data Sheet 7-94, Factory Mutual System, 1969.

Meidl, J. H., *Explosive and Toxic Hazardous Materials*, Glencoe, Beverly Hills, California, 1970.

———, *Flammable Hazardous Materials*, Glencoe, Beverly Hills, California, 1971.

The Condensed Chemical Dictionary, 8th Edition, Van Nostrand, Reinhold, 1971.

Westfield, W. T., "Ignition of Aircraft Fluids on High Temperature Engine Surfaces," *Fire Technology*, Vol. 7, No. 1, Feb. 1971.

Van Dolah, R. W., *et al.*, "Flame Propagation, Extinguishment, and Environmental Effects on Combustion," *Fire Technology*, Vol. 1, No. 2, May 1965.

Basic Facts About Flames and Fire Extinguishment

THE COMBUSTION PROCESS

The combustion process is usually thought of solely as an oxidation process, as indeed it is; however, there are many chemical oxidation processes going on around us that can hardly be called "combustion" processes. For example, our bodies are continually taking in oxygen and, in a series of complicated physio-biological processes, that oxygen is used to "burn" the food we eat, thus giving us energy and sustaining our lives. Iron-metal construction slowly oxidizes from the oxygen in the air and, assisted by moisture, it forms rust (iron oxide). Oil-based paints absorb oxygen from the air and change from a liquid film that is easy to apply to a solid, protecting film that is difficult to remove and that does not return to its liquid state even when solvents are applied to it. Even chlorine in dilute solutions reacts with the substances in paper and wood to change its color (by oxidation) to a dirty yellow shade, a color similar to that obtained after many months of exposure to oxygen and sunlight.

What is Fire?

None of the preceding chemical reactions could be called "fires." The subtle difference between them and fire is the speed with which the oxidation reaction occurs and the amount of heat and accompanying light that issues from the site of the oxidation. Taking these physical properties into consideration, a very simple definition of fire is:

> Fire is a rapid, self-sustaining oxidation process accompanied by the evolution of heat and light of varying intensities.

From the preceding definition, it can be said that in all cases fires are exothermic, and that reaction time is an important consideration.

Oxidation and Reduction

As in all reaction systems, fires involve two widely differing types of components: (1) an oxidizing agent, and (2) a reducing agent. An oxidizing agent is a substance that can combine with, or oxidize, a substance such as a fuel or one of the many elemental metals and nonmetal solids. This oxidation process forms oxides of one type or another. The reaction taking place is an oxidation of a reducing agent, which would be the fuel or elemental metal.

In their fundamental and basic properties, oxidation and reduction processes have been characterized as electron exchange reactions. An oxidizing agent is a substance capable — and eager — to accept electrons, which have a negative charge and a negative sign. A reducing agent is a substance capable of furnishing, or donating, these electrons. In the act of accepting electrons, the oxidizing agent's ability to oxidize is lowered; it is said to be "reduced."

Noting that the elements contain balanced charges, as do compounds formed by their reaction, the formula for the exchange of electrons can be written for the reduction of chlorine (an oxidizing agent) and the oxidation of sodium (a reducing agent) as follows:

$$2Na^\circ + Cl_2 \rightarrow 2Na^+ + 2Cl^-.$$

Sodium does this: $Na^\circ - e^- = Na^{1+}$
(It loses an electron. It is *oxidized.*)
Chlorine does this: $Cl^\circ + e^- \rightarrow Cl^-$
(It gains an electron. It is *reduced.*)

A similar series of electron exchanges can be written for the oxidation of hydrogen by oxygen to water.

In many cases the oxide produced by an oxidation process can also react with a reducing agent, thereby returning to its original form. The reaction of copper to form the red oxide when heated in air, and the ability to return to its original copper color when heated in an atmosphere of hydrogen-containing fuel gas, is typical of the latter case:

copper color		red color		copper color	
$2Cu \rightarrow$	$0 \rightarrow$	$2CuO +$	$2H_2 \rightarrow$	$2Cu$	$+ \; 2H_2O.$
(reducer)	(oxidizer)	(oxide)	(reducer)	(original copper)	(inert water)
oxidation			reduction		

Typical fires, however, react only in the oxidation part of the reactions shown:

(carbonaceous
matter) (air)

$$C + O_2 \rightarrow CO_2$$

or

$$CH_4 + O_2 \rightarrow CO_2 + 2H_2O.$$

(natural gas) (air)

Obviously, fires involve substances that have an affinity for oxygen and that release heat and light in the process. Included in this group are the halogens, which

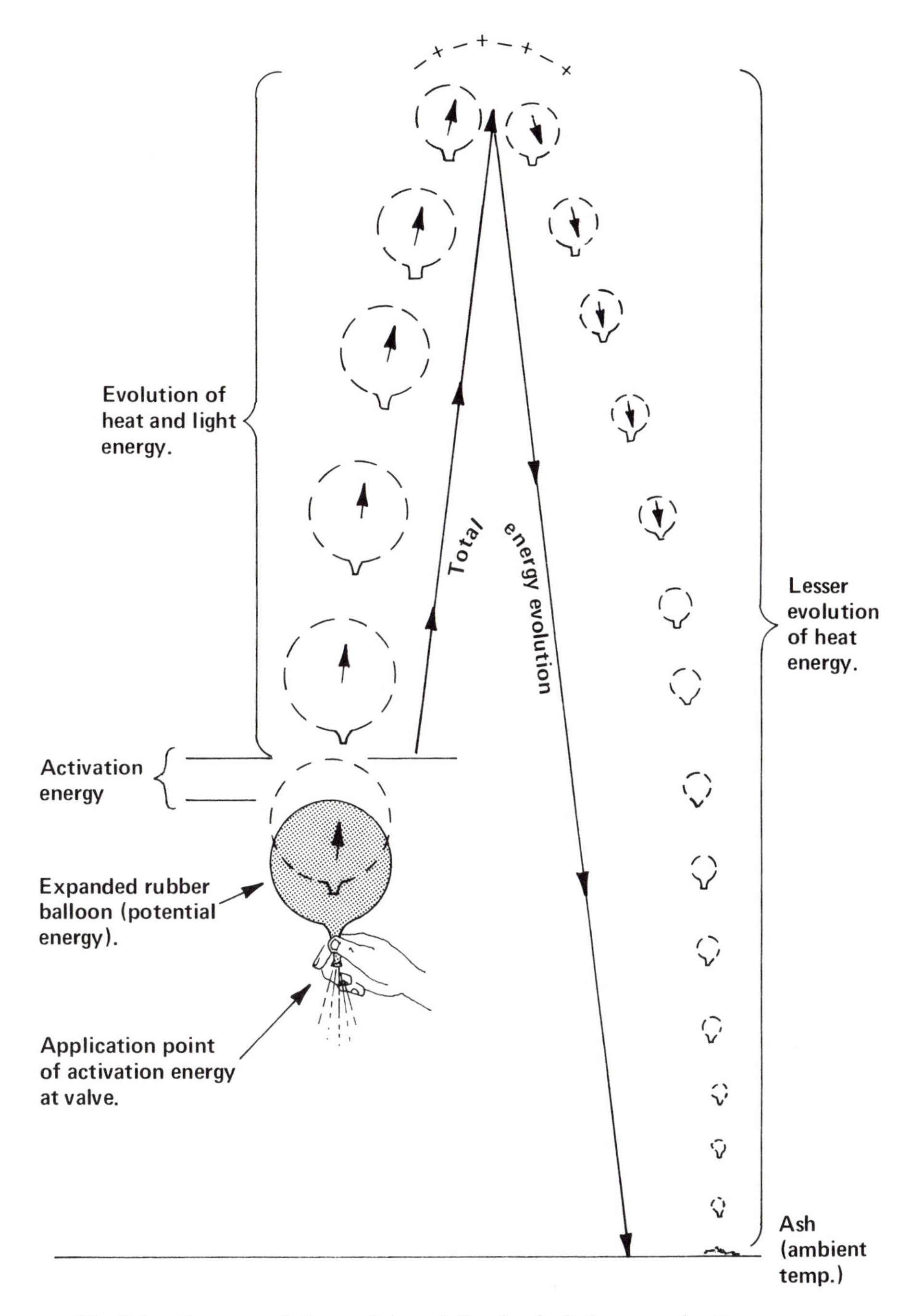

Fig. 8.1. Energy evolution and degradation levels during a combustion process.

are elements that exist in the free state normally as diatomic (two-atom) molecules: chlorine (Cl), bromine (Br), fluorine (F), and iodine (I).

The reaction of oxidation is quite complicated and is variable with substances having different chemical and physical properties. Because of its complexities, scientists have used many types of emphasis in studying oxidation. Flames resulting from different oxidizing agents and reducing agents have been studied thermodynamically by calculation of: (1) their temperature developments at varying sites, (2) their heat release using the energies of dissociation of their chemical entities, and (3) their energies of activation. Chemical-kinetics studies have also been used to bring in the elements of time of reaction and time of exposure of atoms and molecules to oxidizing and other influences.

From an elemental standpoint, all of these studies have been conclusive in determining that combustion and "fire oxidation" requires activation energy, chain reactions involving chain carriers and free radicals in chain reactions, and, in the case of flame extinguishment, free radical quenching reactions.

Activation Energy

Even though molecules of an oxidizing agent and molecules of a reducing agent or fuel can be in intimate contact, no reaction may occur at ordinary temperatures. Fire and rapid oxidation require that a point of high temperature (ignition point) be introduced into the system to start the combination of the two substances and to raise their level of activity to the point at which they will continue to react until all available molecules have reacted. The release of energy (heat and light) when a fire oxidation is activated is similar to the pictorial analogy shown in Figure 8.1 in which an air jet of a rubber balloon, expanded with gas under pressure, is mechanically opened, starting a sudden energy evolution. As shown in Figure 8.1, oxidation reactions (fire) occur rapidly with a large output of heat and light energy, which increases as the temperature increases in the system. When the reactants have completed reacting, there is a slow and lesser evolution of heat energy until ambient temperatures have again been reached. The total energy evolution includes all heat and light generated, and the final level reached is much lower than the level represented by the potential energy of the initial reactants.

This energy level degradation is known as enthalpy, or the heat of reaction, or the heat of formation, of the product of the reaction. For example, the equation for the heat of formation of water is:

$$H_2 + \tfrac{1}{2}O \rightarrow H_2O, \qquad \Delta H^\circ = -68.32 \text{ kcal per mole.}$$

The symbol ΔH° is used to indicate the heat evolved, or formed, by the oxidation. The minus sign indicates that the energy content of the water is 68.32 kcal less than one mole of hydrogen and one-half mole of oxygen.

The Chain Reaction of Combustion

The combustion reaction takes place over a finite time period that varies with respect to the proximity and subdivision of the two components. Even explosions

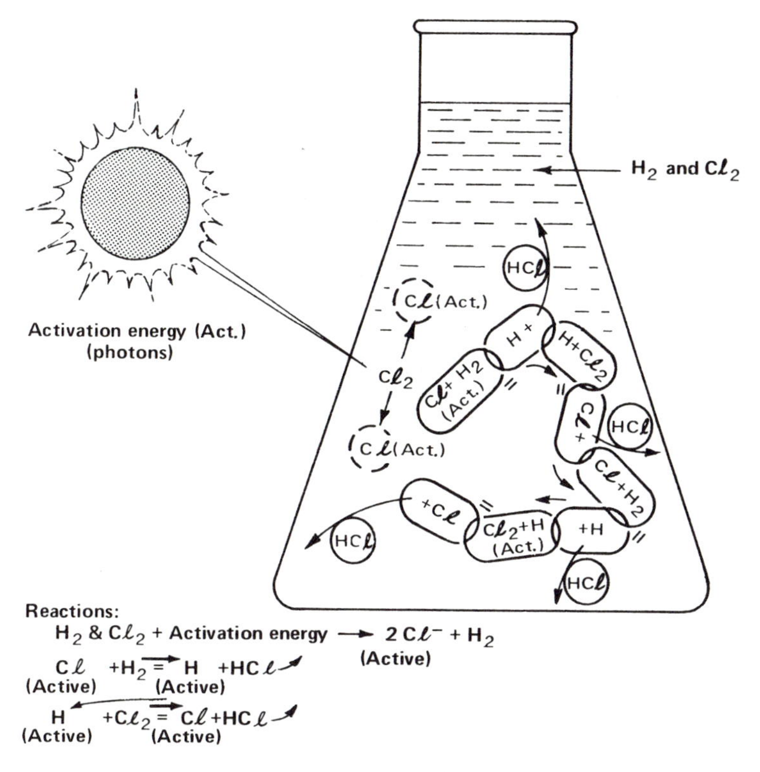

Fig. 8.2. A simple chain reaction.

of a high velocity require time to complete, although they may be measured in nanoseconds (one-billionth of a second).

Combustion is a chain reaction, which is to say that it is a self-sustaining chemical reaction yielding energy or products that cause further reactions of the same kind. To help envision a chain reaction, like a nuclear reaction, picture a closed plastic box with a small hole in the top. On the floor of the box are a large number of set, snap-type mousetraps. Each has a ping-pong ball poised on it, ready to fly up inside the box when the trap is released. When an activating ball is dropped into the box, it bounces around the bottom of the box, contacting and setting off the mousetraps. The poised ping-pong balls are ejected, and each newly released ball activates more mousetraps until all of the balls are ejected in a rapid-fire, spontaneous burst of energy. This is what occurs in a chain reaction.

There are two types of chain reactions of combustion: (1) the unbranched, or simple chain reaction, and (2) the branched, or compound chain reaction. In order to illustrate a simple chain reaction, the oxidation of hydrogen by chlorine will be used. This reaction can also be called chlorination, but because this halogen reaction is similar in its change of energy levels to an oxidation by oxygen, and because it is simpler than the reaction of oxygen with hydrogen, this reaction will be referred to as an oxidation reaction.

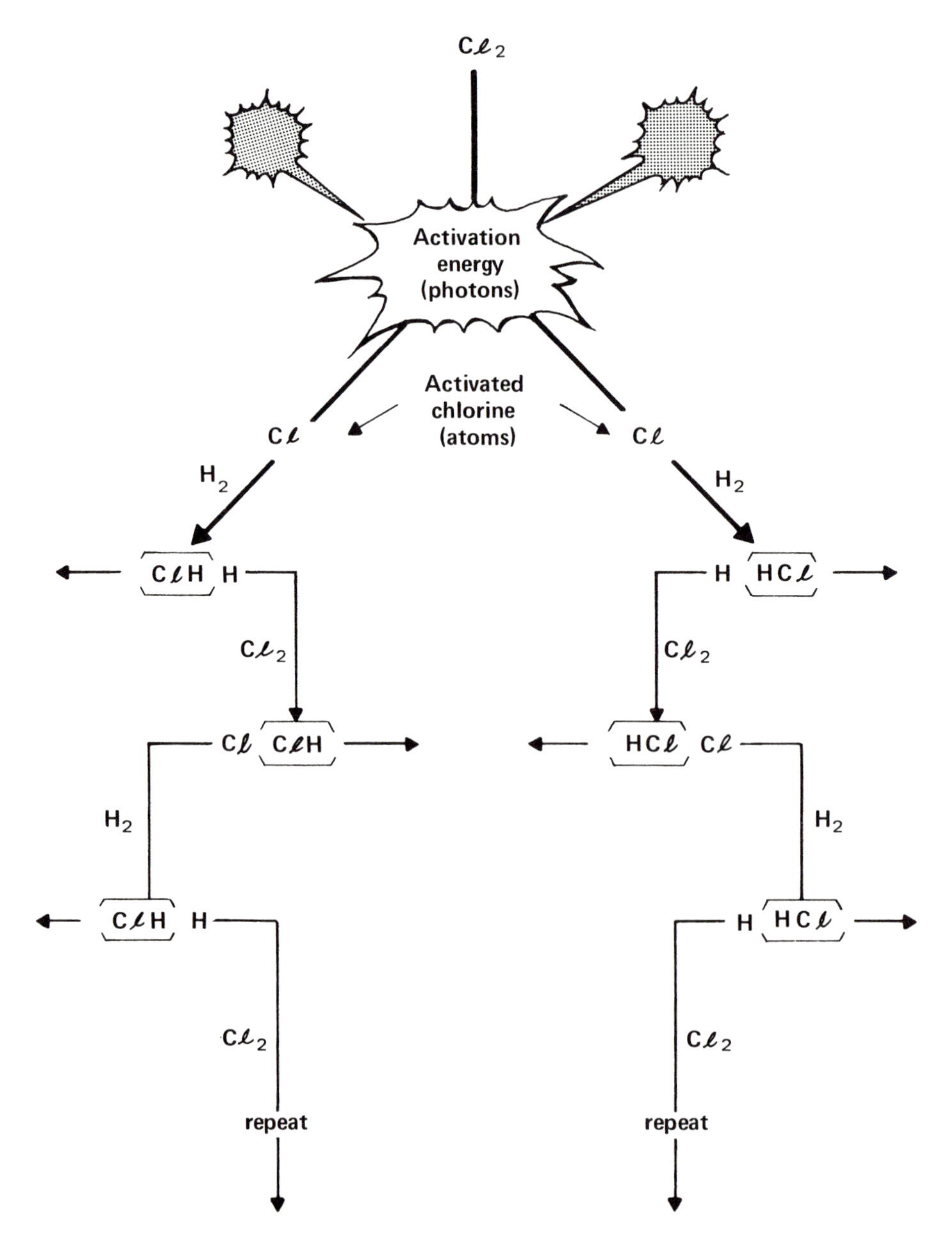

Fig. 8.3. Diagram of the unbranched simple chain reaction of hydrogen and chlorine.

Figures 8.2 and 8.3 illustrate the reaction of hydrogen and chlorine gases when the mixture is activated by sunlight energy. The photons cause the molecular chlorine to become active atoms of chlorine that react with molecular hydrogen. This reaction sets free active atoms of hydrogen capable of reacting with more molecular chlorine which, in turn, sets free atomic chlorine. The chain reaction continues as long as the original gases remain.

In equation form, the reactions are:

$$H_2 + Cl_2 + \text{Activation} \rightarrow 2H + 2Cl$$
$$Cl_2 + H \rightarrow HCl + Cl$$
$$Cl^- + H_2 \rightarrow HCl + H.$$

The branched type of chain reaction involving oxidation is the reaction of hydrogen with oxygen (see Fig. 8.4). It is simply diagrammed in Figure 8.4, even though its action is quite complicated from the point of view of the chemical kinetics involved. The original reactions that occur in this oxidation are simply written in equation steps as follows:

$$H_2 + \text{activation} \rightarrow H + H$$
$$H + O_2 \qquad OH^* + O$$
$$O + H_2 \rightarrow OH^* + H$$
$$OH^* + H_2 \rightarrow H_2O \downarrow + H.$$

These reactions repeat until all reactants have been consumed and have been formed into water.

Note that in all of the preceding chain reactions and their representative equations the activated and atomic elements, H, Cl, O, and the hydroxyl group, OH*, are formed and are immediately reacted with as the oxidations proceed. These are highly active atoms and hydroxyl groups (OH). They are called free radicals and chain carriers, and are extremely important in the extinguishment process by chemical reaction.

The Character of Flames and Their Chemical Reactions

Luminous oxidative flames are a complicated phenomenon. Depending upon the physical conditions pertaining to them and to the fuels and oxidizing agents involved, they vary widely in their chemistry and in the heat and light that they evolve, or the kinetic energy that they generate, as in an explosion, which is merely a flame that usually propagates at supersonic velocities. There are at least four processes that are important to an understanding of flames: (1) convective flow, (2) thermal conduction, (3) molecular diffusion, and (4) chemical reactions occurring under place and time conditions.

Simple flame-reaction chemistry involving known reactants has been well studied, and many reactions can be calculated for their flame temperatures and quantitative radiation from the known enthalpies of their reactants and their concentrations. However, when more complex systems are involved, the study of flames becomes a study of the high-temperature reactions of free radicals such as H, O, OH, (CH_3), (HO_2), and others.

Flames have been separated into two different categories: (1) premixed flames in which the reacting gases have been well mixed prior to ignition and in which combustion occurs without assistance from gases in the surrounding envelope of gases, and (2) diffusion flames in which the fuel gases rely upon processes of

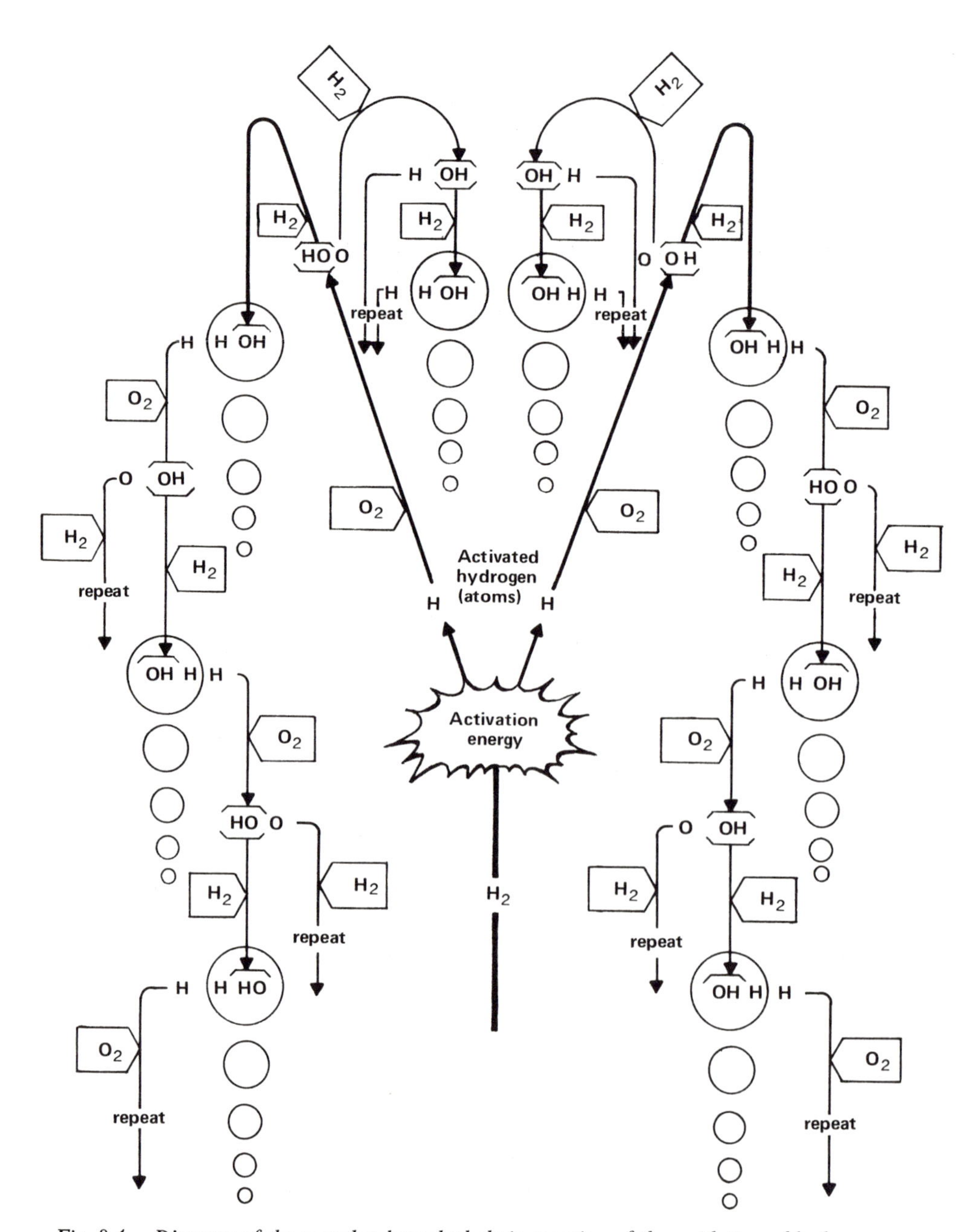

Fig. 8.4. Diagram of the complex branched chain reaction of the oxidation of hydrogen.

admixture and "diffusion" with the oxidizing gases that surround them in the atmosphere.

With the exception of explosions, the fires encountered by fire protection specialists are of the diffusion type. This includes the "glowing" fire where

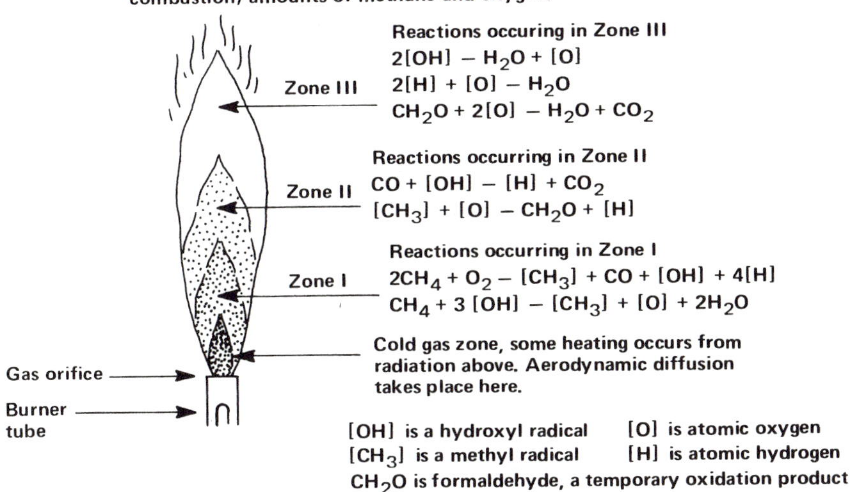

Fig. 8.5. Diagram of a flame.

combustion takes place at lowered velocities and flame lengths, as in the burning of charcoal.

Before reviewing the complexities of diffusion-type combustion, it is of interest to note what the flame-kinetics chemists have discovered concerning the reactions occurring in a "simple" premixed flame where methane, CH_4, is mixed with oxygen and burned from a gas burner. This same reaction occurs in the gas jet burners of an ordinary household gas stove or in a natural gas hot water heater. Figure 8.5 is a self-explanatory pictorial analysis of the events taking place in the oxidation of methane in a premixed flame.[1] These reactions take place rapidly, but not as fast as they would in a contained, correctly balanced (stoichiometric) mixture of methane and oxygen that was ignited by a spark; in such a case, an explosion would occur. Although the same reactions might not occur because of the differing physical conditions influencing them, the end products would be identical.

The complexity of a diffusion flame where gaseous vapors from a fuel are mixed with air and oxygen solely by means of the aerodynamics of the flame and the differing densities of the gaseous components is shown in Figure 8.6.[2]

In this interpretation of a diffusion flame, the term "liquid fuels" denotes almost any of the liquid hydrocarbon fuels that contain carbon and hydrogen in varying proportions, such as gasoline. The example could be used for almost all of the

[1] Fristrom, R. F., "The Mechanism of Combustion in Flames," *Chemical and Engineering News*, October 14, 1963.
[2] Haessler, W. M., *The Extinguishment of Fire*, National Fire Protection Association, Boston, 1974.

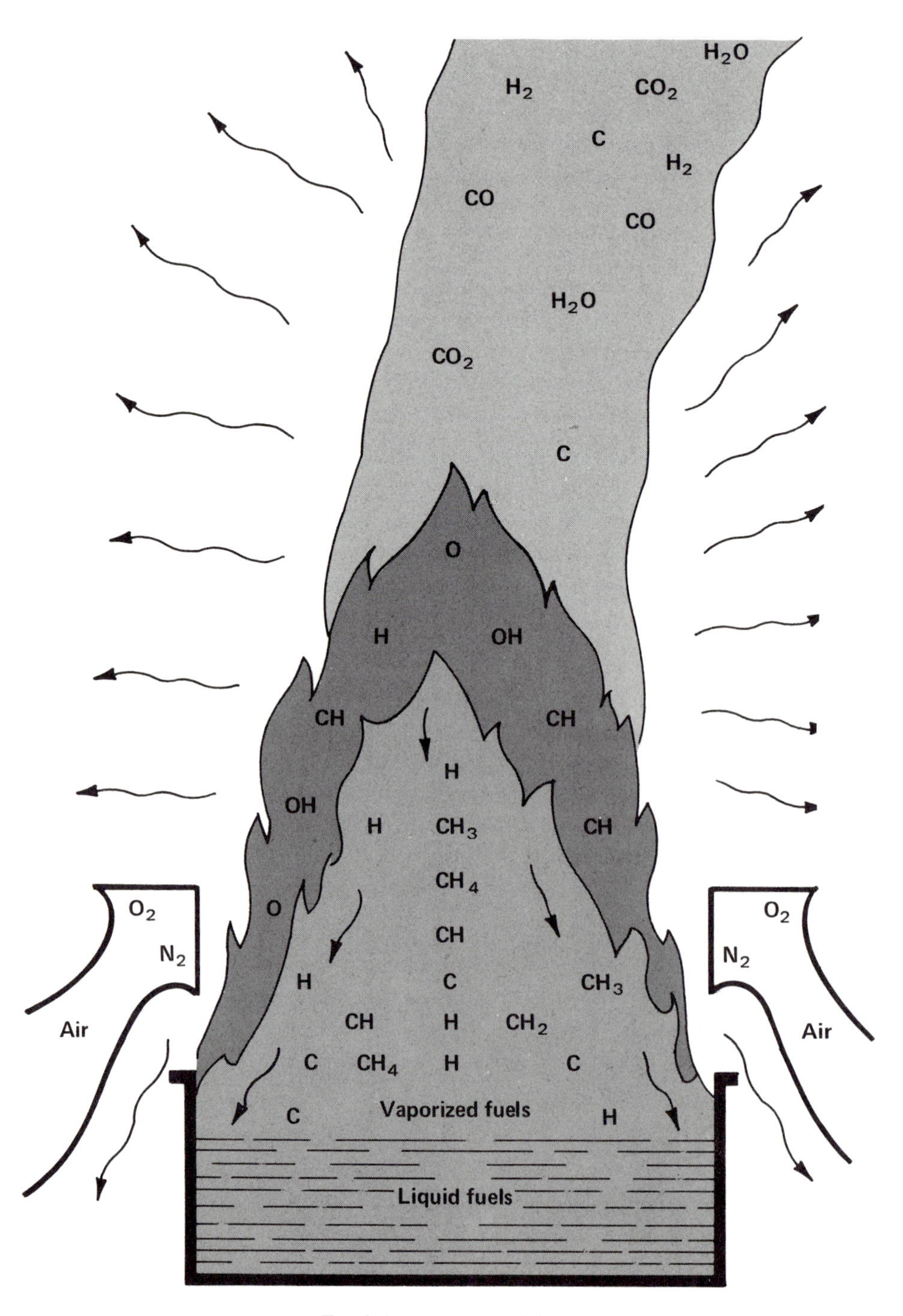

Fig. 8.6. Anatomy of fire.

combustible solids where, as previously emphasized, vapors must be generated in order for flames to propagate.

In Figure 8.6, note the generous evolution of heterogeneous free radicals in the flame matrix; as the stable hydrocarbon vapor breaks down, H, CH, CH_2, CH_3, and C are generated. As oxygen enters into reaction with these, OH and O are generated; finally, we achieve "satisfied," (or inert) carbon dioxide, water, and carbon set free. In Figure 8.6, note also the three levels of reaction in this diffusion flame that relate in a general manner to the three zones of reaction shown in the Figure 8.5 example of a premixed flame.

FREE RADICAL QUENCHING REACTIONS, OR CHEMICAL FIRE EXTINGUISHMENT

At this point we have dissected and categorized in an elementary fashion the basic processes that occur in oxidative flame reactions. To be sure, there are a host of dependent and independent variables that operate in these complicated systems that have not been considered. However, the chain reaction concept and the free radical or activated atomic fragment generation in flames is well substantiated by presently known facts, and these fundamental views are useful in explaining the operation of certain chemical materials that halt flame propagation in complete disproportion to their amounts necessary on any weight or volume basis of consideration.

There have been instances of the use of ordinary baking soda (sodium bicarbonate — $NaHCO_3$) for halting flame propagation, especially from fires in hydrocarbons of one sort or another since Civil War days. (This was at about the same time that the Solvay Process for the cheap production of sodium bicarbonate came into being.) Since then, there have been many theories for the mechanism of the extinguishment of flame by this chemical. The production of carbon dioxide inerting gas and steam when the bicarbonate is heated was the leading contender:

$$2NaHCO_3 + HEAT \rightarrow Na_2CO_3 + CO_2 + H_2O.$$

This was quantitatively disproven by careful experimentation. The dilution of oxygen by the volume of solids injected into the flame by the powder was also shown to be an untenable theory. Evidence of chemical reaction in the flame due to these solids was not shown until after it was discovered that the bicarbonate of potassium was twice as efficient as that of sodium.

In the early 1960s Friedman and Levy published the results of an interesting series of experiments concerning the reaction kinetics of potassium and its compounds in flames.[3] These investigators found that the most likely reactions that account for potassium bicarbonate's extinguishing effectiveness (and similarly, the sodium compound) are due to the compound's ability to break down under

[3] Friedman, R. and Levy, J. B., "Inhibition of Opposed-Jet Methane-Air Diffusion Flames. The Effects of Alkali Metal Vapours and Organic Halides," *Combustion and Flame*, Vol. 7, No. 2, June 1963, p. 195.

heat, yielding potassium hydroxide vapor, which reacts quickly with the OH radical and the activated H* in the branched combustion chain, shown earlier in this chapter.

The chemical reactions probably occurring in this mechanism of free radical quenching are as follows:

$$2KHCO_3 + \text{HEAT} \rightarrow H_2O + CO_2 + K_2O$$
$$K_2O + H_2O \text{ (from the combustion)} \rightarrow 2KOH$$
$$KOH + OH \rightarrow KO + H_2O$$

also: $KOH + H^* \rightarrow K + H_2O$

and a regeneration of KOH occurs as:

$$2K + O \rightarrow K_2O$$
$$K_2O + H_2O \text{ (from the combustion)} \rightarrow 2KOH$$

and the process repeats:

$$KOH + OH \rightarrow KO + H_2O.$$

It can be seen that whenever the vapor form of potassium hydroxide attaches to the free radicals and the chain carriers H* or OH, water is formed (which is inert and passes off as steam, and is no longer reactive) and the free radicals are "quenched," or "inerted," from continuing in the chain reaction.

In support of that reaction mechanism, it is of interest to note that potassium hydroxide solutions in minutely fine spray form have been shown to be even more efficient free radical quenching agents than dry crystalline powdered potassium bicarbonate. (Such solutions are, of course, impractical because of their highly caustic properties.) In the reactions, the hydroxide would bypass the K_2O generation step and its hydrolysis to KOH. It must also be acknowledged that the generation of steam and the evolution of carbon dioxide from the bicarbonate must assist to some extent in the flame extinction mechanism by dilution of oxygen.

Friedman and Levy (as well as other investigators) have also conducted a series of experiments using the "halon" type of extinguishing agents such as methyl bromide and trifluoromonobromomethane (1301). These experiments revealed another type of free radical quenching from the action of the halogen set free by heat from these halogenated hydrocarbons.

When a "halon" type of vaporizing liquid agent is caused to contact the heat of flames and burning vapors or gases, a decomposition of the "halon" takes place and free activated atoms of halogen (Br, Cl, F, etc.) are evolved in the flame matrix. These atoms react quickly with the free radicals OH and H, and a "quenching" takes place (as demonstrated in the earlier example of potassium bicarbonate and sodium).

Using a simple "halon" such as methyl bromide (which is no longer used because of its toxicity), the following chemical reactions, typical of all the "halon" type agents, occur:

$$CH_3Br + \text{HEAT} \rightarrow CH_3 + Br^*$$
$$Br^* + H^* \text{ (from the chain reaction of burning)} \rightarrow HBr$$
$$HBr + OH \text{ (also from the chain reaction)} \rightarrow H_2O + Br^*.$$

Here again we see, as we saw with potassium, a regenerative effect of active bromine following the free radical quenching to inert water or steam during the extinguishment reaction. Obviously, the efficiency of these "chemical" extinguishing agents, when compared to other materials that depend upon cooling or sequestering of fuels for their fire extinguishment abilities, would be much greater on a weight, or even a volume of agent, basis. In many fire suppression experiences, this has turned out to be well-proven over the last twenty-five years.

FOUR BASIC METHODS OF FIRE EXTINGUISHMENT

The preceding paragraphs and the evidence supporting the conclusions made in those paragraphs indicate that there are *four* methods of extinguishment of fire instead of the previously acknowledged age-old three methods: The four methods of extinguishment of fire are as follows:

1. Removal or dilution of air or oxygen to a point where combustion ceases.
2. Removal of fuel to a point where there is nothing remaining to oxidize.
3. Cooling of the fuel to a point where combustible vapors are no longer evolved or where activation energy is lowered to the extent that no activated atoms or free radicals are produced.
4. Interruption of the flame chemistry of the chain reaction of combustion by injection of compounds capable of quenching free radical production during their residence time.

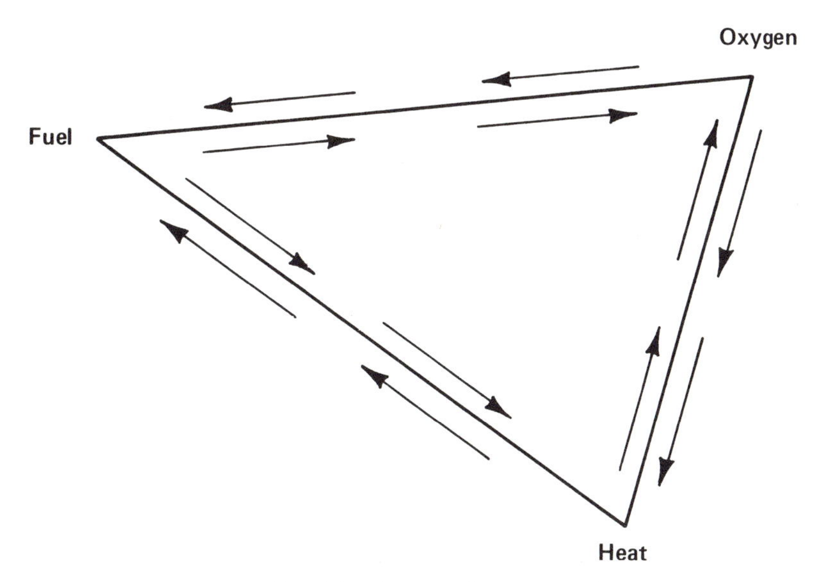

Fig. 8.7. The fire triangle (circa 1920).

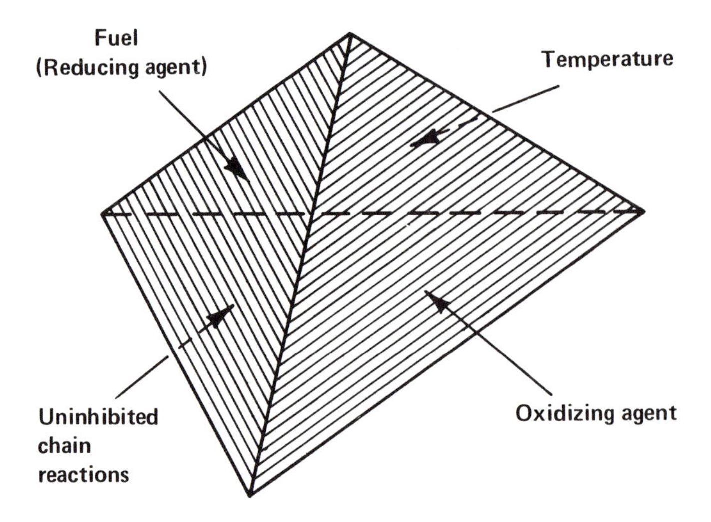

Fig. 8.8. The tetrahedron of fire.

Note that the new fourth method of extinguishment denotes that the action occurs only during contact of the chemical agents with activated groups or with atoms being produced by the combustion process. In a sense this could be seen as a temporary extinguishment process, operating only during the time that the agent particles are present in the flame matrix. If activation energy continues to exist (as with ignition points for vapor or gas reignition) after withdrawal of the agent, the flame reaction will reestablish and will continue.

THE FIRE TRIANGLE AND ITS VARIATIONS

Every student of fire protection technology is well aware of the simple concept of the characterization of fire as a three-component system: fuel, oxygen, and heat. To extinguish fire, only one of these components needs to be removed. The graphical representation of a triangle joining these components is a useful tool for demonstrating the basic actions involved in fire propagation and extinguishment. The illustration of the fire triangle in Figure 8.7 was devised early in the history of fire protection. The arrows were added to indicate interaction of the three components as illustrated in one of the leading fire equipment manufacturer's catalogs dated 1920.

The need for a fourth ingredient to be added to the fire triangle in order to fully describe fire extinguishment mechanisms appeared first in the open literature in 1960 following a series of careful quantitative fire suppression tests by Arthur B. Guise.[4] Immediately following Guise's work, a monograph by Walter M. Haessler in 1961 pioneered the concept of a three-dimensional tetrahedron for describing

[4] Guise, A. B., "The Chemical Aspects of Fire Extinguishment," NFPA *Quarterly*, National Fire Protection Association, Boston, April 1960.

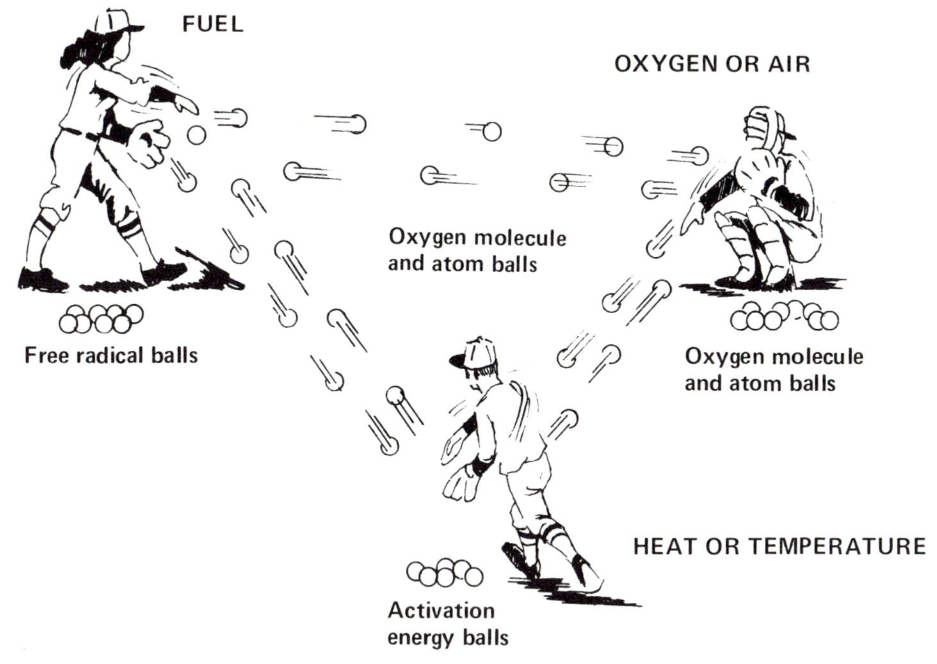

Fig. 8.9. The chain reaction of combustion.

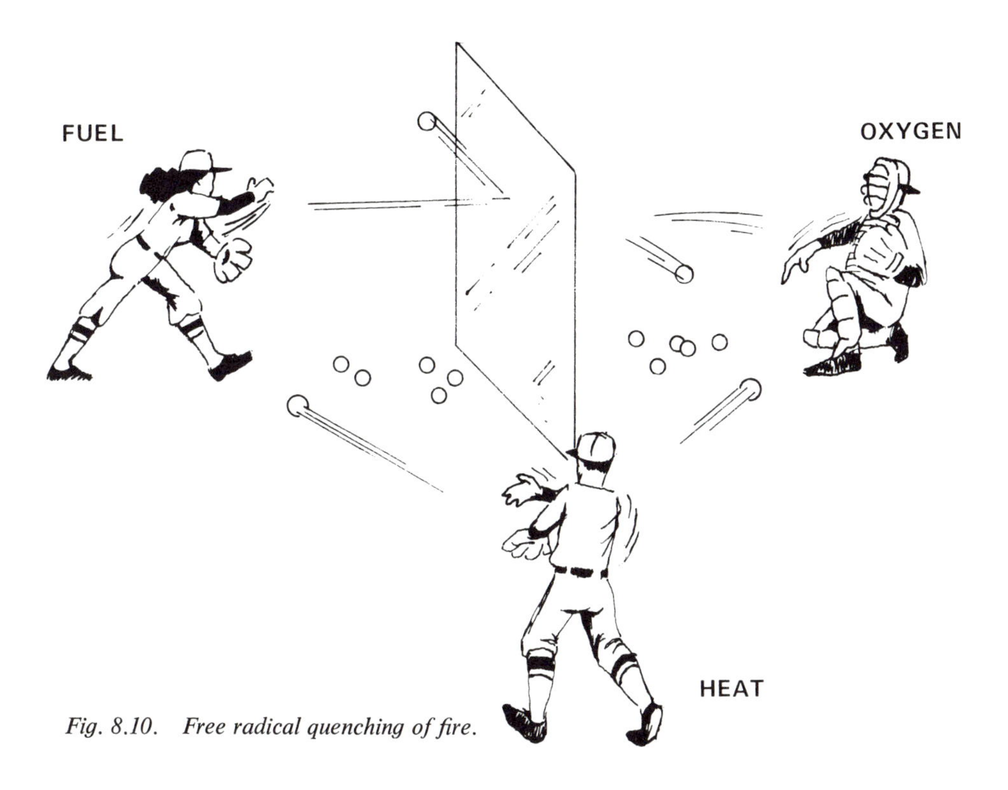

Fig. 8.10. Free radical quenching of fire.

the graphic demonstration of four entities in fire propagation and extinguishment.[5] (See Fig. 8.8.)

The tetrahedron shown in Figure 8.8 indicates the function of the four requirements for fire propagation, along with the idea of removal of any one of the four (shown as upright triangles) to indicate extinguishment. In 1964 this author was not satisfied with the lack of dynamicity shown by graphic portrayals of fire reactions up to this time; therefore, a different, hopefully more dynamic, pictorial version of the fire triangle was evolved.[6] (See Fig. 8.9.)

In explanation of the fourth method of fire extinguishment, a removable barrier was used to indicate chemical extinguishment by free radical quenching, as shown in Figure 8.10.

In summation, the problem of simply, yet effectively, denoting the newer, fourth method of fire extinguishment awaits a fully satisfactory graphical portrayal useful to all of the personnel who are concerned with this type of instructional material. Such a graphical portrayal requires a two-dimensional presentation of a dynamic system showing the newer, fourth method of extinguishment as a possible temporary action, unless the activation energy of the system is lowered to below the reignition level.

ACTIVITIES

1. You have been assigned to tutor a foreign-exchange student who does not understand English. Your first "lecture" is on the chain reaction of combustion. Without using the chain reaction example given in this chapter, describe how you would pictorially show a chain reaction to your student.
2. (a) Why isn't the burning of stored body fat called a fire?
   (b) Is the burning of charcoal briquettes a fire? Why or why not?
3. If molecules of an oxidizing agent come in contact with each other but no reaction takes place, what would the reason be? How could a reaction be started?
4. (a) Does extinguishment of a fire with water differ from free radical quenching reactions?
   (b) Explain why or why not.
5. Identify the basic method of fire extinguishment for each of the following:
   (a) Laying a blanket over a small pile of burning leaves.
   (b) Pouring water over a burning chair.
   (c) Throwing baking soda into a grease fire.
   (d) Spraying a burning rug with a standard hand fire extinguisher.
   (e) Shoveling a ring of dirt around a large-scale forest fire.
   (f) "Dusting" a forest fire with a chemical from an aircraft.
   (g) Throwing sand on a small paper fire.
6. Explain how the fire triangle was updated by the tetrahedron of fire.

[5] Haessler, W. M., *The Extinguishment of Fire*, National Fire Protection Assocation, Boston, 1974.
[6] Tuve, R. L., "A New Look for the Old Fire Triangle," *Fire Engineering*, April 1964.

7. (a) Why are halons an effective form of fire extinguishment?
   (b) How is the halon's free radical quenching extinguishment achieved?
8. From a fire fighter's point of view, discuss with a group of your classmates which of the two types of flames — premixed or diffusion — would be harder to extinguish.
9. All reaction systems involve two widely differing types of components: (1) an oxidizing agent, and (2) a reducing agent.
   (a) What is an oxidizing agent? Explain the oxidation process that takes place at the scene of a fire.
   (b) What is a reducing agent? Explain the reduction process that takes place at the scene of a fire.
10. From what you have learned in this chapter and in your own words, explain what the flame-kinetics chemists have discovered concerning the reactions that occur in a "simple" premixed flame where methane, CH_4, is mixed with oxygen and burned from a gas burner.

SUGGESTED READINGS

Berl, W. G., *et al.*, "An Introduction to Combustion Chemistry," *Fire Research Abstracts and Reviews*, Vol. 13, No. 3, 1971.

Custer, R. L. P., "Oxidation is Only the Beginning," *Fire Engineering*, Vol. 125, No. 7, July 1972.

Emmons, H. W., "Fire and Fire Protection," *Scientific American*, Vol. 231, No. 1, July 1974.

Fire Protection Handbook, 14th Edition, National Fire Protection Association, Boston, 1976.

Fristrom, R. M., *Flame Structure*, Johns Hopkins University, Maryland, February 1974.

Haessler, W. M., *The Extinguishment of Fire*, National Fire Protection Association, Boston, 1974.

Wall, L. A., *The Mechanisms of Pyrolysis, Oxidation and Burning of Organic Materials*, Department of Commerce, National Bureau of Standards, Washington, D.C., June 1972.

Wilson, W. E., Jr., "Structure, Kinetics, and Mechanism of a Methane-Oxygen Flame Inhibited with Methane Bromide," *Tenth Symposium (International) on Combustion*, The Combustion Institute, Pittsburgh, 1965.

Fire Classifications and Water to Foam Agents

THE CLASSIFICATION OF FIRE TYPES AND THEIR RELATIONSHIP TO EXTINGUISHMENT METHODS

Early in the history of organized fire suppression it was realized that fires involving various kinds of fuel differed greatly in their characteristics and required special extinguishing agents and treatment in order to completely and efficiently control them. It was also realized that not all combustible substances could be halted from further burning by the age-old application of water.

The need for agents other than water increased with the discovery of new fire hazardous substances developed by the technical progress of civilization. For example, before plastics came into wide commercial use, bleaches and household cleaners had been stored in glass bottles. As household cleaning problems became more varied, people began to buy plastic-bottled cleansers in large quantities. The storage and handling of these containers became dangerous, and the containers constituted a new hazardous substance that had to be dealt with by agents other than water. Another example is that prior to the invention of an inexpensive method for producing magnesium metal for the fabrication of light and strong construction (as in aircraft parts), there was little need for noting the fact that burning metals required special extinguishing agents or a separate class for their fuel fire hazard: so the NFPA's fourth classification for basic types of fire — Class D fires involving fires in combustible metals — was not officially recognized until about 1960, after many combustible metals other than magnesium had come into general industrial and commercial use.

History does not tell us whether our standard classification of types of fire according to the fuels involved was instituted before or after certain extinguishing agents were developed for specific types of fuel fires. However, modern agents and devices for using them are now categorized according to the class (or classes) of fire for which they are most useful and particularly applicable. Accordingly, today's extinguishing agents are spoken of as Class A, B, C, and D agents in the same manner that we denote rags and paper as Class A combustibles, gasoline and methane as Class B combustibles, computers and transistors as Class C combustibles, and zirconium and potassium as Class D combustibles. Often it is difficult or even impossible to separate the different types of combustible materials involved in a "working" fire; nevertheless, an accurate description of a fire situation requires some identification of the hazards faced in its suppression. The following NFPA classification of types of fires is a basic standard in the United States;[1] over the years this classification has been proven to be both useful and inclusive for categorizing fire hazards, for categorizing agents for combating fires, and for categorizing extinguishing devices:

- **Class A** fires are fires in ordinary combustible materials, such as wood, cloth, paper, rubber, and many plastics.
- **Class B** fires are fires in flammable and combustible liquids, gases, and greases.
- **Class C** fires are fires which involve energized electrical equipment where the electrical nonconductivity of the extinguishing media is of importance. (When electrical equipment is de-energized, extinguishers for Class A or B fires may be used safely.)
- **Class D** fires are fires in combustible metals, such as magnesium, titanium, zirconium, sodium, and potassium.

For purposes of comparison it is interesting to note the British Standard Classification of fires, given in the United Kingdom's *Manual of Firemanship, Book 1*, 1974:

- **Class A:** These are fires involving solid materials normally of an organic nature (compounds of carbon) in which combustion generally occurs with the formation of glowing embers. Class "A" fires are the most common, and the most effective extinguishing agent is generally water in the form of a jet or spray.
- **Class B:** These are fires involving liquids or liquifiable solids. For the purpose of choosing extinguishing agents, flammable liquids may be divided into two groups:
  (1) Those that are miscible with water.
  (2) Those that are immiscible with water.
  Depending on (1) and (2), the extinguishing agents include water spray, foam, "light water," vaporizing liquids, carbon dioxide, and dry chemical powders.
- **Class C:** These are fires involving gases or liquified gases in the form of a liquid spillage, or a liquid or gas leak, and these include methane, propane, butane, etc. Foam or dry chemical powder can be used to control fires involving shallow liquid spills. (Water in the form of spray is generally used to cool containers.)

[1] From NFPA 10, *Standard for Portable Fire Extinguishers*, National Fire Protection Association, Boston, 1975.

- **Class D:** These are fires involving metals. Extinguishing agents containing water are ineffective, and even dangerous; carbon dioxide and the bicarbonate classes of dry chemical powders may also be hazardous if applied to most metal fires. Powdered graphite, powdered talc, soda ash, limestone, and dry sand are normally suitable for Class "D" fires. Special fusing powders (eutectic) have been developed for fires involving some metals, especially the radioactive ones.
- **Electrical Fires:** It is not considered, according to present-day ideas, that electrical equipment must, in fact, be a fire of Class "A," "B" or "D." The normal procedure in such circumstances is to cut off the electricity and use an extinguishing method appropriate to what is burning. Only when this cannot be done with certainty will special extinguishing agents be required which are nonconductors of electricity and nondamaging to equipment; these include vaporizing liquids, dry chemical powders, and carbon dioxide, although the latter's cooling and condensation effects may affect sensitive electronic equipment.

Under "Selection by Hazard," NFPA 10, *Standard for Portable Fire Extinguishers*, includes the following recommendations:

- Extinguishers for protecting Class A hazards shall be selected from among the following: water types, foam, loaded stream, and multipurpose dry chemical.
- Extinguishers for protection of Class B hazards shall be selected from the following: bromotrifluoromethane (Halon 1301), bromochlorodifluoromethane (Halon 1211), carbon dioxide, dry chemical types, foam, and loaded stream.
- Extinguishers for protection of Class C hazards shall be selected from the following: bromotrifluoromethane (Halon 1301), bromochlorodifluoromethane (Halon 1211), carbon dioxide, and dry chemical types.
- Extinguishers and extinguishing agents for the protection of Class D hazards shall be of types approved for use on the specific combustible-metal hazard.

In explanation of the preceding recommendation for Class D Hazards, the NFPA Standard includes the following warning:

- Chemical reaction between burning metals and many extinguishing agents (including water) may range from explosive to inconsequential, depending in part on the type, form, and quantity of metal involved. In general, the hazards from a metal fire are significantly increased when such extinguishing agents are applied.

Portable Extinguisher Classification

As the technology of fire protection grew, there became an increasing awareness of the need to mark portable extinguishers in a distinctive manner in order to indicate the type of fire hazard they had been chosen to protect. In many instances, extinguishers are used by unskilled operators. Often, such use can result in more harm than good because, as extinguishing agents continue to be developed for specific hazards, an unskilled operator might easily apply an unsuitable agent to a particular type of fire. Furthermore, a portable extinguisher is a "first aid" type of device: it must be fully effective and safe for use in its limited quantity. If the extinguisher is not used effectively, a fire emergency of limited scope and size may be enlarged to the point where life and property are imperiled to a high degree.

In addition to the problem of adequately matching the extinguishing agent to the fire hazard to be protected, the question of the quantity of agent contained in a unit as needed to extinguish a fire of a defined size and magnitude was dealt with around 1948. The result of deliberations on these questions of type and quantity of agent in portable extinguishers has been to classify and mark extinguisher containers for type or types of fires on which the extinguishers are recommended for use. (See Fig. 9.1.) In order to assign performance capacities to extinguishers, a number of specification tests are conducted using the types of fires the extinguishers are designed to combat.

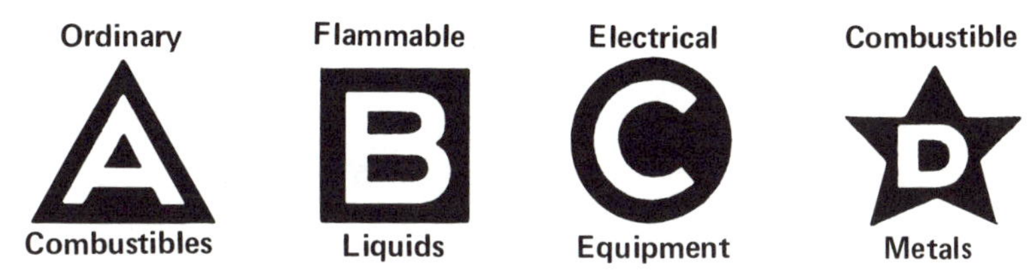

Fig. 9.1. Markings for extinguishers indicating classes of fires on which they should be used. Color coding is part of the identification system. The triangle (Class A) is colored green, the square (Class B) red, the circle (Class C) blue, and the five-pointed star (Class D) yellow. (From *Fire Protection Handbook*, NFPA)

Tests for Class A Ratings

In determining Class A ratings, three tests are used for extinguishers rated 1-A to 6-A. They are the wood-crib fire test, the wood-panel fire test, and the excelsior fire test. For extinguishers having a Class 10-A rating or more, only the wood-crib fire test is used.

The Wood-crib Fire Test: This test consists of layers of nominal 2-by 2-inch or 2-by 4-inch kiln-dried spruce or fir lumber of a specified moisture content arranged as shown in Figure 9.2.

The length, size, and number of individual wood members and their arrangement are varied to test the capability of the extinguishers. For example, a wood-crib test fire for a 1-A rating has 50 wood members, a 2-A has 78, a 4-A has 120, and a 40-A (the largest) has 224. The number of wood members and the nominal size and length of each increase as higher ratings are secured. The crib is ignited by burning n-Heptane in a pan placed symmetrically under the vertical axis of the crib. The size of the square pans and the amount of n-Heptane that is used increase with the size of the cribs (*i.e.*, from one-fourth gallon for a 1-A test to ten gallons for a 40-A test). Freeburn times are standardized as well as methods of recording observations. The tests are conducted indoors in a draft-free room, or outdoors in essentially still air.

Fig. 9.2. A wood crib as used by Underwriters Laboratories for Class A fire tests. (From *Fire Protection Handbook*, NFPA)

The Wood-panel Fire Test: This test is conducted indoors and consists of a solid square wood-panel backing on which two horizontal sections of furring strips are applied. These sections are spaced apart and away from the panel by vertical furring strips. (See Fig. 9.3.) This provides for a large vertical surface area of wood subject to combustion.

The panel sizes vary from 64 square feet for a 3-A, 196 square feet for a 4-A, and 289 square feet for a 6-A. All lumber used is kiln dried to a specified moisture content. In front of the panels, four separate and equal windrows of seasoned basswood, poplar, or aspen excelsior are laid, with the first windrow directly at the base of the test panel and the other three held in reserve about 10 feet from the front of the panel. The amount of excelsior varies from 10 to 60 pounds for the 1-A to 6-A test fires. Fuel oil is used to soak the excelsior, and a small amount of n-Heptane is used as a fuse. The three windrows of excelsior are pushed up to the base of the panel at 45-second intervals during the preburn period. After 3 minutes and 20 seconds, all the remaining excelsior is cleared away. The extinguishant is applied after the furring strips have fallen away from not less than 10 feet of the face of the test panel.

The Excelsior Fire Test: This test utilizes new and seasoned materials in a dry state. These materials are pulled apart and spread evenly and loosely over a prescribed test area, and then packed to a depth of 1 foot (see Fig. 9.4). The area covered varies from 6 pounds of excelsior spread 2 feet 10 inches by 5 feet 8 inches for the 1-A test, to 36 pounds of excelsior spread 6 feet 11

Fig. 9.3. A typical wood panel used by UL to test extinguishers designed for use on Class A (ordinary combustibles) fires. (From *Fire Protection Handbook*, NFPA)

Fig. 9.4. Excelsior prepared for the Class A fire extinguisher test as used by UL. (From *Fire Protection Handbook*, NFPA)

inches by 13 feet 11 inches for the 6-A test. In the excelsior fire test, the material is ignited by a small fuse of n-Heptane and the fire attack is begun from a distance of not less than 15 feet.

Tests for Class B Ratings

In determining Class B ratings, the flammable liquid test fires consist of burning n-Heptane in square steel pans not less than 8 inches in depth and in various sizes. Figure 9.5 illustrates a typical test pan.

Tests for portable extinguishers receiving ratings up to and including 20-B are conducted indoors in an essentially draft-free room. There are five ratings: 1-B, 2-B, 5-B, 10-B, and 20-B. Tests for higher-rated devices are conducted outdoors under conditions of still air and no precipitation. For an extinguisher to become eligible for a given classification and rating, the specified test fires must be extinguished repeatedly and each test conducted by starting with a fully charged extinguisher.

Fig. 9.5. A typical pan arrangement for UL testing of Class B extinguishers. (From *Fire Protection Handbook*, NFPA)

Tests for Class C Ratings

In determining a Class C designation, the extinguishing agent is tested for electrical conductivity only, although no Class C rating is provided unless the extinguisher has an established Class A or Class B, or both, rating. The discharge of the agent from the extinguisher under the test condition must not increase the electrical conductivity, as measured by a milliammeter, through a 10-inch air gap established between an electrically insulated extinguisher and a grounding plate, at a potential of 100,000 volt, 60 cycle alternating current. Figure 9.6 shows the test setup used.

Each extinguisher is operated for 20 seconds, discharging the agent against the target with the potential of 100,000 volts impressed between the extinguisher and the target. The condition is then checked for an additional 15-second discharge with each type of horn or nozzle utilized. In at least one test, the target plate is heated to an initial temperature of 700°F (371°C) prior to the discharge of the extinguisher. All tests must show no meter readings.

Tests for Class D Ratings

In Class D tests, fire extinguishers must be capable of extinguishing combustible metal test fires as designated by the testing laboratory for a given metal. The burning metal must not be scattered beyond the test-bed area during the test. An extinguished test fire must not be subject to reignition after the test, and there must be sufficient unburned combustible metal remaining to show extinguishment by the agent prior to burnout. In addition, studies are made with respect to the toxicity of the agent as well as to the liberated fumes and products of combustion when used on combustible metal fires. Possible reactions that might occur between the burning metals and the agents must be evaluated to ensure safe usage.

Upon completion of the preceding performance tests, a portable extinguisher is rated according to its extinguishing capability and its fitness for operation of the agent it contains for combating fires of the types noted. These ratings are marked on the body of the extinguisher in proximity to the classification mark referred to earlier. Table 9.1 lists some customary extinguisher classifications presently in use.

The proper employment of portable fire extinguishers and their maintenance is a separate study and requires detailed attention. There are a number of excellent

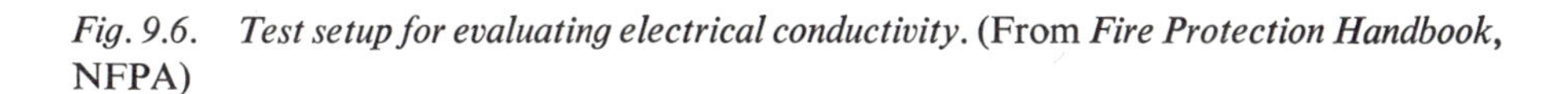

Fig. 9.6. Test setup for evaluating electrical conductivity. (From *Fire Protection Handbook*, NFPA)

Table 9.1 Conversion of Extinguisher Classification*

EXTINGUISHING AGENT	EXTINGUISHER TYPE AND SIZE	CLASSIFICATION 1955 TO JUNE 1, 1969
Chemical solution (soda-acid)	$1\frac{1}{4}$, $1\frac{1}{2}$ gallon	1-A
	$2\frac{1}{2}$ gallon	2-A
	17 gallon	10-A
	33 gallon	20-A
Water	$1\frac{1}{2}$, $1\frac{3}{4}$ gallon (pump or pressure)	1-A
	$2\frac{1}{2}$ gallon (pump or pressure)	2-A
	4 gallon (pump or pressure)	3-A
	5 gallon (pump or pressure)	4-A
	17 gallon (pressure)	10-A
	33 gallon (pressure)	20-A
	5 12-quart or 6 10-quart water-filled pails; 55-gallon water-filled drum with 3 fire pails; 25- to 55-gallon water-filled bucket tank with 5 or 6 fire pails	2-A
Loaded stream	1 gallon	1-A
	$1\frac{3}{4}$ and $2\frac{1}{2}$ gallon	2-A, $\frac{1}{2}$-B†
	33 gallon	20-A
Foam	$1\frac{1}{4}$ and $1\frac{1}{2}$ gallon	1-A, 2-B
	$2\frac{1}{2}$ gallon	2-A, 4-B
	5 gallon	4-A, 6-B
	10 gallon	6-A, 8-B
	17 gallon	10-A, 10-B
	33 gallon	20-A, 20-B
Carbon dioxide	6 or less pounds of carbon dioxide	1-B, C
	$7\frac{1}{2}$ pounds of carbon dioxide	2-B, C
	10 to 12 pounds of carbon dioxide	4-B, C
	15 to 20 pounds of carbon dioxide	4-B C
	25 and 26 pounds of carbon dioxide	6-B, C
	50 pounds of carbon dioxide	10-B, C
	75 pounds of carbon dioxide	12-B, C
	100 pounds of carbon dioxide	12-B, C
Dry chemical	4 to $6\frac{1}{4}$ pounds of dry chemical	4-B, C
	$7\frac{1}{2}$ pounds of dry chemical	6-B, C
	10 to 15 pounds of dry chemical	8-B C
	20 pounds of dry chemical	16-B, C
	30 pounds of dry chemical	20-B, C
	75 to 350 pounds of dry chemical	40-B, C
Wetting agent	10 gallons	6-A
	20 gallons	12-A
	50 gallons	30-A

Note: Carbon dioxide extinguishers with metallic horns do not carry any "C" classification.

† Note: Portable fire extinguishers with fractional ratings do not meet the requirements of the NFPA Extinguisher Standard.

* From *Fire Protection Handbook*, 14th Edition, National Fire Protection Association, Boston, 1976.

treatises on the subject that can be consulted for further information. For example, the 14th Edition of the NFPA's *Fire Protection Handbook* contains a number of chapters devoted to the subject; further references are contained in the *Handbook's* end-of-chapter "Bibliographies."[2] Manufacturer's product catalogs also contain detailed data on specific types of agents and application devices.

WATER AND ITS FORMS

Water — the most abundant and most available substance present on the earth's surface — is the single most useful and most efficient material for controlling and halting combustion and fire. Fortunately, water is found in plentiful amounts almost everywhere. Except in the form of ice and snow, it may be inexpensively and effectively used to halt rapid oxidation reactions. It is almost beyond comprehension to imagine the difficulties society would be confronted with if the physical property of water was different in respect to its ability to retard exothermic reactions by the removal of heat.

Table 9.2 Specific Heats and Heats of Vaporization of Some Common Liquids

COMMON LIQUID	SPECIFIC HEAT		HEAT OF VAPORIZATION	
	BTU/LB.	kJ/kG°K	BTU/LB.	kJ/kG
Acetone	0.514	2.15	223.0	518.0
Ethyl alcohol	0.584	2.44	364.0	846.0
Carbon tetrachloride	0.207	0.86	83.5	194.0
Ether	0.529	2.21	160.0	372.0
Ethylene glycol	0.565	2.36	344.0	800.0
Glycerine	0.627	2.62	419.0	974.0
Hexane	0.541	2.26	157.0	365.0
Toluene	0.410	1.72	156.0	363.0
Water	1.0	418.3	970.3	2,260.0

The Important Properties of Water

Even in the form of a solid (ice), water absorbs heat in the process of changing into a liquid at 32°F (0°C) to the extent of 143.4 Btu per pound (333.2 kiloJoules per kilogram) of water. This, you will remember, is the heat of fusion of ice (water). If this same weight of water's temperature is raised by exposing it to a source of heat, from 32°F (0°C) to the boiling point, 212°F (100°C), the water will absorb 180 Btu (418.3 kiloJoules per kilogram). This is due to the specific heat of water: 1 Btu per °F. The amazing extent of heat absorption occurs when water exceeds the boiling point and becomes steam. This transformation ("heat of vaporization" or "latent heat of evaporation") absorbs 970.3 Btu per pound (2,260 kiloJoules per kilogram). Table 9.2 compares the specific heats and heats of vaporization of some

[2] *Fire Protection Handbook*, 14th Edition, National Fire Protection Association, Boston, 1976.

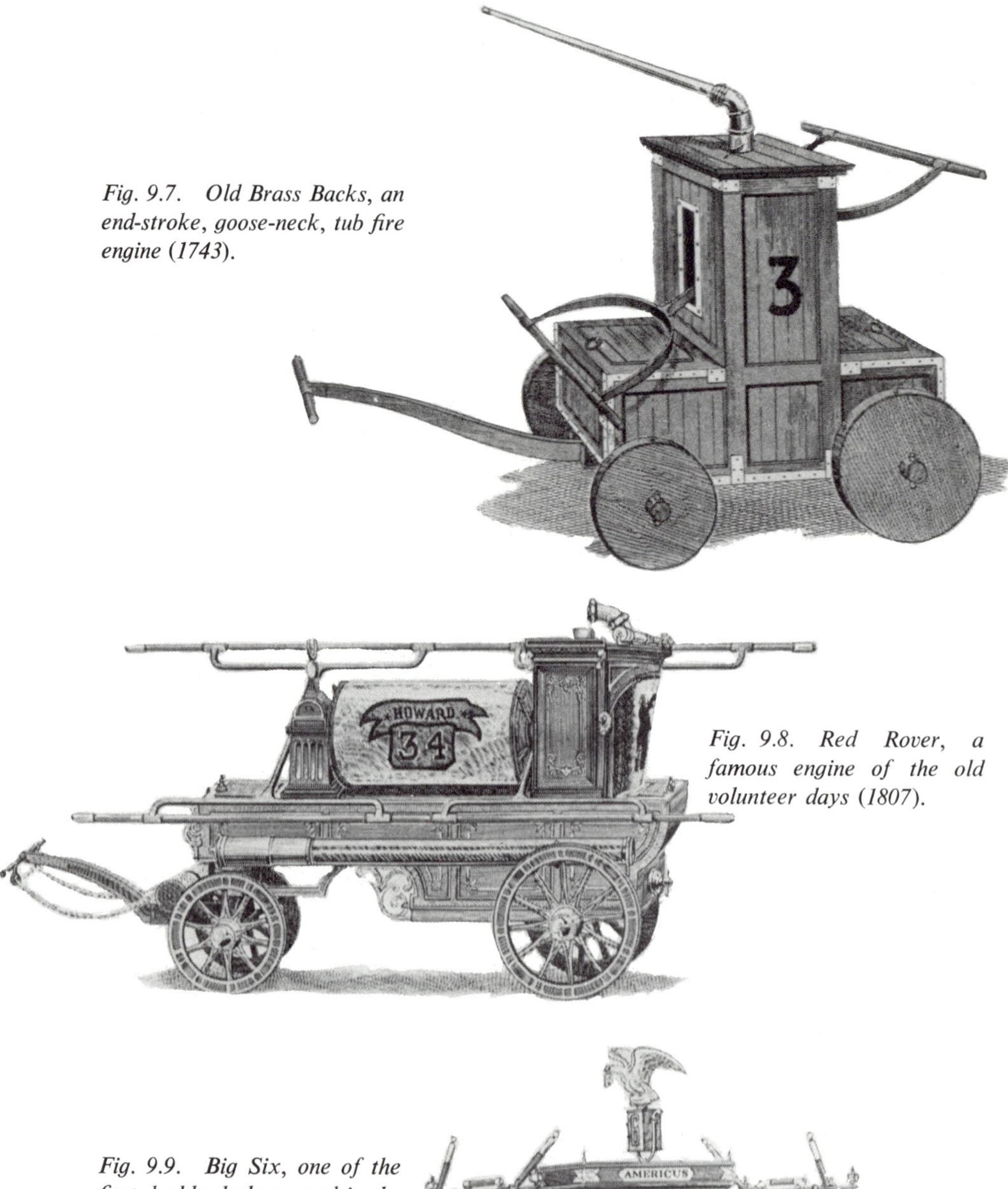

Fig. 9.7. Old Brass Backs, an end-stroke, goose-neck, tub fire engine (1743).

Fig. 9.8. Red Rover, a famous engine of the old volunteer days (1807).

Fig. 9.9. Big Six, one of the first double deckers used in the New York Department (1824).

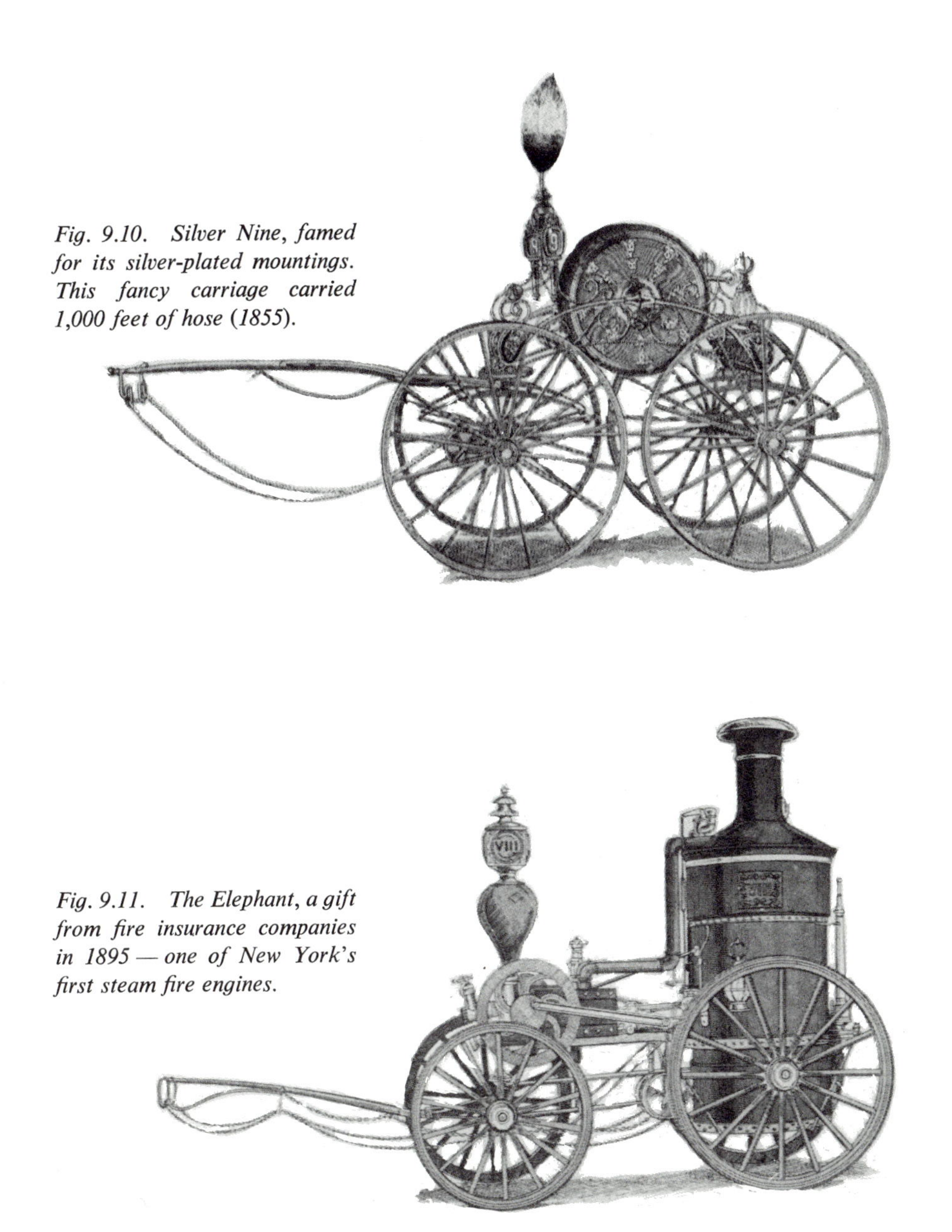

Fig. 9.10. Silver Nine, famed for its silver-plated mountings. This fancy carriage carried 1,000 feet of hose (1855).

Fig. 9.11. The Elephant, a gift from fire insurance companies in 1895 — one of New York's first steam fire engines.

Figs. 9.7–9.11. The illustrations on these two pages depict some of the early methods of carrying water to the site of a fire. These drawings of "fire hoses" and "fire engines" were presented in a monograph that is now out-of-print. However, the originals may be seen in the H.V. Smith Museum of the Home Insurance Company. (From "The Evolution of Fire Hose," *The Boyle Forum*, December 15, 1952) common substances with water; it shows that of all the common liquids, water is the most efficient heat-removing liquid.

Following is an interesting calculation of the heat-removing capacity of a gallon (3.8 litres) of water weighing 8.33 pounds (3.78 kilograms) at 70°F (21.1°C) when it is transformed into steam in a hot fire:

$$\begin{aligned} \text{(Specific Heat Calc.) } 212^\circ - 70^\circ &= 142^\circ \\ 142 \times 8.33 &= 1{,}183 \text{ Btu} \\ \text{(Heat of Vapor Calc.) } 970.3 \times 8.33 &= 8{,}083 \text{ Btu} \\ \text{Total: } & \quad 9{,}266 \text{ Btu/gallon} \\ & \quad (2{,}570 \text{ kJ/litre}). \end{aligned}$$

It follows that a 100 gpm (378.5 litre per minute) water-nozzle pattern directed into a hot fire can remove 926,600 Btu per minute (976,636 kiloJoules per minute). (Contrast this with the fact that the complete combustion of a gallon of gasoline only produces 130,000 Btu.)

There are many other important and valuable characteristics possessed by water, such as:

1. Its solvent ability, which dissolves and washes away many products of combustion (ash) so that the heart of combustion or seat of the fire may be reached.

2. Its small change in viscosity with temperature, so that it may be easily pumped and conducted in hose and pipes from 34°F (1°C) to as high as 210°F (99°C).

3. Its property of flashing into large volumes (an expansion of about 1,700 times the liquid volume) of inerting steam upon exposure to temperatures above its boiling point, thus displacing air and oxygen in the combustion atmosphere (perhaps temporarily).

4. Its rather high surface tension at ordinary temperatures (72 dynes per centimetre), enabling it to issue in a consolidated stream or in discrete water droplets from so-called "fog" nozzles or spraying devices.

5. Its reasonably high density, which confers physical mass to a projected nozzle stream for penetration purposes.

6. Its molecular stability, which enables it to avoid breakdown or dissociation to any appreciable amount in temperatures approaching 3,000°F (1,650°C)—which are higher than ordinary flame matrix temperatures.

The Many Application Methods for Water

The versatility of water for extinguishing fire has been realized since the discovery of fire. Since the discovery of fire, thousands of ways have been devised to easily and effectively bring water to an unwanted fire. Although it is not the purpose of this text to delve into the history of fire protection, it is noteworthy to mention that some of the early means of bringing water to the site of unwanted fires by means of "fire hose" and "fire engines" were presented pictorially in a now "out-of-print" monograph published in 1952.[3] These means are presented as Figures 9.7, 9.8, 9.9, 9.10, and 9.11.

[3] "The Evolution of Fire Hose," *The Boyle Forum*, No. 61, December 15, 1952.

This section will not attempt to cover in detail the entire number of water application devices that are in use today; rather, it will explain some of the operational and fundamental design characteristics of these devices, together with some explanation of their utility for fire protection situations.

Hose Nozzles and Their Variations

Early in our civilization it was found that the use of a smooth tube supplied with water that is pressurized by some means was advantageous for projecting water streams. As a means for delivering water to a distant target, the modern smooth, slightly cone-shaped nozzle has changed little over the years. They are sized according to the internal diameter of the tip of the nozzle tube for their maximum efficient water-delivering capacity at normally available pressures supplied to the nozzle. The hydraulic characteristics of ordinary smooth, conical nozzles and much of the fundamental data now employed in hydraulic work in fire protection were originally developed in a series of extensive investigations made by John R. Freeman in 1888 and 1889.[4] Some of the results of Freeman's measurements on the characteristics of water streams projected from an ordinary nozzle are shown in Figure 9.12.

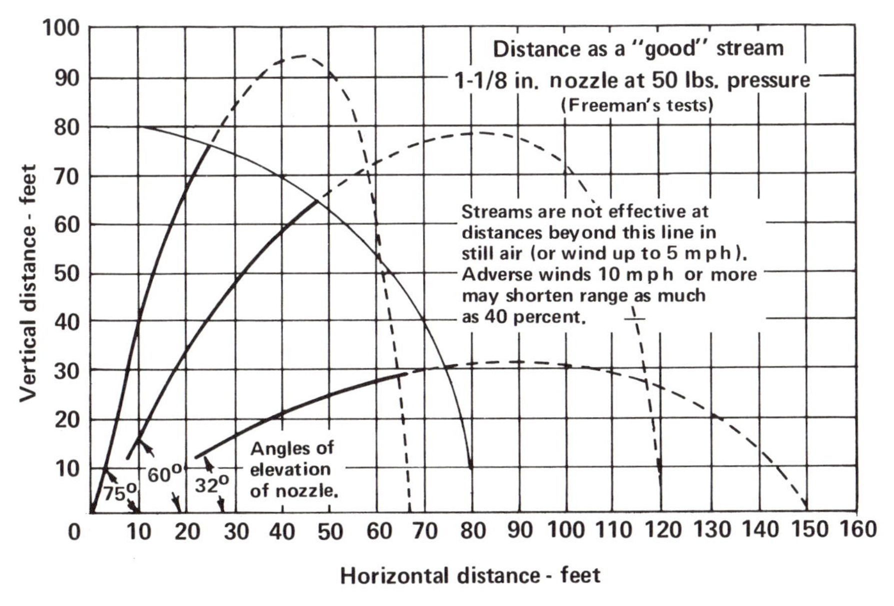

Fig. 9.12. Effective range of a fire stream from a $1\frac{1}{8}$-inch ordinary nozzle. (From *Fire Protection Handbook*, NFPA)

Enlargements in size of the ordinary solid-stream water nozzle are employed in large-scale fire fighting procedures in the form of self-supporting frame standards.

[4] Freeman, J. R., *Transactions of the Society of Civil Engineers*, Vols. XII and XXIV.

These "monitor" nozzles may be strategically placed for unattended operation for purposes of delivering large volumes of water on a burning area.

In the 1930s it was found that far greater efficiencies of water application could be achieved if the water stream was broken up into a wide spray or sprinkler pattern. However, fire fighting operations also required that the projection of a solid stream of water was often needed in addition to the wide coverage pattern of spray. Thus, two types of variable-pattern portable water hoze nozzles were developed. Both of these types afforded the operator a choice of water pattern types. The familiar garden hose "ring" type nozzle was adapted to large water flow rates for adjustable operation to give a "spoiled" solid stream operation and, by merely adjusting the protrusion of the center orifice place clearance, a wide, hollow-cone spray could be selected. Figure 9.13 illustrates the modern design of such an adjustable-pattern portable hose nozzle.

Fig. 9.13. An adjustable portable water nozzle.

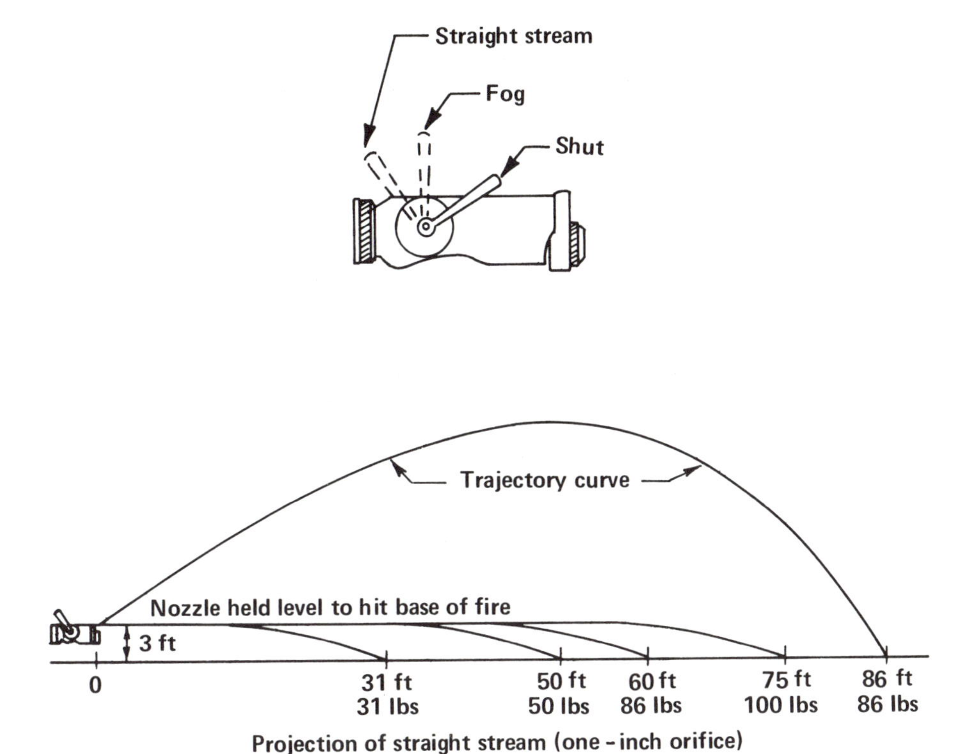

Fig. 9.14. The "All-Purpose" water nozzle and its operating characteristics.

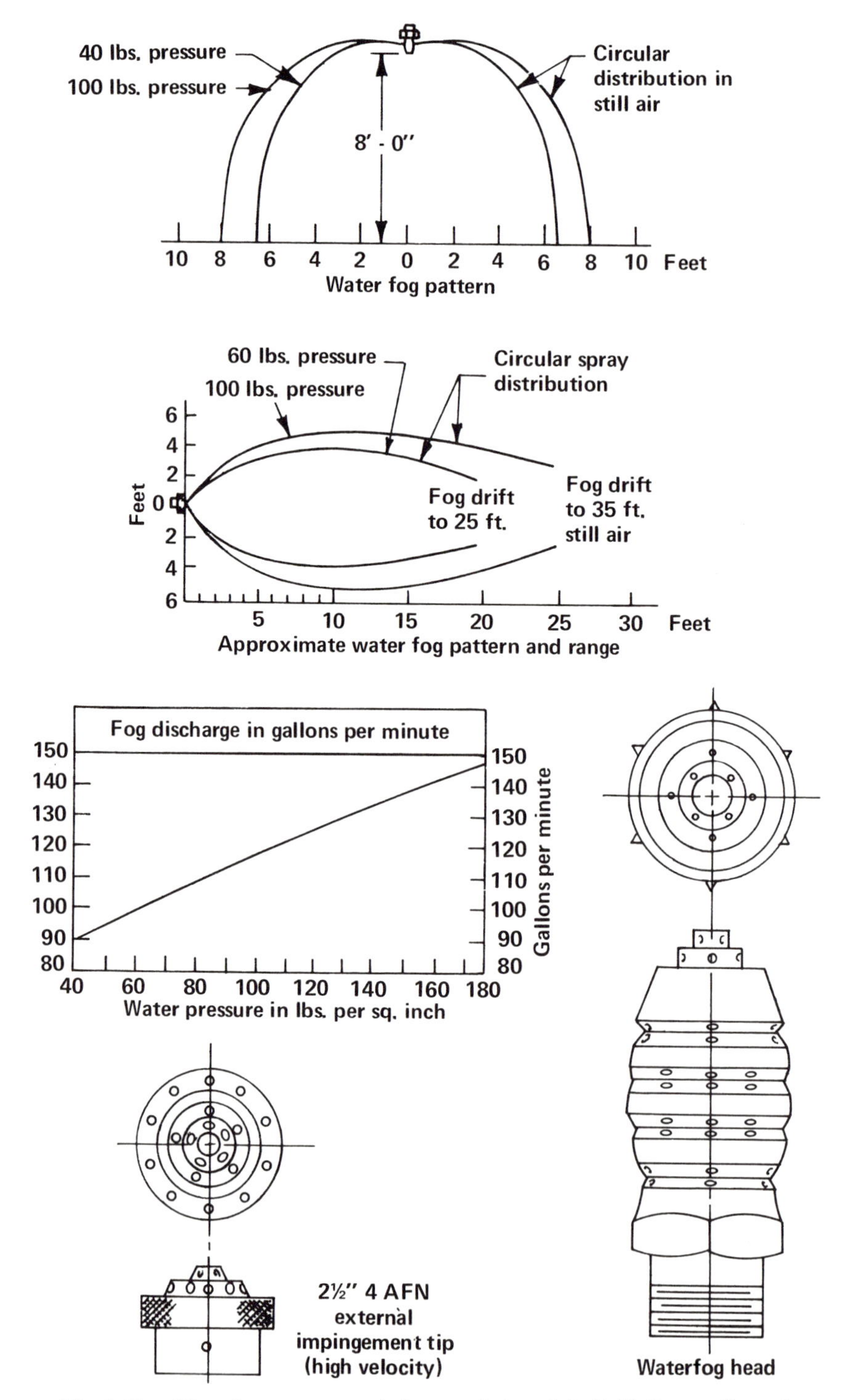

Fig. 9.15. Waterfog patterns and characteristics of the "All-Purpose" nozzle.

At about this same time, another type of adjustable portable hose nozzle was developed with separate waterways. This nozzle delivered water in two ways: (1) either by a well-consolidated stream for projection purposes, or (2) by manipulation of the valve, water was shunted to an "impinging" orifice plate or "head" so that a filled cone spray, or "fog," water pattern was obtained. The details of design of this nozzle are illustrated in Figures 9.14 and 9.15. The "externally" impinging orifice head holder of this "all-purpose" nozzle was also equipped to accommodate an easily substituted "applicator tube" that terminated in an "internally" impinging orifice head (the "L-11") for purposes of providing a wide, bushy "waterfog" pattern. Such a pattern is useful for protection of personnel from flame radiation during fire fighting activities and for full-pattern distribution of water particles. The latter water pattern also shows great effectiveness in the extinguishment of high flash point hydrocarbon fuels.

Other "Water Particle" Producing Devices

The elucidation of the facts surrounding the superior absorption of heat by water in the form of small particles (so-called fog) resulted in the development of a large number of devices for breaking up a pressurized stream of water into discrete particles. These "heads," or nozzles, utilize the kinetic energy and velocity of the water to cause it to shear, or tear, into droplets.

Some designs incorporate a number of orifices placed geometrically to enable the issuing streams to impinge on each other at angles. Maximum velocity of the streams is brought to bear at a point within the envelope of the head, resulting in almost complete dissipation of kinetic energy or velocity in the impingement process. Fine particle sizes of water are produced by this action; however, because their forward movement is nearly lost, they are subject to gravity and wind forces.

Other impingement designs vary the angle of collision of streams from opposing orifices so that some forward velocity of the sheared droplets is retained and a projection of the pattern of water particles is realized. The former impingement methods are loosely called "internal impingement," and the particles that are *projected* are said to be the result of "external impingement" (or "high velocity") designs. (See Figs. 9.14 and 9.15.)

Many spray and small water-particle-producing devices depend upon the ability of water to break into small particles by the sudden release at high velocity of a cohesive thin layer of water into air, causing a shearing action. Spiral passages in the body of some spray devices incorporate centrifugal action to increase the shearing action, as shown in Figure 9.16.

Another type of water spray device employs a spiral of diminishing diameter in the path of a water jet so that a thin layer of water is peeled from the stream, which then breaks up into small particles. This design has the advantage of employing only a single water passage of fairly large diameter, thus preventing clogging by small foreign particles. (See Fig. 9.17.) Other types of water spray nozzles may be constructed along the lines of the standard sprinkler head, where a water jet is delivered to impact a deflector and thus break into small particles.

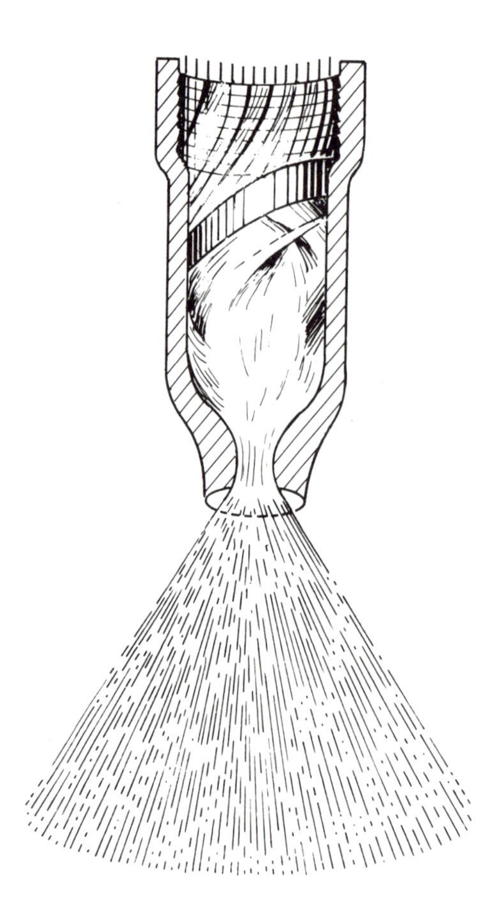

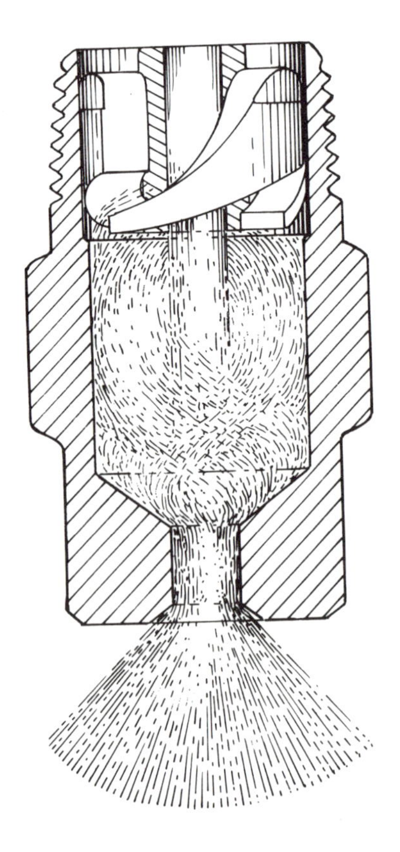

Fig. 9.16. Water spray nozzles having internal spiral water passages. (From *Fire Protection Handbook*, NFPA)

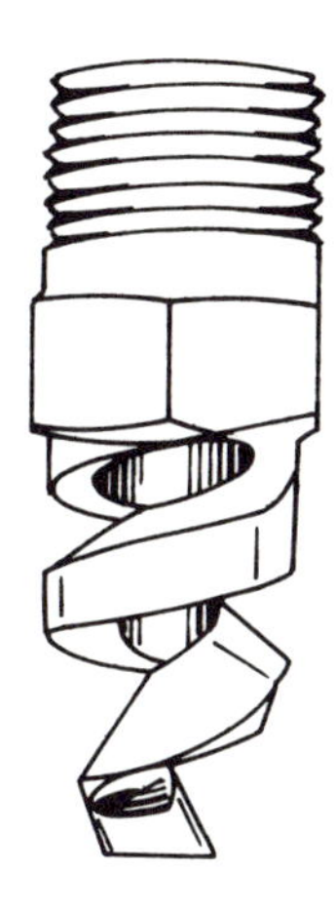

Fig. 9.17. A spiral-type water spray nozzle. (From *Fire Protection Handbook*, NFPA)

When water is caused to break up into small particles by the use of any of the previously described devices or by some other means, it is important to remember that the size of the resulting spray or "fog" particles will vary considerably, depending upon the method of shear or the impingement of stream used. When water streams are directed at high velocity at a target at 90 degrees (approximately) to the axis of the stream (the target may be an opposing stream), breakup of the water attains a maximum amount and a large number of small water droplets are obtained. When water is caused to strike a target at greater angles, larger particles are obtained and some velocity of the stream is preserved so that forward projection of the particles is possible.

Sprinkler Head Nozzles

One of the most widely used and highly effective methods of employment of water for fire protection is the sprinkler system. Sprinkler systems can be installed in most types of occupancies. These systems employ water in spray form, and are so arranged and permanently placed that the area covered by spray from each nozzle or head is cooled. Any fire in that area is controlled and extinguished without human intervention as long as a water supply continues.

In almost all cases, sprinkler systems are automatic in operation, detecting the level of heat developed in the area of each sprinkler head and releasing water spray into that area alone. The automatic sprinkler is unique in this characteristic; it provides its own detector, control valve, and water application pattern, and is not subject to smoke obscuration, toxic gas ingestion, or long time lapse before application of an extinguishing agent after fire detection.

The first perforated pipe sprinkler systems were installed in the United States around 1852. In 1874 Henry S. Parmelee of New Haven, Connecticut, patented the first practical automatic sprinkler head. Parmelee further improved his sprinkler heads, and in 1878 the perforated head was replaced by a rotating slotted turbine. In its crude form, Parmelee's automatic sprinkler head consisted of a brass cap

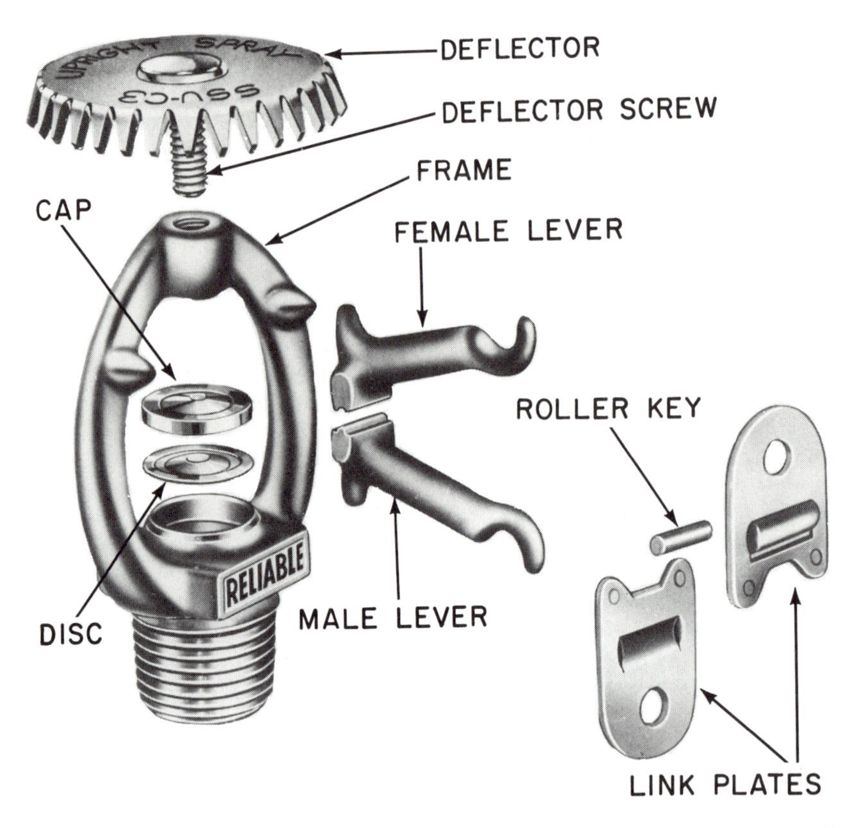

Fig. 9.18. A fusible link type automatic sprinkler. (From *Fire Protection Handbook*, NFPA)

lightly soldered over a perforated oval-shaped distributor. When flames melted the solder, the cap was removed and the water issued from its perforations to about a 270-degree occluded angle.[5]

Many improved designs have been developed for automatic sprinklers since the Parmelee unit. To a large extent, modern devices utilize mechanical arms and linkages to retain the nozzle cap in a closed position. A small amount of a soldering alloy (composed of tin, lead, cadmium, and bismuth), which melts at sharply defined temperatures, is used to maintain the mechanical structure. When the eutectic alloy melts, the structure collapses and opens the nozzle. This construction is detailed in Figure 9.18. The forces exerted on the various components of a sprinkler are shown in Figure 9.19.

There are a number of proprietary designs for automatic sprinkler heads, and there are a variety of designs for special constructions such as window and cornice protection. In general, sprinklers are designed to discharge 15 gpm (56.7 litres/minute) at 7 psi (0.48 bar) to cover a floor area of somewhat more than 100 square feet (9.29 m^2) using an 0.5-inch (12.5-millimetre) orifice. Ordinary sprinklers are

[5] Bryan, John L., *Automatic Sprinkler & Standpipe Systems*, National Fire Protection Association, Boston, 1976, pp. 56–58.

supplied for operation at 135°F (58.5°C), 150°F (62.8°C), 160°F (65.5°C), and 165°F (72.5°C) when a maximum ceiling temperature of 100°F (38°C) is reached. Higher operating temperature sprinklers are available up to 500°F (260.2°C) for special application situations where maximum ceiling temperatures up to 475°F (247°C) are reached.

Automatic sprinkler systems are installed in two basic designs: (1) dry systems having no water in the piping in order to avoid freezing in unattended and unheated spaces (air under pressure is used, which is immediately followed by water if a fire opens a head), and (2) standard wet systems where the piping is filled with water under pressure. The design and maintenance of sprinkler systems is complicated and requires strict adherence to code requirements.

Variations in the Properties of Water for Special Purposes

Water, our most useful and efficient heat-removing agent for fire protection purposes, possesses some physical qualities that restrict its use under certain conditions. These qualities are:

- Water freezes at temperatures frequently found in temperate climates and, as a result, cannot be transported to a fire.
- Water possesses a relatively high surface tension, which restricts it from penetrating and wetting various combustibles where this action is vitally needed for deep-seated fire extinguishment.

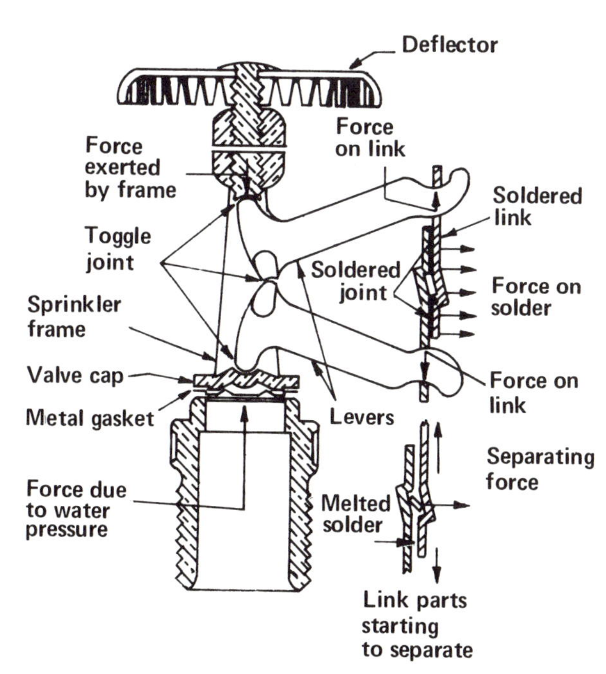

Fig. 9.19 Representative arrangement of the operating parts of a soldered-link automatic sprinkler. (From *Fire Protection Handbook*, NFPA)

- Water's viscosity is relatively low so that it "runs off" readily from a non-horizontal surface, leaving a very thin coating of heat-removing or fire-protecting liquid.
- Water demonstrates a problem of high friction loss in transportation through pipes and hose.
- Water's density and immiscibility with hydrocarbon fuels result in loss of material for heat-removal functions by sinking into the fuel even when employed in fine particle sizes on fires in these fuels.

During the past forty years or so, chemists have succeeded in modifying the preceding qualities to the extent that water can now be employed in fire protection areas not heretofore possible.

Inorganic freezing point depressants are not new in their use for preventing the crystallization of water at low temperatures. However, for many years their major drawback was their corrosivity when used in water solutions contained in metal tanks or pipes. The discovery of the passivating action of the chromate radical has made it possible to store calcium chloride antifreeze solutions for fire protection purposes in metal containers without attack by corrosion during storage. In Table 9.3 the amounts of calcium chloride (in 94 percent pelleted grade) are given for the prevention of freezing of water at the stated temperatures.

Table 9.3 Amounts of Calcium Chloride (94 percent) Needed for Freezing Point Depression of Water

FREEZING POINT		CALCIUM CHLORIDE AMOUNTS*			HYDROMETER READING
°F	°C	POUNDS/GAL	LBS-OZ/GAL	GRAMS/LITER	AT 60/60
30.	−1.1	0.21	0–3.3	25.1	1.021
25.	−3.9	0.73	0–11.6	87.4	1.066
20.	−6.7	1.11	1–1.6	132.9	1.102
15.	−9.4	1.46	1–7.3	174.9	1.128
10.	−12.2	1.74	1–11.8	208.4	1.150
5.	−15.0	1.97	1–15.5	236.0	1.168
0.	−17.8	2.17	2–2.7	260.0	1.183
−5	−20.5	2.38	2–6.0	285.1	1.198
−10	−23.3	2.55	2–8.8	305.5	1.212
−15	−26.1	2.74	2–11.8	328.2	1.225
−20	−29.0	2.90	2–14.4	347.4	1.237
−30	−34.4	3.21	3–3.3	384.5	1.258
−40	−40.0	3.44	3–7.0	412.1	1.274

* Sodium chromate is added for corrosion protection to the extent of 0.5 percent of the calcium chloride used (For example: 0.5 percent of 2.17 lb = 0.17 oz).

Other antifreeze materials such as ethylene glycol, propylene glycol, and glycerine may be used for cold weather protection in wet pipe sprinkler systems under certain conditions that guarantee that the solutions of these chemicals cannot contaminate public water supplies. (These liquids are not recommended for use in portable extinguishers because of the flammability of the residual antifreeze liquid when evaporation of the water occurs in a fire.)

Modern wetting agents have been developed that lower the surface tension of water from its normal level of 73 dynes per centimetre to about 25 dynes per centimetre. This means that the strength of the surface "skin" (that part of a body or particle of water that is exposed to air) of water is dramatically lowered to a point where it is more easily broken or parted by a force acting upon it. When a liquid with a low surface tension comes in contact with a solid surface such as cotton cloth with tiny open spaces, the liquid breaks up and quickly "soaks in." A liquid with a high surface tension might remain on the surface of a cloth in the form of a tight droplet, and may refuse to allow its skin to part so that it could enter the tiny spaces in the cloth.

The effect of wetting agents dissolved in water is to increase the wetting and spreading power of ordinary water; hence, the term "wet water" has been given to these solutions. When wetting agent solutions contact materials such as cotton bales, mattresses, stacked hay or straw, or other tightly consolidated Class A fire hazard materials, the liquid quickly soaks into the material instead of being lost as "run-off." When fire has deeply penetrated these materials, "wet water" will reach the fire by capillary action more quickly and with better efficiency than can be obtained by ordinary water application.

Viscosity-enhancing materials for water have been developed that increase the viscosity and "thickness" of water so that it may be applied to surfaces that are almost vertical in layers up to several times that of plain water. Here again, the function of "run-off" of heat-removing water is diminished and the economy of water application to a fire or a fire hazard is extended.

There are two principal types of water "thickeners." One of these uses organic gel-producing substances to give higher viscosity water solutions to coat burning material or to prevent its ignition; another thickener utilizes bentonite, a powdered mineral that yields a water slurry, or with mixtures of ammonium phosphates and sulfates, also giving a thick slurry for application to fire hazards, particularly in forest fire incidents. The latter "slurry type" mixtures have the added property of retarding flames due to their "solids" content and the capability of ammonium salts to halt flaming and glowing ignition.

Flow property modifiers have recently been developed to lower the friction loss of water during transmission at high velocity through hose. When water is pumped through $1\frac{1}{2}$-inch (nominal 35-millimetre) hose at 100 gpm (378 litres/minute), the pressure loss in 100 feet (30.5 metres) of such hose will be 25 psi (1.72 bar). This pressure loss is due to two factors: (1) the loss due to friction between the water and the walls of the hose (which amounts to about 10 percent of the loss), and (2) the loss due to turbulent flow within the hose at high velocities of transit of the water (which amounts to about 90 percent of the pressure loss).

It has been discovered that dilute solutions of polyethylene oxide (a white powder soluble in water, each molecule of which is made up of thousands of small molecular units strung together in a long chain to form a giant molecule), will cause water to flow in a nonturbulent form (also called Newtonian Flow) in a tube such as a fire hose. The end result of dissolving one gallon (3.81 litres) of polyethylene oxide in 6,000 gallons (22,500 litres) of water is to achieve a 55 to 70 percent increase in flow in a 1½-inch (nominal 35-millimetre) hose so that this size hose delivers as much as 2½-inch (nominal 60-millimetre) size hose. The flow-property-modified solution also doubles the nozzle pressure at the end of a length of hose.

Another name for this polymeric additive for water is "rapid water." There are commercially available systems for injecting this additive in the form of a concentrated slurry into water by automatic means that sense flow rates and add the compound in ratios of 1 to 6,000. Solutions of this material facilitate the delivery of larger amounts of water to a fire hazard using smaller and more manageable hose sizes without sacrificing nozzle pressures or flow volumes.

The modification of the density of water for fire fighting purposes has been accomplished in two ways. One way involves the addition of air to water to form a semi-stable *air foam* that is lighter than almost all flammable and combustible liquids. (See the section of this chapter titled "Foams for Fighting Fires.") The other method involves adding to water an emulsifying agent that is capable of mixing with the top layer of a burning liquid to give a nonflammable floating emulsion of water and fuel.

When small quantities of synthetic detergents are added to water, the surface tension of the water is greatly lowered (as with wetting agents in water solution). The design and purpose of the detergent solutions are different in that they are chosen for their *emulsifying* action instead of for their purely wetting action. When dilute detergent solutions of this type are sprayed or are directed into flammable or combustible fuels, they mix rapidly with the fuel to produce a suspension of the liquid in the detergent solution; this lowers the vapor pressure of the fuel to the point that the amount of combustible vapor coming off is lower than the lower combustible limit in air, and the fuel will no longer burn.

Synthetic detergent emulsifying agents are used in percentages from 0.5 to 2.0 and yield a fuel-solution-air emulsion of lesser density than water or fuel. This emulsion may be very temporary; however, in fires involving high flash point fuels, only a short period of cessation of combustion is sufficient to halt reignition.

FOAMS FOR FIGHTING FIRES

Fire fighting foam is a mass of gas-filled bubbles formed by various methods from aqueous solutions of specially formulated foaming agents. Since foam is lighter than the aqueous solutions from which it is formed and is lighter than flammable liquids, it floats on all flammable or combustible liquids, producing an air-excluding, cooling, continuous layer of vapor-sealing, water-bearing material for purposes of halting or preventing combustion.

Fire fighting foams are formulated in several ways for fire extinguishing action. Some foams are thick and viscous, forming tough heat-resistant blankets over burning liquid surfaces and vertical areas. Other foams are thinner and spread more rapidly. Some are capable of producing a vapor-sealing film of surface-active water solution on a liquid surface, and some are meant to be used as large volumes of wet gas cells for inundating surfaces and for filling cavities. There are various methods of generating and applying foams.

The use of foam for fire protection requires attention to its general characteristics. Foam breaks down and vaporizes its water content under attack by heat and flame. Therefore, it must be applied to a burning surface in sufficient volume and rate to compensate for this loss and to provide an additional amount to guarantee a residual foam layer over the extinguished portion of the burning liquid. Foam is an unstable "air-water emulsion" and may be easily broken down by physical or mechanical forces. Certain chemical vapors or fluids may also quickly destroy foam. When certain other extinguishing agents are used in conjunction with foam, severe breakdown of the foam may occur. Turbulent air or violently uprising combustion gases from fires may divert light foam from the burning area.

THE TYPES OF FOAM CONCENTRATES AND THEIR FOAM CHARACTERISTICS

Protein Foaming Agents

Protein type air foams utilize aqueous liquid concentrates proportioned with water for their generation. These concentrates contain high molecular weight, natural proteinaceous polymers derived from a chemical digestion, and a hydrolysis of natural protein solids. The polymers give elasticity, mechanical strength, and water retention capability to the foams that are generated from them. The concentrates also contain dissolved polyvalent metallic salts that aid the natural protein polymers in their bubble-strengthening capability when the foam is exposed to heat and flame. Organic solvents are added to the concentrates to improve their foamability and foam uniformity as well as to control their viscosity at lowered temperatures. Protein-type concentrates are available for proportioning to a final concentration of either three percent or six percent by volume using either fresh water or sea water. In most instances, these protein-type concentrates produce dense, viscous foams of high stability, high heat resistance, and better resistance to burnback than many other foaming agents.

Fluoroprotein Foaming Agents

While the concentrates utilized for generating fluoroprotein foams are similar in composition to protein foam concentrates, they contain, in addition to natural protein polymers, fluorinated surface active agents that confer a "fuel shedding" property to the foam generated. This makes them particularly effective for fire fighting conditions where the foam becomes coated with fuel, such as in the method of subsurface injection of foam for tank fire fighting, and nozzle or monitor foam applications where the foam may often be plunged into the fuel. Fluoroprotein foams are more effective for in-depth petroleum or hydrocarbon fuel fires than some other foam agents because of this property of "fuel shedding." In addition, these foams demonstrate better compatibility with dry chemical agents than do the protein type foams. Fluoroprotein type concentrates are available for proportioning to a final concentration of either three percent or six percent by volume using either fresh water or sea water.

Aqueous Film-forming Foaming Agents (AFFF)

Aqueous film-forming foam agents are composed of synthetically produced materials that form air foams similar to those produced by the protein-based materials. In addition, these foaming agents are capable of forming water solution films on the surface of flammable liquids; hence, the term "aqueous film-forming foam" (AFFF). AFFF concentrates are available for proportioning to a final concentration of either three percent or six percent by volume with either fresh or sea water.

The air foams generated from AFFF solutions possess low viscosity, have fast spreading and leveling characteristics, and act as surface barriers to exclude air and to halt fuel vaporization just as other foams do. These foams also develop a continuous aqueous layer of solution under the foam. The surface activity of this solution maintains a floating film on hydrocarbon fuel surfaces that helps suppress combustible vapors and cools the fuel substrate. The film, which can also spread over fuel surfaces not fully covered with foam, is self-healing following mechanical disruption; it continues as long as there remains a reservoir of nearby foam for its production. An illustration of the comparison of a steel needle floating on water vs. the mechanism of film formation on a fuel surface by aqueous film-forming foams is shown in Figure 9.20. These both demonstrate surface force phenomena.

The layer of solution draining from the foam mass and floating on the fuel surface has a density equal to ordinary water; it should sink in the liquid fuel below it, which has a lesser density. It does *not* sink until it reaches a limiting thickness (when drops form, it will finally sink). The reasons for this are because of the surface tensions of the liquid solution and of the fuel, and because of the interfacial tension between these two liquids. Also, the property of "amphoteric solubility" of the surfactant in the solution promotes closely packed film formation to a degree that flammable vapors evolving from the fuel cannot penetrate its cohesive character.

Note that in Figure 9.20, the surface tension forces at the air interface with the water act to provide a "skin" on the water capable of supporting the steel needle where it contacts the surface. (A steel razor blade can produce the same effect.) In the lower half of Figure 9.20, the "skin" of a fuel with a density of 0.8 (whether burning or not) supports a thin layer (about 0.0005-inch [.0127-millimetres] thick) of AFFF solution with a density of 1.0 in the same manner as the steel needle is supported. In this instance there also exists interfacial surface tension forces between the two liquids, which bear a certain relationship to the surface tension forces of the single liquids. In order for spreading to occur, it has been found that all of these forces must result in a positive value according to the following relationship:

$$S = ST\ (\text{fuel}) - ST(\text{AFFF solution}) + IT\ (\text{fuel and solution})$$

where: S is the spreading coefficient

ST is the surface tension value of the respective liquids

IT is the interfacial tension value between the two liquids.

Ordinary petroleum fuels demonstrate surface tensions around 24 dynes per centimetre. Interfacial tension between AFFF solution and fuels varies between

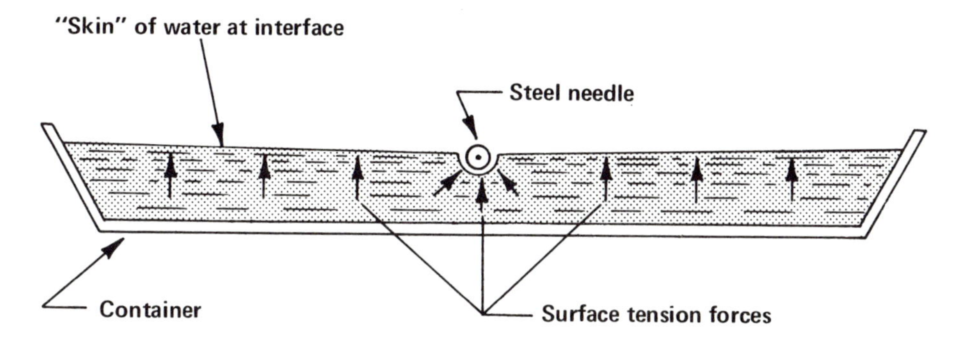

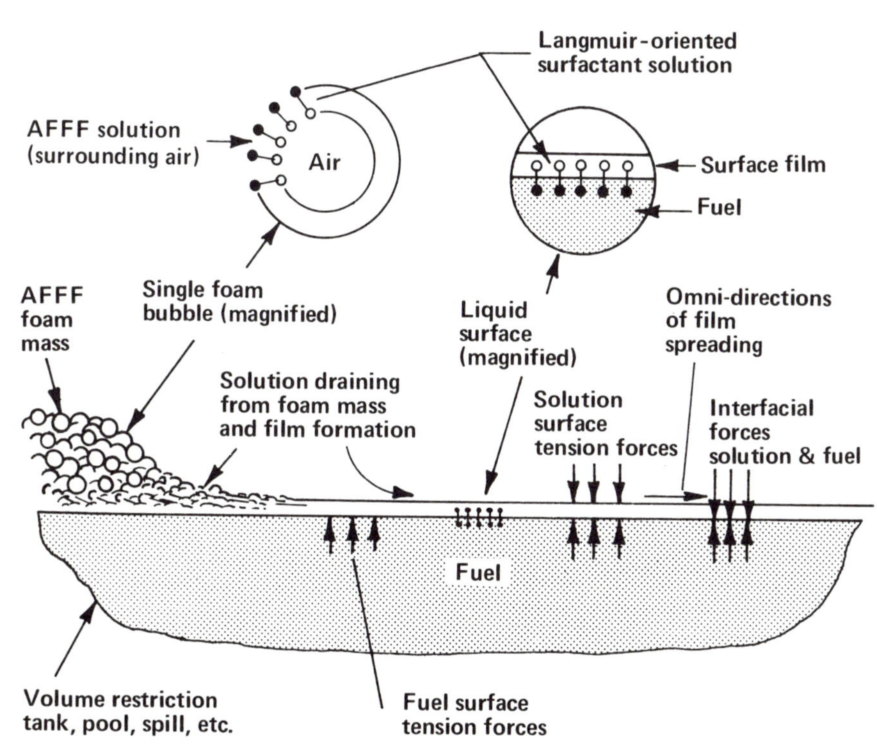

Fig. 9.20. A magnified graphic representation of a steel needle floating on water vs. the mechanism of film formation of a fuel surface by aqueous film-forming foams.

about 1.0 and 6.0 dynes per centimetre. The surface tension of AFFF solutions is about 16.0 dynes per centimetre. (The ST of ordinary protein type foam solutions is about 45 dynes per centimetre, whereas the ST of water is 72 dynes per centimetre.)

In the super-enlarged sections of an AFFF foam bubble and the surface film floating on fuel, the solution is pictured with a succession of open and closed circles connected by a line. Scientist Irving Langmuir originated this concept of surface-active liquids (such as AFFF solution) in order to indicate the way in which the molecules of a surface-active agent orient in water solution. Langmuir postulated that since all surface-active agents consist of two types of compounds in each molecule, one of which is more soluble in water, then the compound could be visualized as a molecule with a head and a tail orienting in the closely packed form visualized in Figure 9.20. The surface-active agent in AFFF solutions is similar in chemical composition to the following organic compound:

$$CF_3(CF_2)_7 \vdots SO_2NH(CH_2)_3N(CH_3)_3I.$$

(This is named: perfluoro-alkylsulfonamide nitrogen-alky trimethyl ammonium iodide.)

In the preceding formula, a dotted vertical line represents the division of the compound into the oil-attracted part on the left and the water-attracted and soluble part on the right. The result of the double action of aqueous film-forming foams is to yield a highly efficient foam extinguishing agent in terms of water and concentrate needed, and the rapidity with which it acts on fuel spills.

All AFFF concentrates contain fluorinated, long-chain hydrocarbons with these particular surface-active properties. Various water-soluble, high-molecular synthetic polymers are added to aid in strengthening the bubble wall and in retarding breakdown. They are nontoxic and biodegradable after dilution. The shelf life of AFFF concentrate compares favorably with other synthetic foam concentrates containing no proteinaceous substances that might change with time.

AFFF can be used as a foam-cover and protecting material for flammable liquids that have *not become* ignited. In some circumstances it may be used for extinguishment of certain water-soluble polar solvents. Because of the extremely low surface tension of the solutions draining from AFFF, these foams may be useful under mixed-class fire situations (Class A and Class B) where deep penetration of water is needed in addition to the surface-spreading action of foam itself.

Foam-generating devices yielding highly stable, homogeneous foams are not necessarily needed in the employment of AFFF. Less sophisticated foaming devices may be used because of the inherent rapid and easy-foaming capability of AFFF solutions. Water spray devices may sometimes be used. AFFF also may be used in conjunction with dry chemical agents without compatibility problems. Although AFFF concentrates *must not* be mixed with other types of foam concentrates, foams made from them do not break down other foams in fire fighting operations.

Synthetic Hydrocarbon Surfactant Foaming Agents

There are many synthetically produced surface-active compounds that foam copiously in water solution. When these compounds are properly formulated, they may be used as fire fighting foams and may be employed in much the same manner as other types of foam.

Hydrocarbon surfactant foam liquid concentrates are employed in one to six percent proportions in water. When these solutions are used in conventional foam-making devices, the resulting air foam possesses low viscosity and fast-spreading qualities over liquid surfaces. Its fire fighting characteristics depend upon the volume of foam layer on the burning surface (which halts access to air and controls combustible vapor production), and upon the minor cooling effect of the water in the foam (which becomes available due to a relatively rapid breakdown of the foam mass). This water solution does *not* possess film-forming characteristics on the flammable liquid surface.

Synthetic hydrocarbon surfactant foams are generally less stable than other types of fire fighting foams. Their water solution content drains away rapidly, leaving a bubble mass that is highly vulnerable to heat or mechanical disruption. To help achieve extinction, these foams must be applied at higher rates than other fire fighting foams. Many formulations of this type of foam concentrate break down other foams if used simultaneously or sequentially.

"Alcohol-type" Foaming Agents

Air foams generated from ordinary agents are subject to rapid breakdown and loss of effectiveness when they are used on fires that involve fuels that are water soluble, water miscible, or of a "polar solvent" type. Examples of this type of liquid are: the alcohols, enamel and lacquer thinners, methyl ethyl ketone, acetone, isopropyl ether, acrylonitrile, ethyl and butyl acetate, and the amines and anhydrides. Even small amounts of these substances mixed with the common hydrocarbon fuels will cause the rapid breakdown of ordinary fire fighting foams.

Certain special foaming agents have, therefore, been developed. These agents are called "alcohol-type" concentrates. Some of these concentrates must be foamed and applied to the burning surface almost immediately after they are proportioned into water. Solutions of this type cannot be pumped long distances because their maximum "transit times" (the time required for foam solutions to travel from the eductor, or foam generator, to the discharge outlet), before foam application, are short. They would be ineffective if this time was to be exceeded.

High Expansion Foaming Agents

"High expansion" foam is an agent for control and extinguishment of Class A and Class B fires and is particularly suited as a flooding agent for use in confined spaces. The foam is an aggregation of bubbles mechanically generated by the passage of air or other gases through a net, a screen, or other porous medium that is wetted by an aqueous solution of surface-active foaming agents. Under proper conditions, fire fighting foams of expansions from 100 to 1 (100) up to 1,000 to 1 (1,000) can be generated.

High expansion foam is a unique vehicle for transporting wet foam masses to inaccessible places, for total flooding of confined spaces, and for volumetric displacement of vapor, heat, and smoke. Tests have shown that under certain circumstances high expansion foam, when used in conjunction with water from automatic sprinklers, will provide more positive control and extinguishment than will either extinguishing agent by itself. (High-piled storage of rolled paper stock is an example.) Optimum efficiency in any one type of hazard is dependent upon the rate of application and the foam expansion and stability.

Liquid concentrates for producing high expansion foams consist of synthetic hydrocarbon surfactants of a type that will foam copiously with a small input of turbulent action. They are usually used in about two percent proportion in water solution.

High expansion foam is particularly suited for indoor fires in confined spaces. Its use outdoors may be limited because of the effects of weather. High expansion foam has several effects on fires:

- When generated in sufficient volume it can prevent air, necessary for continued combustion, from reaching the fire.
- When forced into the heat of a fire, the water in the foam is converted to steam, reducing the oxygen concentration by dilution of the air.
- The conversion of the water to steam absorbs heat from the burning fuel. Any hot object exposed to the foam will continue the process of breaking down the foam, converting the water to steam, and of being cooled.
- Because of its relatively low surface tension, solution from the foam that is not converted to steam will tend to penetrate Class A materials. However, deep-seated fires may require overhaul.
- When accumulated in depth, high expansion foam can provide an insulating barrier for protection of exposed materials or structures not involved in a fire, thereby preventing fire spread.

Research has shown that using air from inside a burning building for generating high expansion foam has an adverse effect on the volume and stability of the foam produced. Combustion and pyrolysis products can reduce the volume of foam produced and can increase the drainage rate when they react chemically with the foaming agent. The high temperature of the air breaks down the foam as it is being generated. Physical disruption, apparently caused by vapor and by solid particles from the combustion process, also takes place. Each of these factors causes foam breakdown, and each may be compensated for by higher rates of foam generation.

Foam that is generated from the gases of combustion becomes toxic, and entry into a foam-filled passage must not be attempted without self-contained breathing apparatus. The foam mass also obscures vision, and life-lines must be used when entering such a passage.

Chemical Foam Agents

These foam-producing materials have become obsolete because of the superior economics and ease of handling of the liquid foam-forming concentrates previously discussed. Chemical foam is formed from the temperature-sensitive chemical reaction in aqueous solution between aluminum sulfate ("A") (acidic) and sodium bicarbonate ("B") (basic), which also contains proteinaceous foam stabilizers. In wet systems, these chemicals are stored in solution in large separate tanks. In powder systems and in all portable methods for its use, the chemicals are added through "hoppers" into flowing water streams. The foam is formed by the generation of carbon dioxide gas from the chemical reaction of the two solutions of compounds. Although chemical foam is quite stable and heat resistant, it is generally stiff and slow-moving; it "bakes" under flame attack, and will form open fissures in the foam layer that exposes the underlying fuel. Chemical foam "fixed" systems require constant maintenance. Portable devices for use of such systems are difficult to operate at a fire.

CONCENTRATE PROPORTIONING

The process of producing and applying fire fighting air foams to hazards requires three separate operations, each of which consumes energy. They are: (1) the proportioning process, (2) the foam generation phase, and (3) the method of distribution.

In general practice, the functions of air foam generation and distribution take place nearly simultaneously within the same device. There are also many types of proportioning and foam-generating equipment.

In certain portable devices all three functions are combined into one device. The design and performance requirements of foam systems dictate the choice of types of proportioning, generating, and distributing equipment for the protection of specific hazards.

Foam Concentrate Proportioners

In order that a predetermined volume of liquid foam concentrate may be taken from its source and placed into a water stream to form a foam solution of fixed concentration, two general method classifications are made:

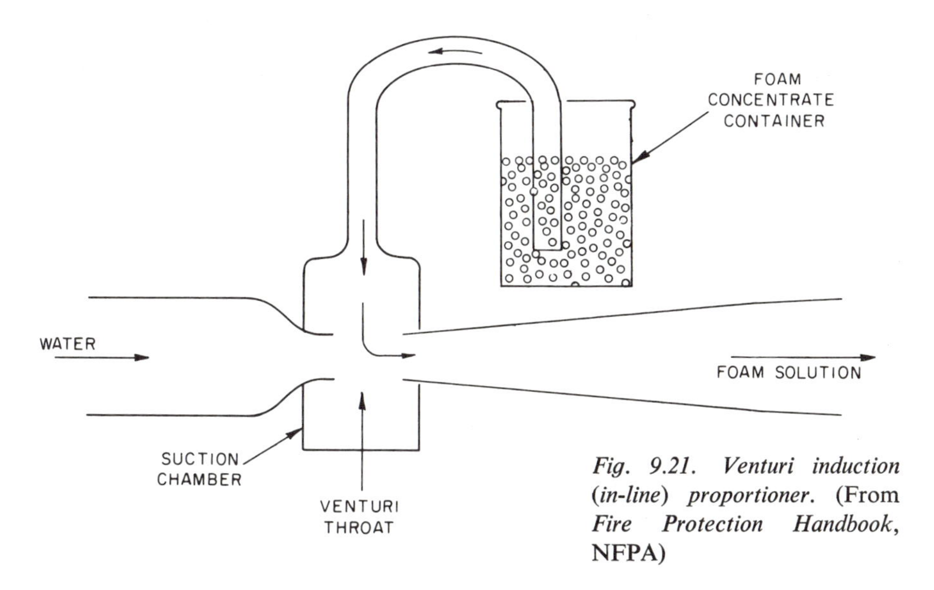

Fig. 9.21. Venturi induction (in-line) proportioner. (From *Fire Protection Handbook,* NFPA)

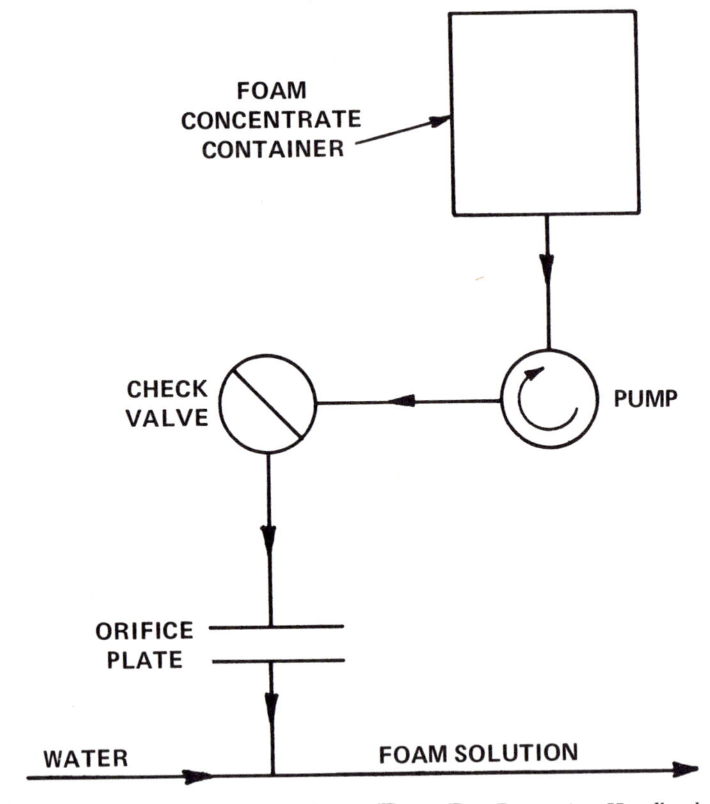

Fig. 9.22. Concentrate pump proportioner. (From *Fire Protection Handbook*, NFPA)

1. Methods that utilize the pressure energy of the water stream by venturi action and orifices to induct concentrate. (In general, such devices impose a 35 percent pressure drop on the water stream.)

2. Methods that utilize external pumps or pressure heads to inject concentrate into the water stream at a fixed ratio to flow.

Figures 9.21 and 9.22 illustrate the general principles of the two different methods of proportioning. These (and other) foam concentrate proportioning methods may be easily arranged to produce foam solutions of fixed concentrations of three percent or six percent by volume of the liquid foam concentrate in the water stream.

AIR FOAM GENERATING METHODS

Air foam nozzles, foam tubes, and foam makers are the devices that mix air with proportioned foam solution to form finished air foam for application to a hazard. The most widely used type of foam makers are those in which the foam is generated by inspiration of air into the device by means of a venturi nozzle. Air foam generators can be classified into five general categories, depending upon the amount of energy that is used in the foam-making process. These categories are:

1. Non-aspirating Methods: Devices such as water sprinklers or water spray nozzles produce a watery foam-froth. Use of such devices for foam production is confined to special applications because of the poor quality, sloppy foam that results. They do not actually inspirate air for foam production, but rely upon droplet collision and air turbulence for foam-froth production. The resulting froth is relatively unstable. Very little energy is required for this method of foam making, but effective AFFF foams may be produced in these devices.

2. Air Aspirating Foam Nozzles, Foam Tubes, and Ordinary Foam Makers: These constitute the majority of commercial foam-making devices for portable use or for fixed installation. They generate foam by venturi action and by aspiration of air into a turbulent foam solution stream. Figure 9.23 illustrates the basic principles of the air aspirating method of foam production.

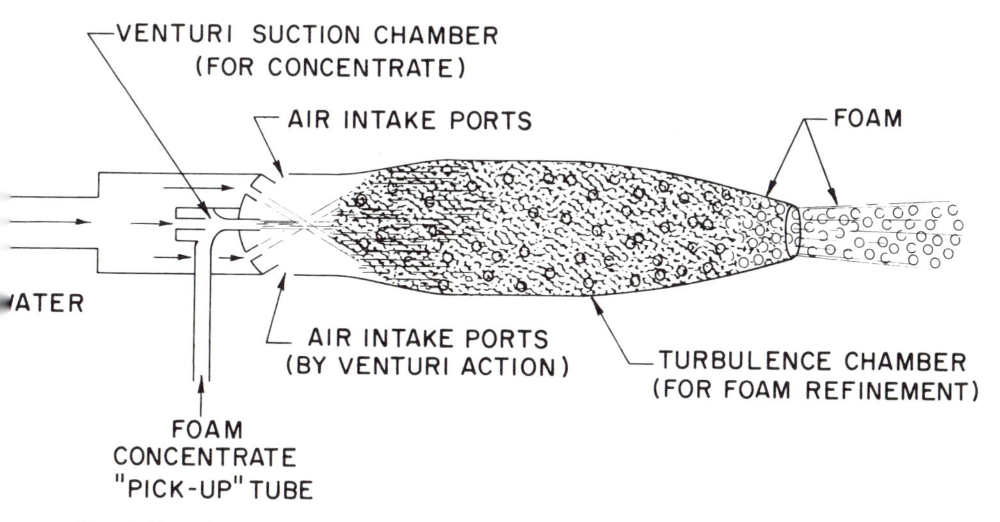

Fig. 9.23. Cross-section of an aspirating type foam maker with a concentrate "pick-up" tube. (From *Fire Protection Handbook*, NFPA)

Approximately 90 percent of the kinetic energy of the incoming foam solution is exhausted in this type of device, and the air foam issuing from it may be distributed only in ways that do not require large pressure differentials.

3. "High Back-Pressure" Foam Makers: These modified venturi devices are especially designed to conserve foam pressure energy. They operate at relatively high back-pressures for purposes of discharging foam through extended lengths of pipe or hose, and for subsurface injection of foam for fuel tank fire fighting. They are also useful for converting old chemical foam systems to air foam systems. Although they operate at higher pressures than ordinary venturi devices, the foam produced by "high back-pressure" foam makers retains some (approximately 25 percent) residual pressure. (See Fig. 9.24.)

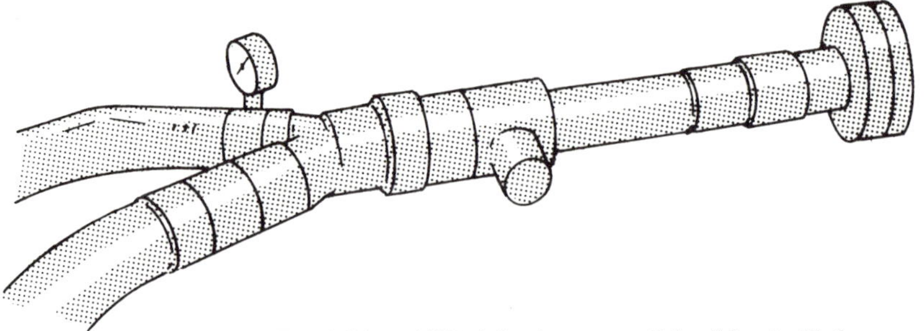

Fig. 9.24. A "high back-pressure" (or "forcing") foam maker. (From *Fire Protection Handbook*, NFPA)

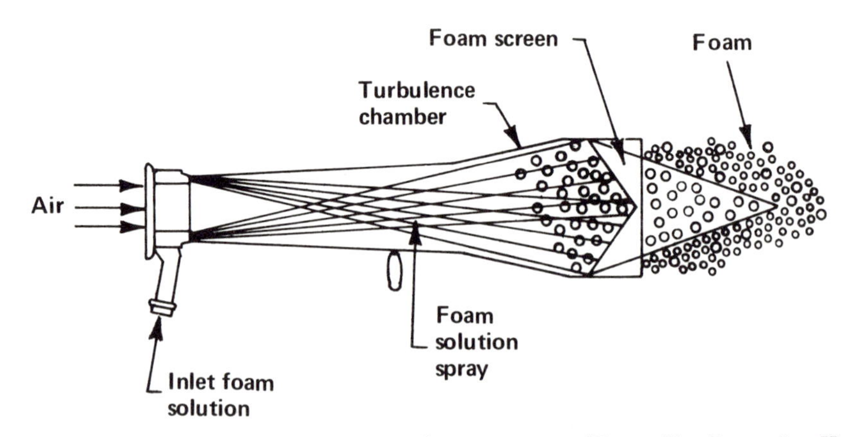

Fig. 9.25. Aspirating type high expansion foam generator. (From *Fire Protection Handbook*, NFPA)

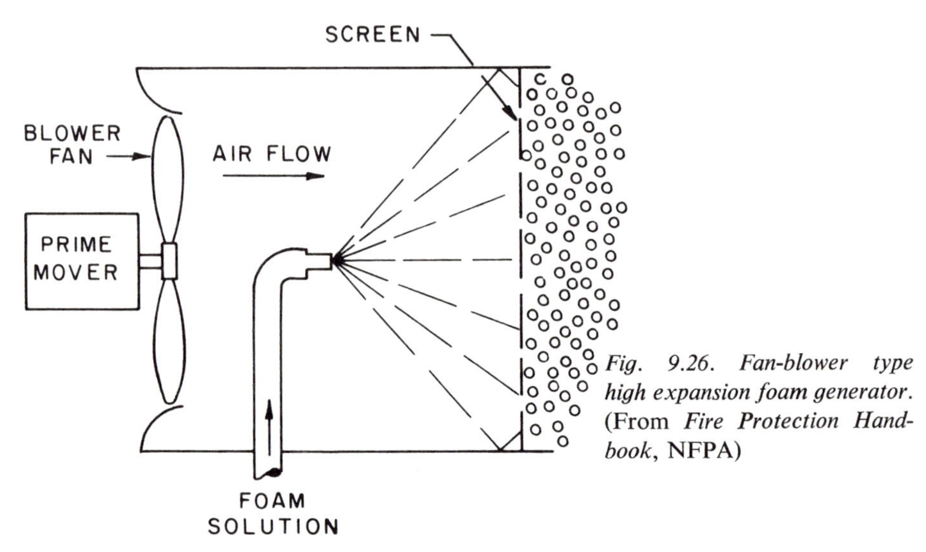

Fig. 9.26. Fan-blower type high expansion foam generator. (From *Fire Protection Handbook*, NFPA)

4. Pumped Foam Devices: In these systems compressed air is injected or pumped into the foam solution under pressure. They are inherently more expensive, and have limited commercial or industrial use. Pumped foam devices require additional increments of power in order to inject air, and the resulting homogeneous foam retains some kinetic energy.

5. High Expansion Foam Generating Devices: There are two principal methods used for the generation of this type of fire fighting foam. One of these utilizes a modified venturi action with very turbulent flow; the other requires input of energy to form the finished foam. The latter system results in high expansion foam containing enough residual kinetic energy to enable it to be forced through large tubes and passageways. Figures 9.25 and 9.26 illustrate the operating principles of high expansion foam generating devices.

Chemical Foam Generating Methods

Except for the generation of chemical foam from wet systems in which the simple mixing of the two solutions ("A" and "B" solutions) will produce foam, it is necessary to pour two chemical powders into a water stream in a fixed ratio to produce the foam. The "A" and "B" chemicals may be previously mixed and stored together in tightly sealed containers. Figure 9.27 illustrates the equipment used to generate foam using previously mixed "A" and "B" powders.

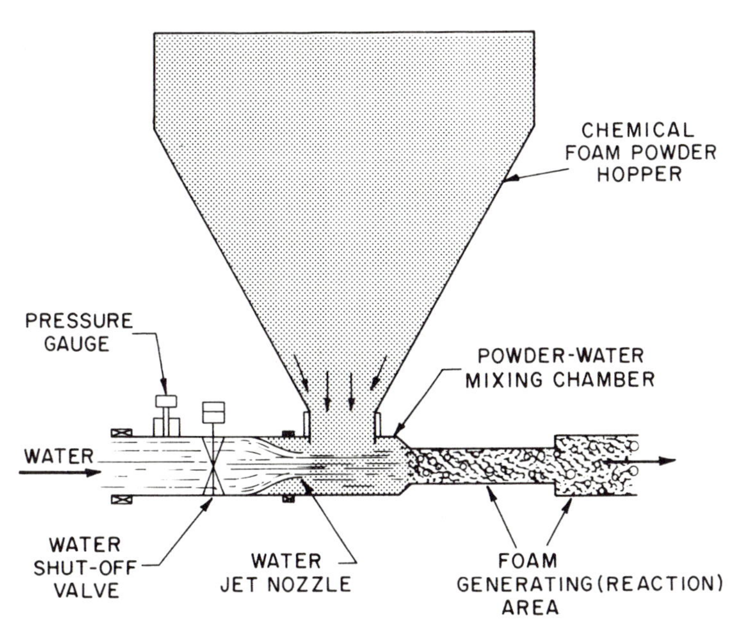

Fig. 9.27. A chemical foam hopper generator for a single (previously mixed "A" and "B") powder system. When dual powders are used, two hoppers are piped in parallel to a single water source. Dual piping is used from their outlets. (From *Fire Protection Handbook*, NFPA)

The admixture of solids into water must be done carefully in order to prevent clogging of the mixing throat. Back-up of water into the cone prevents free flow of the powders into the cone's apex. If this should occur — as often happens during fire fighting — the operation must be stopped, and all moisture must be removed from the walls of the device.

FOAM GENERATING EQUIPMENT AND SYSTEMS

The Hoseline Foam Nozzle

This is the most universally used portable, air-aspirating, foam fire fighting device for flammable liquid fire operations. It is manufactured in a variety of operating capacities for one-person handling. Supplied with foam solution from a proportioner or by means of a pickup tube, it successfully combats fires resulting from spills or leaks of flammable liquid or fires of average proportion in tanks or fuel pits. In order to provide the variety of foam stream patterns that may be needed for successful extinguishment, these nozzles usually contain built-in devices that allow continuous foam pattern variation from a solid straight stream to an inverted filled umbrella shape. (*N.B.:* As with water streams, foam is employed in: (1) a solid straight stream for range or reach, (2) a flat, wide, bushy shape for gentle "snowstorm" application on the burning fuel surface, and (3) a very wide, circular, bushy shape for radiance shielding of the operator during fire extinguishment or penetration into the fire area.) Larger capacity hose foam nozzles (250 gpm [946 litres per minute] and more) are designed for one- or two-person operation at a solid stream foam pattern only. (See Fig. 9.28.)

Fig. 9.28. Large capacity air foam hose nozzle. (From *Fire Protection Handbook*, NFPA)

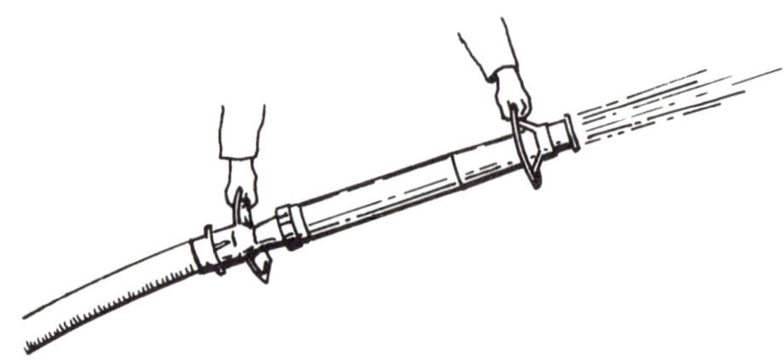

The "Fog-foam" Nozzle

Many situations in fuel fire extinguishment do not require foams of particular physical characteristics such as high stability or homogeneous bubble size. Where easy nozzle handling and simple foam pattern requirements are important, the discharge from the "fog-foam" type nozzle offers superior portability and high foam rate discharge. (See Fig. 9.29.) The open ports in the back side of this nozzle induct air by means of a crude venturi action that occurs in the short cavity. This

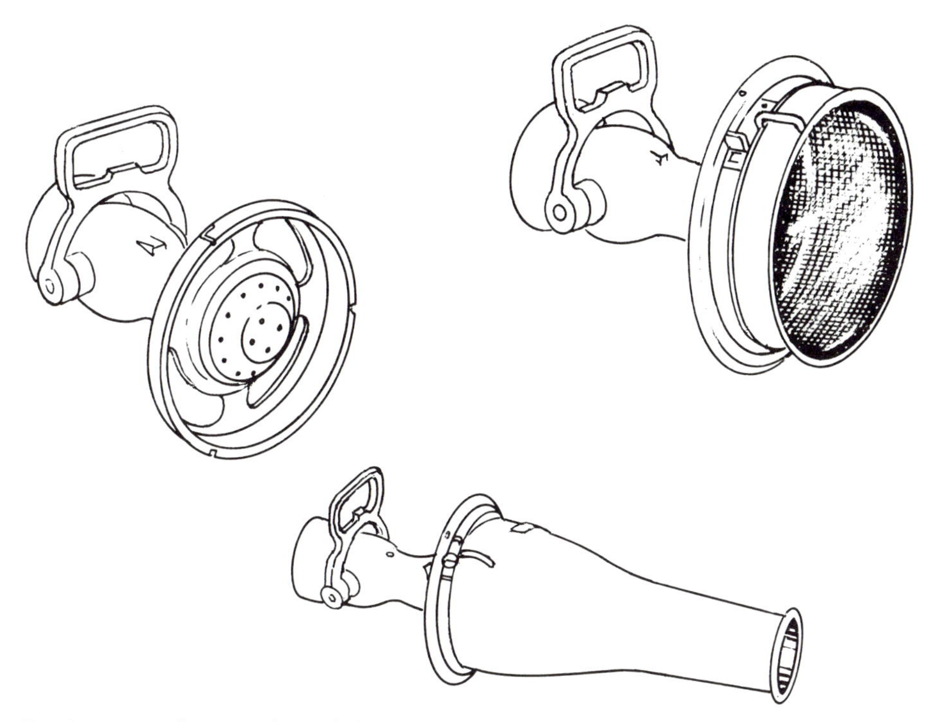

Fig. 9.29 Fog-foam nozzles with foam-spray pattern screen and stream shaper attachments. (From *Fire Protection Handbook*, NFPA)

air induction is followed by turbulence at the front face screen that forms foam that is issued in a wide, bushy pattern. With the screen in place, a wide, fully filled foam pattern is obtained for gentle "snowstorm" application to a burning surface. The "stream-shaper" attachment affords a consolidated foam stream of greater reach. By removal of all attachments from the internally impinging jet head of the nozzle, it may be used as a high velocity water fog nozzle with a wide water spray pattern.

Water Fog or Spray Nozzles for Foam

There are several types of water fog or spray nozzles for portable use that provide an acceptable fire fighting foam of adequate characteristics when supplied with certain foam concentrates. The most universal design for this type is shown in Figure 9.30.

These portable water fog nozzles can be used with aqueous film-forming foam solutions (AFFF) for combating fuel spill fires in shallow depths. They are sometimes used in this manner on crash-rescue mobile vehicles. The foam product that results from the discharge of AFFF solution devices that do not aspirate air is generally fast draining and does not impart the same degree of burnback resistance as the foam product that results from AFFF agents produced from foam-generating devices.

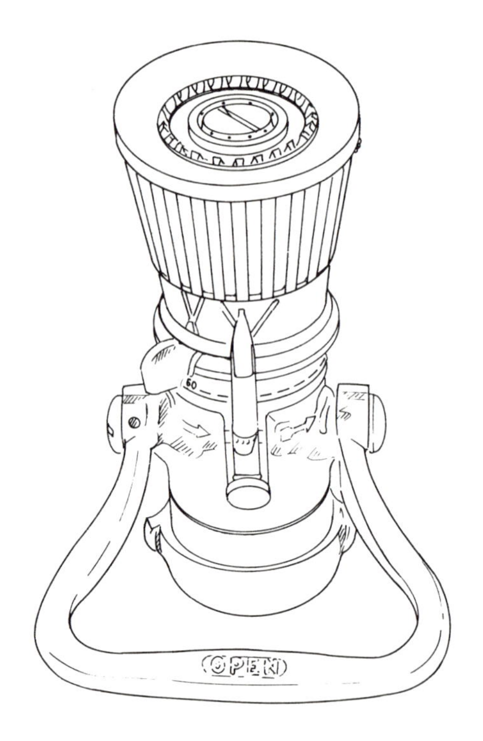

Fig. 9.30. Variable pattern water fog nozzle for use with AFFF. (From *Fire Protection Handbook*, NFPA)

Fixed Foam Generating Systems

Where flammable liquid fire protection is required for permanently installed hazards such as fuel storage tanks containing flammable or combustible liquids, air foam generating and distributing devices are installed integrally with the hazard. These devices, which are piped to a source of water, foam concentrate, or foam solution, may be arranged for automatic activation by fire detectors in the event of fire, or may be manually controlled.

The fire protection of exterior fuel storage tanks requires that several foam makers be installed at equally spaced positions on the top periphery of the tank. These are connected to a line (or lines) on the ground that simultaneously supplies foam solution to each foam maker in case of ignition of the flammable contents of the tank. Frangible seals at the discharge outlet of the foam maker prevent vapor loss from the tank. The seals are designed to burst when foam pressure is applied. The air inlet to the aspirating foam maker is provided with a screen to prevent clogging from foreign matter, such as birds' nesting materials. A universal or swing-pipe joint is installed (see Fig. 9.31) in the foam solution inlet pipe to prevent tank top distortion from fracturing the supply piping during the fire.

Spray Foam Nozzles

In commercial plants that handle flammable or combustible liquids, there are many situations where accidental spillage or flow of fuels may occur at floor or ground levels. The utility of a fixed foam-making device situated above the hazardous area and capable of instantaneously and completely covering that area can readily be seen. The spray deflector shown in Figure 9.32 accomplishes this action when installed from an overhead piping grid with nozzle outlets arranged so that their foam pattern overlaps each nearest nozzle pattern. The nozzle is supplied with foam solution, and foam is produced by air aspiration at the head just prior to the deflector construction.

Subsurface Foam Injection Systems

The problems inherent with the application of foam from above the burning surface (topside) are sometimes difficult to combat. The obvious solution to these problems is to apply foam from the underside of the fire by causing it to come up through the contents of the tank. The subsurface foam injection system accomplishes this by injecting foam under the pressure of the head of fuel in the tank, using the high back-pressure foam maker.

Entry of foam may be provided at several points at the base of the tank (base injection), or it may be accomplished by means of the product line. Where large tanks are involved, a branched pipe foam distributor may be installed on or slightly above the floor of the tank, connected to a central foam injection point

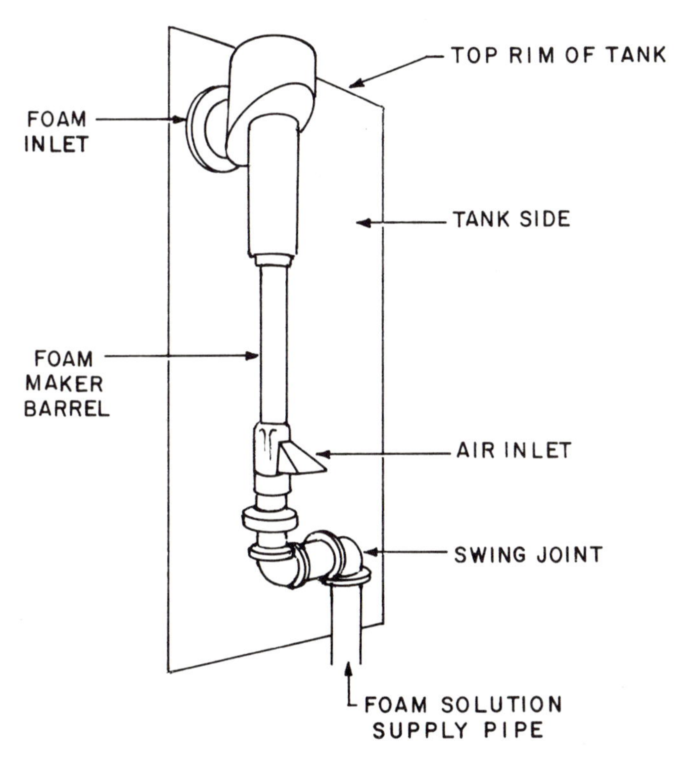

Fig. 9.31. Air foam maker at top of storage tank. (From *Fire Protection Handbook*, NFPA)

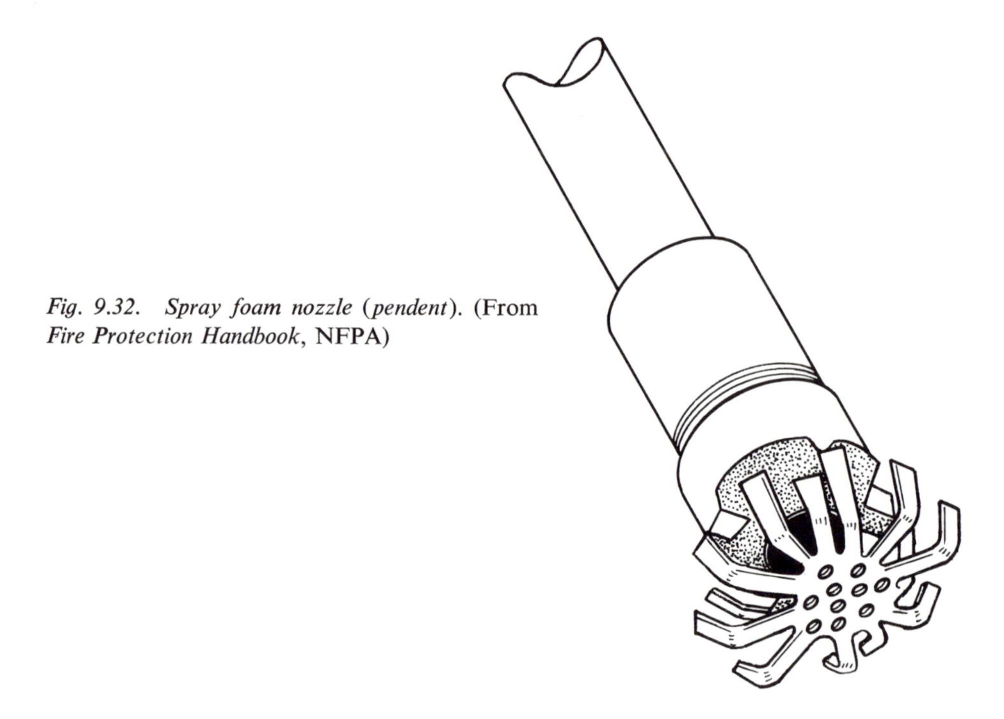

Fig. 9.32. Spray foam nozzle (pendent). (From *Fire Protection Handbook*, NFPA)

outside the tank. Mobile foam concentrate proportioning equipment is used with subsurface foam injection systems, pumping from a protected position outside the dike or berm around the tank.

Foam-water Sprinkler and Spray Systems

The advantages of permanently installed water sprinkler systems for the protection of many types of occupancies in terms of the saving of lives and property is well known. In areas where flammable and combustible liquids are processed, stored, or handled, a water discharge is often ineffective for controlling or extinguishing fires. The evolution of aspirating type foam-making sprinklers and spray nozzles has been successful in replacing water sprinkler nozzles for such systems so that fires in these occupancies may be controlled and property may be safeguarded.

When supplied with foam solution, sprinkler system piping "grids" provided with foam-water nozzles generate air foam in essentially the same "water-sprinkler" pattern as does water when water is discharged from the same nozzle. This dual capability affords the system with the advantages of both a Class A and a Class B hazard extinguishment ability.

Foam-water spray nozzles also discharge water or foam, and are used in similar sprinkler system designs where a special directional discharge pattern of foam or water is desired. These are used in fewer instances than the foam-water nozzle.

Mobile Fire Protection Vehicles Using Foam

Mobile foam trucks developed by the U.S. Air Force, the U.S. Navy, and municipal authorities in the United States and in some other countries, are large, custom-designed vehicles with oversize running gear that allows the vehicles to travel over all types of airport terrain. The trucks carry their own water supplies and foam concentrate (usually for six percent proportioning). Although their supply of foam concentrate and water is limited, it is sufficient to enable fire fighters to form a rescue passageway to a burning aircraft.

High Expansion Foam Generating Equipment

High expansion foam systems are basically used to control or extinguish fires involving surface fires in flammable and combustible liquids and solids, and deep-seated fires involving solid materials subject to smoldering. Three-dimensional fires in flammable liquids (falling or flowing under pressure) having flash points below 100°F (37.7°C) generally cannot be extinguished with this technique, although such fires may be kept under control by application of high expansion foam. Key factors to consider in determining the design adequacy of a high expansion foam system are:

- The quality and adequacy of water supply, the adequacy of supply of foam liquid concentrate, and the source of the air supply.
- Suitability of the generator and foam delivery system (piping, fittings, valves, ducts).
- Needed submergence volume and time as influenced by the space being protected, the nature of the hazards involved, leakage of the foam, and similar factors.

Combined Agent or "Twinned" Equipment

The superior capability of dry chemical agent (especially potassium salt types) for very fast flame control and three-dimensional flowing fuel extinguishment is well documented. Its fire fighting deficiency of reflash protection over fuel surfaces that have been extinguished is also well known. With the advent of dry chemical compatible aqueous film-forming foam (AFFF) concentrates, it became possible to almost simultaneously apply a coating of vapor-securing foam to a burning fuel surface that has been freshly extinguished by the chemical action of dry chemical discharges.

The logical extension of these facts has been incorporated into portable and mobile devices for one-person operation of dual trigger-valve controlled nozzles and monitor nozzles discharging AFFF and dry chemical, but linked together (twinned) for coordinated action and control. Figure 9.33 illustrates the twinned pistol-grip, trigger-valve dual nozzle device for portable dual hoseline use. The use of this combined agent attack allows flammable liquid fire extinguishment with both speed and freedom from reignition.

Fig. 9.33. Twinned nozzle applicator for combined agent use — AFFF nozzle in operator's left hand and dry chemical nozzle in right hand. (From *Fire Protection Handbook*, NFPA)

PRECAUTIONS AND REQUIREMENTS FOR THE USE OF FIRE FIGHTING FOAMS

Because all air foams consist of a delicate emulsion of air and foam-forming concentrate solution, fire service personnel must be aware of the conditions under which this agent is used. Thus, the following requirements are important:

- The character and physical qualities of the foam must be adequate for the fire protection task it is to be used upon.
- The characteristics of the foam equipment used must be correctly chosen.
- The application rates of foam to a burning liquid surface must be no less than 0.1 gpm/square foot (4.17 litres per minute, per square metre).
- Foams applied to burning liquid surfaces that destroy them by surface active incompatibility must be applied at much greater rates.
- Foams (and water sprays) applied to burning liquids that are above 212°F (100°C) will result in temporary expansion (slopover) of the hot liquid, steam, and air.
- Foams will be destroyed if applied simultaneously with many types of other extinguishing agents.

ACTIVITIES

1. (a) If a grease fire started in a deep-frying skillet on your stove, would you use water to extinguish the flames? Why or why not?
   (b) Describe another method of extinguishment that could be used if no extinguishing agent were available. (HINT: Formulate your answer from what you have learned in this text about combustion.)
2. (a) Explain the difference in operation between a wet pipe and a dry pipe sprinkler system.
   (b) What particular types of occupancies would use a dry pipe system? Why?
3. (a) How could a burning rag constitute not only a Class A fire, but also a Class B fire? How could a plastic bottle constitute both a Class A and a Class B fire?
   (b) Explain how you would extinguish the preceding two types of fires.
4. Besides water, the earth provides us with other natural "extinguishers." If a minor paper fire was started on the street, how might you "naturally" extinguish the fire if no water was available?
5. (a) With a group of your classmates, discuss the positive characteristics of water that make it an effective means of fire extinguishment.
   (b) Next, compile your own list of the instances in which water would be a negative means of fire extinguishment.
   (c) Then, compare your list with the lists compiled by your classmates.
6. Discuss the various methods of impinging water on a fire. Explain why these different methods are needed.
   (a) What are the advantages of an automatic sprinkler system?
   (b) What are the disadvantages of an automatic sprinkler system?
7. (a) Discuss with the rest of the class the various types and uses of foam fire extinguishers.
   (b) Compile a list of the fire types that would be handled effectively by a particular foam, including specific locations.
8. (a) Compile a list of hazardous substances present in the storage area(s) of your residence (*i.e.*, basement, garage, hall closet, etc.).
   (b) Include in your listing the classifications of these hazardous substances according to the standard classification of combustible materials.
9. When fire fighting foams are utilized, there are certain basic precautions and requirements that fire fighting personnel should be aware of. Outline what you believe are the most important of these precautions and requirements. Discuss your reasoning with a group of your classmates.

SUGGESTED READINGS

Alvares, N. J. and Lipska, A. E., "The Effect of Smoke on the Production and Stability of High-Expansion Foam," *Journal of Fire and Flammability*, Vol. 3, April 1972.

"Aqueous Foams — Their Characteristics and Applications," *Industrial and Engineering Chemistry*, Vol. 48, November 1956.

Averill, C. F., " 'Super Garages' for Jumbo Jets Require Super Fire Protection Systems," *Fire Journal*, Vol. 66, No. 2, March 1972.

Ballas, Thomas, "Electric Shock Hazard Studies of High Expansion Foam," *Fire Technology*, Vol. 5, No. 1, February 1969.

Breen, D. E., "Hangar Fire Protection with Automatic AFFF Systems," *Fire Technology*, Vol. 9, No. 2, May 1973.

Brodey, Milton, "Slippery Water Reduces Friction Loss," *Fire International*, 1972.

Bryan, J. L., *Fire Suppression and Detection Systems*, Glencoe, Beverly Hills, California, 1974.

Burford, R. R., "The Use of AFFF in Sprinkler Systems," *Fire Technology*, Vol. 12, No. 1, February 1976.

UL 711, *Classification, Rating, and Fire Testing of Class A, B, C, and D Extinguishers*, Underwriters Laboratory, April 1965.

Clough, T. C., "Research on Friction Reducing Agents," *Fire Technology*, Vol. 9, No. 1, February 1973.

Cray, E. W., "High-Expansion Foam," NFPA *Quarterly*, Vol. 58, No. 1, July 1964.

Di Maio, L. R., "Characteristics of Foams and Use on Polar Solvents," *Fire Engineering*, June 1975.

Evans, E. M. and Nash, P., "The Base Injection of Foams Into Fuel Storage Tanks," *Fire Prevention and Technology*, No. 14, March 1976.

Fire Protection Manual for Hydrocarbon Processing Tanks, 2nd Edition, Gulf, Houston, 1973.

Fire Service Hydraulics, 2nd Edition, R. H. Donnelley, New York, 1970.

Fittes, D. W., Griffiths, D. J., Nash, P., "The Use of Light Water for Major Aircraft Fires," *Fire Technology*, Vol. 5, No. 4, November 1969.

Gassman, J. J. and Hill, R. G., "Fire Extinguishing Methods for New Passenger-Cargo Aircraft," National Aviation Facilities Experimental Center, Atlantic City, N.J.

Geyer, George B., "Extinguishing Agents for Hydrocarbon Fuel Fires," *Fire Technology*, Vol. 5, No. 2, May 1969.

Guides for Fighting Fires In and Around Petroleum Storage Tanks, American Petroleum Institute, National Fire Protection Association, 1974.

Handbook of Tables for Applied Engineering Sciences, CRC Press, Chemical Rubber Company, Cleveland, Ohio.

Hird, D., "The Use of Foaming Agents for Aircraft Crash Fires," *Fire*, September 1974.

Hird, D., Rodriguez, A., Smith, D., "Foam — Its Efficiency in Tank Fires," *Fire Technology*, Vol. 6, No. 1, February 1970.

Kalelkar, A. S., "Understanding Sprinkler Performance: Modeling of Combustion and Extinction," *Fire Technology*, Vol. 7, No. 4, November 1971.

Meldrum, D. N., "Aqueous Film-Forming Foam — Facts and Fallacies," *Fire Journal*, Vol. 66, No. 1, January 1972.

Nash, P., "Fire Protection of Flammable Liquid Storages by Water Spray and Foam," Building Research Establishment Current Paper No. CP 42/75.

———, "The Extinction of Aircraft Crash Fires," *Fire Prevention*, No. 112, January 1976.

Nash, P. and Young, R. A., "The Performance of the Sprinkler in the Extinction of Fire," Building Research Establishment Current Paper No. CP 67/75, July 1975.

"Navy Tests 'Light Water' Airborne Fire-Fighting System," *Aviation Week and Space Technology*, August 23, 1965.

Stechishen, E., *Measurement of the Effectiveness of Water as a Fire Suppressant*, Forest Fire Research Institute, Ottawa, Ontario, Canada, May 1970.

Thorne, P. F., "Drag Reduction in Fire-Fighting," Building Research Establishment Current Paper No. CP 70/75, August 1975.

10

Gas to "Halon" Extinguishing Agents

INERTING GAS EXTINGUISHING AGENTS

To say that almost all fires can be prevented or halted by sufficient dilution of the oxygen surrounding the combustible material may sound repetitious. However, that is exactly the mode of action of inerting gases.

Inerting gases are a number of gases that do not react with combustible materials, and that are capable of simple dilution of oxygen or air atmospheres to the point where oxidation (or combustion) either cannot start or cannot continue. Since these gases are nonreactive, they are classified as *inert* gases. The atmosphere they produce is inert and no longer capable of promoting combustion. Helium, argon, krypton, nitrogen, and carbon dioxide are used to produce inert atmospheres. Flue gases or products of combustion gases such as those from oil burners are also inerting gases since they consist largely of carbon dioxide and nitrogen (with perhaps five percent oxygen) and may be piped to a hazard as an inerting gas.

Carbon dioxide gas and nitrogen are the principal inert gases in modern use. Several of the halogenated hydrocarbon gases and vapors are also employed as inert gases. However, these compounds also exert a free radical quenching action in the presence of flames. These will be dealt with in more detail later in this chapter in a section titled "Halogenated Extinguishing Agents."

Carbon Dioxide Inert Gas

Carbon dioxide gas has a distinguished history of use as a fire extinguishing and inerting gas. It possesses many favorable physical characteristics and is only mildly toxic (it warns the person breathing it by increasing the respiratory rate and leaving no lasting toxic effect at five percent concentration, thus making it possible for the person breathing it to escape before suffocation can occur at concentrations above nine percent).

Among the inerting gases, carbon dioxide is the only one where large volumes of gas may be liquefied and maintained in that state in a small container volume at ordinary temperatures and at not extremely high pressures. Its temperature-pressure relationships and physical states are given in Figure 10.1. The explanation for the production of carbon dioxide particles in a solid state "snow" during discharge is entirely connected with the cooling effect on the stored liquid during release of pressure.

Carbon dioxide gas is extremely heavy in comparison to air or to the gases of combustion; because of this, it quickly sinks to the level of the point of combustion or base of the flames, replacing air and oxygen and maintaining a flame-smothering atmosphere. Compared to air, its density is 1.5 (1.5 times as heavy).

Almost the entire effect of fire extinguishment by carbon dioxide can be attributed to its dilution of oxygen to achieve an inert atmosphere. It has been shown that the cooling effect of the extremely cold "snow" mixture discharged from an extinguisher amounts to only about 120 Btu per pound (279 kiloJoule per kilogram) of gas. This can be considered to be diminutive when compared to water

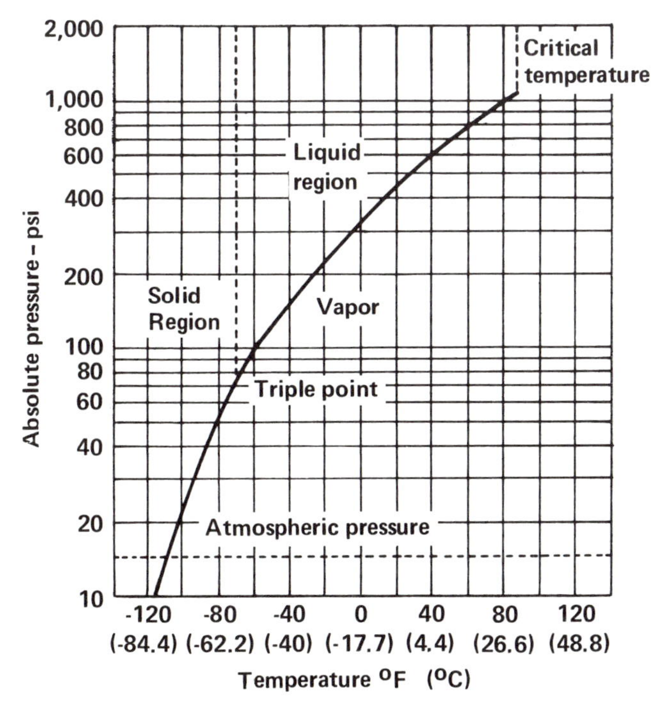

Fig. 10.1 Effect of pressure and temperature change on the physical state of carbon dioxide at constant volume. Above the critical temperature (87°F [31°C]) it is entirely gas, irrespective of pressure. Between 87.8°F (31°C) and −69°F (−56.6°C) — temperature of the triple point — in a closed container, it is part liquid and part gas. Below the triple point it is either a solid or a gas, depending on the pressure and temperature. (From *Fire Protection Handbook*, NFPA)

with its total cooling capability of 1,150 Btu per pound (2,672.6 kiloJoule per kilogram) when it becomes steam.

Except for a few anomalies, carbon dioxide extinguishes flammable and combustible liquid and gas fires when concentrations of about 30 to 38 percent carbon dioxide are reached. Certain gases and vapors with wide flammable limits require higher gas concentrations to extinguish (see Table 10.1).

Table 10.1 Minimum Carbon Dioxide Concentrations for Extinguishment

MATERIAL	THEORETICAL MIN. CO_2 CONCENTRATION (PERCENT)
Acetylene	55
Acetone	26*
Benzol, benzene	31
Butadiene	34
Butane	28
Carbon disulfide	55
Carbon monoxide	53
Coal gas or natural gas	31*
Cyclopropane	31
Dowtherm	38*
Ethane	33
Ethyl ether	38*
Ethyl alcohol	36
Ethylene	41
Ethylene dichloride	21
Ethylene oxide	44
Gasoline	28
Hexane	29
Hydrogen	62
Isobutane	30*
Kerosene	28
Methane	25
Methyl alcohol	26
Pentane	29
Propane	30
Propylene	30
Quench, lubricating oils	28

Note: The theoretical minimum extinguishing concentrations in air for the above materials were obtained from Bureau of Mines, Bulletin 503. Those marked * were calculated from accepted residual oxygen values.

Carbon dioxide is classified as a nonpermanent or "temporary" extinguishing agent. When ventilation of an atmosphere of carbon dioxide gas occurs and oxygen again becomes available to a combustible material, it can be reignited easily unless it has cooled below either its ignition temperature or its fire point. This characteristic is particularly true where Class A fires and materials are concerned, or where deep-seated glowing type combustion may continue after

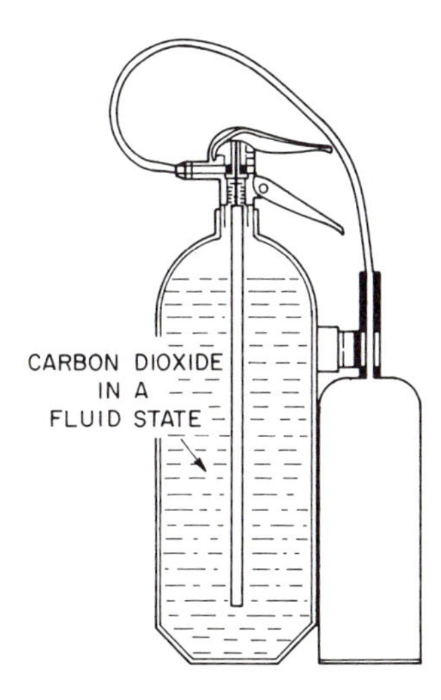

Fig. 10.2. Carbon dioxide fire extinguisher. (From *Fire Protection Handbook*, NFPA)

flame control has been accomplished with carbon dioxide application. The gas diffuses very slowly through materials that are difficult to penetrate. Therefore, cooling of these materials by time or some other means, with a sustained atmosphere of about 38 percent or more concentration of carbon dioxide is necessary.

There are two principal types of storage and containment units for carbon dioxide fire protection systems. One employs a steel container at fairly high pressures of about 850 psi (58.6 bar) where the liquid and gas phase are in equilibrium. The other consists of a low pressure storage unit maintained at 300 psi (20.7 bar) and kept at 0°F (−18.5°C) by a refrigeration unit. The latter containment allows large amounts of gas (largely in liquid form) in containers that need not consist of heavy walls. (See Fig. 10.2.)

The familiar portable CO_2 extinguisher utilizes the high pressure container. Large high pressure "bottles" are also employed in total flooding systems for inerting large closed spaces. These latter systems are connected to either automatic or manual detecting and activating devices, and have special piping and nozzles to direct the gas to the hazards it protects. In spaces where personnel might be concerned, the sudden gas evolution is accompanied by an alarm system so that human exposure is avoided. In systems utilizing the low pressure refrigerated storage for CO_2, the piping and nozzle arrangements are similar to those of the high pressure designs. (See Figs. 10.3 and 10.4.)

Nitrogen Inerting Gas

Nitrogen, the other principal gas contained in air, could well be used as an oxygen diluent and thus halt combustion and produce an inert atmosphere. Because nitrogen is not an easily condensible gas and must be stored at high pressures of about 2,000 psi (137.9 bar) in heavy cylinders in the gas phase, it does not find wide use as a fire extinguishant or inerting gas. It is slightly lighter than air; for this reason, it is difficult to direct or maintain nitrogen in the "seat of

Fig. 10.3. The hand lines being used in this demonstration of fighting an aircraft fire are discharging carbon dioxide at the rate of 450 pounds per minute. (From *Fire Protection Handbook*, NFPA)

burning" or at the base of flames where it can be most effective. Liquefied nitrogen has been used and experimented with for fires. However, the liquid must be applied in deluge amounts.

HALOGENATED EXTINGUISHING AGENTS

As previously discussed, the halogens — fluorine (F), chlorine (Cl), bromine (Br), and Iodine (I) — are vigorous oxidizing agents. When combined with the metallic and nonmetallic elements with one or more atoms of oxygen, the resulting "ionic" halogen compounds attack and oxidize many compounds. When the lower aliphatic series — methane (CH_4), ethane (C_2H_6), propane (C_3H_8), and butane (C_4H_{10}) — are observed, it can be seen that hydrocarbons (and those above them in the series, including octane, etc.), are highly flammable and readily capable of rapid oxidation. If one of the highly oxidizing halogens is combined chemically with a readily oxidizable hydrocarbon by means of special methods that utilize high pressures that are devoid of oxygen, the result is a vapor (gas) or liquid that is an organic compound capable of decomposition and, as such, is considered to be an efficient fire extinguishing agent.

The old and now forbidden carbon tetrachloride "Pyrene" extinguisher fluid is the earliest halogenated extinguishing agent. Carbon tetrachloride (CCl_4) is fully chlorinated methane and was used in portable pump-tank fire extinguishers as early as 1907. In the 1930s a number of accidental deaths and serious injuries were attributed to the ingestion of vapors of this compound or its breakdown products, and by 1950 legislation was passed to halt its use.

In the 1930s another halogenated agent, methyl bromide (CH_3Br), began to be used, predominantly in the United Kingdom. A "sister" compound to carbon tetrachloride, methyl bromide was more effective than carbon tetrachloride; however, its toxicity militated against its wide use, except in areas where no human exposure was involved (such as in aircraft piston engine nacelles). Methyl bromide, which is a gas at ordinary temperatures and which has a boiling point of 40°F (4°C), will itself burn. Containing three atoms of hydrogen and one atom of carbon in its molecule, methyl bromide possesses a narrow flammable range between 13.5 and 14.5 percent in air. Above and below this range it is an efficient fire extinguishant.

During World War II, a compound containing both chlorine and bromine substituted in methane — chlorobromomethane (CH_2ClBr) — began to be used. Its principal use was on German war vessels where it was utilized for fire protection

Fig. 10.4. A carbon dioxide fire extinguishing system for a restaurant kitchen range and exhaust system. Outlets are so arranged that discharge of carbon dioxide under the hood floods the surfaces of the cooking equipment and enters the exhaust system to extinguish grease fires in the duct system. (From *Fire Protection Handbook*, NFPA)

by means of flooding engine and machinery compartments where diesel fuel was stored or employed. It was also used as a secondary agent by the U.S. Air Force in 1950. This compound produced slightly less toxic effects than the previously used halogenated agents.

The "Halon" Extinguishing Agents

In 1946 it became obvious that of all the halogens, bromine substitution in the lower aliphatic hydrocarbons resulted in a powerful extinguishing agent; however, it also produced toxic vapors. It was also noticed that the fluorine bond to carbon was very stable (note the nontoxic, highly stable, "Freon" refrigerants), and if a compound containing fluorine and bromine in some new ratio to carbon could be synthesized, perhaps the result would be less toxic. This theory was proposed by the author to the Army Engineers Fire Research Unit, and chemists at the Purdue University were put to work on an organic synthesis problem. The result is the modern series of "Halon" halogenated agents. The name "Halon," which was derived from obvious chemical roots, was proposed by James Malcolm at the U.S. Army's Corps of Engineers Laboratory in 1950.

The ponderous chemical names for each of the halogenated extinguishing agents has been simplified by the use of digits in the following "Halon" system for naming these compounds:*

> The first digit of the number represents the number of carbon atoms in the compound molecule; the second digit, the number of fluorine atoms; the third digit, the number of chlorine atoms; the fourth digit, the number of bromine atoms; and the fifth digit, the number of iodine atoms (if any). Hydrogen is not numbered.
>
> For example:
>
> The chemical formula for carbon tetrachloride is CCl_4.
>
> Its Halon number is "104."
>
> The formula for methyl bromide is CH_3Br.
>
> Its Halon number is "1001."
>
> The formula for bromochlorodifluoromethane is $BrCClF_2$.
>
> Its Halon number is "1211."

The British Fire Service adheres to the initial capital "alphabet" system where:

Bromochlorodifluoromethane ($BrCClF_2$) is "BCF."
And: Bromotrifluoromethane ($BrCF_3$) is "BTM."

* The halogenated extinguishing agents are currently known as Halons, and the "Halon System" for naming the halogenated hydrocarbons was devised by the U.S. Army Corps of Engineers. This simplified system of nomenclature describes the chemical composition of the materials without the use of chemical names or possibly confusing abbreviations (*i.e.*: "BT" for bromotrifluoromethane, and "DDM" for dibromodifluoromethane).

In the ten-year period from 1964 to 1974, the relatively large number of halogenated agents (they have also been called vaporizing liquids and halogenated hydrocarbons) was narrowed down to the volume usage of only two materials. Problems of toxicity of the undecomposed and the fire-decomposed liquids and their vapors, difficulties with dispensing equipment design and fabrication, and overall cost effectiveness were the most probable reasons for the limited usage.

Table 10.2 lists the important characteristics of the currently important Halons 1211 (bromochlorodifluoromethane) and 1301 (bromotrifluoromethane) in comparison with Halon 1011 (bromochloromethane). Halon 1011 was used to some extent in the United States until 1970, at which time its use was discontinued after learning of unfavorable German experiences with the compound. In the United States in 1976 the only two materials recognized in the National Fire Protection Association's Standards were bromochlorodifluoromethane (Halon 1211) and bromotrifluoromethane (Halon 1301).

Table 10.2 indicates that each of these agents is quite heavy; thus, their vapors will readily sink and stay in the region of the base of the flames for efficient flame extinction. Halons 1211 and 1301 do not differ widely in their weight efficiency of extinguishment, and their high "lethal concentration" values show lesser toxicity problems than the older German-originated Halon 1011. Carbon tetrachloride is roughly equivalent to Halon 1011 in this respect. (The higher the concentration of substance it takes to be dangerous to humans, the less toxic the substance is.)

Extinguishment Mechanism Chemistry

It would be valuable to know the exact and unvarying ways in which the halogenated extinguishing agents operate in preventing flame propagation or halting it in the case of a "working" fire. Unfortunately, this is not possible because of the limited extent of currently available knowledge about combustion kinetics and the influence of the actions and reactions of the halogenated materials. It is known that chemical reactions are involved in this situation because these agents are considerably more effective than processes of heat removal or smothering can account for. Therefore, they must act by removing the active chemical species involved in the chain reactions of flame propagation.

Interestingly, it has been shown over the years that the fire fighting efficiency of extinguishing agents containing the halogens is opposite in succession to their efficiency as oxidizing agents. When the halogens act as oxidizing materials, fluorine is the most active oxidizer; fluorine is followed by chlorine, bromine, and iodine, the latter being the least effective. When the halogens are coupled with methyl or ethyl groups in an organic compound, their efficiencies as fire extinguishing agents show that iodine compounds such as methyl iodide (a highly toxic material, consequently of no usefulness in fire protection) is the most effective, followed by bromine, chlorine, and fluorine compounds, the latter having the least effectiveness for flame extinction.

Chemical kineticists and other investigators have found that Halon 1301 and Halon 1211 function in different ways, depending upon the fire conditions under which they are used. Thus, the use of these Halons as inerting agents involves different chemical and physical reactions than when they are employed in an on-going, hot fire, or than when they are directed into a fire that is just getting started. However, it does seem certain that the basic mode of action of the Halons in fire situations is that when they break down under the temperature of a flame their products of decomposition have an affinity for uniting with the highly active free radicals O, H, and OH produced by burning fuels, subsequently removing them from further flame propagation.

When considering the reactions that take place when Halon 1301 breaks down under heat attack, the following flame matrix reactions seem to occur most often and in most cases:

1. $$BrCF_3 + \text{Heat} \rightarrow \underbrace{Br + CF_3}_{\text{(Active)}}$$

Subsequently, the following reaction can occur:

2. $$CF_3 + \text{Heat} \rightarrow \underset{\text{(Active)}}{F} + \text{(Other CF compounds)}.$$

If CF_4 (methane) was the fuel, the following would probably occur during oxidation:

3. $$2CH_4 + O_2 \rightarrow \underbrace{4H + OH}_{\text{(Active)}} + CO + \text{(Other CH compounds)}.$$

The active Br and F from equations 1 and 2 could also attack the native fuel molecule in the following fashion:

4. $$CH_4 + Br \rightarrow HBr + \text{(Other CH compounds)}.$$

5. $$CH_4 + F \rightarrow HF + \text{(Other CH compounds)}.$$

And the active atomic H from equation 3 would also react with the active Br as follows:

6. $$H + Br \rightarrow HBr.$$

Regeneration of Br and F is now possible by the following reactions with the hydroxyl radical from the oxidation reaction of equation 3:

7. $$OH + HBr \rightarrow \underset{\text{(Active)}}{Br} + \underset{\text{(Inert)}}{H_2O}$$

8. $$OH + HF \rightarrow \underset{\text{(Active)}}{F} + \underset{\text{(Inert)}}{H_2O}.$$

In summary, it is more than likely that the efficiency of halogenated agents is due to the following significant reactions from the preceding scheme:

- Capture of active atomic hydrogen by halogen and removal of it from further heat-producing reactions.
- Reaction of active free radical hydroxyl groups from oxidation of fuels and subsequent removal by halogen hydride reaction resulting in inert water.

Table 10.2 Characteristics of Important Halogenated Agents

	AGENT		
	BROMO-CHLORO-METHANE	BROMO-CHLORO-DIFLUORO-METHANE	BROMO-TRIFLUORO-METHANE
Chemical formula	$BrCH_2Cl$	$BrCClF_2$	$BrCF_3$
Halon no.	1011	1211	1301
UK name	CB	BCF	BTM
Type of agent	Liquid	Liquid Gas	Liquid Gas
Boiling point:			
°F	151.0	25.0	−72.0
°C	66.0	−4.0	−58.0
Density	1.93	1.83	1.57
Efficiency: (Wt. required per unit B Rating)			
Lbs.	2.0	0.6	0.6
Kilo	0.85	0.25	0.25
Lethal concentration (ppm)			
Vapor	65,000	324,000	832,000
Fire decomp. vapor	4,000	7,650	14,000

- Regeneration of active atomic halogen by the preceding reaction so that the processes of capture and removal of atomic and free radical groups may be repeated.[1]

The flame extinguishment efficiencies of Halon 1301 and 1211 on a basis of volume of vapor (gas) are almost equal, with Halon 1301 showing a slightly superior performance. In accurate tests there seems to be considerable variation in the percentage of extinguishment vapor necessary; however, it may be concluded that volume concentrations of five percent of Halon 1301 will extinguish fires in most hydrocarbon fuels, whereas six percent of Halon 1211 is required. There are several anomalies to these figures in the list of all flammable liquids and gases. For example, hydrogen gas fires require very high concentrations of Halons for extinguishment (up to 27 percent of Halon 1211 in some tests). In order to effect extinguishment, oxygenated fuels such as dimethyl ether or methyl alcohol also require Halon concentrations up to ten or eleven percent. The Halons require higher concentrations to *inert* a combustible mixture than when they are used to extinguish flames. Although the conditions under which the extinguishing agent is judged affect the outcome considerably, it may be concluded that concentrations

[1] Ford, C. L., "An Overview of Halon 1301 Systems," a paper in "Halogenated Fire Suppressants," *American Chemical Society Symposium Series* 16, 1975, pp. 1–63.

of eight or nine percent are required in order to prevent flame propagation in combustible mixtures of ordinary fuels in air.

Successful extinguishment of solid combustibles by flooding with Halon vapor where the fire may have penetrated the mass of material requires a "soaking time" in order for the vapor to reach the seat of the fire. This "soaking time" varies greatly with the composition and variables of geometry of the solid combustible. Paper products in stacks, thermoplastics, and ordinary wood construction where fires cannot become deep-seated can be fully extinguished with Halon 1301 concentrations of five to six percent with "soaking times" of ten minutes. Deep-seated fires in corrugated paper or wood storages in rooms will require concentrations of up to thirteen percent with "soaking times" up to twenty minutes or longer. In general, Halon 1301 requires less vapor for fire extinguishment than when Halon 1211 is used; in the case of flooding or inerting extinguishment, about ten percent less Halon 1301 is needed. Usually, Halon 1301 and Halon 1211 are equivalent to sodium bicarbonate dry chemical in ordinary flame extinguishment by local application procedures. On a weight basis, Halon 1301 is approximately 2.5 times more effective than carbon dioxide.

In fire protection, it is important to consider the anesthetic or narcotic effects of the Halons. Neither Halon 1301 nor Halon 1211 has shown any residual poisoning effects, the effect of high concentration being of a temporary nature. Experiments on animals have shown that the toxic effects of the Halons are twofold: (1) they can either stimulate or depress the central nervous system to produce effects ranging from tremors and convulsions to lethargy and unconsciousness, and (2) they can sensitize the heart to adrenalin by causing disorders that range from a few isolated abnormal beats to the complete disorganization of normal rhythm. Tests on humans have shown that high concentrations of both agents lead to dizziness, impaired coordination, and reduced mental acuity; also, it is possible that prolonged exposure at high concentrations could lead to unconsciousness and even death. It is necessary to consider this possibility in relationship to the concentrations of each agent as required to extinguish fire.

Halon 1301 is the least toxic gaseous extinguishant, and for most total flooding applications it can be discharged into occupied spaces. Most applications call for extinguishing concentrations ranging from four to six percent volume, and it has been established that there is minimal, if any, effect on the central nervous system with concentrations up to seven percent volume. Between seven and ten percent volume concentration, light-headedness and reduced dexterity have been noted. However, it is only above ten percent by volume that toxic effects can be considered as being potentially serious enough to be avoided.

Halon 1211 has a lower threshold safety limit than Halon 1301, and all evidence seems to indicate that the hazard for concentrations of equivalent fire extinguishment capability is comparable to that of carbon dioxide. This would indicate safe application from portable extinguishers, but a restricted use in fixed systems. Concentrations up to four percent volume have been shown to be tolerable for a time period up to one minute (compared to ten percent for Halon 1301). Halon 1211 has been used effectively as a fire fighting agent by railroad systems, airlines and airports, telephone companies, and computer users on a world-wide basis.

Application Equipment for the Halons

Portable extinguishers have been devised for using both Halon 1301 and Halon 1211 for fire hazards in a number of instances. It is generally claimed that because of its higher boiling point and stream projection, the Halon 1211 compound is more universally acceptable in such hand application.

Details of the design engineering of installed or "fixed" systems for these agents are beyond the scope of this text. However, it is important to realize that there are two types of systems available. They are:

1. Total flooding systems for hazard protection in enclosed hazards, such as magnetic tape storage vaults and processing areas for paints, etc.
2. Local application systems for discharging Halons in such a manner that the stationary burning hazard is surrounded by a high concentration of agent to extinguish a fire.

There are, of course, special systems designed for specific hazard protection. Such special systems combine the preceding two conditions.

DRY CHEMICAL EXTINGUISHING AGENTS

Dry chemical agents are mixtures of powdered chemical compounds of varying composition that constitute our most efficient flame-halting materials. They are stored in moisture-proof containers so that they may be pneumatically blown into a fire by means of pressurized gas discharges that convey the solid powder from its container into the combustion zone. The extinguishment properties of dry chemicals, referred to as "dry powders" in the United Kingdom (not to be confused with "dry chemical foam powders" or other dry extinguishing powders for combustible metal fire control), do not include cooling, smothering, or inert gas evolution. They are "temporary" extinguishants since they only provide instantaneous and momentary flame extinction.

The history of dry chemical agents for quelling flames goes back many years, probably to the time of the Civil War days, at which time Solvay perfected the process for the cheap production of sodium bicarbonate. The pneumatic gas expulsion of the powder originated in Germany in 1913. In the 1920s, American railroad coaches that were lighted by kerosine lamps carried several long red tubes of "bicarbonate of soda." These tubes, located at the ends of each coach, were for kerosine lamp fire protection. The instructions for the use of the tubes were: "Remove cap and sling powder with a rapid swinging motion at the flames."

Sodium bicarbonate powder and pressurized extinguishers for its use went through many developmental stages. In 1959 a major change was the discontinued use of borax as a part of the mixture. Borax was discontinued because it could not be shown to be effective. Moisture and caking problems of the powder were solved by the use of zinc stearate (like baby powder) for coating the crystals. Today's formulations employ silicones for this purpose. Experimentation with grinding and sizing operations were conducted to obtain optimum "throw" of the powder along with maximum flame-quenching ability. It was found that a particle size range from about 5 to 75 microns was best, with most particles in the range of 20 to 25 microns.

In 1959, under this author's leadership, a United States Navy Fire Research Team originated the potassium bicarbonate formula for dry chemical. Named "Purple K" for the purple flame spectra it demonstrates when it encounters flame temperatures and the "K" symbol for potassium, it has twice the flame-quenching capacity per pound when compared with the sodium compound.

In about 1960 a new German-developed dry chemical became available in the United States. This new dry chemical contained monoammonium phosphate (chemically: ammonium dihydrogen phosphate, $NH_4H_2PO_4$), which is an acidic salt (in contrast to the bicarbonate types that are basic and that react with the acidic types). This compound shows about the same extinguishing power for Class B fuels as is obtained with sodium bicarbonate. Also, it breaks down in the presence of heat to yield a glassy coating of metaphosphoric acid, HPO_3. When monoammonium phosphate dry chemical is used on Class A fuels such as wood, the glassy residue on the solid combustible halts further combustion. Thus, the material can be used for extinguishment of both Class A and Class B fires. Dry chemical extinguisher discharges can be used for Class C electrical fires because the powder particulates do not conduct electricity back to the operator. Therefore, because the new powder possessed a Class "ABC" capability, it became known as "Multi-purpose" dry chemical.

In about 1964 another potassium-based dry chemical consisting of potassium chloride was developed. This became "Super-K." Its efficiency of flame extinction is slightly less than that of potassium bicarbonate.

In approximately 1967 a British chemical company developed still another dry chemical with much greater extinguishment efficiency than was previously obtainable from any other dry chemical. This dry chemical consists of an addition product of urea with potassium bicarbonate. (The chemistry of this material will be dealt with in a later paragraph.) The new compound has been named "Monnex" and, on a weight of agent basis, is more than twice as efficient as potassium bicarbonate for flame extinction.

The Chemistry and Extinguishing Mechanisms of Dry Chemicals

As with the halogenated agents, the exact kinetics and thermochemistry that account for all of the reactions occurring when dry chemicals extinguish flames is not completely known. One problem in researching and studying this subject is that the accurate and precise transport of dry crystals (of varying sizes) into the flame matrix is a very difficult mechanical task. Gases and vapors are much easier to inflict into the combustion zone for study, as can be seen by the number of investigations that have been made over the years in the flame studies of the vapors of halogenated agents.

The tenable conclusion that free radical quenching must be the principal avenue of mechanism of dry chemical extinguishing agents can be made, as per the following evidence: the high weight efficiencies of dry chemicals when compared to other extinguishing agents; the negative results of investigations endeavoring to show that replacement of oxygen by solids content of the crystals is the mechanism;

that evolution of carbon dioxide is responsible; that heat capacity transferral to the solids is sufficient to halt flame propagation.

There is little conflicting evidence against the concept that H and OH free radical active species exist in flames of burning hydrocarbon fuels. These are vulnerable for even short lengths of time to combination and removal as inert or nonreactive water molecules.

Some evidence has shown that the sodium bicarbonate dry chemical extinguishment mechanism proceeds as follows:[2]

$$2NaHCO_3 + \text{Heat} \rightarrow CO_2 + H_2O + Na_2CO_3$$
$$Na_2CO_3 + \text{Heat} \rightarrow CO_2 + Na_2O$$
$$Na_2O + H_2O \text{ (In flame matrix)} \rightarrow 2NaOH$$
$$NaOH + H \text{ (Active, in flame)} \rightarrow Na + H_2O \text{ (Inert)}$$

and:

$$NaOH + OH \text{ (Active, in flame)} \rightarrow NaO + H_2O \text{ (Inert)}.$$

The reactive compounds Na and NaO, resulting from the latter two equations, regenerate NaOH by reaction in the flame matrix to repeat the water-producing equations.

Another mechanism involving H and OH free radicals has been postulated by Haessler as follows:[3]

$$NaHCO_3 + \text{Heat} \rightarrow CO_2 + OH + Na$$
$$OH + H \text{ (Active, in flame)} \rightarrow H_2O$$
$$Na + H \text{ (Active)} \rightarrow NaH$$
$$OH \text{ (Active, in flame)} + NaH \rightarrow Na + H_2O.$$

The potassium bicarbonate dry chemical extinguishment mechanism would follow either of the sodium ones, with one important consideration:

In all chemical reactions comparing sodium to potassium compound activities, the tendency toward reactivity and activity, potassium is higher in its ability to unite and undergo combination or reaction. This may be due to its lower ionization level than sodium or due to its lower decomposition temperature. These facts have not been confirmed by experimentation. At any rate, the fire extinguishing efficiencies of the potassium compounds are superior in all respects to those of sodium.

According to Haessler, potassium chloride exerts a similar action to that ascribed to potassium bicarbonate:[3]

$$KCl \rightarrow K + Cl$$
$$Cl + H \text{ (Active)} \rightarrow HCl$$
$$K + H \text{ (Active)} \rightarrow KH$$
$$KH + OH \text{ (Active)} \rightarrow K + H_2O \text{ (Inert)}.$$

The process repeats with the regenerated potassium.

---

[2] Friedman, R. and Levy, J. B., "Inhibition of Opposed-Jet Methane—Air Diffusion Flames. The Effects of Alkali Metal Vapors and Organic Halides," *Combustion and Flame*, Vol. 7, No. 2, June 1963, p. 165.

[3] Haessler, W. M., *The Extinguishment of Fire*, National Fire Protection Association, Boston, 1974.

Evidence exists to show that the British "Monnex" dry chemical, which is the addition product of urea and potassium bicarbonate, is unique in chemical fire extinguishment reactions. Urea, or carbamide, $CO(NH_2)_2$, is capable of forming adducts, or inclusion complexes, under the correct processing conditions. Such complexes are not true compounds, but are crystalline mixtures in which the molecules of one of the components are contained within the crystal lattice framework of the other compound. When certain physical pressures (such as heat) are exerted on the complex, it may separate into its original components. Apparently "Monnex" dry chemical is such an inclusion complex with urea bound up with potassium bicarbonate.

A number of years ago it was proven that an extremely finely powdered dry chemical was a more efficient flame-quenching agent than the same powder with a coarser particle-size distribution. The fine powder was incapable of projection to any distance, thereby rendering it impractical for use.[4]

It has been claimed that the "Monnex" dry chemical consists of particles of the normal and customary size distribution for efficient projection from a nozzle under gas pressure. However, when the crystals are acted upon by heat in a fire a breakup, or shattering, of crystals occurs that yields very fine particles and more efficient flame quenching by the potassium bicarbonate than is customarily obtained. The functions of decomposition and chain breaking by the potassium bicarbonate content in the "Monnex" powder are presumably identical but superior in effect to those attributed to potassium bicarbonate, as previously described.

Haessler describes the action of "Monnex" dry chemical in a different manner. Concluding that the urea-potassium bicarbonate material is actually potassium carbamate (KNH_2COO), he outlines the decomposition of this compound in the flame matrix including the evolution of gaseous ammonia (NH_3) after union with active hydrogen atoms.

The mechanism of flame quenching by monoammonium phosphate (multipurpose), with concurrent deposition on Class A material of a coating of glassy metaphosphoric acid, probably involves union of active H atoms or OH radicals with NH_3 radicals during dissociation of the NH_4 evolved upon decomposition of the compound in the flame. The phosphoric acids remaining from the decomposition may then dehydrate fully to the anhydrous glassy metaphosphoric acid, HPO_3. This postulated mode of action has not been fully investigated.

Equipment for Application of Dry Chemical Extinguishing Agents

The sodium bicarbonate nontoxic dry chemical agents have enjoyed a wide popularity and usage since about 1946, especially for extinguishing flammable and combustible liquid fires. Prior to this time they were available for many fire protection purposes, and were packaged in many forms.

The gas propulsion form of agent discharge to a hazard has been found to be exceedingly practical and reliable. In some devices the gas is kept separate from the chemical until the agent is needed, at which time the pressurized gas bottle is

[4] Dolan, J. E. and Dempster, P. B., "The Suppression of Methanol-Air Ignitions by Fire Powders," *Journal of Applied Chemistry*, Vol. 5, 1955, p. 510.

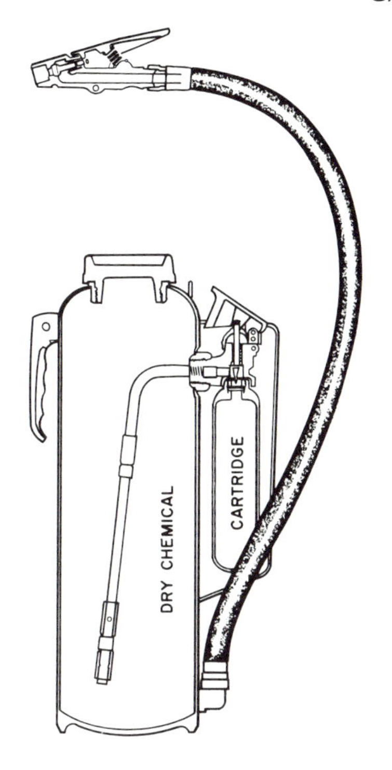

Fig. 10.5. Dry chemical fire extinguisher, cartridge operated. (From *Fire Protection Handbook*, NFPA)

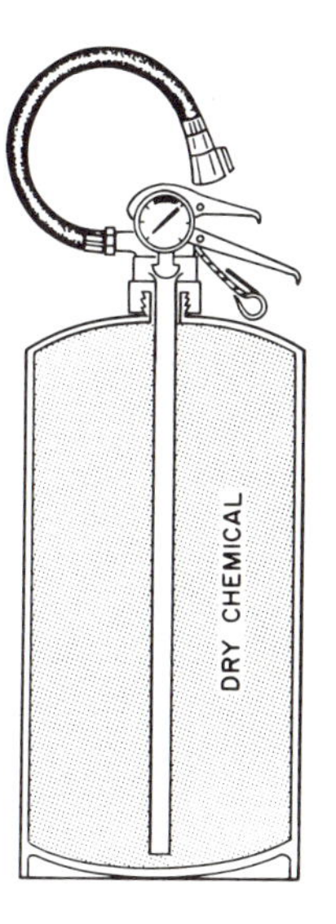

Fig. 10.6. Dry chemical fire extinguisher, stored pressure with disposable shell. (From *Fire Protection Handbook*, NFPA)

opened to the bulk of powder in a lower pressure container, creating a stirring action on the powder and promoting a powder-laden gas discharge. In another type of device the powder container is pressurized with gas to a moderately low pressure, and turbulent stirring occurs when the pressure is released through an appropriately designed powder delivery nozzle. A third type of container consists of a ball that is rotated 180° when gas is released to charge the unit for use. Both carbon dioxide and nitrogen gases are used as a propulsion gas for all of these devices, both in portable and fixed equipment. (See Figs. 10.5 and 10.6.)

Using the preceding principles, equipment has been designed for piping dry chemical agent to hazards, or in portable hose and nozzle systems for manual application to hazards. The design engineering for piped dry chemical systems is particularly critical because of the fact that flow requirements involve an unstable and temporary mixture of a gas and a solid, quite different than solid-fluid flow or gas flow alone.

The Evolution of A Dual Action Extinguishing Method Using Dry Chemical Agents

It can be seen from the foregoing paragraphs of this section that dry chemical extinguishing agents, especially those containing potassium compounds, constitute the most powerful flame extinction materials per unit of weight that are available in a practical sense. However, as with inerting and halogenated agents, dry chemical agents are temporary extinguishants. They cannot be relied upon to maintain permanent or lasting extinguishment if sources of reignition are present or if combustible vapors may be expected to continue to evolve from the extinguished fuel, and thus constitute a continuing fire hazard.

During the late 1940s attempts were made to utilize the flame-quenching abilities of dry chemical agents before, during, and following the application of protein foams to gasoline fuel fires. It was found that an almost magical foam blanket breakdown took place when ordinary sodium bicarbonate dry chemicals were discharged near or on the foam, or into the fuel in contact with the protein air foam.

Considerable research and testing resulted in the interesting conclusion that the zinc or magnesium stearate used in the then current dry chemical mixtures (to prevent caking of the powder and promote flow properties) was undergoing reactions to produce stearic acid in the presence of heat and moisture. Through surface chemical causes, stearic acid is a powerful foam breaker. For these reasons, around 1954 the metal stearates were removed from dry chemical formulations. Instead, to avoid caking, they were processed with silicone compounds and became compatible agents.

With the 1959 advent of the more active potassium bicarbonate dry chemical formulations, the problem of protein foam breakdown increased in severity; this

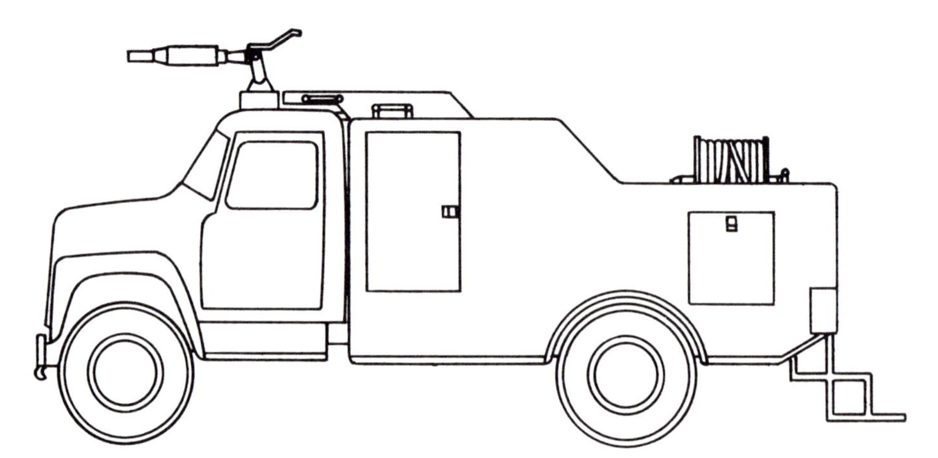

Fig. 10.7. Combined agent discharge (aqueous film-forming foam and potassium dry chemical) from fire fighting vehicle. (From Ansul Company)

was evident, even when the application of silicone treatment made the powder as compatible as possible. The solution to this dilemma would have to lie in the direction of improvement of the nature of the foam so that resistance to the highly efficient potassium agent would be secured.

In 1962 the United States Navy Fire Research Team originated the fluorocarbon foam concentrate now known as AFFF foam (see "Foams for Fighting Fires" in Chapter 9 of this text), which possessed a phenomenally low surface tension and was impervious to attack by all dry chemical agents. Following this progress in attainment of dry chemical agent and foam compatibility, it was possible to combine the two extinguishing agents into a single operation by: (1) using powerful potassium bicarbonate to halt flame propagation, and (2) almost simultaneously applying AFFF foam to maintain permanent extinguishing by cooling and preventing flammable vapor evolution from hydrocarbon fuels. (See "Foam Generating Equipment and Systems" in Chapter 9 of this text. See also Fig. 10.3.)

The combined agent attack on hydrocarbon fuel fires has produced high fire extinguishing efficiencies (up to twelve times more than the early attempts to use similar techniques). The combined agent attack on hydrocarbon fuel fires is in use throughout the world today.

COMBUSTIBLE METAL EXTINGUISHING AGENTS

Another class of fire, Class D, is an important consideration in any discussion of combustible materials and suitable extinguishants for fires in them. There are only a few elemental metals or metal alloys that burn in air. Many of these metals require large inputs of heat before they will reach ignition temperatures, while others ignite spontaneously upon exposure to air. Small amounts of burning metals cause only minor concern; but, when masses of burning metal are involved, the need for halting the vigorous and high temperature oxidation becomes major.

The flame zone in the combustion of metals is a high temperature, highly reactive area. Despite a large amount of investigation on the subject, no successful material or agent has been developed to halt the oxidation reactions of burning metals. Reliance must be placed upon: (1) withdrawal of oxygen (smothering or inerting), or (2) cooling in some way, be it through radiation or heat removal with a cooling agent, in order to lower the temperature of the mass below its ignition point. The following extinguishing agents and methods operate in either or both of these modes: G-1 Powder, Met-L-X Powder, Na-X Powder, inert gas blanketing, Ternary Eutectic Chloride powder, and water.

G-1 Powder is a proprietary mixture composed of graphite and foundry coke, carefully sized to facilitate easy handling and packing when carefully applied to the burning surface with a hand scoop or shovel. An organic phosphate is added to the mixture. The graphite is ignited with difficulty; it is a good heat conductor for absorbing heat and for radiating heat from the burning area. The organic phosphate decomposes to yield a noncombustible smoke that penetrates the spaces between the graphite particles, thus excluding oxygen. The mixture is reducing in action. G-1 Powder is moderately effective for all metal fires.

Met-L-X Powder is also a proprietary mixture consisting largely of carefully sized sodium chloride with added tricalcium phosphate powder and zinc stearate to facilitate discharge from a pressurized dry chemical type extinguisher and delivery nozzle. A small amount of powdered thermoplastic material is added to the mixture so that when the powder is carefully discharged on the burning metal surface in some bulk, the thermoplastic melts and holds the sodium chloride particles together for efficient cooling and smothering of the burning area. It may be used on all metal fires.

Na-X Powder is similar in action to Met-L-X. Its main ingredient is sodium carbonate for a greater resistance to higher temperatures. Na-X Powder is especially designed for sodium metal fires.

Inert gas blanketing of burning metals is an effective extinguishing method when it can be performed. Only the completely inert gases — argon, helium, neon, etc., — may be safely employed in inert gas blanketing.

Ternary Eutectic Chloride powder is of British origin. It contains proportions of potassium chloride, sodium chloride, and barium chloride. This mixture, which melts at the temperature of burning metals, covers the burning surface and excludes oxygen.

Water, the most effective cooling agent, must be applied judiciously only in the cases of burning magnesium and titanium. Water will dissociate into hydrogen and oxygen at the burning metal surface, leading to a more rapid burning. Water must never be used on sodium and similar metals.

ACTIVITIES

1. Explain why each of the following statements is correct or incorrect:
   (a) Water can effectively be used to quench a sodium metal fire.
   (b) Met-L-X Powder can be used on all metal fires.
   (c) G-1 Powder, when used on metal fires, is moderately effective but yields a highly combustible smoke.
   (d) Na-X Powder contains sodium carbonate and is highly effective on sodium metal fires.
   (e) *All* inert gases can be used for safe inert gas blanketing.
   (f) Ternary Eutectic Chloride is composed of a mixture of elements that coats the burning metal, but does not melt.
2. Review the dangers of the toxicity levels of extinguishing agents, paying particular attention to the adverse effects of carbon tetrachloride.
   (a) Prepare a written statement that explains how you feel the adverse effects from the once predominant use of carbon tetrachloride could have been prevented.
   (b) If you had just discovered a new extinguishing agent, how might you go about testing it to prevent any accidental deaths or injuries from its use?
3. Locate the inerting gases in the "Perodic Table of the Elements" in Chapter 1 of this text, and list the main characteristics of each. Then write an explanatory description of the quality of inert gases that makes them so effective in halting combustion.

4. Why is carbon dioxide inert gas classified as a nonpermanent, or "temporary," extinguishing agent? What changes would be necessary to make it a "permanent" extinguishing agent?
5. Describe the two principal types of storage and containment units for carbon dioxide fire protection systems.
6. The gas propulsion form of agent discharge to a hazard has been found to be both practical and reliable. Describe the operation of at least two types of devices that utilize this form of agent discharge. Include in your description the names of gases used in such devices, both in portable and fixed equipment.
7. Explain why the design engineering for piped dry chemical systems is particularly critical.
8. Write, in your own words, an explanation of the "Halon System" of nomenclature for naming the halogenated hydrocarbons.
9. Continuing research efforts and investigation have failed to develop a successful material or agent for halting the oxidation reactions of burning metals. Thus, the lowering of the temperature of a mass of burning metal to below its ignition point is dependent upon what two methods?
10. The combined agent attack on hydrocarbon fuel fires has produced high fire extinguishing efficiencies. Describe how dry chemical agent and foam are combined into a single operation.

SUGGESTED READINGS

Cawood, E., "Stored Pressure: What are the Advantages?," *Fire*, March 1971.

Cholin, R., "How Deep is Deep?," *Fire Journal*, Vol. 66, No. 2, March 1972.

Creitz, E. C., "Extinction of Fires by Halogenated Compounds — A Suggested Mechanism," *Fire Technology*, Vol. 8, No. 2, May 1972.

Fire Protection Handbook, 14th Edition, National Fire Protection Association, Boston, 1976.

Gann, R. G., *Halogenated Fire Suppressants*, ACS Symposium Series No. 16, American Chemical Society, Washington, D.C., 1975.

Guise, A. B., "Potassium Bicarbonate — Base Dry Chemical," NFPA *Quarterly*, July 1962.

OSHA Synopsis: *Fire Extinguisher Requirements for General Fire Protection*, National Association of Fire Equipment Distributors, Chicago.

Richardson, J. F., "General Principles of Fire Extinction," *The Institution of Fire Engineers Quarterly*, Vol. 32, No. 86, June 1972.

Robertson, A. F. and Rappaport, M. W., "Fire Extinguishment in Oxygen Enriched Atmospheres," National Aeronautics and Space Administration Report CR-121150, February 1973.

Symposium on An Appraisal of Halogenated Fire Extinguishing Agents, National Academy of Sciences, Washington, D.C.

Thorne, P. F., "The Anaesthetic Hazard in Total Flooding Systems Using BCF and BTN," *Fire*, April 1971.

Williamson, H. V., "Halon 1301 — Minimum Concentrations for Extinguishing Deep-Seated Fires," *Fire Technology*, Vol. 8, No. 4, November 1972.

APPENDIX I
CONVERSION FACTORS*

As this text is being prepared, many industrial firms, governmental agencies, and other organizations are beginning the long and complicated task of converting to the metric system. The United States, Canada, and a few other smaller countries will soon adopt the International System of Units (termed "SI"), thereby joining other industrialized nations that have already adopted this system. When the SI system is officially adopted, all present U.S. measurement units will be replaced by metric designations.

Obviously, standards-making organizations will need to include metric measurements and terms in their standards and related literature. Manufacturers' data sheets and descriptive literature, governmental publications, and even verbal communications concerning measurements will have to include metric terms. Of equal importance is the fact that many products will be changed in size and design to produce optimum performance under the metric system. Within the next decade it is likely that a large percentage of fire apparatus and fire protection equipment will be considerably altered to meet the requirements of metrical measurements.

At this time it is not possible to predict how the numerous dimensions and calculations relating to fire protection will be changed, but the basic equivalents of metric and English systems can be summarized for easier mathematical conversions. This appendix includes the terms and quantities most likely to be used in fire protection calculations.

The SI system consists of seven base units and two supplementary units.

Base Units.

Quantity	Unit	SI Symbol
length	metric	m
mass	kilogram	kg
time	second	s
electric current	ampere	A
thermodynamic temperature	kelvin	K
luminous intensity	candela	cd
amount of substance	mole	mol

Supplementary Units.

Quantity	Unit	SI Symbol
plane angle	radian	rad
solid angle	steradian	sr

Derived Units. All of the additional units with a few exceptions are derivations of the base and supplementary units.

---

* From *Annual Fire Protection Reference Directory*, National Fire Protection Association, Boston, 1975–1976.

CONVERSION FACTORS (Continued)

SI Prefixes. The usual practice is to use not more than 4 digits and then use the next higher prefix. If a table or chart would be clearer using more than 4 digits, this can be waived.

Name	Symbol	Multiplication Factor
tera	T	10^{12}
giga	G	10^{9}
mega	M	10^{6}
kilo	k	10^{3}
milli	m	10^{-3}
micro	μ	10^{-6}
nano	n	10^{-9}
pico	p	10^{-12}
femto	f	10^{-15}
atto	a	10^{-18}

Units and Conversions. The conversions listed have been rounded off to three places to the right of the decimal for numbers greater than (0.1), and to a maximum of three significant figures for conversions less than (0.1). Exact conversions are noted. If more accurate conversions are required, refer to ANSI Z210.1–1973, Metric Practice Guide, available from American National Standards Institute.

Area.

Name of Unit			Conversion
U.S.	SI	SI Symbol	U.S. to SI
square inch	square millimetre	mm^2	1 $in.^2$ = 645.160 mm^2*
square foot	square metre	m^2	1 ft^2 = 0.0929 m^2
square yard	square metre	m^2	1 yd^2 = 0.836 m^2

* = exact

Density.

Name of Unit			Conversion
U.S.	SI	SI Symbol	U.S. to SI
pound per cubic foot	kilogram per cubic metre	kg/m^3	1 lb/ft^3 = 16.018 kg/m^3
pound per gallon+	kilogram per cubic metre	kg/m^3	1 lb/gal = 119.827 kg/m^3

+U.S. liquid measure

CONVERSION FACTORS (Continued)

Electricity.

Name of Unit U.S.	SI	SI Symbol	Conversion U.S. to SI
ampere	ampere	A	same
cycle per second	hertz	Hz	1 cycle/sec. = 1 Hz
farad	farad	F	same
gauss	tesla	T	1 gauss = 0.0001 T
oersted	ampere per metre	A/m	1 oersted = 79.577 A/m
ohm	ohm	Ω	same
volt	volt	V	same
watt	watt	W	same

Heat and Heat Transfer.

Name of Unit U.S.	SI	SI Symbol	Conversion U.S. to SI
British Thermal Unit	kiloJoule	kJ	1 Btu = 1.055 kJ
calorie	Joule	J	1 cal = 4.187 J
calorie/sq. cent.	kiloJoule per square metre	kJ/m^2	1 cal/cm^2 = 41.868 kJ/m^2
Btu/hr	watt	W	1 Btu/hr = 0.293 W
Btu/lb	kiloJoule per kilogram	kJ/kg	1 Btu/lb = 2.326 kJ/kg
Btu/ft^3	kiloJoule per cubic metre	kJ/m^3	1 Btu/ft^3 = 37.259 kJ/m^3
Btu/ft^2 hr	watt per square metre	W/m^2	1 Btu/ft^2 hr = 3.155 W/m^2

The values of Btu and calorie are taken from the International Table.

Illumination.

Name of Unit U.S.	SI	SI Symbol	Conversion U.S. to SI
footcandle	lux	lx	1 footcandle = 10.764 lx

The lux is equal to $\frac{cd \cdot sr}{m^2}$. The term cd · sr (candela · steradian) is the lumen. The footcandle is equal to the illumination of one lumen per square foot.

Length.

Name of Unit U.S.	SI	SI Symbol	Conversion U.S. to SI
inch	millimetre	mm	1 in. = 25.400 mm*
inch	centimetre	cm	1 in. = 2.540 cm*

CONVERSION FACTORS (Continued)

Length. ***(Continued)***

foot	metre	m	1 ft = 0.305 m
mile	kilometre	km	1 mile = 1.609 km

(*N.B.*: mm, m, and km are the preferred units. Millimetres (mm) should be used instead of centimetre [cm].)

* = exact

Mass and Force.

Name of Unit			Conversion
U.S.	SI	SI Symbol	U.S. to SI
1 pound (mass)	kilogram	kg	1 pound (mass) = 0.454 kg
1 ounce (mass)	gram	g	1 ounce (mass) = 28.350 g
1 pound (force)	newton	N	1 pound (force) = 4.448 N

A newton (N) is the force that gives a body one kilogram in weight an acceleration of 1 metre per second every second $\frac{(1\ \text{kg} \cdot \text{m})}{\text{s}^2}$.

Volume.

Name of Unit			Conversion
U.S.	SI	SI Symbol	U.S. to SI
gallon+	litre	l	1 gal = 3.785 l
gallon+	cubic metre	m^3	1 gal = 0.00379 m^3
quart+	litre	l	1 qt = 0.946 l
pint+	litre	l	1 pt = 0.473 l
fluid ounce+	millilitre	ml	1 fluid oz = 29.574 ml
cubic foot	cubic metre	m^3	1 ft^3 = 0.0283 m^3
cubic yard	cubic metre	m^3	1 yd^3 = 0.765 m^3

\+ Gallon, quarts, pints, and fluid ounces are U.S. liquid measures.
(*N.B.*: 1 cubic diametre = 1 litre approximately.)

Water Density and Flow Rate.

Name of Unit			Conversion
U.S.	SI	SI Symbol	U.S. to SI
gallon per minute+	litre per minute	l/min	1 gpm = 3.785 l/min
gallon per minute+ per square foot	litre per minute per square metre	$l/\text{min.m}^2$	1 gpm/ft^2 = 40.746 $l/\text{min.m}^2$

\+U.S. liquid measure

CONVERSION FACTORS (Continued)

Pressure and Stress.

Name of Unit U.S.	SI	SI Symbol	Conversion U.S. to SI
pounds per square inch	pascal	Pa	1 psi = 6894.757 Pa

Pressures are usually given in kPa and stresses are usually given in MPa. (*N.B.*: 1 bar = 10^5 Pa.)

Temperature.

Name of Unit U.S.	SI	SI Symbol	Conversion U.S. to SI
degree Fahrenheit	degree Celsius	°C	5/9 (°F-32) = °C

The degree Celsius is the same as the degree Centigrade. The official SI unit for temperature is the Kelvin (K. K = °C + 273.15) but the use of C can be continued where absolute temperatures are not used.

APPENDIX II
TEMPERATURE CONVERSION TABLES*

The column of figures headed "Reading in °F or °C to be converted" refers to the temperature, either in degrees Fahrenheit or degrees Centigrade, which is to be converted. If converting from Fahrenheit degrees to Centigrade degrees, the equivalent temperature will be found in the column headed "°C"; if converting from degrees Centigrade to degrees Fahrenheit, the equivalent temperature will be found in the column headed "°F." This arrangement is very similar to that of Sauveur and Boylston, copyrighted 1920, and is published with their permission.

°F	Reading in °F or °C to be converted	°C	°F	Reading in °F or °C to be converted	°C	°F	Reading in °F or °C to be converted	°C
....	−458	−272.22	....	−368	−222.22	....	−278	−172.22
....	−456	−271.11	....	−366	−221.11	....	−276	−171.11
....	−454	−270.00	....	−364	−220.00	....	−274	−170.00
....	−452	−268.89	....	−362	−218.89	−457.6	−272	−168.89
....	−450	−267.78	....	−360	−217.78	−454.0	−270	−167.78
....	−448	−266.67	....	−358	−216.67	−450.4	−268	−166.67
....	−446	−265.56	....	−356	−215.56	−446.8	−266	−165.56
....	−444	−264.44	....	−354	−214.44	−443.2	−264	−164.44
....	−442	−263.33	....	−352	−213.33	−439.6	−262	−163.33
....	−440	−262.22	....	−350	−212.22	−436.0	−260	−162.22
....	−438	−261.11	....	−348	−211.11	−432.4	−258	−161.11
....	−436	−260.00	....	−346	−210.00	−428.8	−256	−160.00
....	−434	−258.89	....	−344	−208.89	−425.2	−254	−158.89
....	−432	−257.78	....	−342	−207.78	−421.6	−252	−157.78
....	−430	−256.67	....	−340	−206.67	−418.0	−250	−156.67
....	−428	−255.56	....	−338	−205.56	−414.4	−248	−155.56
....	−426	−254.44	....	−336	−204.44	−410.8	−246	−154.44
....	−424	−253.33	....	−334	−203.33	−407.2	−244	−153.33
....	−422	−252.22	....	−332	−202.22	−403.6	−242	−152.22
....	−420	−251.11	....	−330	−201.11	−400.0	−240	−151.11
....	−418	−250.00	....	−328	−200.00	−396.4	−238	−150.00
....	−416	−248.89	....	−326	−198.89	−392.8	−236	−148.89
....	−414	−247.78	....	−324	−197.78	−389.2	−234	−147.78
....	−412	−246.67	....	−322	−196.67	−385.6	−232	−146.67
....	−410	−245.56	....	−320	−195.56	−382.0	−230	−145.56
....	−408	−244.44	....	−318	−194.44	−378.4	−228	−144.44
....	−406	−243.33	....	−316	−193.33	−374.8	−226	−143.33
....	−404	−242.22	....	−314	−192.22	−371.2	−224	−142.22
....	−402	−241.11	....	−312	−191.11	−367.6	−222	−141.11
....	−400	−240.00	....	−310	−190.00	−364.0	−220	−140.00
....	−398	−238.89	....	−308	−188.89	−360.4	−218	−138.89
....	−396	−237.78	....	−306	−187.78	−356.8	−216	−137.78
....	−394	−236.67	....	−304	−186.67	−353.2	−214	−136.67
....	−392	−235.56	....	−302	−185.56	−349.6	−212	−135.56
....	−390	−234.44	....	−300	−184.44	−346.0	−210	−134.44
....	−388	−233.33	....	−298	−183.33	−342.4	−208	−133.33
....	−386	−232.22	....	−296	−182.22	−338.8	−206	−132.22
....	−384	−231.11	....	−294	−181.11	−335.2	−204	−131.11
....	−382	−230.00	....	−292	−180.00	−331.6	−202	−130.00
....	−380	−228.89	....	−290	−178.89	−328.0	−200	−128.89
....	−378	−227.78	....	−288	−177.78	−324.4	−198	−127.78
....	−376	−226.67	....	−286	−176.67	−320.8	−196	−126.67
....	−374	−225.56	....	−284	−175.56	−317.2	−194	−125.56
....	−372	−224.44	....	−282	−174.44	−313.6	−192	−124.44
....	−370	−223.33	....	−280	−173.33	−310.0	−190	−123.33

(Continued)

* From *Lange's Handbook of Chemistry*, by Norbert Adolph Lange, 11th Edition, McGraw-Hill, New York, 1972.

Temperature Conversion Tables (*Continued*)

°F	Reading in °F or °C to be converted	°C	°F	Reading in °F or °C to be converted	°C	°F	Reading in °F or °C to be converted	°C
−306.4	−188	−122.22	−79.6	−62	−52.22	+82.4	+28	−2.22
−302.8	−186	−121.11	−76.0	−60	−51.11	+84.2	+29	−1.67
−299.2	−184	−120.00	−72.4	−58	−50.00	+86.0	+30	−1.11
−295.6	−182	−118.89	−68.8	−56	−48.89	+87.8	+31	−0.56
−292.0	−180	−117.78	−65.2	−54	−47.78	+89.6	+32	±0.00
−288.4	−178	−116.67	−61.6	−52	−46.67	+91.4	+33	+0.56
−284.8	−176	−115.56	−58.0	−50	−45.56	+93.2	+34	+1.11
−281.2	−174	−114.44	−54.4	−48	−44.44	+95.0	+35	+1.67
−277.6	−172	−113.33	−50.8	−46	−43.33	+96.8	+36	+2.22
−274.0	−170	−112.22	−47.2	−44	−42.22	+98.6	+37	+2.78
−270.4	−168	−111.11	−43.6	−42	−41.11	+100.4	+38	+3.33
−266.8	−166	−110.00	−40.0	−40	−40.00	+102.2	+39	+3.89
−263.2	−164	−108.89	−36.4	−38	−38.89	+104.0	+40	+4.44
−259.6	−162	−107.78	−32.8	−36	−37.78	+105.8	+41	+5.00
−256.0	−160	−106.67	−29.2	−34	−36.67	+107.6	+42	+5.56
−252.4	−158	−105.56	−25.6	−32	−35.56	+109.4	+43	+6.11
−248.8	−156	−104.44	−22.0	−30	−34.44	+111.2	+44	+6.67
−245.2	−154	−103.33	−18.4	−28	−33.33	+113.0	+45	+7.22
−241.6	−152	−102.22	−14.8	−26	−32.22	+114.8	+46	+7.78
−238.0	−150	−101.11	−11.2	−24	−31.11	+116.6	+47	+8.33
−234.4	−148	−100.00	−7.6	−22	−30.00	+118.4	+48	+8.89
−230.8	−146	−98.89	−4.0	−20	−28.89	+120.2	+49	+9.44
−227.2	−144	−97.78	−0.4	−18	−27.78	+122.0	+50	+10.00
−223.6	−142	−96.67	+3.2	−16	−26.67	+123.8	+51	+10.56
−220.0	−140	−95.56	+6.8	−14	−25.56	+125.6	+52	+11.11
−216.4	−138	−94.44	+10.4	−12	−24.44	+127.4	+53	+11.67
−212.8	−136	−93.33	+14.0	−10	−23.33	+129.2	+54	+12.22
−209.2	−134	−92.22	+17.6	−8	−22.22	+131.0	+55	+12.78
−205.6	−132	−91.11	+19.4	−7	−21.67	+132.8	+56	+13.33
−202.0	−130	−90.00	+21.2	−6	−21.11	+134.6	+57	+13.89
−198.4	−128	−88.89	+23.0	−5	−20.56	+136.4	+58	+14.44
−194.8	−126	−87.78	+24.8	−4	−20.00	+138.2	+59	+15.00
−191.2	−124	−86.67	+26.6	−3	−19.44	+140.0	+60	+15.56
−187.6	−122	−85.56	+28.4	−2	−18.89	+141.8	+61	+16.11
−184.0	−120	−84.44	+30.2	−1	−18.33	+143.6	+62	+16.67
−180.4	−118	−83.33	+32.0	±0	−17.78	+145.4	+63	+17.22
−176.8	−116	−82.22	+33.8	+1	−17.22	+147.2	+64	+17.78
−173.2	−114	−81.11	+35.6	+2	−16.67	+149.0	+65	+18.33
−169.6	−112	−80.00	+37.4	+3	−16.11	+150.8	+66	+18.89
−166.0	−110	−78.89	+39.2	+4	−15.56	+152.6	+67	+19.44
−162.4	−108	−77.78	+41.0	+5	−15.00	+154.4	+68	+20.00
−158.8	−106	−76.67	+42.8	+6	−14.44	+156.2	+69	+20.56
−155.2	−104	−75.56	+44.6	+7	−13.89	+158.0	+70	+21.11
−151.6	−102	−74.44	+46.4	+8	−13.33	+159.8	+71	+21.67
−148.0	−100	−73.33	+48.2	+9	−12.78	+161.6	+72	+22.22
−144.4	−98	−72.22	+50.0	+10	−12.22	+163.4	+73	+22.78
−140.8	−96	−71.11	+51.8	+11	−11.67	+165.2	+74	+23.33
−137.2	−94	−70.00	+53.6	+12	−11.11	+167.0	+75	+23.89
−133.6	−92	−68.89	+55.4	+13	−10.56	+168.8	+76	+24.44
−130.0	−90	−67.78	+57.2	+14	−10.00	+170.6	+77	+25.00
−126.4	−88	−66.67	+59.0	+15	−9.44	+172.4	+78	+25.56
−122.8	−86	−65.56	+60.8	+16	−8.89	+174.2	+79	+26.11
−119.2	−84	−64.44	+62.6	+17	−8.33	+176.0	+80	+26.67
−115.6	−82	−63.33	+64.4	+18	−7.78	+177.8	+81	+27.22
−112.0	−80	−62.22	+66.2	+19	−7.22	+179.6	+82	+27.78
−108.4	−78	−61.11	+68.0	+20	−6.67	+181.4	+83	+28.33
−104.8	−76	−60.00	+69.8	+21	−6.11	+183.2	+84	+28.89
−101.2	−74	−58.89	+71.6	+22	−5.56	+185.0	+85	+29.44
−97.6	−72	−57.78	+73.4	+23	−5.00	+186.8	+86	+30.00
−94.0	−70	−56.67	+75.2	+24	−4.44	+188.6	+87	+30.56
−90.4	−68	−55.56	+77.0	+25	−3.89	+190.4	+88	+31.11
−86.8	−66	−54.44	+78.8	+26	−3.33	+192.2	+89	+31.67
−83.2	−64	−53.33	+80.6	+27	−2.78	+194.0	+90	+32.22

(Continued)

Temperature Conversion Tables (*Continued*)

°F	Reading in °F or °C to be converted	°C	°F	Reading in °F or °C to be converted	°C	°F	Reading in °F or °C to be converted	°C
+195.8	+91	+32.78	+309.2	+154	+67.78	+422.6	+217	+102.78
+197.6	+92	+33.33	+311.0	+155	+68.33	+424.4	+218	+103.33
+199.4	+93	+33.89	+312.8	+156	+68.89	+426.2	+219	+103.89
+201.2	+94	+34.44	+314.6	+157	+69.44	+428.0	+220	+104.44
+203.0	+95	+35.00	+316.4	+158	+70.00	+431.6	+222	+105.56
+204.8	+96	+35.56	+318.2	+159	+70.56	+435.2	+224	+106.67
+206.6	+97	+36.11	+320.0	+160	+71.11	+438.8	+226	+107.78
+208.4	+98	+36.67	+321.8	+161	+71.67	+442.4	+228	+108.89
+210.2	+99	+37.22	+323.6	+162	+72.22	+446.0	+230	+110.00
+212.0	+100	+37.78	+325.4	+163	+72.78	+449.6	+232	+111.11
+213.8	+101	+38.33	+327.2	+164	+73.33	+453.2	+234	+112.22
+215.6	+102	+38.89	+329.0	+165	+73.89	+456.8	+236	+113.33
+217.4	+103	+39.44	+330.8	+166	+74.44	+460.4	+238	+114.44
+219.2	+104	+40.00	+332.6	+167	+75.00	+464.0	+240	+115.56
+221.0	+105	+40.56	+334.4	+168	+75.56	+467.6	+242	+116.67
+222.8	+106	+41.11	+336.2	+169	+76.11	+471.2	+244	+117.78
+224.6	+107	+41.67	+338.0	+170	+76.67	+474.8	+246	+118.89
+226.4	+108	+42.22	+339.8	+171	+77.22	+478.4	+248	+120.00
+228.2	+109	+42.78	+341.6	+172	+77.78	+482.0	+250	+121.11
+230.0	+110	+43.33	+343.4	+173	+78.33	+485.6	+252	+122.22
+231.8	+111	+43.89	+345.2	+174	+78.89	+489.2	+254	+123.33
+233.6	+112	+44.44	+347.0	+175	+79.44	+492.8	+256	+124.44
+235.4	+113	+45.00	+348.8	+176	+80.00	+496.4	+258	+125.56
+237.2	+114	+45.56	+350.6	+177	+80.56	+500.0	+260	+126.67
+239.0	+115	+46.11	+352.4	+178	+81.11	+503.6	+262	+127.78
+240.8	+116	+46.67	+354.2	+179	+81.67	+507.2	+264	+128.89
+242.6	+117	+47.22	+356.0	+180	+82.22	+510.8	+266	+130.00
+244.4	+118	+47.78	+357.8	+181	+82.78	+514.4	+268	+131.11
+246.2	+119	+48.33	+359.6	+182	+83.33	+518.0	+270	+132.22
+248.0	+120	+48.89	+361.4	+183	+83.89	+521.6	+272	+133.33
+249.8	+121	+49.44	+363.2	+184	+84.44	+525.2	+274	+134.44
+251.6	+122	+50.00	+365.0	+185	+85.00	+528.8	+276	+135.56
+253.4	+123	+50.56	+366.8	+186	+85.56	+532.4	+278	+136.67
+255.2	+124	+51.11	+368.6	+187	+86.11	+536.0	+280	+137.78
+257.0	+125	+51.67	+370.4	+188	+86.67	+539.6	+282	+138.89
+258.8	+126	+52.22	+372.2	+189	+87.22	+543.2	+284	+140.00
+260.6	+127	+52.78	+374.0	+190	+87.78	+546.8	+286	+141.11
+262.4	+128	+53.33	+375.8	+191	+88.33	+550.4	+288	+142.22
+264.2	+129	+53.89	+377.6	+192	+88.89	+554.0	+290	+143.33
+266.0	+130	+54.44	+379.4	+193	+89.44	+557.6	+292	+144.44
+267.8	+131	+55.00	+381.2	+194	+90.00	+561.2	+294	+145.56
+269.6	+132	+55.56	+383.0	+195	+90.56	+564.8	+296	+146.67
+271.4	+133	+56.11	+384.8	+196	+91.11	+568.4	+298	+147.78
+273.2	+134	+56.67	+386.6	+197	+91.67	+572.0	+300	+148.89
+275.0	+135	+57.22	+388.4	+198	+92.22	+575.6	+302	+150.00
+276.8	+136	+57.78	+390.2	+199	+92.78	+579.2	+304	+151.11
+278.6	+137	+58.33	+392.0	+200	+93.33	+582.8	+306	+152.22
+280.4	+138	+58.89	+393.8	+201	+93.89	+586.4	+308	+153.33
+282.2	+139	+59.44	+395.6	+202	+94.44	+590.0	+310	+154.44
+284.0	+140	+60.00	+397.4	+203	+95.00	+593.6	+312	+155.56
+285.8	+141	+60.56	+399.2	+204	+95.56	+597.2	+314	+156.67
+287.6	+142	+61.11	+401.0	+205	+96.11	+600.8	+316	+157.78
+289.4	+143	+61.67	+402.8	+206	+96.67	+604.4	+318	+158.89
+291.2	+144	+62.22	+404.6	+207	+97.22	+608.0	+320	+160.00
+293.0	+145	+62.78	+406.4	+208	+97.78	+611.6	+322	+161.11
+294.8	+146	+63.33	+408.2	+209	+98.33	+615.2	+324	+162.22
+296.6	+147	+63.89	+410.0	+210	+98.89	+618.8	+326	+163.33
+298.4	+148	+64.44	+411.8	+211	+99.44	+622.4	+328	+164.44
+300.2	+149	+65.00	+413.6	+212	+100.00	+626.0	+330	+165.56
+302.0	+150	+65.56	+415.4	+213	+100.56	+629.6	+332	+166.67
+303.8	+151	+66.11	+417.2	+214	+101.11	+633.2	+334	+167.78
+305.6	+152	+66.67	+419.0	+215	+101.67	+636.8	+336	+168.89
+307.4	+153	+67.22	+420.8	+216	+102.22	+640.4	+338	+170.00

(Continued)

Temperature Conversion Tables (*Continued*)

°F	Reading in °F or °C to be converted	°C	°F	Reading in °F or °C to be converted	°C	°F	Reading in °F or °C to be converted	°C
+644.0	+340	+171.11	+870.8	+466	+241.11	+1097.6	+592	+311.11
+647.6	+342	+172.22	+874.4	+468	+242.22	+1101.2	+594	+312.22
+651.2	+344	+173.33	+878.0	+470	+243.33	+1104.8	+596	+313.33
+654.8	+346	+174.44	+881.6	+472	+244.44	+1108.4	+598	+314.44
+658.4	+348	+175.56	+885.2	+474	+245.56	+1112.0	+600	+315.56
+662.0	+350	+176.67	+888.8	+476	+246.67	+1115.6	+602	+316.67
+665.6	+352	+177.78	+892.4	+478	+247.78	+1119.2	+604	+317.78
+669.2	+354	+178.89	+896.0	+480	+248.89	+1122.8	+606	+318.89
+672.8	+356	+180.00	+899.6	+482	+250.00	+1126.4	+608	+320.00
+676.4	+358	+181.11	+903.2	+484	+251.11	+1130.0	+610	+321.11
+680.0	+360	+182.22	+906.8	+486	+252.22	+1133.6	+612	+322.22
+683.6	+362	+183.33	+910.4	+488	+253.33	+1137.2	+614	+323.33
+687.2	+364	+184.44	+914.0	+490	+254.44	+1140.8	+616	+324.44
+690.8	+366	+185.56	+917.6	+492	+255.56	+1144.4	+618	+325.56
+694.4	+368	+186.67	+921.2	+494	+256.67	+1148.0	+620	+326.67
+698.0	+370	+187.78	+924.8	+496	+257.78	+1151.6	+622	+327.78
+701.6	+372	+188.89	+928.4	+498	+258.89	+1155.2	+624	+328.89
+705.2	+374	+190.00	+932.0	+500	+260.00	+1158.8	+626	+330.00
+708.8	+376	+191.11	+935.6	+502	+261.11	+1162.4	+628	+331.11
+712.4	+378	+192.22	+939.2	+504	+262.22	+1166.0	+630	+332.22
+716.0	+380	+193.33	+942.8	+506	+263.33	+1169.6	+632	+333.33
+719.6	+382	+194.44	+946.4	+508	+264.44	+1173.2	+634	+334.44
+723.2	+384	+195.56	+950.0	+510	+265.56	+1176.8	+636	+335.56
+726.8	+386	+196.67	+953.6	+512	+266.67	+1180.4	+638	+336.67
+730.4	+388	+197.78	+957.2	+514	+267.78	+1184.0	+640	+337.78
+734.0	+390	+198.89	+960.8	+516	+268.89	+1187.6	+642	+338.89
+737.6	+392	+200.00	+964.4	+518	+270.00	+1191.2	+644	+340.00
+741.2	+394	+201.11	+968.0	+520	+271.11	+1194.8	+646	+341.11
+744.8	+396	+202.22	+971.6	+522	+272.22	+1198.4	+648	+342.22
+748.4	+398	+203.33	+975.2	+524	+273.33	+1202.0	+650	+343.33
+752.0	+400	+204.44	+978.8	+526	+274.44	+1205.6	+652	+344.44
+755.6	+402	+205.56	+982.4	+528	+275.56	+1209.2	+654	+345.56
+759.2	+404	+206.67	+986.0	+530	+276.67	+1212.8	+656	+346.67
+762.8	+406	+207.78	+989.6	+532	+277.78	+1216.4	+658	+347.78
+766.4	+408	+208.89	+993.2	+534	+278.89	+1220.0	+660	+348.89
+770.0	+410	+210.00	+996.8	+536	+280.00	+1223.6	+662	+350.00
+773.6	+412	+211.11	+1000.4	+538	+281.11	+1227.2	+664	+351.11
+777.2	+414	+212.22	+1004.0	+540	+282.22	+1230.8	+666	+352.22
+780.8	+416	+213.33	+1007.6	+542	+283.33	+1234.4	+668	+353.33
+784.4	+418	+214.44	+1011.2	+544	+284.44	+1238.0	+670	+354.44
+788.0	+420	+215.56	+1014.8	+546	+285.56	+1241.6	+672	+355.56
+791.6	+422	+216.67	+1018.4	+548	+286.67	+1245.2	+674	+356.67
+795.2	+424	+217.78	+1022.0	+550	+287.78	+1248.8	+676	+357.78
+798.8	+426	+218.89	+1025.6	+552	+288.89	+1252.4	+678	+358.89
+802.4	+428	+220.00	+1029.2	+554	+290.00	+1256.0	+680	+360.00
+806.0	+430	+221.11	+1032.8	+556	+291.11	+1259.6	+682	+361.11
+809.6	+432	+222.22	+1036.4	+558	+292.22	+1263.2	+684	+362.22
+813.2	+434	+223.33	+1040.0	+560	+293.33	+1266.8	+686	+363.33
+816.8	+436	+224.44	+1043.6	+562	+294.44	+1270.4	+688	+364.44
+820.4	+438	+225.56	+1047.2	+564	+295.56	+1274.0	+690	+365.56
+824.0	+440	+226.67	+1050.8	+566	+296.67	+1277.6	+692	+366.67
+827.6	+442	+227.78	+1054.4	+568	+297.78	+1281.2	+694	+367.78
+831.2	+444	+228.89	+1058.0	+570	+298.89	+1284.8	+696	+368.89
+834.8	+446	+230.00	+1061.6	+572	+300.00	+1288.4	+698	+370.00
+838.4	+448	+231.11	+1065.2	+574	+301.11	+1292.0	+700	+371.11
+842.0	+450	+232.22	+1068.8	+576	+302.22	+1295.6	+702	+372.22
+845.6	+452	+233.33	+1072.4	+578	+303.33	+1299.2	+704	+373.33
+849.2	+454	+234.44	+1076.0	+580	+304.44	+1302.8	+706	+374.44
+852.8	+456	+235.56	+1079.6	+582	+305.56	+1306.4	+708	+375.56
+856.4	+458	+236.67	+1083.2	+584	+306.67	+1310.0	+710	+376.67
+860.0	+460	+237.78	+1086.8	+586	+307.78	+1313.6	+712	+377.78
+863.6	+462	+238.89	+1090.4	+588	+308.89	+1317.2	+714	+378.89
+867.2	+464	+240.00	+1094.0	+590	+310.00	+1320.8	+716	+380.00

(Continued)

Temperature Conversion Tables (*Continued*)

°F	Reading in °F or °C to be converted	°C	°F	Reading in °F or °C to be converted	°C	°F	Reading in °F or °C to be converted	°C
+1324.4	+718	+381.11	+1551.2	+844	+451.11	+1778.0	+970	+521.11
+1328.0	+720	+382.22	+1554.8	+846	+452.22	+1781.6	+972	+522.22
+1331.6	+722	+383.33	+1558.4	+848	+453.33	+1785.2	+974	+523.33
+1335.2	+724	+384.44	+1562.0	+850	+454.44	+1788.8	+976	+524.44
+1338.8	+726	+385.56	+1565.6	+852	+455.56	+1792.4	+978	+525.56
+1342.4	+728	+386.67	+1569.2	+854	+456.67	+1796.0	+980	+526.67
+1346.0	+730	+387.78	+1572.8	+856	+457.78	+1799.6	+982	+527.78
+1349.6	+732	+388.89	+1576.4	+858	+458.89	+1803.2	+984	+528.89
+1353.2	+734	+390.00	+1580.0	+860	+460.00	+1806.8	+986	+530.00
+1356.8	+736	+391.11	+1583.6	+862	+461.11	+1810.4	+988	+531.11
+1360.4	+738	+392.22	+1587.2	+864	+462.22	+1814.0	+990	+532.22
+1364.0	+740	+393.33	+1590.8	+866	+463.33	+1817.6	+992	+533.33
+1367.6	+742	+394.44	+1594.4	+868	+464.44	+1821.2	+994	+534.44
+1371.2	+744	+395.56	+1598.0	+870	+465.56	+1824.8	+996	+535.56
+1374.8	+746	+396.67	+1601.6	+872	+466.67	+1828.4	+998	+536.67
+1378.4	+748	+397.78	+1605.2	+874	+467.78	+1832.0	+1000	+537.78
+1382.0	+750	+398.89	+1608.8	+876	+468.89	+1850.0	+1010	+543.33
+1385.6	+752	+400.00	+1612.4	+878	+470.00	+1868.0	+1020	+548.89
+1389.2	+754	+401.11	+1616.0	+880	+471.11	+1886.0	+1030	+554.44
+1392.8	+756	+402.22	+1619.6	+882	+472.22	+1904.0	+1040	+560.00
+1396.4	+758	+403.33	+1623.2	+884	+473.33	+1922.0	+1050	+565.56
+1400.0	+760	+404.44	+1626.8	+886	+474.44	+1940.0	+1060	+571.11
+1403.6	+762	+405.56	+1630.4	+888	+475.56	+1958.0	+1070	+576.67
+1407.2	+764	+406.67	+1634.0	+890	+476.67	+1976.0	+1080	+582.22
+1410.8	+766	+407.78	+1637.6	+892	+477.78	+1994.0	+1090	+587.78
+1414.4	+768	+408.89	+1641.2	+894	+478.89	+2012.0	+1100	+593.33
+1418.0	+770	+410.00	+1644.8	+896	+480.00	+2030.0	+1110	+598.89
+1421.6	+772	+411.11	+1648.4	+898	+481.11	+2048.0	+1120	+604.44
+1425.2	+774	+412.22	+1652.0	+900	+482.22	+2066.0	+1130	+610.00
+1428.8	+776	+413.33	+1655.6	+902	+483.33	+2084.0	+1140	+615.56
+1432.4	+778	+414.44	+1659.2	+904	+484.44	+2102.0	+1150	+621.11
+1436.0	+780	+415.56	+1662.8	+906	+485.56	+2120.0	+1160	+626.67
+1439.6	+782	+416.67	+1666.4	+908	+486.67	+2138.0	+1170	+632.22
+1443.2	+784	+417.78	+1670.0	+910	+487.78	+2156.0	+1180	+637.78
+1446.8	+786	+418.89	+1673.6	+912	+488.89	+2174.0	+1190	+643.33
+1450.4	+788	+420.00	+1677.2	+914	+490.00	+2192.0	+1200	+648.89
+1454.0	+790	+421.11	+1680.8	+916	+491.11	+2210.0	+1210	+654.44
+1457.6	+792	+422.22	+1684.4	+918	+492.22	+2228.0	+1220	+660.00
+1461.2	+794	+423.33	+1688.0	+920	+493.33	+2246.0	+1230	+665.56
+1464.8	+796	+424.44	+1691.6	+922	+494.44	+2264.0	+1240	+671.11
+1468.4	+798	+425.56	+1695.2	+924	+495.56	+2282.0	+1250	+676.67
+1472.0	+800	+426.67	+1698.8	+926	+496.67	+2300.0	+1260	+682.22
+1475.6	+802	+427.78	+1702.4	+928	+497.78	+2318.0	+1270	+687.78
+1479.2	+804	+428.89	+1706.0	+930	+498.89	+2336.0	+1280	+693.33
+1482.8	+806	+430.00	+1709.6	+932	+500.00	+2354.0	+1290	+698.89
+1486.4	+808	+431.11	+1713.2	+934	+501.11	+2372.0	+1300	+704.44
+1490.0	+810	+432.22	+1716.8	+936	+502.22	+2390.0	+1310	+710.00
+1493.6	+812	+433.33	+1720.4	+938	+503.33	+2408.0	+1320	+715.56
+1497.2	+814	+434.44	+1724.0	+940	+504.44	+2426.0	+1330	+721.11
+1500.8	+816	+435.56	+1727.6	+942	+505.56	+2444.0	+1340	+726.67
+1504.4	+818	+436.67	+1731.2	+944	+506.67	+2462.0	+1350	+732.22
+1508.0	+820	+437.78	+1734.8	+946	+507.78	+2480.0	+1360	+737.78
+1511.6	+822	+438.89	+1738.4	+948	+508.89	+2498.0	+1370	+743.33
+1515.2	+824	+440.00	+1742.0	+950	+510.00	+2516.0	+1380	+748.89
+1518.8	+826	+441.11	+1745.6	+952	+511.11	+2534.0	+1390	+754.44
+1522.4	+828	+442.22	+1749.2	+954	+512.22	+2552.0	+1400	+760.00
+1526.0	+830	+443.33	+1752.8	+956	+513.33	+2570.0	+1410	+765.56
+1529.6	+832	+444.44	+1756.4	+958	+514.44	+2588.0	+1420	+771.11
+1533.2	+834	+445.56	+1760.0	+960	+515.56	+2606.0	+1430	+776.67
+1536.8	+836	+446.67	+1763.6	+962	+516.67	+2624.0	+1440	+782.22
+1540.4	+838	+447.78	+1767.2	+964	+517.78	+2642.0	+1450	+787.78
+1544.0	+840	+448.89	+1770.8	+966	+518.89	+2660.0	+1460	+793.33
+1547.6	+842	+450.00	+1774.4	+968	+520.00	+2678.0	+1470	+798.89

(Continued)

Temperature Conversion Tables (*Continued*)

°F	Reading in °F or °C to be converted	°C
+2696.0	+1480	+804.44
+2714.0	+1490	+810.00
+2732.0	+1500	+815.56
+2750.0	+1510	+821.11
+2768.0	+1520	+826.67
+2786.0	+1530	+832.22
+2804.0	+1540	+837.78
+2822.0	+1550	+843.33
+2840.0	+1560	+848.89
+2858.0	+1570	+854.44
+2876.0	+1580	+860.00
+2894.0	+1590	+865.56
+2912.0	+1600	+871.11
+2930.0	+1610	+876.67
+2948.0	+1620	+882.22
+2966.0	+1630	+887.78
+2984.0	+1640	+893.33
+3002.0	+1650	+898.89
+3020.0	+1660	+904.44
+3038.0	+1670	+910.00
+3056.0	+1680	+915.56
+3074.0	+1690	+921.11
+3092.0	+1700	+926.67
+3110.0	+1710	+932.22
+3128.0	+1720	+937.78
+3146.0	+1730	+943.33
+3164.0	+1740	+948.89
+3182.0	+1750	+954.44
+3200.0	+1760	+960.00
+3218.0	+1770	+965.56
+3236.0	+1780	+971.11
+3254.0	+1790	+976.67
+3272.0	+1800	+982.22
+3290.0	+1810	+987.78
+3308.0	+1820	+993.33
+3326.0	+1830	+998.89
+3344.0	+1840	+1004.4
+3362.0	+1850	+1010.0
+3380.0	+1860	+1015.6
+3398.0	+1870	+1021.1
+3416.0	+1880	+1026.7
+3434.0	+1890	+1032.2
+3452.0	+1900	+1037.8
+3470.0	+1910	+1043.3
+3488.0	+1920	+1048.9
+3506.0	+1930	+1054.4
+3524.0	+1940	+1060.0
+3542.0	+1950	+1065.6
+3560.0	+1960	+1071.1
+3578.0	+1970	+1076.7
+3596.0	+1980	+1082.2
+3614.0	+1990	+1087.8
+3632.0	+2000	+1093.3
+3650.0	+2010	+1098.9
+3668.0	+2020	+1104.4
+3686.0	+2030	+1110.0
+3704.0	+2040	+1115.6
+3722.0	+2050	+1121.1
+3740.0	+2060	+1126.7
+3758.0	+2070	+1132.2
+3776.0	+2080	+1137.8
+3794.0	+2090	+1143.3
+3812.0	+2100	+1148.9
+3830.0	+2110	+1154.4
+3848.0	+2120	+1160.0
+3866.0	+2130	+1165.6
+3884.0	+2140	+1171.1
+3902.0	+2150	+1176.7
+3920.0	+2160	+1182.2
+3938.0	+2170	+1187.8
+3956.0	+2180	+1193.3
+3974.0	+2190	+1198.9
+3992.0	+2200	+1204.4
+4010.0	+2210	+1210.0
+4028.0	+2220	+1215.6
+4046.0	+2230	+1221.1
+4064.0	+2240	+1226.7
+4082.0	+2250	+1232.2
+4100.0	+2260	+1237.8
+4118.0	+2270	+1243.3
+4136.0	+2280	+1248.9
+4154.0	+2290	+1254.4
+4172.0	+2300	+1260.0
+4190.0	+2310	+1265.6
+4208.0	+2320	+1271.1
+4226.0	+2330	+1276.7
+4244.0	+2340	+1282.2
+4262.9	+2350	+1287.8
+4280.0	+2360	+1293.3
+4298.0	+2370	+1298.9
+4316.0	+2380	+1304.4
+4334.0	+2390	+1310.0
+4352.0	+2400	+1315.6
+4370.0	+2410	+1321.1
+4388.0	+2420	+1326.7
+4406.0	+2430	+1332.2
+4424.0	+2440	+1337.8
+4442.0	+2450	+1343.3
+4460.0	+2460	+1348.9
+4478.0	+2470	+1354.4
+4496.0	+2480	+1360.0
+4514.0	+2490	+1365.6
+4532.0	+2500	+1371.1
+4550.0	+2510	+1376.7
+4568.0	+2520	+1382.2
+4586.0	+2530	+1387.8
+4604.0	+2540	+1393.3
+4622.0	+2550	+1398.9
+4640.0	+2560	+1404.4
+4658.0	+2570	+1410.0
+4676.0	+2580	+1415.6
+4694.0	+2590	+1421.1
+4712.0	+2600	+1426.7
+4730.0	+2610	+1432.2
+4748.0	+2620	+1437.8
+4766.0	+2630	+1443.3
+4784.0	+2640	+1448.9
+4802.0	+2650	+1454.4
+4820.0	+2660	+1460.0
+4838.0	+2670	+1465.6
+4856.0	+2680	+1471.1
+4874.0	+2690	+1476.7
+4892.0	+2700	+1482.2
+4910.0	+2710	+1487.8
+4928.0	+2720	+1493.3
+4946.0	+2730	+1498.9
+4964.0	+2740	+1504.4
+4982.0	+2750	+1510.0
+5000.0	+2760	+1515.6
+5018.0	+2770	+1521.1
+5036.0	+2780	+1526.7
+5054.0	+2790	+1532.2
+5072.0	+2800	+1537.8
+5090.0	+2810	+1543.3
+5108.0	+2820	+1548.9
+5126.0	+2830	+1554.4
+5144.0	+2840	+1560.0
+5162.0	+2850	+1565.6
+5180.0	+2860	+1571.1
+5198.0	+2870	+1576.7
+5216.0	+2880	+1582.2
+5234.0	+2890	+1587.8
+5252.0	+2900	+1593.3
+5270.0	+2910	+1598.9
+5288.0	+2920	+1604.4
+5306.0	+2930	+1610.0
+5324.0	+2940	+1615.6
+5342.0	+2950	+1621.1
+5360.0	+2960	+1626.7
+5378.0	+2970	+1632.2
+5396.0	+2980	+1637.8
+5414.0	+2990	+1643.3
+5432.0	+3000	+1648.9
+5450.0	+3010	+1654.4
+5468.0	+3020	+1660.0
+5486.0	+3030	+1665.6
+5504.0	+3040	+1671.1
+5522.0	+3050	+1676.7
+5540.0	+3060	+1682.2
+5558.0	+3070	+1687.8
+5576.0	+3080	+1693.3
+5594.0	+3090	+1698.9
+5612.0	+3100	+1704.4

APPENDIX III*

TABLE OF HAZARDOUS MATERIALS AND CHEMICAL INDEX**

Column

a — Hazards of substances are coded: Ex Explosive; F Flammable; P Poison; C Corrosive or absorbed by skin; I Irritant; V Vapors, fumes or dusts hazards; U Unstable; Ox Oxidizing agent. (Principal hazard is listed under "Class.")

b — Flash point, in degrees Fahrenheit unless otherwise indicated.

c — Toxicity, giving the maximum allowable concentration in parts per million in air for an eight hour exposure, where this is known.

d — Suitable extinguishing agent, as listed by the Factory Mutual Engineering Division: 1. Water, 2. Foam, 3. CO_2 or dry chemical, 4. Shut off the flow of gas or vapor if possible before using CO_2 or dry chemical, 5. Special extinguishing agent for metal fires.

e — Solubility and other properties or reactions.

Other abbreviations are as follows:

Comb. — combustibles
Decomp. — decomposes
Det. — detonates
Ig. — ignites
Mixt. — mixture
Sol. — soluble in water
Sp. fl. — spontaneously flammable
Viol. — violently

CHEMICAL	(a) CLASS	(b) F.P. (°F)	(c) TOX. (PPM)	(d) EXT. AGENT	(e) REMARKS
Acetal	F	−5	Moderate	3	Slightly soluble
Acetaldehyde	F	−36	200	1,3,4	UI Soluble forms sp. ex. peroxides
Acetic acid, glacial	F	109	10		C Soluble
Acetic anhydride	F	129	5	1,2,3	Reacts vigorously with water
Acetone	F	0	1000	1,3	V I
Acetone cyanohydrin	F	165	V	2,3	Decomposes forms highly toxic HCN
Acetonitrile	F	42	40	2,3	V Extremely toxic soluble
Acetyl chloride	F	40	V	3	Reacts violently with water
Acetylene	F	gas		4	U Lighter than air. Can detonate under high pressures unless dissolved in acetone
Acetylene dichloride	F	39	200	2,3	V I
Acetylene tetrabromide	P		1		V Decomposes to yield Fl. and highly toxic vapors at 374 °F
Acetylene tetrachloride	P liq.		5		V I
Acetyl peroxide	Ex & Ox	F solid	Explodes by heat or shock		V I slightly soluble
Alcohol—See specific type					
Acrolein	F	−15	.5	3	U P V I Sol.
Acrylic acid	F	130	V		Extreme irritant; U may viol. polymerize at high temp.
Acrylonitrile	F	32	20	3	U P V Mild odor Sol.
Adiponitrile	F	200	Highly	1,2,3	
Aldrin (pesticide)			Highly	1,2,3	May be in powder form or comb. solution
Alkylaluminums	F	Sp. fl. in air			C V Reacts violently with water
Allyl alcohol	F	70	5	3	Sol. P V C I
Allylamine	F	−20	V	2,3	Toxic I
Allyl chloride	F	−25	V	3,4	Not sol. P I
Allylchloroformate	F	88	V	2,3	I Decomposes to yield phosgene with heat
Aluminum borohydride	F	Sp. fl. in air			C V
Aluminum bromide	solid		V I		Reacts violently with water

* From *Fire Officer's Guide to Emergency Action,* Charles W. Bahme, National Fire Protection Association, Boston, January 1974.
** This table does not list every known hazardous substance, but owing to space consideration sets forth about 300 of the most hazardous materials.

CHEMICAL	(a) CLASS	(b) F.P. (°F)	(c) TOX. (PPM)	(d) EXT. AGENT	(e) REMARKS
Aluminum carbide		Decomp. water to produce methane			
Aluminum chloride	solid		V I		Reacts violently with water
Aluminum triethyl	F	Sp. fl.			Reacts violently with water to produce ethane
Aluminum trimethyl	F	Sp. fl.			Reacts with water to produce methane
Amatol	Ex	Mixture of TNT and ammonium nitrate			
Ammonal	Ex	Powerful and brisant high explosive			
Ammonia (gas)	F		50	1,3,4	V I Sol.
Ammonium bichromate	F solid & oxidizer		V		C I
Ammonium bromate	Ex & Ox	Explodes by deton. in compressed state			
Ammonium chlorate	Ex & Ox	Shock hazardous Ex; also by heat			
Ammonium dichromate	F & Ox solid				Toxic; combustible
Ammonium hydroxide (aqua ammonia)			V		C I
Ammonium nitrate	Ex & Ox	May decompose and explode at elevated temp. in pure state, and by heat or shock if contam. with oil or combustibles. Decomposes to yield toxic oxides of nitrogen			
Ammonium nitrite	Ex & Ox	Explosive by heat above 158 °F or by shock			
Ammonium perchlorate	Ex & Ox	May explode in fire; forms HE with combustibles			
Ammonium permanganate	Ex & Ox	Explodes by shock; powerful oxidizer			
Ammonium persulfate	Ex & Ox	May be explosive on contact with combustibles			
Ammonium picrate, dry	Ex & Ox	Explodes by spark, friction or shock			
Amylamine	F	30	Moderate	2,3	I
Amyl mercaptan	F	65	20		I
Aniline, oil	F	158	5	2,3	C Slightly sol. P V
Antimony pentasulfide	F	solid			Yields toxic sulfur dioxide when burning
Arsenic trichloride	P				Water reactive
Azides, most	Ex		variable		Ex with shock or heat
Barium chlorate	Ox			1	Dust is irritant; forms explosive mixture with comb. materials. Yields toxic fumes when involved in fire
Barium nitrate	Ox				See sodium nitrate
Barium peroxide	Ox				See sodium peroxide
Benzene (Benzol)	F	12	25	2,3	I C P V Not sol.
Benzotrifluoride	F	54		3	V Reacts with water or heat to yield toxic and irritant fumes
Benzoyl chloride	F	162	Yields phosgene with heat	2,3	I C V Decomp. with water
Benzoyl peroxide, dry	Ex & Ox F solid	Ex by friction, low heat or shock			V
Benzyl chloride	F	153	1	1,2,3	V C U Not sol.
Beryllium (powder)	F solid				Extremely toxic respiratory poison
Black powder	Ex	Functions by burning			
Blasting agents	Ex				Ammonium nitrate with oil or carbonaceous mixture. Det. in fire
Boron	F solid	Sp. fl. in air			
Boron hydrides	F	Sp. fl. in air	1		P I U Ex with water
Boron trichloride	gas		V		C I Reacts with water
Boron triethyl	F	Sp. fl. in air			Reacts violently with water to produce toxic fumes
Boron trifluoride	Ox gas				Reacts with water to produce toxic fumes
Boron trimethyl	gas	Sp. fl. in air			
Bromates	Ox solid		V		
Bromine	Ox liq.		1		V I C
Bromine trifluoride	Ox liq.				C Decomp. violently with water
Bronze (powder)	F solid	if contains aluminum powder			Minor irritant
Butadiene	F	gas	1000	4	U Not sol. Forms Ex peroxides in air
Butane	F	−76		4	Slightly sol.
N-Butyl acetate	F	72	150	3	I Slightly sol.
Butyl alcohol	F	84	100	1,3	I Sol.
N-Butylamine	F	10		2,3	I
Butyl ether	F	77		2,3	I
Butyl hydroperoxide	Ex & F liq.				U shock and heat sensitive ox mat.; likely to explode if involved in fire

CHEMICAL	(a) CLASS	(b) F.P. (°F)	(c) TOX. (PPM)	(d) EXT. AGENT	(e) REMARKS
Butyllithium (in petroleum solvents)	F	−40 to 25	C	5	Reacts with water and comb. mat. to cause fire
Butyl peracetate (75% sol, in solvent)	Ex & F liq.		V		U shock and heat sensitive contact with comb. mat. can cause fire or explosion. Decomp. at 200 °F
Butyl perbenzoate	Ex & F liq.				U Heat sensitive; contact with comb. mat. can cause fire or explosion. Decomp. at 140 °F
Butyl peroxypivalate	Ex & F liq.				U Shock and heat sensitive. Decomp. at 85 °F. Contact with comb. can cause fire or explosion
Butyraldehyde (iso 2nd normal)	F	−40 & 20		2,3	I
Butyric acid	F	161	toxic	1,2,3	I
Cadmium compounds	P		V		
Calcium carbide				3	Yields acetylene with water
Calcium chlorate	Ox solid				See chlorates
Calcium cyanamide	P		V		Yields ammonia 2nd acetylene with water
Calcium cyanide	P		V		I Yields HCN with acid or water
Calcium hypochlorite	Ox solid		Yields chlorine and oxygen with heat		Containers may explode in fire
Calcium hypophosphite		Yields poisonous phosphine with heat			
Calcium oxide (Quick lime)		Gives off heat with water			C
Calcium peroxide	Ox solid				C I Ex at 527 °F
Calcium phosphide			Yields sp. fl. and poisonous phosphine with water		
Carbon bi (or di) sulfide	F	−22	20	1,3	Vapor-air mixt. det. with shock
Carbon monoxide	P gas		50	2	
Carbon tetrachloride			10		
Caustics—See Sodium hydroxide					
Cellulose nitrate					
dry cotton	Ex				Ex with heat or shock
wet cotton	F solid				Used to make lacquer. Yields toxic products in fire
nitrated over 12.2%	Ex (any form—wet or dry)				Used in smokeless powder
Chloracetylene	Ex				U Sp. Ex in sunlight
Chlorates	Ox solid				May Ex in fire
Chlorinated diphenyl oxide			.5		V I
Chlorine (procedures for leaks)	Ox gas		1		V I P
Chlorine dioxide	Ex		very		V I P U Extremely Ex gas
Chlorine heptoxide	Ex		very		Decomp. easily and Ex
Chlorine trifluoride	Ox gas		.1		Ig. comb. reacts with water to prod. toxic and C fumes
Chlorobenzene	F	85	75	2,3	V I Narcotic; Not sol.
Chloroform			50		V I Anaesthetic
Chloroisocyanuric acid	P & Ox solid		I		Reacts with water to yield toxic fumes
Chlorophenol	F	250	V		P C absorbed through skin
Chloropicrin	I				Powerful irritant
Chlorosulfonic acid	C & Ox liq.		V		I Reacts with water to yield toxic fumes. Can ig. comb. mat.
Chromic acid, dry	Ox solid		P C I		Reacts with organic matter to produce fire. Contents may explode in fire
Cordeau detonant	Ex	May Ex if involved in fire			"Primacord"
Cresol	F liq. or solid		5	1,2,3	C V P
Crotoanldehyde	F	55	toxic	2,3	V I U Polymerization may occur in fire and violently rupture container
Cumene hydrophenoxide	Ex & F liq.		toxic		U P Powerful oxidizer; may explode in fire or in contact with comb. mat.
Cyanides In fumigation In heat treating	P		V		I Yields HCN

CHEMICAL	(a) CLASS	(b) F.P. (°F)	(c) TOX. (PPM)	(d) EXT. AGENT	(e) REMARKS
Cyanoacetic acid	F solid			3	V Yields toxic and irritant vapor with heat
Cyanogen gas	F	gas	Very pois. haz. like HCN		Prevent shock
Cyclohexane	F	−4	300	2,3	Not sol.
Cyclohexanol	F	154	50	1,2,3	V I Slightly sol.
Cyclohexanone	F	111	50	3	V I
Cyclohexylamine	F	90		2,3	V I Severe respiratory and eye irritant
Cyclopropane	F	gas	400	4	Anaesthetic not sol.
Decaborane	F solid	176	.05	1,3	V Forms shock-sensitive ex. with halogenated mat. yields hydrogen with water
Deuterium	F gas	Sp. fl. in air		4	Heavy hydrogen
Diamylamine	F	124		2,3	I C
Diazo compounds	Ex				Most are sensitive explosives
Diborane	F gas	Sp. fl. in air	.1	4	V I Extremely toxic. Reacts with water to cause ignition
N-Dibutylamine	F	125		1,2,3	V
Dibutyl ether-n	F	77	15	3	V Not sol.
Dibutyl peroxide	F, Ex & Ox	65	V I		Explodes above 228 °F
3, 4-Dichloroaniline	F solid			3	P Highly toxic. Melts at 161 °F
Dichlorobenzene	F	151	50	1,2,3	V I
1, 4-Dichlorobutane	F	126		1,2,3	Moderately toxic
Dichlorodifluoromethane	Non. flam.		1000		Freon 12
Dichloroethylene-1, 1	F	0	100	4	U V I Not sol.
Dichloroisocyanuric acid	Ox solid			I	V I Contact with water yields toxic gases; with organic materials may cause fire
Dichloromonofluoromethane	Non. flam.		1000		Freon 21
1-Dichloro-1-nitroethane	F	168	10	1,3	Not sol.
Dichlorotetrafluoroethane	Non. flam.		1000		Freon 114
Diethylaluminum chloride	F	Sp. fl. in air			Reacts violently with water C I
Diethylamine	F	−0	25	3	Sol. C I
Diethylchlorosilane	F	90		2,3	V I C
Diethylenetriamine	F	215		1,2,3	I
Diethyl ether (Ether ethyl)	F	−49	400	3	Forms sp. Ex peroxides. Anaesthetic
Diethyl sulfate	F	220		1,2,3	I Decomp. with heat to yield toxic sulfur oxides
Diethylzinc	F	Sp. fl. in air		3	Reacts violently with water
Diisopropylamine	F	30		3	I Sol.
Diisopropyl Peroxydicarbonate	Ex & Ox solid		V I		Decomp. above 53 °F. Shock and heat will det.
Diketene	F	93			U V I
Dimethyl aniline (Xylidine)	F	185	5	1,2,3	Haz. similar to aniline
Dimethyl ether	F	gas		3,4	Sol. Haz. similar to Diethyl ether
Dimethyl phosphine	F	Sp. fl. in air		2,3	V P I
Dimethyl sulfate	F	182	1	1,2,3	Slightly sol.
Dimethyl sulfide	F	below 0	very	2,3	V Slightly sol.
Dinitroaniline — 2,4	Ex	435	1	1,3	P I V
Dinitrobenzol (Dinitrobenzene)	Ex	302	1	1,3	P I V Det. by shock or heat
Dinitrochlorobenzene	Ex	382		1,3	V P C I Det. at 300 °F
Dinitrochlorohydrin	Ex	0.2			V P I C
Dinitroglycol	Ex				In a class with nitroglycerine
Dinitronaphthol salts	Ex	Explodes when warmed	More toxic than aniline		P V
Dinitrophenol	Ex	With low heat	More toxic than aniline		P V
Dinitrotoluene (Dinitrotoluol)	Ex	404	1.5	1,3	May det. in a fire P V
Dioxane	F	54	100	3	V I Sol. forms Ex peroxides
Divinyl benzene	F	169	moderately	1,2,3	V U Polymerizes at high temp. with possible violent rupture of container
Dynamite	Ex	May explode while burning	0.2		V Not sol.
Endrin		Powder dicomp. at 392			V P I Not sol.

CHEMICAL	(a) CLASS	(b) F.P. (°F)	(c) TOX. (PPM)	(d) EXT. AGENT	(e) REMARKS
Epichlorohydrin (EPI)	F	105	5	1,2,3	V P I U sol.
Ethanolamine	F	185	3	1,2,3	I
Ethyl acetate (Acetic ether)	F	24	400	3	Slightly sol. V I
Ethyl acrylate	F	60	25	2,3	V U I Slightly sol.
Ethyl alcohol (Ethanol)	F	55	1000	3	Sol.
Ethylamine	F	Below 0	10	2,3,4	C I Sol.
N-Ethylaniline	F	185	very	1,2,3	P V Not sol.
Ethylbenzene	F	59	200	2,3	V I Not sol.
Ethyl bromide (Bromoethane)	F	under −4	200	1,3	V I Not sol.
Ethyl chloride (Chloroethane)	F	−58	1000	3	I Anaesthetic Not sol.
Ethylene	F	gas		4	Anaesthetic sol.
Ethylene cyanohydrin	F	265		1,2,3	V Decomp. above 442 °F to yield toxic cyanides
Ethylene dibromide	Non. flam. liq.		25		Usually in highly flam. diluent. P V I
Ethylene dichloride	F	56	50	2,3	P V I Not sol.
Ethyleneglycoldinitrate	Ex		0.2		Same class as nitroglycerine
Ethyleneimine	F	12	.5	2,3	U P V I
Ethylene oxide	F	Below 0	50		Contents may explode in fire. Needs no air for comb. U V I Sol.
Ethyl ether—See Diethyl ether					
Ethyl formate (Formic ether)	F	−4	100	2,3	V Slightly sol.
Ethyl mercaptan (Ethanethiol)	F	−40	10	3	Not sol. Odorant for LPG
Ethyl methacrylate monomer	F	68			V U I
Ethyl methyl ether	F	−35		3,4	Sol. U
Ethyl methyl ketone	F	21	200		Sol.
Ethyl nitrate (Nitric ether)	Ex at 185	50		2,3	
Ethyl nitrite (Nitrous ether)	Ex	−31		2,3,4	Sp. decomp. at 194 °F
Ethyltrichlorosilane	F	72		2,3	U P V I
Ethyl vinyl ether	F	0			U
Fluoride compounds	P				V I
Fluorine	Ox gas	Sp. fl. with anything on expos. 0.1			Viol. reactive V I C
Formaldehyde, solution	F about 150		5	3	P V I Sol.
Formic acid	F	156	5	1,2,3	V I C
Freons, refrigerant gases					
Fulminates, e.g., silver, gold or mercury	Ex	High explosive (Primary class)			
Furfuryl	F	140	5	1,2,3	V I
Furfuryl alcohol	F	167	50	3	Sol. U
Glacial acetic acid—See Acetic acid					
Gold carbide	Ex	Explodes by percussion, friction or rapid heating			
Heptane-iso	F	0	500	2,3	Not sol.
Hexachlorethane-mixt.	F solid		1	3	Reacts violently and ig. with water
Hexane (Hexyl hydride)	F	−7	500	3	Not sol.
Hexanitrocarbanilide	Ex	Explodes spont. at 653; also by shock			
Hexanitro-oxanilide	Ex	About as powerful as TNT			
Hexogen (Cyclonite)	Ex	1½ times more powerful than TNT; very sensitive			
Hydrazine, anhydrous diamine	F	100	1	1,3	Highly spont. Ex with Ox
Hydrazoic acid (Aziomide)	Ex				
Hydrides			variable		
Hydrogen	F gas			4	Slightly sol.
Hydrogen chloride (Hydrochloric acid—Muriatic acid)	C	non fl.	5		Can dilute spill with water V I C
Hydrogen cyanide (Hydrocyanic acid—HCN)	Ex	0	10	4	Can be det. by heavy initiation. F P U V
Hydrogen fluoride (Hydrofluoric acid)	C		3	1	V I C
Hydrogen peroxide	Ox. liq.		1 (90%)		Strong concen. sp. ig. with combust. V I C U sol. May det. at elevated temp.
Hydrogen phosphide	See phosphine		0.3		
Hydrogen sulfide	F gas		10	4	V I Sewer gas
Hydroxylamine (Oxammonium)	Ex	Explodes at 265			Sol. V
Isophorone	F	184	25	1,2,3	V
Isoprene	F	−65		2,3	U V

CHEMICAL	(a) CLASS	(b) F.P. (°F)	(c) TOX. (PPM)	(d) EXT. AGENT	(e) REMARKS
Isopropyl amine	F	−35	5	3	
Isopropyl ether	F	−18	500	2,3	U anaesthetic
Isopropyl formate	F	22		1,2,3	I V
Ketine	F gas		very		U I
Lauroyl peroxide	Ex & Ox & F	Same haz. as Benzoyl peroxide			
Lead azide	Ex				
Lead picrate	Ex	Det. from fire or shock			
Lindane	solid				V I Yields toxic phosgene and HCL with heat
Liquefied petroleum gas	F		1000	4	See Butane and Propane
Lithium	F solid	Decomp. water liberating hydrogen			U
Lithium aluminum hydride	solid			3	U Decomp. yielding hydrogen above 257. Water reactive
Lithium aluminum hydride, ethereal	F	Sp. fl. in air		3	U Very reactive to water yielding hydrogen
Lithium borohydride	solid			3	U Violent reaction with water releasing hydrogen
Lithium carbide	solid			3	U Decomp. by water yielding acetylene
Lithium hydride	solid			5	U May ig. spont. in air or with moisture; decomp. by water yielding hydrogen
Magnesium—powder, ribbon or chips	F solid	Yields hydrogen when damp		5	Staring at fire injures eyes
Magnesium ethyl	solid	Sp. fl. in air		3	Yields ethane with water
Magnesium hypophosphite	solid	Yields phosphine at 212			
Magnesium methyl	solid	Sp. fl. in air		3	Yields methane with water
Magnesium perchlorate	Ox solid	Yields oxygen with water; forms explosive with comb. mat.			
Maleic anhydride	F solid	215		3	C I
Mannite hexanitrite	Ex	Explodes by shock			
Mannitol hexanitrate	Ex	Explodes by shock or low heat			
Mercury	P		.1		V
Mercury bichloride	P				V
Mesityl oxide	F	87	25	3	V I
Methyl acetate	F	14	200	3	Narcotic I
Methacrylic acid	F	171		1,2,3	U I V
Methylacrylate	F	27	10	2,3	U I
Methyl alcohol (Methanol)	F	52	200	3	P C V Sol.
Methylamines	F gases		10	4	V I Sol.
Methyl bromide (Bromomethane)	Barely F gas	ig. at 999	20	4	Fumigant P V I
Methyl butyl ketone (Propylacetone)	F	95	100	2,3	
Methyl chloride (Chloromethane)	F gas		100	4	Refrigerant Sol.
Methyl cyclohexanol	F	149	100	2,3	
Methyl cyclohexanone	F	118	100	2,3	
Methylene chloride (Carrene)	F	212	500	1,2,3	
Methyl ether (Dimethyl ether)	F	925		4	U Sol.
Methyl ethyl ether	F	−35		2,3	U Sol. anesthetic
Methyl ethyl ketone	F	21	200		Sol.
Methyl ethyl ketone peroxide (40% diluted in solvents)	Ex	over 100	V I		Explodes at 230 °F or by shock
Methyl formate	F	−2	100	3,4	Fumigant
Methyl isobutyl ketone	F	73	100	2,3	I Sol.
Methyl methacrylate	F	50	100	2,3	U
Methyl nitrate	Ex	Explodes at 302 or by percussion			
Methyl nitrite	Ex	Explodes at 149 or by percussion			
Methyl parathion (in 20% xylene solution)	F	115	.1	Highly toxic	May explode in a fire
Methylstyrene-a	F	129	100	2,3	U
Methyl vinyl ether	F	925		4	U
Nitrate baths					
Nitrate film					
Nitrate fires					
Nitrates					
Nitric acid	Ox liq.		2		V I
			Gives off nitric oxides		Sol. C
Nitric oxide	Ox gas		5		P V
Nitroaniline	F solid	390	1	1,2,3	P V
Nitrobenzene	F	190	1	1,2,3	P V C Not sol.

CHEMICAL	(a) CLASS	(b) F.P. (°F)	(c) TOX. (PPM)	(d) EXT. AGENT	(e) REMARKS
Nitro-carbo-nitrate	Ex & Ox solid				Mixt. of Ammonium nitrate, dinitrocotton, etc. Used as blasting agent. Not a sensitive Ex
Nitrochlorobenzene, meta or para	F solid	261	1	1,3	P Not sol.
Nitroethane	Ex	82	100	3	Rapid heating to high temp. May cause detonation
Nitrogen bromide	Ex	Explodes on slightest provocation			
Nitrogen chloride	Ex	Explodes on slightest provocation			
Nitrogen pentoxide	Ex	Highly toxic P. Liberates oxygen with violence at 122.			
Nitrogen tetroxide	Ox gas		5		V Highly toxic
Nitroglycerine	Ex	518	.2		Explodes at 502 °F V P I
Nitromethane	Ex	112	100	3	Ex by shock or at 550 °F
Nitrophenol, mono	F solid			Highly toxic	C V
Nitropropane	F	120	25	3	U May explode on heating
M-Nitrotoluene	Ex & F solid	223	5	3	V I Melts at 61 °F
O-Nitrotoluene	Ex & F	223	5	3	V I U Melts at 25 °F
Nitrous oxide	Ox gas				Laughing gas
Organic phosphates					Pesticides P V
Oxalic acid					P I V
Oxidizers					
Oxygen	Ox gas				
Oxygen liquefied	Ox liq.	Forms HE with comb. mat.			Frostbite hazard
Ozone, liq.	Ex		.1		V I
Paraldehyde	F	63		1,3	V
Parathion	F		.1		Fatal by skin contact
Pentaborane	F	Sp. fl. in air	.005		Extremely toxic
Pentachlorophenol	P solid		.5		C I V
Pentaerythrite tetranitrate		Explodes by shock or low heat			High explosive
Peracetic acid	Ex	105	V		C I Explodes at 230 °F and by shock
Perchlorethylene	See Tetrachloroethylene				
Perchloric acid—to 72% strength	C	Flam. of explosive with org. matter			C V I
Perchloric acid (dehydrated)	Ex	Explodes by heat, shock or with org. matter			
Persulfuric acid	Ox liq.	Yields oxygen when warmed; strong oxidizing agent			C
Phenol	F	175	5	1,3	P C V Sol.
Phenylamine—See Aniline					
Phosgene	P gas		.1	Highly toxic	
Phosphides	solids		Yields phosphine with water		
Phosphine	F gas	Sp. fl. in air	.3	4	P
Phosphorus, red	F solid	Ig. by friction	Very toxic	1	Ex with Ox materials
Phosphorus, white or yellow	F solid	Sp. fl. in air	.1	1	Kept under water P C
Phosphorus heptasulfide	solid	Ig. by friction			With water, yields hydrogen sulfide
Phosphorus oxychloride	liq.	Ig. with comb.			V C I reacts violently with water
Phosphorus pentasulfide	F solid	Sp. fl. in moist air	1	3	V Yields hydrogen sulfide with water
Phosphorus sesquisulfide	F solid	Ig by friction		1	Moderately toxic
Phosphorus tribromide	liq.		Very toxic		Decomp. by water with evolution of heat
Phosphorus trichloride	liq.		.5	V	Reacts violently with water
Picramates	Ex	Explosive when warmed			
Picramic acid (Dinitroaminophenol)	Ex				
Picric acid (Trinitrophenol), liq.	F	302	V	1	May explode in a fire
Picric acid, crystals	Ex	Explodes by shock or at 502			
Potassium	F solid	Sq. ig. with air	V	3,5	C Yields hydrogen with water which ig. sp. Haz. similar to sodium
Potassium arsenate	P		highly		Dust P V
Potassium bichromate	Ox solid				Dust P V I
Potassium bifluoride	P		highly	Decomp. by heat to yield fluroine V	
Potassium bromate	Ox solid	Forms explosive with comb. mat.			
Potassium chlorate	Ox solid	May explode with org. mat. and acids			Containers may explode in fire. Sol.
Potassium cyanide	F solid		highly	3	P Yields HCN with acid or water

CHEMICAL	(a) CLASS	(b) F.P. (°F)	(c) TOX. (PPM)	(d) EXT. AGENT	(e) REMARKS
Potassium dichloroisocyanurate	Ex solid	V I U			Decomp. violently with heat; yields toxic gases with water
Potassium hydride	F solid	Sp. fl. in air		3	
Potassium hydroxide	C	Solid or liq.	2		Lye. Heats with water. V I
Potassium hyponitrite	Ex				Det. at low temp.
Potassium hypophosphite	solid	Decomp. when heated to form phosphine			
Potassium nitrate	Ox solid				See sodium nitrate
Potassium nitrite	Ox solid				See sodium nitrite
Potassium nitrocyanide	Ex		highly		Explodes at 752
Potassium nitromethane	Ex				
Potassium percarbonate	Ox solid				
Potassium perchlorate	Ox solid	Will det. when heavily confined and exposed to heat			I
Potassium permanganate	Ox solid		V	Sp. fl.	With glycerine or ethylene glycol
Potassium peroxide	Ox solid			V I	With comb. ig. easily by friction or with water
Potassium persulfate	Ox solid	Lib. oxygen at 212 or at ord. temp. in water			
Potassium picrate	Ex		highly		
Potassium sulfide	F solid		V I	3	Avoid water
"Primacord"	Ex	(Cordeau detonant)			May Ex in fire
Propane—See also LP-Gas	F gas		1000	4	Not sol.
Propareyl bromide	Ex & F	64	V I		Explodes by shock or heat
Propellants, missile					
Propionic acid	F	130	V	1,2,3	C I
Propionic anhydride	F	145	V	1,3	C I Reacts with water to form acid
Propyl amine	F	−35	V	1,2,3	I
Propylene	F gas			4	Anesthetic sol.
Propylene dichloride	F	60	75	2,3	U reacts with metals to produce heat
Propylene oxide	F	−35	100	1,2,3	U sol.
Propyl ether-iso	F	−18		2,3	U
Propyl nitrate	Ex & F	68	25	1,3	Monopropellant. Likely to detonate in fire V
Pyridine	F	68	5	2,3	V I
Pyrophoric iron	solid	Sp. fl. in air			
Pyrophoric nickel	solid	Sp. fl. in air			
Red fuming nitric acid	Ox. liq		2		P I C V
Rocket fuels					
Rubidium	F solid				Ig. when moist
Selenium	F solid		V		P
Silicides (of light metals)	solid	Decomp. water to yield hydrogen			Tox-variable
Silicon hexachloride	F	Vapors sp. ig. in air at high temp.			Reacts with water to highly toxic gas
Silicon hydride	F gas	Sp. fl. in air at ord. temp.			
Silver nitrate	Ox solid				P C
Silver nitrite	Ex	Explodes by percussion or warming			
Silver oxalate	Ex		highly		Explodes when heated
Silver oxide	Ox solid				C
Smokeless powder	Ex				
Sodium	F solid		V	3,5	C Decomp. water forming hydrogen which ig. sp.
Sodium alloys	F solid	Sp. fl. in air	V	3,5	Yields hydrogen when wet which igs. sp.
Sodium amide (Sodamide)	Ex		V		May explode if stored for long periods
Sodium bichromate	Ox solid				C V I Decomp. at 212
Sodium bromate	Ox solid				
Sodium carbide	solid			3	Yields acetylene with water
Sodium chlorate	Ox solid	Forms HE with comb.			May explode in fire
Sodium chlorite	Ox solid	Ex with comb. on heat, friction or impact			U May explode in fire
Sodium cyanide	P		highly		Yields HCN rapidly with acids and slowly with water
Sodium dichloroisocyanurate	Ox solid		V	1,2,3	Ig. with org. matter. Yields dense toxic fumes with heat
Sodium fluoride	P				
Sodium hydride, crystals	F	Sp. fl. in air	highly	3	Decomp. by water or heat
Sodium hydrosulfite	F solid			3	With water, yields SO_2 and may ig. comb. mat.
Sodium hydroxide	C	Hot solns. may corrode metals forming hydrogen			I V Sol.

CHEMICAL	(a) CLASS	(b) F.P. (°F)	(c) TOX. (PPM)	(d) EXT. AGENT	(e) REMARKS
Sodium hypochlorite	Ox solid	With heat, yield O_2 and chlorine			Cont. may explode in fire
Sodium hypophosphite	solid	Yields phosphine when heated			Decomp. products sp. fl. and toxic
Sodium nitrate	Ox solid	Mixt. with org. mat. ig. by friction			Yields toxic fumes when involved in fire
Sodium perborate	Ox solid				
Sodium perchlorate	Ox solid	May explode when mixed with comb. mat. by heat or friction			
Sodium peroxide	Ex	Explodes by shock or at 680			
Sodium sulfide	F solid		toxic	1	Yields toxic and flam. hydrogen sulfide when heated
Strontium peroxide	Ox solid	Reacts explosively with water			
Styrene	F	90	100	2,3	U I Cont. may violently rupture in fire
Sulfur	F solid			I	Yields toxic sulfur dioxide when burning
Sulfur chloride	F	245	1	1,3	Reacts violently with water
Sulfur dioxide	gas	non-fl.	5		V I P
Sulfuric acid	C		1		Concen. acid reacts violently with water
Sulfur trioxide	C gas		V		Reacts violently with water P I
Tetrachloroethane (Acetylene tetrachloride)	P		5		V I
Tetrachloroethylene	V		100		I Cleaning solvent
Tetraethyllead (TEL Compound)	F & Ex	200	.075		Det. in a fire. Very toxic
Tetrafluroethylene	F & Ex	Yields toxic vapors in a fire			Explosive decomp. with heat. Shock sensitivity minimal
Tetrahydrofuran	F	6	200	3	U forms org. peroxides unless inhibited
Tetramethyllead (TML Compound)	F	100	.075		Can det. above 212 °F. Very toxic
Tetranitroaniline (TNA)	Ex	Explosive at 437 °F	highly		
Tetrazene	Ex				
Tetryl	Ex	Det. by spark or shock			
Thallium	F solid	Kept immersed in water		P	
Thionyl fluoride	gas		V I		Water reactive to yield hydrofluoric acid
Thorium, powder	F solid				Radioactive element
Titanium, powder	F solid				
Titanium tetrachloride	C liq.		V I		Reacts with water to yield hydrochloric acid
Toluene (Toluol)	F	40	200	2,3	
Toluidines (O & P)	F solid		5	1,2,3	Vapors ig. at 185 and are highly toxic P
Triamylamine	F	215	V I	1,2,3	
Tributylamine	F	187	V I	1,2,3	
Trichloroethylene (Perchloroethylene)			100		Practically non-comb.
Trichloroisocyanuric acid	Ox solid		V I	1	May ig. with comb. mat. Yields toxic fumes when exposed to water or fire
Trichlorosilane (Silicon chloroform)	F	Sp. fl. in air		3	Violent reaction with water
Triethylamine	F	20			Not sol.
Trinitroaniline (Picramide)	Ex		V		
Trinitroanisol (Methyl picrate)	Ex				
Trinitrobenzaldehyde	Ex				
Trinitrobenzene (TNB)	Ex		V		Highly sensitive to shock or heat
Trinitrophenol—See Picric acid					
Trinitrotuluene (TNT)	Ex		V		Det. at 464 °F or strong shock
Tritium (Hydrogen 3)	F gas				Radioactive isotype of hydrogen
UMDH (Unsymmetrical dimethylhydrazine)	F	5	.5	1,3	V I
Uranium powder	F solid		highly	Radioactive	
Urea nitrate	Ex & F solid				
Urea peroxide	F & Ox solid				May ig. by friction
Vinyl acetate	F	18		2,3	U I Sol. Cont. may violently rupture in fire

CHEMICAL	(a) CLASS	(b) F.P. (°F)	(c) TOX. (PPM)	(d) EXT. AGENT	(e) REMARKS
Vinyl bromide	F gas		500	4	Yields toxic fumes with heat
Vinyl chloride	F gas		500	3	U Yields toxic fumes with heat
Vinyl cyclohexane	F	61			U
Vinyl ether (Divinyl ether)	F	−22	anesthetic	3	U Yields toxic fumes with heat
Vinyl ethyl ether	F	−50			U Yields toxic fumes with heat. Not soluble
Vinylidene chloride	F	0			U Yields toxic fumes with heat
Vinyl isobutyl ether	F	15			U
Vinyl isopropyl ether	F	−26			U
Vinyl methyl ether	F gas			4	U
2-Vinylpyridine	F				U
Vinyltrichlorosilane	F	70	V		P I Yields toxic fumes with heat
Vinyltoluene	F	127		1,2,3	I U
Xylene (Xylol) (O,M,P)	F	81–90	100	2,3	V I
Xylidine-o (o-Dimethylaniline)	F	260	5	1,2,3	P V Twice as toxic as aniline with no warning
Zinc chlorate	Ox solid	Mixt. with org. explodes by friction or shock			
Zinc ethyl	F	Sp. fl. in air		3	Violent reaction with water
Zinc, molten					
Zinc phosphide	F solid			3	Yields phosphine gas
Zinc picrate	Ex	Explodes at 592. Primary explosive			
Zinc propyl, liq.	F	Sp. fl. in air. Yields ethane with water			
Zirconium, powder	F	May ig. sp. in air		5	
Zirconium tetrachloride	C solid	Reacts with water to yield irrit. fumes			

Subject Index

D

H

I

M

O

P

Q

T

U

V

X

Y

Z

Consolatore

The Simon & Schuster

SHORT PROSE READER

The Simon & Schuster

SHORT PROSE READER

Fourth Edition

ROBERT W. FUNK
Eastern Illinois University

SUSAN X DAY
Iowa State University
University of Houston

ELIZABETH MCMAHAN
Illinois State University

LINDA S. COLEMAN
Eastern Illinois University

Upper Saddle River, New Jersey 07458

Library of Congress Cataloging-in-Publication Data

The Simon & Schuster short prose reader / [compiled by] Robert W. Funk . . . [et al.].— 4th ed.
p. cm.
Includes bibliographical references and index.
ISBN 0-13-192589-X
1. College readers. 2. English language—Rhetoric—Problems, exercises, etc. 3. Report writing—Problems, exercises, etc. I. Title: Simon and Schuster short prose reader. II. Funk, Robert.

PE1417.S453 2005
808'.0427—dc22

2004029331

Editorial Director: Leah Jewell
Senior Acquisitions Editor: Craig Campanella
Assistant Editor: Jennifer Conklin
Editorial Assistant: Joan Polk
Production Liaison: Joanne Hakim
Marketing Manager: Kate Stewart
Marketing Assistant: Mariel DeKranis
Manufacturing Buyer: Benjamin Smith
Cover Art Director: Jayne Conte
Cover Design: Bruce Kenselaar
Cover Illustration/Photo: Catherine Secula/Photolibrary.com
Director, Image Resource Center: Melinda Reo
Manager, Rights and Permissions: Zina Arabia
Manager, Visual Research: Beth Brenzel
Manager, Cover Visual Research & Permissions: Karen Sanatar
Image Permission Coordinator: Craig Jones
Photo Researcher: Diana Gongora
Composition/Full-Service Project Management: Karen Berry/Pine Tree Composition
Printer/Binder: Phoenix Book Tech Park

Credits and acknowledgments borrowed from other sources and reproduced, with permission, in this textbook appear on pages 397–399.

Pearson Education LTD.
Pearson Education Singapore, Pte. Ltd
Pearson Education, Canada, Ltd
Pearson Education–Japan
Pearson Education Australia PTY, Limited
Pearson Education North Asia Ltd
Pearson Educación de Mexico, S.A. de C.V.
Pearson Education Malaysia, Pte. Ltd
Pearson Education, Upper Saddle River, New Jersey

10 9 8 7 6 5 4 3 2 1
S.E. ISBN 0-13-192589-X
A.I.E. ISBN 0-13-192591-1

In memory of our great and good friend

Deborah Wiatt

Contents

Active Reading 1

2

The Reading-Writing Connection 15

3

Strategies for Conveying Ideas: *Narration* and *Description* 30

Strategies for Making a Point: *Example* and *Illustration* 72

7

Strategies for Examining Two Subjects: *Comparison* and *Contrast* 189

11

Combining Strategies: Further Readings 365

Thematic Contents

Family

Sports

Health

Science and Nature and Technology

Morals and Ethics

Education and Learning

Language and Writing

Minority Perspectives

Gender

Human Behavior

Social Issues

Work

Humor

Entertainment and Leisure

Editing Skills: Contents

Chapter 7 Strategies for Examining Two Subjects: *Comparison* and *Contrast*

Chapter 8 Strategies for Explaining How Things Work: *Process* and *Directions*

Chapter 9 Strategies for Analyzing Why Things Happen: *Cause* and *Effect*

Chapter 10 Strategies for Influencing Others: *Argument* and *Persuasion*

Preface

Good readers are usually good writers, and good writers are always good readers. Researchers tell us that reading and writing are complementary processes that involve the use of language to create meaning. This text is designed to reinforce this relationship and to encourage reading by students who want to improve their writing. The selections are short and lively, not too difficult, but rich enough to provide ideas for thought and discussion. The instructional apparatus accompanying each reading has two main goals:

1. to encourage students to use writing as a means of exploring the readings, and
2. to point out strategies used in the essays that students can employ in their own compositions.

The Simon & Schuster Short Prose Reader is a flexible resource. The numerous readings, activities, and writing topics give instructors the freedom to select from a broad range of assignments and approaches.

NEW TO THE FOURTH EDITION

Features new to this edition include the following:

- **Images and Ideas** activities to accompany each rhetorical strategy, with an engaging visual accompanied by questions that prompt students to examine the image and respond to it in discussion and writing.
- Increased emphasis on **Combining Strategies,** including boxed questions for a selected essay in each chapter that direct students to look closely at how writers use more than one rhetorical strategy to develop their essays. Chapter 11 has also been revised to include an annotated essay and more questions about combined strategies.
- Additional writing prompts that focus specifically on **Collaborative Writing, Using the Internet,** and **Writing about Reading.**
- Expanded treatment of **audience** in Chapter 2 and throughout the text.
- Added discussion of the **similarities and differences** format for comparison/contrast writing.
- More detailed instruction on three specific approaches to argument: **counterargument, problem-solution,** and **pro-and-con.**
- Updated material for the **debates** on the pros and cons of the death penalty and same-sex marriage (Chapter 10).

- **Sixteen new readings,** including selections by Katha Pollitt, Elizabeth Berg, Evan Thomas, Juleyka Lantigua, Leonard Pitts, Rick Reilly, Greg Critser, Emily Nelson, David Myers, and Cynthia Crossen, along with **three new student essays.** Many favorites by Russell Baker, Judith Cofer, Brent Staples, Gloria Naylor, Isaac Asimov, Judith Viorst, Suzanne Britt, Mark Twain, Dave Barry, Wayson Choy, Stephen King, Jade Snow Wong, Arthur Ashe, Barbara Huttmann, Bill Bryson, Langston Hughes, and Richard Selzer have been held over from previous editions.

These additions strengthen the key features that have made the first three editions of *The Simon & Schuster Short Prose Reader* a popular and successful text.

THE READING SELECTIONS

The readings are brief, accessible, and easy to teach. They cover a wide range of topics and viewpoints to involve students with ideas and issues that relate to their own experience but also expand their understanding of people, places, and ideas. A special effort has been made to appeal to a cross-section of students by including a number of works by women and writers from diverse cultural backgrounds. Many of the selections are standard pieces that have been used successfully in writing classes; others are new readings that have never been anthologized before.

ORGANIZATION

The readings are grouped according to their major pattern of organization and presented as strategies for approaching a given writing task. The introduction to each strategy explains the point, the principles, and the pitfalls of using this particular pattern. An Images and Ideas page at the end of each introduction includes a relevant visual accompanied by questions that draw connections between the image and the chapter's rhetorical strategy. This opening material is followed by four published essays (eight in Chapter 10) and a student essay that further illustrate the strategy. Chapter 11 provides five additional essays that combine strategies in a variety of ways.

INSTRUCTIONAL FEATURES

Two **introductory chapters** present a concise explanation of the interrelated processes of reading and writing. Chapter 1 gives specific directions for learning how to become active readers, including a sample reading that has been annotated by an active reader. Chapter 2 describes the process of writing in response to reading. It includes a sample student essay and a brief reading to respond to and write about.

The **pre-reading apparatus** includes three instructional aids: (1) a brief thinking/writing activity (Preparing to Read) that gets students ready to read by evoking thoughts and feelings on the subject of the reading; (2) a short biographical headnote that provides a context for reading; and (3) a list of Terms to Recognize that defines potentially unfamiliar words in the selection.

The **post-reading apparatus** includes a selection of activities that instructors can assign, as needed, to help students increase their skills in reading and writing:

1. **Responding to Reading**—a journal-writing or possible listserve assignment that asks students to record their reactions to an issue or idea in the selection they have just read. This brief activity promotes fluency and may be used as the basis for the essay assignments that follow.
2. **Gaining Word Power**—an exercise that draws words from the reading and helps students add them to their active vocabulary.
3. **Considering Content**—a series of questions that assist students in becoming focused, attentive readers. Answering these questions assures basic comprehension.
4. **Considering Method**—several questions that help students to identify successful strategies in the reading and to examine rhetorical choices that the author made.
5. **Writing Step by Step**—a sequence of specific directions that guide students in writing a short essay imitating the reading's structure and purpose. This directed writing can be used to provide inexperienced writers with a successful composing experience.
6. **Other Writing Ideas**—additional writing assignments that relate to the rhetorical mode or subject matter of the reading. These assignments, which include a mixture of personal and academic topics, focus on Collaborative Writing, Using the Internet, and Writing about Reading.
7. **Editing Skills**—an exercise that helps students to check and improve the essay they have just written. Each editing section focuses on a different skill, one that pertains to some grammatical, mechanical, or rhetorical feature of the reading selection.

This extensive apparatus gives teachers and students a wide variety of choices for exploring the reading-writing connection.

OTHER FEATURES

To help instructors who want to correlate reading assignments or organize their courses according to issue-centered units, the **Thematic Contents** groups the reading selections according to several common themes. The text also includes a **Glossary** of useful rhetorical terms and a guide to **Editing Skills** and their locations within chapters. With these references, instructors can direct individual students to sections addressing their particular editing problems.

ACKNOWLEDGMENTS

We want to extend appreciation and thanks to the many people who have helped us in producing this book, especially our editors and the editorial staff at Prentice Hall: Leah Jewell, Craig Campanella, Joan Polk, and Jennifer Conklin, and to our production editor at Pine Tree Composition, Karen Berry. We are indebted to the students in our writing classes for field-testing the assignments and for providing us with inspiration and insights as well as sample essays. We are also grateful for the excellent ideas and suggestions provided by our reviewers: Gail Upchurch, Olive Harvey College; Karen Hackley, Houston Community College; Deonne Kunkel, Diablo Valley College; Anne Wilbourne, Southeastern Louisiana University; Kerry Cantwell, Alamance College; and Susan McDermott, Hudson Valley Community College. And of course we want to thank Bill, Brian, Danny, and Casey for their patience, support, and encouragement.

Robert W. Funk
Susan X Day
Elizabeth McMahan
Linda S. Coleman

The Simon & Schuster

SHORT PROSE READER

Active Reading

Most people who write well also read well—and vice versa. The two skills are so intertwined that they are often taught together, as we do in this textbook. Reading gives you not only information and amusement but also a sense of how sentences work and how paragraphs are best organized. Most of the time, you get this sense without really paying attention: it just seeps into your mind with the rest of the material. In this textbook, we ask you to make the reading-writing connection more consciously than you may have done before. By looking carefully at good writing, you will better understand the content as well as the techniques the writers use. And by being aware of yourself as a reader, you will develop useful strategies for attracting and keeping the attention of your own readers.

LEARNING TO BE AN ACTIVE READER

Did you ever finish reading something, look up from the page, and realize that you didn't take in anything at all? That you passed your eyes over the print, but you might as well have stared out the window? At such times, you know that you have been an extremely passive reader. On the opposite end of the spectrum, you've probably had the experience of being swept away from reality while reading, so involved in the printed word that the rest of your world fades. Much of your college reading won't be able to carry you off that completely. By learning to be an **active reader**, though, you will be able to handle your reading assignments competently—remembering more and using what you have learned more effectively. The main idea is to stay involved with the reading through interaction—bringing mental and emotional energy to the task.

KEEPING A JOURNAL

One good way to become an interactive reader is to keep a journal about your reading. In a journal, you can experiment with ideas and express your responses with greater freedom than you can in formal writing assignments. Before each reading selection in most chapters of this book, we ask you a Preparing to Read question, which you may answer in your journal as well as in class discussions. This activity starts you thinking about the ideas you will find in the essay that follows. After each selection, we give you a Responding to Reading suggestion to consider in your journal, encouraging you to write about your personal reactions. You will see as this chapter goes along how your two journal entries fit into your role as an active reader.

PREVIEWING THE READING

If you're the type who simply plunges into a reading assignment, you're missing something. Study skills experts emphasize the value of previewing the reading, getting your mind ready for full comprehension. **Previewing** involves more than merely counting the number of pages you have to go: no one needs to remind you to do that! The trick is to develop a mental set that makes your brain most receptive to the material.

Title

Try stopping after you read the **title** and asking yourself what it suggests to you. The sample essay we use in this chapter is named "Handled with Care." Where have you heard such a phrase before? What image does it bring to your mind? The title often gives you a clue about what's ahead.

Author and Other Publication Facts

With some assignments, you will recognize the author's name. Bob Greene is the writer of "Handled with Care," and you may know that he was a longtime columnist for the *Chicago Tribune*. You may also know that his columns often comment on culture and politics and the connections between them. Again, the author's name provides clues about the essay that follows. You may know something about the time period when it was written (as you would if the byline reads "Mark Twain"). Similarly, if the byline reads "Dave Barry," you can expect something humorous about modern life.

Even when you don't recognize the author, you can take note of other publication facts. The date of publication, if you have it, gives you an idea of how current the information is. If the reading is reprinted, as the ones in this textbook are, consider where it originally appeared and whether that means anything to you. If an essay first showed up in *U.S. News and World Report,* you can assume it will have a conservative political slant; if it came from *The Nation,* it will probably have a liberal viewpoint. If it came from a city newspaper, like the *Chicago Tribune,* you can't be so sure about the political perspective, since most big newspapers attempt to cover the spectrum.

Visual Features and Supplements

Page through the reading, notice its design, and look at parts that stand out from the ordinary print, such as headnotes, headings, photos, diagrams, boxed material, summaries, and questions after the reading. Your textbooks, like most of the successful web pages you visit, are specially designed to include lots of these helpful materials. Unfortunately, many students skip them, thinking they're not as important as the rest. Actually, they are there to focus attention on what *is* important in the reading. Or they may give you information that assists you in understanding the reading; for example, the headnote paragraphs before essays in this textbook give a little biographical information about the writer. The Terms to Recognize section lists some difficult words from the essay and their definitions, so you won't have to look them up in the dictionary right away. In this list, we provide only the definition that fits the way the term is used in the reading; the term probably has other meanings as well.

Some reading material doesn't include any obvious help. It's just straight print. But you can benefit from the only visual clue: paragraph indentations. Read the first sentence of each paragraph. This survey will probably give you ideas about the content and organization of the reading.

Responses and Predictions

"This preview stuff just slows me down," you may be thinking. "I don't have time for it." Let us assure you, the first stage does seem slow, but it makes later stages faster and more efficient. It also markedly increases your memory for what you read.

The main thing you're doing as you preview is responding to clues and making predictions about what the reading contains. It may seem like a guessing game, but actually you are clearing the brush from pathways in your brain, making the way easy for the information to get through. Through guessing what to expect, you are directing your attention, focusing on the material so that it won't be fighting through a tangle of thoughts about why you squabble with your roommate and what you'll have for lunch.

Instead, let your mind wander through the associations and experiences you already have with the material suggested by your preview. What people, events, and feelings in your own life did you stumble across in your preview? If you are from Chicago, you are able to visualize the setting of the essay and may even be able to predict Bob Greene's overall style. Or the title, "Handled with Care," may have reminded you of a delicate package that happened to reach you intact, miraculously, through the mail.

A FIRST READING

Now is the time to plunge in. Try to place yourself in a setting that aids concentration. This setting varies from person to person, and you probably know your ideal situation. You can't always get it, but at least don't undermine yourself by choosing a spot where you know you'll be distracted or where you know you'll be lulled to sleep. Sitting up at a table or desk is a good idea because the position suggests that you are going to work.

You'll need a pencil, pen, or pad of Post-its to read interactively. Make it a habit. Mark words and terms that you need to look up later. Write questions—or just question marks—in the margins near material you don't quite understand. Write your spontaneous responses (*Yes! No! Reminds me of Aunt Selma! What!? Prejudiced crap!*) in the margins. Underline sentences that you think may contain the main ideas and phrases that impress you with the way they are worded. We provide one reader's markings of "Handled with Care" in this chapter.

STAYING AWARE OF CONVENTIONS

Conventions are the traditional ways of doing things: for example, we have conventional ways of beginning and ending telephone conversations, and we expect everyone to follow them. If a friend closed a phone conversation by saying, "Pick me up at 9:15," and immediately hung up, you would think it strange not to have any of the usual sign-off words. The same type of expectation goes for writing **conventions**, some of which we outline here. Stories and poems don't have to follow these conventions, but nonfiction works like **essays**, textbooks, and manuals do. You can enhance your reading by looking for conventional features as you go along. They provide clues to the author's purpose and help to break down complicated readings into manageable chunks.

Subject

Each piece of writing is expected to deal with one subject or topic. This should be fairly clear to you near the beginning, perhaps even from the title ("Why We Crave Horror Movies," in Chap. 9, is about exactly that). Once you identify the subject, you can be pretty sure the whole reading will stay on that topic.

Main Idea or Thesis

We expect a reading not only to have a subject but also to say something *about* the subject. That is the **thesis** or main idea. Frequently, the main idea comes up early in the reading, clearly expressed in one sentence. At other times, you must put together the main point piece by piece as you read. It may finally be stated in a sentence at the end, or it may not be stated directly at all. As you read, underline sentences that seem to add to your understanding of the author's main point.

Supporting Material

Writers must prove their main points by providing convincing **supporting material**. This can be in the form of logical reasoning, emotional appeal, examples, evidence from experts, specific details, facts, and statistics. Different main points and different readers lend themselves to different types of supporting material: for example, your math textbook uses mostly logical reasoning and examples, while an essay about capital punishment might use all the forms we listed. While you read an assignment, ask yourself: *What forms of supporting material does this author use and are they appropriate for the intended audience?*

Patterns of Organization

The conventions of subject, thesis, and supporting material deal with the content of the reading. You also need to look at *how* the content is presented. After each selection in the following chapters, you will see questions called Considering Content and Considering Method. The method questions ask you to consider the techniques the author used to present material, including organization.

We organize the chapters in this book according to **patterns of development** that are the most conventional ways to organize writing: patterns such as comparison and contrast, cause and effect, and narration. Each chapter explains one basic pattern, though in practice, most authors combine several strategies to support and develop their central pattern. As you read an essay, notice how the writer uses patterns to arrange ideas.

Paragraphs

As a reader, you will also notice how writers package their meaning in units of thought called **paragraphs**. Paragraph indention usually signals the introduction of a new topic or a new aspect of a current topic. In other words, writers start new paragraphs to show that they are moving on to another topic or subtopic. Paragraphs in prose essays tend to be longer than paragraphs in newspapers and magazines, where journalists break their paragraphs frequently to make narrow columns of type easier to read.

Transitions

Another element of a writer's method involves how he or she makes connections between ideas. These connections are called **transitions**. A common place for a transition is between two paragraphs (at the end of one and/or at the beginning of the next one), where the author shows the logical relationship between them. Recognizing transitions helps you direct your thought process in the way the author wants you to. For example, a paragraph that begins "Furthermore, . . ." lets you know that you should expect material that adds to and agrees with the material before it. A paragraph that begins "On the other hand, . . ." lets you know that you should expect material that contradicts or shows the opposite of the material before it. By noticing transitions, you prepare the appropriate mindset for understanding what comes next.

A SAMPLE ESSAY

Handled with Care

BOB GREENE

The day the lady took her clothes off on Michigan Avenue, people were leaving downtown as usual. The workday had come to an end; men and women were heading for bus and train stations, in a hurry to get home. 1

She walked south on Michigan; she was wearing a white robe, as if she had been to the beach. She was blond and in her thirties. As she passed the 2

Radisson Hotel, Roosevelt Williams, a doorman, was opening the door of a cab for one of the hotel's guests. The woman did not really pause while she walked; she merely shrugged the robe off, and it fell to the sidewalk. She was wearing what appeared to be the bottom of a blue bikini bathing suit, although one woman who was directly next to her said it was just underwear. She wore nothing else.

3 Williams at first did not believe what he was seeing. If you hang around long enough, you will see everything: robberies, muggings, street fights, murders. But a naked woman on North Michigan Avenue? Williams had not seen that before and neither, apparently, had the other people on the street.

4 It was strange; her white robe lay on the sidewalk, and by all accounts she was smiling. But no one spoke to her. A report in the newspaper the next day quoted someone: "The cars were stopping, the people on the buses were staring, people were shouting, and people were taking pictures." But that is not what other people who were there that afternoon said. The atmosphere was not carnival-like, they said. Rather, they said, it was as if something very sad was taking place. It took only a moment for people to realize that this was not some stunt designed to promote a product or a movie. Without anything telling them, they understood that the woman was troubled, and that what she was doing had nothing to do with sexual titillation; it was more of a cry for help.

5 The cry for help came in a way that such cries often come. The woman was violating one of the basic premises of the social fabric. She was doing something that is not done. She was not shooting anyone, or breaking a window, or shouting in anger. Rather, in a way that everyone understood, she was signaling that things were not right.

6 The line is so thin between matters being manageable and being out of hand. One day a person may be barely all right; the next the same person may have crossed over. Here is something from the author John Barth:

> 7 She paused amid the kitchen to drink a glass of water; at that instant, losing a grip of fifty years, the next-room-ceiling plaster crashed. Or he merely sat in an empty study, in March-day glare, listening to the universe rustle in his head, when suddenly a five-foot shelf let go. For ages the fault creeps secret through the rock; in a second, ledge and railings, tourists and turbines all thunder over Niagara. Which snowflake triggers the avalanche? A house explodes; a star. In your spouse, so apparently resigned, murder twitches like a fetus. At some trifling new assessment, all the colonies rebel.

8 The woman continued to walk past Tribune Tower. People who saw her said that the look on her face was almost peaceful. She did not seem to think she was doing anything unusual; she was described as appearing "blissful." Whatever the reaction on the street was, she seemed calm, as if she believed herself to be in control.

She walked over the Michigan Avenue bridge. Again, people who were there report that no one harassed her; no one jeered at her or attempted to touch her. At some point on the bridge, she removed her bikini bottom. Now she was completely undressed, and still she walked. "It was as if people knew not to bother her," said one woman who was there. "To tell it, it sounds like something very lewd and sensational was going on. But it wasn't like that at all. It was as if people knew that something very . . . fragile . . . was taking place. I was impressed with the maturity with which people were handling it. No one spoke to her, but you could tell that they wished someone would help her." 9

Back in front of the Radisson, a police officer had picked up the woman's robe. He was on his portable radio, advising his colleagues that the woman was walking over the bridge. When the police caught up with the woman, she was just standing there, naked in downtown Chicago, still smiling. The first thing the police did was hand her some covering and ask her to put it on; the show was over. 10

People who were there said that there was no reaction from the people who were watching. They said that the juvenile behavior you might expect in such a situation just didn't happen. After all, when a man walks out on a ledge in a suicide attempt, there are always people down below who call for him to jump. But this day, by all accounts, nothing like that took place. No one called for her to stay undressed; no one cursed the police officers for stopping her. "It was as if everyone was relieved," said a woman who saw it. "They were embarrassed by it; it made them feel bad. They were glad that someone had stopped her. And she was still smiling. She seemed to be off somewhere." 11

The police charged her with no crime; they took her to Read Mental Health Center, where she was reported to have signed herself in voluntarily. Within minutes things were back to as they always are on Michigan Avenue; there was no reminder of the naked lady who had reminded people how fragile is the everyday world in which we live. 12

....

MARKING THE TEXT

Here is an example of how a student reader marked the Bob Greene essay.

Handled with Care

like a package with breakables in it

BOB GREENE

—Emotion— shock value!

The day the lady took her clothes off on Michigan Avenue, people were leaving downtown as 1 usual. The workday had come to an end; men and women were heading for bus and train stations, in a hurry to get home.

Time of day important?

2 She walked south on Michigan; she was wearing a white robe, as if she had been to the beach. She was blond and in her thirties. As she passed the Radisson Hotel, Roosevelt Williams, a doorman, was opening the door of a cab for one of the hotel's guests. The woman did not really pause while she walked; she merely shrugged the robe off, and it fell to the sidewalk. She was wearing what appeared to be the bottom of a blue bikini bathing suit, although one woman who was directly next to her said it was just underwear. She wore nothing else.

Why?

What difference would that make?

3 Williams at first did not believe what he was seeing. If you hang around long enough, you will see everything: robberies, muggings, street fights, murders. But a naked woman on North Michigan Avenue? Williams had not seen that before and neither, apparently, had the other people on the street.

Contrast with other outrageous events

Subject—an unusual event

4 It was strange; her white robe lay on the sidewalk, and by all accounts she was smiling. But no one spoke to her. A report in the newspaper the next day quoted someone: "The cars were stopping, the people on the buses were staring, people were shouting, and people were taking pictures." But that is not what other people who were there that afternoon said. The atmosphere was not carnival-like, they said. Rather, they said, it was as if something very sad was taking place. It took only a moment for people to realize that this was not some stunt designed to promote a product or a movie. Without anything telling them, they understood that the woman was troubled, and that what she was doing had nothing to do with sexual titillation; it was more of a cry for help.

Papers were wrong—misreported the event!

Main point?

People had sympathy

5 The cry for help came in a way that such cries often come. The woman was violating one of the basic premises of the social fabric. She was doing something that is not done. She was not shooting anyone, or breaking a window, or shouting in

Transition: expand on "cry for help"

more overt "violations"?

anger.] Rather, in a way that everyone understood, she was signaling that things were not right.

6 The line is so thin between matters being manageable and being out of hand. One day a person may be barely all right; the next the same person may have crossed over. Here is something from the author John Barth: *Thesis?* *Who is?*

> 7 She paused amid the kitchen to drink a glass of water; at that instant, losing a grip of fifty years, the next-room-ceiling plaster crashed. Or he merely sat in an empty study, in March-day glare, listening to the universe rustle in his head, when suddenly a five-foot shelf let go. For ages the fault creeps secret through the rock; in a second, ledge and railings, tourists and turbines all thunder over Niagara. Which snowflake triggers the avalanche? A house explodes; a star. In your spouse, so apparently resigned, murder twitches like a fetus. At some trifling new assessment, all the colonies rebel.

Supporting material from expert— more examples *?* *?*

8 The woman continued to walk past Tribune Tower. People who saw her said that the look on her face was almost peaceful. She did not seem to think she was doing anything unusual; she was described as appearing "blissful." Whatever the reaction on the street was, she seemed calm, as if she believed herself to be in control. *Because she was acting out her cry for help?*

9 She walked over the Michigan Avenue bridge. Again, people who were there report that no one harassed her; no one jeered at her or attempted to touch her. At some point on the bridge, she removed her bikini bottom. Now she was completely undressed, and still she walked. "It was as if people knew not to bother her," said one woman who was there. "To tell it, it sounds like something very lewd and sensational was going on. But it wasn't like that at all. It was as if people knew that something very . . . fragile . . . was taking place. I was impressed with the maturity with which people were handling it. No one spoke to *Def. of maturity*

her, but you could tell that they wished someone would help her."

Back in front of the Radisson, a police officer had picked up the woman's robe. He was on his portable radio, advising his colleagues that the woman was walking over the bridge. When the police caught up with the woman, she was just standing there, naked in downtown Chicago, still smiling. The first thing the police did was hand her some covering and ask her to put it on; the show was over. 10

A quiet response!

People who were there said that there was no reaction from the people who were watching. They said that the juvenile behavior you might expect in such a situation just didn't happen. After all, when a man walks out on a ledge in a suicide attempt, there are always people down below who call for him to jump. But this day, by all accounts, nothing like that took place. No one called for her to stay undressed; no one cursed the police officers for stopping her. "It was as if everyone was relieved," said a woman who saw it. "They were embarrassed by it; it made them feel bad. They were glad that someone had stopped her. And she was still smiling. She seemed to be off somewhere." 11

Contrast w/ what was expected

People identified with her? Feared for her? for themselves?

The police charged her with no crime; they took her to Read Mental Health Center, where she was reported to have signed herself in voluntarily. Within minutes things were back to as they always are on Michigan Avenue; there was no reminder of the naked lady who had reminded people how fragile is the everyday world in which we live. 12

Because? Surprising? Main point? See ¶ 6.

CLARIFYING MEANING

Put yourself in the student reader's place to see what happens next after reading and marking the text.

Using the Dictionary

First, use a dictionary to look up terms and words you marked as unfamiliar on the first reading. In this textbook, some will be defined just before the selection. Be sure to look up even words you *think* you know but are a bit fuzzy on. What exactly does *assessment* mean in paragraph 7? Though we often use the word to mean *evaluation,* in this case it means *taxation,* which makes more sense in the context.

You may use specialized dictionaries to look up unfamiliar references in the selection. For example, Greene writes that John Barth is an author. If you looked in a biographical or literary dictionary, you would find that Barth is a contemporary author who sees the world as absurd, as making no real sense. This detail helps you understand the quotation. Many writers make references to names from mythology, philosophy, and literature that you will need to look up. The reference librarian will show you where the specialized dictionaries are kept.

Reading Aloud

Return to the spots where you drew a question mark, and read those passages slowly aloud. Hearing your voice find the proper way to read a sentence may shed the necessary light on its meaning.

Discussing

Having a conversation about your reading will usually help you understand it. Another person who has read the same selection will probably have different reactions and may be able to clarify points that stumped you. Even someone who has not read the selection may be a good sounding board to discuss the ideas with.

Rereading

At some point, you will need to go back and reread the whole assignment, especially if you are going to be tested on it or intend to write a formal essay about it. With difficult material, the second read through will be more comfortable and will allow you to notice things you missed the first time.

MAKING INFERENCES AND ASSOCIATIONS

Bob Greene's essay is a good example of one that isn't completely spelled out for you. You have to make judgments about what he writes, inferences or conclusions about the meaning.

Reading between the Lines

You can train yourself to infer knowledge that lies below the surface meaning of the words. To *infer* means to arrive at an idea or a conclusion through reasoning. When you infer, you balance what the writer says with your own ideas and hunches about what is left unsaid. This process may sound difficult, but making **inferences** is a skill that can be learned.

Developing Inference Skills

Here are some suggestions for improving your ability to read between the lines:

Read beyond the words. Fill in details and information to complete the writer's suggestions. Use the writer's hints to discover the meanings that often lie beneath the surface. But don't go too far: you should be able to point to words and phrases that support what you have inferred.

Question yourself as you read and after you finish. You might use questions like these: Why did the author include these details? What does this example mean? How am I supposed to react to this sentence?

Draw conclusions and speculate on outcomes. In reflecting on Bob Greene's essay, for example, you might ask yourself these questions: Is the article saying that people are more sensitive than we usually assume—or less sensitive? Is the message of the selection positive or negative? Would people have reacted differently if the woman had not been young, blonde, and attractive? What truths about our society does this incident suggest?

Make associations between the reading and your own experience. For instance: Have you ever witnessed a "cry for help"? Was it like the one Greene describes, or different? How did you and other people respond?

Your own observations and reflections add richness to the selection's meaning. Our Responding to Reading exercises will assist you in developing your personal reaction.

WRITING TO UNDERSTAND AND RESPOND

If you write out your Responding to Reading assignment, you have already begun the interactive process that will set the selection firmly in your memory. Study skills experts point out that we have four modes of verbal communication: we listen, we read, we speak, and we write. Different people learn best through different modes, but college learning often emphasizes only the first two: listening to course lectures and reading textbooks. When you add the other two modes to your study habits, you more than double your learning potential. Speaking in class and discussing the material with friends and classmates are important. Writing about what you have heard and read is equally important. Here are some ways to write about a reading assignment:

1. Without looking at the reading, write a summary, 100 to 200 words long, of the selection. As you write, you will develop a sense of which parts of the selection are unclear in your mind. These will be the parts you find it hard to express. Compare your summary to the original, and revise your summary to make it as accurate as possible. This summary will be a fine study aid if you are going to be tested.

2. You can make another study aid by constructing an outline of the important points. This outline can be a simple list of key thoughts in the order they appeared in the essay, like this:

A. A woman removed her clothes while walking down Michigan Avenue.
B. The witnesses said the crowd reacted in a sad way, not in the excited, noisy way one might expect.
C. The woman was crying out for help by violating conventional behavior.
D. The episode reflected the thin line between ordinary and shocking events.
E. The crowd identified with the woman instead of looking at her as a freak.
F. After the woman was taken away, Michigan Avenue quickly went back to its usual state.

3. In your journal, write a letter to the author of your selection. What would you say to him or her if you could? Do you have any questions? These might be brought up in class discussion if you have them on hand.

4. You will also benefit from writing out answers to the Considering Content and Considering Method questions, which are designed to help you focus on meaning and technique.

The writing you have done so far will be of great help to you when you need to draft an assigned essay of your own based on your reading. This process is the subject of the next chapter.

WEB SITES

www.csbsju.edu/academicadvising/helplist.htm

Visit this site to gather additional strategies for effective reading and studying.

www.campaignformentalhealth.org/site/PageServer?pagename=share_your_story

The "Share Your Story" link at "Campaign for Mental Health.org" offers a variety of stories by people who live with mental illness.

Chapter 2

The Reading-Writing Connection

The connections between reading and writing are strong: in both activities you use language to create meaning. In Chapter 1, you learned how writing can help you to understand and remember what you read; in this chapter, you will learn how reading and responding to essays can help you to improve your writing.

WRITING IN RESPONSE TO READING

The basic principle of this book is that reading and writing go together. Each chapter follows a four-part pattern that you will discover works well in many of your college classes: (1) read a selection, (2) examine the content, (3) analyze the techniques, and (4) write something of your own that relates to the reading.

When you read, you get ideas for your own writing. Reading can supply you with topics to write about and show you how to write about them. Even when you already have a topic, reading can help you to come up with material to develop that topic. Reading will also provide you with models to follow. By studying the strategies and techniques that professional writers use, you can learn methods and procedures for writing effectively on many different subjects and in many different writing situations.

BUILDING AN ESSAY

Writing an essay is a lot like building a house. A writer fits separate pieces of meaning together to make an understandable statement. If you want to write well, you need to learn the basic skills of constructing an essay.

Despite differences in education and personality, most writers follow a remarkably similar process of *prewriting, planning, writing, revising,* and *editing.* Whether building a single paragraph or a ten-page article, successful writers usually work through several stages, which go roughly like this:

1. Find a subject; gather information. (Prewriting)
2. Focus on a main idea; map out an approach. (Planning)
3. Prepare a rough draft. (Writing)
4. Rework and improve the draft. (Revising)
5. Correct errors. (Editing)

If you follow these operations, you will learn to write more productively and more easily. But keep in mind that this sequence is only a general guide. The steps often overlap and loop around. The important point to remember is that writing is done in stages; successful writers take the time to build their essays step by step and to polish and finish their work the way a good carpenter sands rough surfaces.

Source: © 1999 United Feature Syndicate, Inc.

Finding Ideas

One of the most difficult challenges of writing is coming up with a topic. Even when you are responding to a reading, you still have to decide what to say about it. In this textbook and in most classes, you will be given some direction toward a topic; the job from there on is up to you. Rather than wait for inspiration to strike, you can go after the ideas you need by doing some **prewriting**. Here are three methods that experienced writers use to generate material for writing.

1. **Freewriting.** Write without stopping for five or ten minutes. Don't pause to consider whether your ideas are any good or not; just get down as many thoughts as you can within the time limit. If you're freewriting on a computer, turn down the contrast to write without seeing the screen, or turn the monitor off. After the time is up, read through your freewriting and highlight anything that strikes you as interesting or important. Then do some more freewriting on one or two of these points. Here is an example of freewriting done by student Tara Coburn in response to "Handled with Care," the article by Bob Greene that you read in Chapter 1:

> Basic themes of Greene's essay. People watched a break in normalcy with maturity, saw the fragility of life, a lesson with tones of sadness, embarrassment, sympathy, but then forgot it. Compared with the lesson of the fragility of life when the situation is personal. The reaction of the people on Michigan avenue was formed by their detachment from the woman. If they'd known her, someone would have spoken to her, helped her, the world would not have gone back to normal, the incident would not have been forgotten. Their lesson was fleeting but when there is a tear in the social normalcy of your daily life and someone you love shows you the fragility of life, it is more real.

At this point, Tara stopped and looked at what she had written. She liked the idea that surfaced in the last sentence and decided that she had a topic she could develop into an essay. She put the freewriting aside for a while—to let the ideas work around in her mind before she moved on to the next stage in the process.

2. **Brainstorming.** As an alternative to freewriting, you can ask yourself a question and list as many answers as you can. For example: *What have I done that's unexpected or out of the ordinary?* or *When is it all right to get involved with someone in trouble?* Challenge yourself to make the list as long as you can. If necessary, ask yourself a new question: *When is it a bad idea to help someone in trouble?* In making this list you have already started writing. Think of it as a bank of raw material on which you can draw.

In order to develop ideas for her topic, Tara Coburn posed this question to herself: *When did the fabric of my life begin to tear?* And then she brainstormed a list of responses to that question:

```
part of my world, my fabric, was belief in my dad
thought he could do anything
helped us build a snowman
finished the ice cream
fix anything
even our cat brought him things
build anything—bed, chairs, room
not a book smart guy
tear happened gradually
getting older, tired, looking older
can't fix everything—computer
as strong as everything seemed, it was fragile
not a cry for help, more a sign that he can't help
anymore
like the people on Mich. Ave., it's a sign it's my turn
to come forward
```

3. **Questioning.** Write a broad topic—such as *Helping People in Trouble*—at the top of a sheet of paper. Then write the headings *Who? What? When? Where? Why? How?* down the page. Fill in any thoughts about the topic that occur to you under these headings. If you can't think of anything for one heading, go to the next, but try to write something under each heading. The goal is to think creatively about the topic as you try to come up with material to use in writing.

Devising a Working Thesis

At this stage, you need to collect the ideas that you came up with in prewriting and organize them. One way to focus your material is to ask yourself: *What point do I want to make?* The answer to that question will lead you to your main idea, or working **thesis**. Once you decide what point you want to make, then you can go through your prewriting material and decide which details to use and which ones to toss.

As you learned in Chapter 1, a thesis says something *about* the subject of a reading. As a reader, your job is to discover the writer's thesis; as a writer, your job is to provide a clear thesis for your readers. Look at the difference between a **subject** and a thesis in these examples:

Subject: Helping strangers
Thesis: I think we are responsible for helping people in trouble.

Subject: Doing something socially unacceptable
Thesis: As soon as I got to college, I set out to prove that I was an adult and beyond the control of my parents.

Here is an example of the thesis statement that Tara Coburn devised from her freewriting and her brainstorming list:

```
     When I saw that my dad's strength was limited, I
began to realize that life was fragile and fleeting and
that I would have to learn to be strong and do things for
him and for myself.
```

You may change or refine your thesis as the paper develops, but having an idea of what you want to say makes the actual writing considerably easier.

Making a Plan

Having a plan to follow makes you less likely to wander from your main point. An outline of your major points will provide you with a framework for your first draft; it can help you shape and arrange your thoughts and keep you from making organizational missteps. There is no need for complete sentences or balanced headings in your outline. Just make a list of your main points in the order that you plan to cover them. The following brief plan is based on the thesis you just read:

```
                    SOMEONE TO HELP

 1. Opening
      Yarn ankle bracelet—first clue that life is fragile
      My social fabric—my belief in my dad
 2. Always thought Dad could do anything
      Helped us with the giant snow bunny
      Lifted its head—not as easy as it looked then
 3. Dad always did the hard work
      Cranked the ice cream
      Even the cat came to him
      Built furniture, almost everything around me
 4. Realized he was getting older
      Hair and beard turned white
      Couldn't fix the computer—but I could
      I became the fixer—my chance to grow stronger
```

```
5. Closing
    People on Michigan Ave—strangers didn't know what
    to do
    I love my dad—I can step forward to help him
    Helping makes me grow and become stronger
```

Composing a Draft

If you have an outline or plan to work from, you shouldn't have any trouble producing the first draft of your paper. Don't fret about trying to write a brilliant **introduction**. Skip it if you can't come up with anything inspired, and start right in on the first main point. You can always come back and add an introduction when you revise.

Some people write the first draft from start to finish without bothering to search for the best word or the right phrase. If that's your method, fine. But many successful writers stop to reread what they have written, especially when they get stuck; rereading helps them to regain momentum and bring back fluency. The main goal is to get your ideas down on paper in a reasonably complete draft. Then you are ready for the important next step: revising.

Improving the Draft

Set your first draft aside, at least overnight, so you can look at it in a new light. This process of looking at your draft *again* is called **revising**, and it literally means "re-seeing." In fact, you want to try to see your work now with different eyes—the eyes of a reader.

When revising your draft, concentrate on making major improvements in content and organization. Such improvements might include enlarging or narrowing the thesis, adding more examples or cutting irrelevant ones, and reorganizing points to improve logic or gain emphasis. Tackling the simple problems first may seem reasonable, but you will find that dealing with a major difficulty may eliminate some minor problems at the same time—or change the way you approach them. If you try to do the fine-tuning and polishing first, you may burn up valuable time and energy and never get around to the main problems.

Targeting the Readers

Most of the writing you do has an **audience**—the person or group of people who will read what you've written. Keeping these readers in mind will help you decide not only what to say but also how best to say it.

You already have considerable reader awareness: you would not write a letter applying for a job in the same way you write a thank-you note to dear Aunt Marie. But writing a class assignment presents a special problem in audience. You know you're writing for your instructor and perhaps other members of the class, but you also know that they are going to read your paper no matter what. In order to develop your writing abilities, you need to get beyond the captive, limited audience of an instructor and your classmates and learn how to write for readers who have different expectations and interests.

Although they often have specific, well-defined audiences, many successful writers still find it useful to imagine an audience that is reasonably informed and generally attentive, readers who will keep reading so long as the writing is interesting and worthwhile. For example, when we write the chapters of this book, we think about the students who have been in our classes over the years. If you can picture such an audience, you'll have an easier time deciding what to revise and how to improve your paper. Having other people read your drafts and give you their reactions will also teach you a lot about the importance of audience.

Getting Feedback

Writers routinely seek the help of potential readers to find out what is working and what is not working in their drafts. Someone else can often see places where you *thought* you were being clear but were actually filling in details in your head, not on the page.

The ideal people to help you evaluate your first draft are the members of your own writing class. They will be familiar with the assignment and will understand why you are writing the paper and for whom. In many writing classes, students work together on their papers. Meeting in small groups, they read photocopies of each other's drafts and respond to them; sometimes they post their drafts on a class Web site or submit them electronically on a computer bulletin board. If your instructor doesn't set up a peer review system, try to get several readers' reactions to your drafts. You can meet together outside of class or use an Internet mailing list. Here are some questions to use in asking for feedback:

Have I made my thesis clear to the reader?

Does the introduction get the reader's attention?

Are there any points that the reader might not understand?

Do I need to give the reader more reasons and examples?

Have I shown the reader how every point relates to the thesis?

Does the **conclusion** tie everything together for the reader?

Polishing the Final Draft

When you are satisfied with the changes you've made to improve content and organization, you can move on to matters of spelling, word choice, punctuation, capitalization, and mechanics. This is the **editing** stage, and you cannot skip it. Readers become quickly annoyed by writing that is full of errors.

Source: For Better of For Worse © 1995 Lynn Johnston Productions. Dist. by Universal Press Syndicate. Reprinted with permission. All Rights Reserved.

Here are some additional tips that will help you polish and correct your final draft:

1. Let your work sit for a day to clear your head and increase your chances of spotting errors.
2. Read your draft out loud, listening for anything that sounds unclear or incomplete or awkward.
3. Don't try to do everything at once; save time to take a break when you need one.
4. Slow down when you edit: look at each word and punctuation mark individually, and watch for mistakes that you know you usually make.
5. Ask a reliable reader to check over your draft one more time before you turn it in.

SAMPLE STUDENT ESSAY

The essay that follows was written by Tara Coburn, a first-year student at Eastern Illinois University. She was responding to some of the ideas expressed in "Handled with Care" by Bob Greene. The comments in the margin call your attention to important features of organization and development.

Someone to Help

1 When I was eight, I had a favorite red yarn ankle bracelet that I wore for an entire year. I tugged on it one day to show that it was strong enough to last another year, and it broke. It was a clue that the things we count on in life are fragile. In his essay "Handled with Care," Bob Greene describes how the sudden appearance of a naked woman on busy Michigan Avenue taught the people walking in the street a similar lesson—that "the line is so thin between matters being manageable and being out of hand." Part of my world—my "social fabric"—was based on my childhood belief that my father could do anything. When I began to see that my dad's strength was limited, his weakness reminded me that life is fragile and fleeting and that I would have to learn to be strong on my own.

Introductory paragraph: opens with a specific example, relating main idea to the reading and moving into the thesis.

Thesis: last sentence of first paragraph.

2 As a child, no matter how big the problem, I always thought, "Dad can do it." One winter, the other neighborhood children and I had a grand plan to build a giant snow bunny. We somehow managed to lift the bottom and middle balls into place, but the head was far too heavy for us to boost up six feet onto a half-finished snow bunny. My very first thought was to ask my dad, and he tramped out into the snow to check the situation. At the time, he seemed to lob the snow bunny's head effortlessly into place, but thinking back I realize how heavy that big ball of snow must have been.

First body paragraph begins with topic sentence and develops it with an extended example.

3 Dad was always the one we called on to do the hard work. When the homemade

Second body paragraph begins with topic sentence and develops it with several examples.

ice cream was getting to the final stages of freezing, we called Dad to kneel down and crank the final turns that no one else could manage. His reputation as a Mr. Fix-it was renowned, even to our cat, Mousie. When she had played too roughly with the mouse or snake from the backyard, she'd drop the dead animal at his feet with a look that said, "Daddy, I broke my toy. Will you make it play again?" My dad built much of the world I lived in, from the chairs in the living room to the bed I slept in. Being surrounded by things he built is probably the reason my ideas about him were so important to me.

Paragraph ends by summing up the significance of the examples.

4 The tears in my social fabric appeared gradually as I realized that my dad was getting older. When I came home from college for the first time, I noticed that his beard had started to turn from red to white; he was becoming an old man. Not only was his appearance changing, but the arrival of our household's first computer also revealed that my dad cannot fix everything. Now when he says to me, "The computer's little doodad is spinning and I can't make it stop!" his frustration is obvious. But his inability has given me the chance, for once, to be the fixer. I always relied on him, but as he gets older and I learn more, I can let him rely on me. The fragility I now see in my dad has given me the chance to be stronger.

Third body paragraph begins with topic sentence and develops it with two specific examples.

End of paragraph sums up the point and connects it to the main thesis.

5 The people on Michigan Avenue did not step forward to help the naked woman. One passerby remarked, "No one spoke to her, but you could tell that they wished someone would help her." The

Conclusion returns to Greene's essay to make a point through contrast.

```
crowd felt sorry for the woman, but
they were strangers and didn't know how
to respond. My dad's turn from strong
to fragile was not a cry for help; it
was just a sign that he cannot always
help me as he once did. But because I
know and love my father, I can step
forward to help him. And by helping him
I have grown strong and have learned to
help myself.
```

Closing sentences reinforce the thesis.

RESOURCES FOR WRITERS ON THE INTERNET

You can find a lot of advice on the writing process at Web sites and online services. They vary widely in quality, presentation, and amount of detail. The following are some of the most helpful and usable:

- **http://web.uvic.ca/wguide/**
  The University of Victoria's Hypertext Writer's Guide will help you through the basics of the writing process and answer questions about essays, paragraphs, sentences, words, and documentation.
- **www.powa.org/**
  Paradigm Online Writing Assistant offers useful advice on writing various types of papers. It contains sections on discovery, organization, editing, and other topics.
- **http://owl.english.purdue.edu**
  Purdue University's Online Writing Lab has printer-friendly handouts, interactive Power-Point presentations, and hypertext workshops about the process and mechanics of writing—one of the most extensive collections of advice about writing on the Web.

RESPONDING TO A READING

Now that you have seen samples of close reading and of writing in response to reading, it's time to try these skills yourself. Use the advice in Chapters 1 and 2 as you practice.

Preparing to Read

Are you a successful writer? Do you like to write? Did you have any experiences in English or other classes that affected your attitude toward writing? What were they?

Learning to Write

Russell Baker

The winner of a Pulitzer Prize for journalism, Russell Baker began his career as a writer for the Baltimore Sun *and moved to the* New York Times *in the 1950s, where he wrote the "Observer" column from 1962 to 1998. He currently hosts the* Masterpiece Theater *series on PBS. Baker is known for his humorous observations of everyday life, but in this excerpt from his autobiography* Growing Up *(1982), his lighthearted tone gives way to a serious description of an important personal event.*

TERMS TO RECOGNIZE

notorious *(para. 1)*	known widely and usually unfavorably, disreputable
prim *(para. 1)*	formal and neat, lacking humor
listless *(para. 2)*	without energy, boring
ferocity *(para. 2)*	fierce intensity, savagery
irrepressible *(para. 2)*	impossible to control or hold back
essence *(para. 3)*	the most important ingredient or element, fundamental nature
antecedent *(para. 4)*	the word that a pronoun refers to
exotic *(para. 6)*	rare and unusual
reminiscence *(para. 8)*	a thing remembered, a memory
contempt *(para. 10)*	scorn, disrespect
ridicule *(para. 10)*	mockery, teasing
ecstasy *(para. 11)*	bliss, delight, joy

When our class was assigned to Mr. Fleagle for third-year English I anticipated another grim year in that dreariest of subjects. Mr. Fleagle was notorious among City students for dullness and inability to inspire. He was said to be stuffy, dull, and hopelessly out of date. 1

To me he looked to be sixty or seventy and prim to a fault. He wore primly severe eyeglasses, his wavy hair was primly cut and primly combed. He wore prim vested suits with neckties blocked primly against the collar buttons of his primly starched white shirts. He had a primly pointed jaw, a primly straight nose, and a prim manner of speaking that was so correct, so gentlemanly, that he seemed a comic antique.

I anticipated a listless, unfruitful year with Mr. Fleagle and for a long time was not disappointed. We read *Macbeth.* Mr. Fleagle loved *Macbeth* and wanted us to love it too, but he lacked the gift of infecting others with his own passion. He tried to convey the murderous ferocity of Lady Macbeth one day by reading aloud the passage that concludes 2

> . . . I have given suck, and know
> How tender 'tis to love the babe that milks me.
> I would, while it was smiling in my face,
> Have plucked my nipple from his boneless gums. . . .

The idea of prim Mr. Fleagle plucking his nipple from boneless gums was too much for the class. We burst into gasps of irrepressible snickering. Mr. Fleagle stopped.

"There is nothing funny, boys, about giving suck to a babe. It is the—the very essence of motherhood, don't you see." 3

He constantly sprinkled his sentences with "don't you see." It wasn't a question but an exclamation of mild surprise at our ignorance. "Your pronoun needs an antecedent, don't you see," he would say, very primly. "The purpose of the Porter's scene, boys, is to provide comic relief from the horror, don't you see." 4

Later in the year we tackled the informal essay. "The essay, don't you see, is the. . . ." My mind went numb. Of all forms of writing, none seemed so boring as the essay. Naturally we would have to write informal essays. Mr. Fleagle distributed a homework sheet offering us a choice of topics. None was quite so simpleminded as "What I Did on My Summer Vacation," but most seemed to be almost as dull. I took the list home and dawdled until the night before the essay was due. Sprawled on the sofa, I finally faced up to the grim task, took the list out of my notebook, and scanned it. The topic on which my eye stopped was "The Art of Eating Spaghetti." 5

This title produced an extraordinary sequence of mental images. Surging up out of the depths of memory came a vivid recollection of a night in Belleville when all of us were seated around the supper table—Uncle Allen, my mother, Uncle Charlie, Doris, Uncle Hal—and Aunt Pat served spaghetti for supper. Spaghetti was an exotic treat in those days. Neither Doris nor I had ever eaten spaghetti, and none of the adults had enough experience to be good at it. All the good humor of Uncle Allen's house reawoke in my 6 mind as I recalled the laughing arguments we had that night about the socially respectable method for moving spaghetti from plate to mouth.

7 Suddenly I wanted to write about that, about the warmth and good feeling of it, but I wanted to put it down simply for my own joy, not for Mr. Fleagle. It was a moment I wanted to recapture and hold for myself. I wanted to relive the pleasure of an evening at New Street. To write it as I wanted, however, would violate all the rules of formal composition I'd learned in school, and Mr. Fleagle would surely give it a failing grade. Never mind. I would write something else for Mr. Fleagle after I had written this thing for myself.

8 When I finished it the night was half gone, and there was no time left to compose a proper, respectable essay for Mr. Fleagle. There was no choice next morning but to turn in my private reminiscence of Belleville. Two days passed before Mr. Fleagle returned the graded papers, and he returned everyone's but mine. I was bracing myself for a command to report to Mr. Fleagle immediately after school for discipline when I saw him lift my paper from his desk and rap for the class's attention.

9 "Now, boys," he said, "I want to read you an essay. This is titled 'The Art of Eating Spaghetti.'"

10 And he started to read. My words! He was reading *my words* out loud to the entire class. What's more, the entire class was listening. Listening attentively. Then somebody laughed, then the entire class was laughing, and not in contempt and ridicule, but with openhearted enjoyment. Even Mr. Fleagle stopped two or three times to repress a small prim smile.

11 I did my best to avoid showing pleasure, but what I was feeling was pure ecstasy at this startling demonstration that my words had the power to make people laugh. In the eleventh grade, at the eleventh hour as it were, I had discovered a calling. It was the happiest moment of my entire school career. When Mr. Fleagle finished he put the final seal on my happiness by saying, "Now that, boys, is an essay, don't you see. It's—don't you see—it's of the very essence of the essay, don't you see. Congratulations, Mr. Baker."

WEB SITE

www.albany.edu/writers-inst/olv6n2.html#baker

Read a transcript of a 1991 interview with Russell Baker, who talks about his life and his writing career.

Suggestions for Writing

1. Baker's experience in eleventh-grade English changed the way he thought about himself. Have you ever had such an eye-opening ex-

perience—some event you really want to write about, some incident you want to "recapture and hold" for yourself? Write an essay in which you describe what happened.

2. Write an essay about the most important thing that happened to you in school. It could be either a positive or negative experience—one that taught you something about yourself or about school or about the subject you were studying.
3. USING THE INTERNET. Take a look at the list of common myths about writing at **http://school.newsweek.com/resources/writing.php**. Then get together with a small group of classmates and talk about how each of you feels about writing. Compare notes on what you like and don't like about writing. Which of the ten myths do you still believe in? Discuss the kinds of writing that you have done, and tell each other about previous writing experiences. Then write an essay explaining your thoughts and feelings about writing. If you changed your attitude (as Baker did), tell about that change.
4. Write an essay describing your writing process: how you get started, how you go about getting ideas, what you like to write about, where you like to write, whether you like background music or need quiet, whether you make an outline or plunge right in, how long it takes to complete an assignment, whether you work in stages, how many drafts you write, whether you write by hand or use a computer, how much correcting and recopying you do, and so on. Think carefully about the way you write, and describe it in as much detail as you can. Conclude your essay by explaining what you think you could do to make your writing process more efficient.

Chapter 3

Strategies for Conveying Ideas: *Narration* and *Description*

He falls back upon the bed awkwardly. His stumps, unweighted by legs and feet, rise in the air, presenting themselves. I unwrap the bandages from the stumps, and begin to cut away the black scabs and the dead, glazed fat with scissors and forceps. A shard of white bone comes loose. I pick it away. I wash the wounds with disinfectant and redress the stumps.

—Richard Selzer, "The Discus Thrower"

That powerful paragraph, written by a surgeon good with words as well as with scalpels, combines description and narration. Dr. Selzer, narrating an experience in treating a terminally ill patient, uses description to make his account of this brief event vividly, compellingly real. Simply put, a narrative is a story; **narration** is the telling of a story. The preceding passage is taken from a longer narrative (with a beginning, a middle, and an end) that makes a point. As you can see, the strength of narrative and descriptive writing lies in the use of vivid language and in the selection of precise details.

THE POINT OF NARRATION AND DESCRIPTION

Although the selections in this chapter combine several writing strategies, they are primarily narratives that make a point using descriptive details to add clarity, liveliness, and interest. You will notice that most of the readings throughout this text use both narration and description to help them develop many different kinds of essays.

Using Narratives

Consider how often we use narrative in everyday speech. If you want to convince your daughter to avoid becoming pregnant as a teenager, you'll probably tell her the story of your high school friend who made that mistake and missed her chance to become the architect she always wanted to be. We tell stories because they are convincing—they have the ring of truth to them—and most people are able to learn from the experience of others. So, narrative is a good strategy to consider if you are writing to persuade, to make a point.

If you are going to write a narrative essay, be sure your story has a point. You wouldn't tell a joke without a punch line, and only mothers and close friends will hold still for stories without a purpose.

Much more frequently, you will use short narratives as part of a longer essay. Notice in your reading how often writers begin essays with a brief narrative to catch our interest and lead us into the topic. Once the writers have our attention, they may move to other strategies to develop their ideas and present the material, but a good story makes an effective lure.

Using Description

Essays of pure description are rare, but descriptive details provide one of the most common ways of adding interest and clarity to your writing. **Description** can put a picture in the minds of readers, helping them to see what you mean. Most writing would seem dull and lifeless without descriptive details, like this sentence:

> The firefighter rescued a child.

Although the action referred to is exciting, the sentence is blah. But add some descriptive details, and the sentence gains meaning and interest:

> The exhausted firefighter, a rookie on the force, staggered from the flames with the limp body of an unconscious child cradled in his arms.

You can, of course, include too much description. But use any details that come to you while writing your first draft. Then, when you revise, decide whether you've gone too far, and eliminate any that seem overdone or unnecessary. Of course, add more details if you think you have too few.

THE PRINCIPLES OF NARRATION AND DESCRIPTION

Good narrative and descriptive writing depends, as does most writing, on making choices. In narratives, the main choices involve deciding which events to include and which ones to leave out. In descriptions, the choices involve selecting words that appeal to the senses—usually to sight, but also to touch, taste, hearing, and smell.

Organizing the Events

If you've ever listened to a boring storyteller, you know how important a concise and effective framework is to a narrative. Organizing a narrative seems easy because the story almost always proceeds in **chronological order** (according to the time in which events happened). But a poor storyteller (or a writer who fails to revise) will get things out of order and interrupt the tale with "Oh, I forgot to mention that Marvin got fired just before his cat got lost and his dog was run over." Or the storyteller will get hung up on a totally unimportant detail:

> The moment I saw him—I think it was at the senior prom—or was it at the homecoming dance—or it could have been the party at Yolanda's—

> oh, wait, I don't think it was at a party at all, it was at a football game—or no, a basketball game. . . .

even though the point of the story has nothing to do with where or when it happened.

Get your story straight in your mind before you begin, or else straighten it out when you revise. Eliminate any dull or unnecessary material, and then go to work on making it interesting.

Including Specific Details

All good writing is full of specific **details**, but a narrative will simply fall flat without them. Recall the paragraph we quoted by Dr. Richard Selzer. We could summarize that paragraph in a single sentence:

> After the patient fell back awkwardly in bed, I removed the bandages, cleaned, disinfected, and rebandaged the ends of his amputated legs.

What did we leave out? The details—in this case mostly descriptive details—and what a difference they make in allowing us to visualize the doctor's performance.

Selecting Descriptive Words

Effective description depends heavily on the use of specific details, but also crucial are the words you choose in presenting those details. Consider this sentence quoted from Dr. Selzer's paragraph:

> I unwrap the bandages from the stumps, and begin to cut away the black scabs and the dead, glazed fat with scissors and forceps.

Look at what happens when we substitute less specific, less descriptive words in that same sentence:

> I take the bandages off the amputated legs, and begin to remove the dead tissue with my instruments.

The meaning is the same, and most of us would probably have written it that way, but it's clear that the force of Dr. Selzer's sentence lies in his word choice: *unwrap, stumps, cut away, black scabs, glazed fat, scissors, forceps.*

Good description stems from close observation, paying close attention to the sights, sounds, textures, tastes, or smells around you. Then you must search for exactly the right words to convey what you experience to your readers. Try to think of words that go beyond the general to the specific:

GENERAL ←——————→ SPECIFIC

a drink	a soft drink	a diet cola	a stale diet Coke
move	run	run fast	race headlong
weather	rain	cold rain	cold, blowing rain

You get the idea. As you read, be alert for words that convey **images**, that let you know how something looks or feels or moves or tastes or smells or sounds.

When you revise, try to replace the following useful but run-of-the-mill verbs with words that are more specific and more interesting:

is (are, was, were, etc.)	go	get	has
come	move	do	make

Notice the difference a livelier word choice makes:

O'Malley moved on to second base.
O'Malley slid into second base.

I'm going to do my homework.
I'm going to struggle with my homework.

Lapita made a chocolate mousse.
Lapita whipped up a luscious chocolate mousse.

Keep a vocabulary list and try to use the new words in speaking and writing. The theory is that if you use a word three times, it enters your vocabulary. Part of becoming a good writer (as well as a good speaker) depends on increasing the number of words you have at your command, so get to work on it.

THE PITFALLS OF NARRATION AND DESCRIPTION

The narratives included in this chapter, written by experienced authors, will not show you the many things that can go wrong in this kind of writing. Descriptions and narratives are among the easiest kinds of writing to do, but they are probably the most difficult to do well. You need to find someone who will read your draft, not only for enjoyment, but also with the promise of helping you improve it. It's always hard to see the flaws in your own writing. It's especially hard with narratives and descriptions. But you can do a good job if you are willing to work at it and can find a reliable person to help you in revising.

As you rewrite, ask yourself—and ask your helper to answer—the following questions:

1. Is the point of my narrative clear? It may not be stated directly, but readers should be able to see *why* I'm telling this story.
2. Are the events in order? Are there any gaps? Any backtracking?
3. Are there enough details to be clear and interesting?
4. Are there any unnecessary details that I should take out?
5. Do the descriptions provide an image (a picture, a sound, a smell, a taste, an atmosphere, an action)?

Respond to the questions as honestly as you can—and have your helper do the same. Then keep revising until you are both happy with the results.

WHAT TO LOOK FOR IN NARRATION AND DESCRIPTION

As you read the essays in this chapter, pay attention to these characteristics:

1. *Look at the way the essay is put together.* Probably the events are told chronologically, as they happened, but if not, try to figure out why the writer departed from the usual method of organization.
2. *Decide what point the narrative makes.* Its purpose may be simply to entertain the readers, but more often it will illustrate or make a point. Look for an underlying meaning that you discover as you think about the story and the author's reasons for telling it.
3. *Consider the elements of the narrative.* Think about why these particular events, people, and descriptions are included. If they are not crucial, try to decide what they add to the story and what would be lost if they were left out.
4. *Notice the descriptive details.* Underline any sentences or phrases that appeal to your senses or put a picture in your mind.
5. *Pick out good descriptive words.* Most of these you will already have underlined. Add them to your vocabulary list.

IMAGES AND IDEAS

Source: AP Wide World Photos (left); Robert Funk (right).

For Discussion and Writing

Look at the two photographs reproduced here. First, give each one a title that you think expresses the mood or atmosphere of the picture. The title will probably be a general word or phrase that describes an emotion or state of mind.

Next, choose one of the pictures to write about. Your purpose is to present a detailed description of the picture to someone who has never seen it. The description should carefully differentiate this photo from other ones like it—that is, if your readers were to look at five winter scenes, they could pick out this particular one. In your description, get across the feeling or state of mind you used in your title. Be sure to use words that are very specific and have strong associations. Your description should be at least one page long. You could also make up a narrative to convey your description.

√ Preparing to Read

Have you ever been involved in a serious accident or caught in a burning building or been especially frightened by something? Describe how you felt at the time. Were you terrified—or did everything happen too fast? Record the thoughts and feelings you experienced once the incident was over.

Rain of Fire

Evan Thomas

Evan Thomas, a graduate of Harvard and the University of Virginia Law School, has been assistant managing editor of Newsweek *magazine since 1991 and the lead writer on many major news stories. His vivid reporting on the September 11th disaster helped win for* Newsweek *the National Magazine Award for General Excellence in 2002. The following selection appeared in* Newsweek *(Dec. 31, 2001/Jan. 7, 2002) as Part IV of an extensive article entitled "The Day That Changed America." If you want to find out how Virginia DiChiara made it down those seventy-eight flights of stairs, read the whole gripping article.*

Terms to Recognize

moseying *(para. 1)*	moving slowly and aimlessly
translucent *(para. 4)*	clear, letting light through
prosaic *(para. 6)*	everyday, ordinary
suffused *(para. 8)*	spread throughout
cacophony *(para. 8)*	harsh, nerve-jangling sound

A self-described workaholic, Virginia DiChiara was normally out of her house and on her way to work by 7 A.M. But the morning of September 11 was so brilliantly beautiful that DiChiara decided to dawdle. She let her two golden retrievers play in the yard, cooked herself some eggs, poured herself a cup of coffee. "I was just moseying along," she says. "I didn't feel like rushing." She left her Bloomfield, N.J., home at 7:40, a 40-minute delay that would end up saving her life. 1

It was a little after 8:40 when she entered the lobby of the North Tower of the World Trade Center. Together with a Cantor Fitzgerald co-worker, she rode the elevator up to the 78th floor, where she crossed a lobby to take a second elevator the rest of the way to her office on the 101st floor. The elevator door opened and she pressed the button for 101. It was 8:46 A.M. 2

As the elevator doors closed, Flight 11 plowed into the northern face of Tower I some 20 floors above. The elevator went black and "bounced around like a ball," DiChiara recalls. "I remember seeing two lines shooting around the top of the elevator"—electrical cables that had come loose and were spitting current—and "everybody started screaming." In front of her was a man named Roy Bell, who later said that the sound of impact was "deafening," like someone banging a 2-by-2-foot sheet of aluminum with a hammer "six inches from your head." The right wall of the elevator car crashed into Bell, breaking several of his fingers and flinging him to the left side. Miraculously, the elevator doors remained open about a foot. Within seconds, Bell "just sprinted" out of the elevator, he recalls. "Inside was not where you wanted to be." 3

DiChiara had crouched down behind Bell. She saw Bell go through, and thought, "I don't hear any screaming, so I know he's not on fire . . . I'm outta here." She decided to go for it. But as she gathered herself, huge blue flames—translucent teardrops of fire, a foot in diameter—began falling in a steady curtain. DiChiara dropped her bag, covered her face with her palms and squeezed through the door, her elbows pushing the black rubber guards on the elevator doors. Left behind was her Cantor co-worker. DiChiara never saw her again; at times she feels guilty that she made it out and her co-worker did not. 4

DiChiara was aflame when she emerged from the elevator. "I remember hearing my hair on fire," she says. (She later joked, "I must have put on some extra hair spray.") With her hands she tapped out the fire. "I got it out, I got it," she said to herself. Then, feeling something else, she looked back and saw flames rising from her shoulder. In that instant, she remembered the old lesson from grade school: stop, drop and roll. She threw herself to the carpeted floor and rolled over and over, frantically patting out the flames. "I remember getting up and just looking at myself," she says. "'OK, everything's out.' And then sort of laughing, almost like a hysteria, like a little giggle, like, 'Oh my God, let me do it again just in case I missed it.' I was so scared, like there was an ember on my body that was still going to go up." 5

DiChiara crawled some 20 feet down the hallway and sat with her back propped against a wall. She was wearing a sleeveless cotton shirt that day, and her arms and hands were seared with third-degree burns. In shock, she did not feel the pain—yet. Improbably prosaic thoughts crossed her mind. In the briefcase she'd left on the elevator were some airplane tickets recently purchased for a vacation jaunt to the Florida Keys, as well as a wad of cash. Should she go back and retrieve it? "No," she thought to herself. "Just stay right where you are." 6

Then she spotted a co-worker, Ari Schonbrun, head of global accounts receivable at Cantor. "Ari!" she called out. He turned around and looked at her. "Virginia! Oh my God!" he said. "Ari, I'm badly burned," Virginia 7 told him. She was gradually realizing how grave her condition truly was, and beginning to feel it as well. "I'm in so much pain," she said. Schonbrun was horrified. "The skin was peeled off her arms," he says. "You knew this woman was in trouble." DiChiara read his expression. "I knew by the look on his face that I was bad," she says. Schonbrun told her, "Virginia, take it easy. We're going to get help. Don't worry. You're going to be fine." She begged, "Whatever you do, don't leave me." Schonbrun reassured her, "I'm not going to leave you. I'll be with you."

The hallways were smoky, suffused with the nauseating smell of burned jet fuel, littered with debris, and completely dark save for some outdoor light filtering in from windows at the end of the hall. Schonbrun gently guided DiChiara toward a small security office behind the elevator banks where the lights still worked. About a dozen people were huddled there, including two security guards. DiChiara lay down on the floor, on the verge of passing out. A woman sitting nearby was crying. One of the security guards was furiously dialing for help on the office phone but couldn't get through. The other guard had a radio, but she was paralyzed, crying. Schonbrun told her, "You've got to calm down. You've got to get on that radio and get us help." The guard tried, but the only sound coming over the radio was a cacophony of screams. 8

Singed by his narrow escape through the elevator doors, Bell had also made it to the security office. His doctors would later tell him that the few seconds between his exit and DiChiara's made the difference between his second-degree and her third-degree burns. Suddenly, a man appeared saying that he was a fire warden. There was a stairwell in the middle of the tower that they could use, he announced. Schonbrun leaned over DiChiara and laid out the options: either they could wait for someone to rescue them, or they could start heading down by foot. 9

For DiChiara, there was no choice. No way was she going to sit there and wait. Gritting her teeth, she got up and headed for the stairwell. 10

RESPONDING TO READING

Everyone has seen on television the visual images of the horrific events of September 11, 2001. The previous selection recounts the experience of one woman caught up in that tragedy. What did you get from the reading that you did not get from seeing the visuals?

GAINING WORD POWER

Thomas uses a number of precise descriptive word choices in this narrative. Several appear in the Terms to Recognize section preceding the selection. Here are some others:

DiChiara decided to *dawdle.* (para. 1)

Flight 11 *plowed* into the northern face of Tower I. (para. 3)

DiChiara was *aflame.* (para. 5)

Her arms and hands were *seared.* (para. 6)

One of the security guards was *furiously* dialing for help. (para. 8)

We repeat these sentences below, minus the italicized words. Fill in each blank with your own definition of the missing word.

DiChiara decided to __________.

Flight 11 __________ into the northern face of Tower I.

DiChiara was __________.

Her arms and hands were __________.

One of the security guards was __________ dialing for help.

What is lost or changed in the second set of sentences? Which sentences do you prefer?

CONSIDERING CONTENT

1. Why was Virginia DiChiara late going to work on September 11, 2001? Was she a person who was often late?
2. How did she know her hair was on fire? How did she extinguish the flames on her clothing?
3. How did DiChiara discover how severe her burns were? Why do you suppose she did not immediately know how gravely she was injured?
4. What help did the survivors get from the security guard's radio?
5. Why did DiChiara decide to struggle down seventy-eight flights of stairs instead of waiting for rescue?
6. What does Thomas show us about her character, throughout the narrative, that aids her in making that life-saving decision?

CONSIDERING METHOD

1. How is this selection organized?
2. What is the author's **purpose**? Who is the intended **audience**?
3. At the end of paragraph 2, Thomas includes this small detail in a sentence all by itself: "It was 8:46 A.M." Why do you think he made that stylistic choice?
4. In paragraph 7, Thomas writes, "In shock, she did not feel the pain—yet." What does he achieve with the use of that dash?

5. How does the writer engage our emotions in this narrative?
6. The piece is packed with specific details. Point out three or four that you find particularly effective and explain why you think Thomas included them.
7. What makes the concluding paragraph effective?

WRITING STEP BY STEP

Think about a catastrophe that you witnessed or were involved in—a flood, a fire, an auto wreck, an earthquake, a tornado, or a hurricane. Jot down as many specific details as you can remember. Then write a short narrative describing the experience. Use "I" in referring to yourself and "we" if others join you in the action. Or, write about the experience of someone else caught up in a major ordeal, the way Evan Thomas writes about Virginia DiChiara's struggles in the previous essay. Collect material through interviewing the other person, with your narrative purposes in mind.

A. Begin, as Thomas does, with a brief explanation of how you got involved—what you were doing, who (if anyone) was with you, and what led up to the catastrophe.

B. Then recount the events from beginning to end. Before you begin, make a detailed list of everything that happened and all the specific details you can recall. Next, decide which actions to include and which to leave out.

C. You do not need a thesis statement in a personal narrative. Let your readers infer what your point is.

D. Include any interesting thoughts that went through your mind (as Thomas does in para. 6). Use quotation marks, since you are, in effect, quoting yourself. If other people are involved, quote directly anything memorable that they said.

E. Use plenty of description (like "lines shooting around the top of the elevator," "translucent teardrops of fire, a foot in diameter," "frantically patting out the flames"), as well as pertinent specific actions ("She let her two golden retrievers play in the yard, cooked herself some eggs, poured herself a cup of coffee," "One of the security guards was furiously dialing for help on the office phone but couldn't get through"). Put pictures in your readers' minds. Lively verbs help create striking images—*moseying, plowed, bounced, shooting, banging.*

F. Avoid telling your readers what to feel. Let them experience your emotions through your forceful recounting of the events.

G. Experiment with a couple of very short sentences for emphasis.

H. Write a concluding sentence or a short concluding paragraph that brings the account to a definite close. Don't let your narrative just trail off at the end.

OTHER WRITING IDEAS

1. In his account of the survivors trapped on the 78th floor, Evan Thomas shows them experiencing many varying emotions—fear, panic, hysteria, courage, self-discipline, determination, resolve. Think of a time when you felt one of these emotions and write a narrative account analyzing the experience. Describe where you were, what happened, and how you reacted. You might conclude by explaining how you feel about the occurrence now, as you reflect on it.
2. COLLABORATIVE WRITING. Write a narrative to support or disprove some familiar proverb or saying, such as "Home is where the heart is," "Crime doesn't pay," or "Virtue is its own reward." Get together with several classmates before you begin writing to help one another decide what saying would make a good choice and what story might provide convincing illustrations.
3. USING THE INTERNET. Visit the Chicago Historical Society's Web site on the Great Chicago Fire and read the narrative by Bessie Bradwell Helmer (**www.chicagohs.org/fire/witnesses/bradwell.html**). Using the information you've acquired, write a narrative recounting your imaginary escape from the Great Chicago Fire of 1871. Or, if you prefer, follow the progress of a single lucky person who fled the conflagration and survived. You can follow the chronology of Bessie Helmer's account, but try for the immediacy that Evan Thomas captures in his reporting of Virginia DiChiara's miraculous escape. Supply your own specific events, vivid descriptions, and action verbs.
4. WRITING ABOUT READING. Analyze the writing strategies used in Thomas's essay. Use the five questions on page 35 in the introduction of this chapter to guide you in your analysis.

EDITING SKILLS: SENTENCE COMBINING

Writing lots of short sentences with little variety produces a choppy effect that sounds a bit childish. In order to make your writing fluent, you need to take those choppy sentences and combine them into a single sophisticated one. It works like this:

Choppy sentences:

She had two golden retrievers.
She let them play in the yard.
She cooked herself some eggs.
She poured herself a cup of coffee.

If you combine those sentences, you get one of Evan Thomas's well-crafted sentences:

> She let her two golden retrievers play in the yard, cooked herself some eggs, poured herself a cup of coffee.

Here is another example:

Choppy sentences:

She threw herself to the floor.
The floor was carpeted.
She rolled over and over.
She was frantic.
She patted out the flames.

Thomas wrote his sentence this way:

> She threw herself to the carpeted floor and rolled over and over, frantically patting out the flames.

EXERCISE

Now you try it. Here are some short sentences to combine into a single effective one. Begin the first one with *if.*

You may like to travel.
You may want to travel alone.
You are a woman.
I have some tips for you.

My friend is a woman
She travels alone a lot.
She has given me good advice.

You may be on a train or plane.
You choose your departure times.
You choose your arrival times.
Check the time of your arrival at your destination.
Be sure your arrival will be during daylight hours.

The station or airport may be far from your hotel.
You need to find the best way to get there.
Taxis are expensive.

You should travel light.
You should take only two bags.
One bag should be on wheels.
You will probably have to carry your own luggage.

Now examine the narrative you just wrote. If you find limp or choppy sentences, combine them into pleasing ones.

WEB SITE

http://911digitalarchive.org/

Click on "Stories" to read compelling personal narratives written by people who were involved in the tragedy of September 11th.

PREPARING TO READ

Did you ever have an experience that changed the way you look at the world—perhaps while traveling, or getting to know a stranger, or suffering a serious illness, or being inspired by a teacher?

Jackie's Debut: A Unique Day

MIKE ROYKO

Born in Chicago, Mike Royko (1932–1997) attended Wright Junior College there. He wrote a syndicated column for the Chicago Tribune *and was awarded the Pulitzer Prize. His columns often ridiculed human greed, vanity, and stupidity. The piece reprinted here is perhaps not typical, since it focuses on an uplifting event—a step toward reversing the usual pattern of ignorant prejudice. It appeared in the* Chicago Daily News *on Wednesday, October 15, 1972, the day after Jackie Robinson died. If you enjoy this article about Jackie Robinson, you might want to read Robinson's autobiography,* I Never Had It Made.

TERMS TO RECOGNIZE

scalpers *(para. 3)*	people selling tickets at higher than regular prices
Ls *(para. 4)*	elevated trains
caromed *(para. 11)*	hit and bounced
chortling *(para. 12)*	chuckling and snorting with laughter

All that Saturday, the wise men of the neighborhood, who sat in chairs on the sidewalk outside the tavern, had talked about what it would do to baseball. I hung around and listened because baseball was about the most important thing in the world, and if anything was going to ruin it, I was worried. 1

Most of the things they said, I didn't understand, although it sounded terrible. But could one man bring such ruin? They said he could and would. And the next day he was going to be in Wrigley Field for the first time, on the same diamond as Hack, Nicholson, Cavarretta, Schmidt, Pafko, and all my other idols. I had to see Jackie Robinson, the man who was going to somehow wreck everything. So the next day, another kid and I started walking to the ball park early. 2

We always walked to save the streetcar fare. It was five or six miles, but I felt about baseball the way Abe Lincoln felt about education. Usually, we 3 could get there just at noon, find a seat in the grandstands and watch some batting practice. But not that Sunday, May 18, 1947. By noon, Wrigley Field was almost filled. The crowd outside spilled off the sidewalk and into the streets. Scalpers were asking top dollar for box seats and getting it.

I had never seen anything like it. Not just the size, although it was a new record, more than 47,000. But this was 25 years ago, and in 1947 few blacks were seen in the Loop, much less up on the white North Side at a Cub game. That day, they came by the thousands, pouring off the northbound Ls and out of their cars. They didn't wear baseball-game clothes. They had on church clothes and funeral clothes—suits, white shirts, ties, gleaming shoes, and straw hats. I've never seen so many straw hats. Big as it was, the crowd was orderly. Almost unnaturally so. People didn't jostle each other. 4

The whites tried to look as if nothing unusual was happening, while the blacks tried to look casual and dignified. So everybody looked slightly ill at ease. For most, it was probably the first time they had been that close to each other in such great numbers. 5

We managed to get in, scramble up a ramp and find a place to stand behind the last row of grandstand seats. Then they shut the gates. No place remained to stand. 6

Robinson came up in the first inning. I remember the sound. It wasn't the shrill, teen-age cry you now hear, or an excited gut roar. They applauded, long, rolling applause. A tall middle-aged black man stood next to me, a smile of almost painful joy on his face, beating his palms together so hard they must have hurt. 7

When Robinson stepped into the batter's box, it was as if someone had flicked a switch. The place went silent. He swung at the first pitch and they erupted as if he had knocked it over the wall. But it was only a high foul that dropped into the box seats. I remember thinking it was strange that a foul could make that many people happy. When he struck out, the low moan was genuine. 8

I've forgotten most of the details of the game, other than that the Dodgers won and Robinson didn't get a hit or do anything special, although he was cheered on every swing and every routine play. But two things happened I'll never forget. Robinson played first, and early in the game a Cub star hit a grounder and it was a close play. Just before the Cub reached first, he swerved to his left. And as he got to the bag, he seemed to slam his foot down hard at Robinson's foot. It was obvious to everyone that he was trying to run into him or spike him. Robinson took the throw and got clear at the last instant. 9

I was shocked. That Cub, a home-town boy, was my biggest hero. It was not only an unheroic stunt, but it seemed a rude thing to do in front of people who would cheer for a foul ball. I didn't understand why he had done it. It wasn't at all big league. I didn't know that while the white fans were 10 relatively polite, the Cubs and most other teams kept up a steady stream of racial abuse from the dugout. I thought all they did down there was talk about how good Wheaties are.

11 Later in the game, Robinson was up again and he hit another foul ball. This time it came into the stands low and fast, in our direction. Somebody in the seats grabbed for it, but it caromed off his hand and kept coming. There was a flurry of arms as the ball kept bouncing, and suddenly it was between me and my pal. We both grabbed. I had a baseball.

12 The two of us stood there examining it and chortling. A genuine major-league baseball that had actually been gripped and thrown by a Cub pitcher, hit by a Dodger batter. What a possession! Then I heard a voice say: "Would you consider selling that?" It was the black man who had applauded so fiercely. I mumbled something. I didn't want to sell it.

13 "I'll give you $10 for it," he said.

14 Ten dollars. I couldn't believe it. I didn't know what $10 could buy because I'd never had that much money. But I knew that a lot of men in the neighborhood considered $60 a week to be good pay. I handed it to him, and he paid me with ten $1 bills. When I left the ball park, with that much money in my pocket, I was sure that Jackie Robinson wasn't bad for the game.

15 Since then, I've regretted a few times that I didn't keep the ball. Or that I hadn't given it to him free. I didn't know, then, how hard he probably had to work for that $10. But Tuesday I was glad I had sold it to him. And if that man is still around, and has that baseball, I'm sure he thinks it was worth every cent.

RESPONDING TO READING

Read again Royko's last paragraph. Explain in your journal why he wishes, as an adult, that he had given the baseball to the man who bought it from him. Do you think you would you feel the same way?

GAINING WORD POWER

Following is the vocabulary entry for the word *unique,* which appears in Mike Royko's title. It comes from the *Webster's New World Dictionary,* 3rd college edition, published by Simon & Schuster. Can you make sense of it?

> **u | nique** (yo͞o nēk′) ***adj.*** [[Fr < L *unicus,* single < ONE]] 1 one and only; sole *[a unique* specimen*]* 2 having no like or equal; unparalleled *[a unique* achievement*]* 3 highly unusual, extraordinary, rare, etc.: a common usage still objected to by some—**u | nique′ly *adv.*—u | nique′ness *n.***

All dictionaries are unique, but here's how to read this entry from the *New World.*

1. The boldfaced word itself tells you how to spell it, and the thin line (between the *u* and the *n*) shows where it divides into syllables. Some dictionaries use centered dots instead.
2. The syllables inside the parentheses (some dictionaries use slash marks) let you know how to pronounce the word. If you don't understand the symbols, check inside the front or back cover or at the bottom of the page.
3. The boldfaced abbreviation tells you the part of speech. Notice that further down in the entry you'll see other forms of the same word that are different parts of speech (*uniquely* ***adv.*** and *uniqueness* ***n.***). Sometimes those other parts of speech have different meanings listed.
4. Those weird notations inside the double brackets give the *etymology* of the word—that is, they tell us how it came into our language. The entry reads this way: "The word *unique* comes to us from the French, derived from the Latin *unicus,* meaning *single,* derived from *unus,* meaning *one.*" Check the explanatory notes in your dictionary to learn how to figure out the etymologies.
5. Various meanings are numbered. Some dictionaries list the most common meaning first; others begin with the oldest meaning, which would often be the least common. So, again, check your dictionary's explanatory notes. The *New World* lists meanings from oldest to newest.
6. All dictionaries give warnings about usage. If you look up the word *ain't,* you will find it labeled *slang* or *nonstandard,* perhaps even with a warning that some people have strong objections to it. In the entry for *unique,* the third meaning of "highly unusual, extraordinary, rare, etc." is followed by a caution that this is "a common usage still objected to by some." Those *some* often include English teachers, so avoid writing *most unique* or *quite unique* or *very unique.* Remember the etymology.

Now, to practice what you just learned, look up the word *debut* from Royko's title, and answer the following questions:

1. How many syllables does it have?
2. What different ways can it be correctly pronounced?
3. What parts of speech can it be?
4. What language did it come from?
5. What are two meanings of the word?
6. Are there any usage labels or warnings?

CONSIDERING CONTENT

1. What made the old men of the neighborhood say that Jackie Robinson would ruin the game of baseball?
2. How did young Royko and his friend get to Wrigley Field, and why did he choose that way to get there?
3. What does he mean when he says, "I felt about baseball the way Abe Lincoln felt about education" (para. 3)?
4. Why was he surprised to see so many black people at the game?
5. Why were the black people wearing their good clothes?
6. When the young narrator says, "It wasn't at all big league" (para. 10), what does he mean?

CONSIDERING METHOD

1. How does Royko get the readers' attention in the opening paragraph?
2. How does he let us know that the "wise men of the neighborhood" are not truly wise—that he is being sarcastic? What small detail in the sentence gives us the clue?
3. Why does Royko tell most of the narrative through the thoughts of himself as a young boy?
4. In paragraph 3, why does he give the exact date—Sunday, May 18, 1947?
5. When you think about all the details he could have included in describing the crowd at the baseball game, why do you think he chose to relate how the black people were dressed?
6. The **point of view** shifts briefly in paragraph 10, as Royko tells us something he learned later. How does he handle this shift so that we scarcely notice it?
7. In paragraphs 12 and 13, why does he give us the exact words of the man who wants to buy the baseball?
8. Explain why the final paragraph makes a good conclusion.

WRITING STEP BY STEP

Using Mike Royko's column as an example, write a narrative essay about a childhood experience that suddenly let you see some less than admirable aspect of the adult world. Royko, after making it clear that he grew up in a white section of the city, lets us see how the vicious action of his former hero on the Cub team opened his eyes to racial prejudice.

Think of a similar experience in your own past that will allow you to show the unfairness of some human behavior or the pain caused by some human weakness. Perhaps you could tell about your accidental discovery of disloyalty or cheating by someone you admired and trusted. Or you could tell the story of the first time you observed adult violence or adult cruelty. Write this assignment as a cautionary tale for an audience of high school seniors.

A. Think about the story and the insight you gained from it. Write out the meaning of the incident as a thesis statement, but do not include it in the essay. Just keep it in mind as a guide in selecting details.
B. Jot down the events you want to cover, leaving lots of space between them. Next, go over the events, and fill in the spaces with all the details you can think of about each event. Then, go through the whole sheet again, and carefully decide just which events and which details you want to use in your narrative. Select only the most descriptive details that will allow your readers to experience the event as you did.
C. Use *I* in telling your story. Consider narrating it from the point of view of yourself as a child, as Royko does. Try to remember how you actually saw things when you were young and innocent and tell the story that way. See Royko's paragraphs 1, 10, and 14 for good examples.
D. Begin, as Royko does, by briefly setting the scene. Be sure to work in the time, either here or later (as Royko does in para. 3).
E. Don't give away the point of your story, but try to work in a teaser (or a *delay*) as he does in his opening. The men are talking about what "it" would do to baseball, but we readers have no clue what "it" means and read on to find out.
F. In your conclusion, try to let your readers understand what you learned from the incident, but don't tell them straight out. See Royko's last three paragraphs, in which he tells us how he felt as a child and how he feels now as a man—and both are positive feelings. He lets us see how wrong racial prejudice is and at the same time reminds us of how proud blacks could feel about Jackie Robinson's success.
G. If exactly what people say is important to the story, put the speech in direct quotations, as Royko does in paragraphs 12 and 13.

OTHER WRITING IDEAS

1. Did you ever do something quite wrong in response to peer pressure? Tell the story of how this happened and how you felt about it at the time—and how you feel about it now.
2. USING THE INTERNET. Enter the name of Rosa Parks into your search engine, and you will find the story of another African American who, like Jackie Robinson, led the way in the fight against racial segregation. After reading about her, write a narrative for a group of

middle school students telling the story of her quiet, courageous stand against injustice.

3. COLLABORATIVE WRITING. With a group of classmates, have a brainstorming session in which you discuss difficult ethical and moral decisions each of you has faced and how you responded. For many, deciding whether to cheat, steal, or tattle creates their first moral crisis. Then, tell the story of a tough decision you had to make and what happened as a result.
4. WRITING ABOUT READING. What difference does it make that the story is told by an adult looking back? Write an essay about the use of memory and point of view in this selection. Think of how you would tell a story from your childhood now, in contrast with how you would've told it at the time.

EDITING SKILLS: PUNCTUATING CONVERSATION

When adding conversation to your narrative, you know, of course, to use quotation marks around other people's words. But you also need to notice how other marks are used with those quotation marks. Look at the punctuation in these sentences from Royko's essay:

> Then I heard a voice say: "Would you consider selling that?"
>
> "I'll give you $10 for it," he said.

The words telling who is talking—called a tag—need to be separated from the quotation. When the tag comes *before* the quotation, you can use either a comma or a colon to separate. When the tag comes *after* the quoted words, use a comma.

Sometimes you may want to put the tag in the middle:

> "I don't want to sell it," I mumbled; "it's mine!
>
> "Go away!" I yelled, "or I'll call a cop."
>
> "I'm going home," I told Joe, "before I lose this ball."

If you start a new sentence after the tag, you need a semicolon or a period, just as you would in any other writing.

And don't stack up punctuation. If you use an exclamation mark or a question mark, omit the comma or period.

Notice that periods and commas go before the ending quotation marks. But with question marks (and exclamation marks), you have to decide whether the tag is a question (or an exclamation) or whether the quoted words are.

"Oh, well," Jamal said, "it takes all kinds!"

Can you believe Jamal said, "It takes all kinds"?

Finally, whenever you change speakers, begin a new paragraph.

As you read your way through the selections in this text, pay attention to the way quoted material is punctuated.

EXERCISE

Now, for practice, put the necessary punctuation in the following sentences:

Oh, heaven help us Marvin exclaimed I forgot to do our income taxes

We'd better get started on them fast then responded Rosa

Marvin groaned You get the receipts together while I try to find the calculator

How am I supposed to know where to find the receipts asked Rosa

Surely roared Marvin you've been keeping them all together some place

Rosa was silent during a long pause, then asked Was I supposed to

Marvin slapped his palm against his forehead and yelled We're doomed

Don't worry, honey soothed Rosa they never send you to Leavenworth on a first offense

Go back and look at the essay you just wrote: Did you punctuate the conversations accurately? Check all your quotations carefully, and make any necessary corrections.

WEB SITES

www.negroleaguebaseball.com/

The online home of Negro League history

www.blackbaseball.com/

The Negro League's Web Site

Both sites provide historical background and perspectives that relate to the topic of Royko's essay.

Preparing to Read

How do you picture a typical prison guard? Have your ideas been shaped by movies and television? In your journal, write a brief description of how you think a prison guard would look, talk, and act.

A Guard's First Night on the Job

William Recktenwald

William Recktenwald is a journalist who once served as a guard in a maximum security prison in Pontiac, Illinois. The following firsthand account first appeared in the St. Louis Globe Democrat *in 1978.*

TERMS TO RECOGNIZE

orientation *(para. 1)*	an introduction to and explanation of an activity or a job
contraband *(para. 2)*	smuggled goods
cursory *(para. 2)*	hasty, not complete or thorough
apprehensive *(para. 9)*	worried, anxious, uneasy
virtually *(para. 9)*	for all practical purposes
ruckus *(para. 9)*	noisy disturbance
din *(para. 10)*	loud, continuous noise
equivalent *(para. 15)*	the equal of, the same as

When I arrived for my first shift, 3 to 11 P.M., I had not had a minute of training except for a one-hour orientation lecture the previous day. I was a "fish," a rookie guard, and very much out of my depth. A veteran officer welcomed the "fish" and told us: "Remember, these guys don't have anything to do all day, 24 hours a day, but think of ways to make you mad. No matter what happens, don't lose your cool. Don't lose your cool!" 1

I had been assigned to the segregation unit, containing 215 inmates who are the most trouble. It was an assignment nobody wanted. To get there, I passed through seven sets of bars. My uniform was my only ticket through 2 each of them. Even on my first day, I was not asked for any identification, searched, or sent through a metal detector. I could have been carrying weapons, drugs, or any other contraband. I couldn't believe this was what's meant by a maximum-security institution. In the week I worked at Pontiac, I was subjected to only one check, and that one was cursory.

3 The segregation unit consists of five tiers, or galleries. Each is about 300 feet long and has 44 cells. The walkways are about 3½ feet wide, with the cells on one side and a rail and cyclone fencing on the other. As I walked along one gallery, I noticed that my elbows could touch cell bars and fencing at the same time. That made me easy pickings for anybody reaching out of a cell.

4 The first thing they told me was that a guard must never go out on a gallery by himself. You've got no weapons with which to defend yourself, not even a radio to summon help. All you've got is the man with whom you're working. My partner that first night was Bill Hill, a soft-spoken six-year veteran who immediately told me to take the cigarettes out of my shirt pocket because the inmates would steal them. Same for my pen, he said—or "They'll grab it and stab you."

5 We were told to serve dinner on the third tier, and Hill quickly tried to fill me in on the facts of prison life. That's when I learned about cookies and the importance they have to the inmates. "They're going to try and grab them, they're going to try and steal them any way they can," he said. "Remember, you only have enough cookies for the gallery, and if you let them get away, you'll have to explain to the guys at the end why there weren't any for them."

6 Hill then checked out the meal, groaning when he saw the drippy ravioli and stewed tomatoes. "We're going to be wearing this," he remarked, before deciding to simply discard the tomatoes. We served nothing to drink. In my first six days at Pontiac, I never saw an inmate served a beverage.

7 Hill instructed me to put on plastic gloves before we served the meal. In view of the trash and waste through which we'd be wheeling the food cart, I thought he was joking. He wasn't. "Some inmates don't like white hands touching their food," he explained.

8 Everything went routinely as we served the first 20 cells, and I wasn't surprised when every inmate asked for extra cookies. Suddenly, a huge arm shot through the bars of one cell and began swinging a metal rod at Hill. As he ducked away, the inmate snared the cookie box. From the other side of the cart, I lunged to grab the cookies—and was grabbed in turn. A powerful hand from the cell behind me was pulling my arm. As I jerked away, objects began crashing about, and a metal can struck me in the back.

9 Until that moment I had been apprehensive. Now I was scared. The food cart virtually trapped me, blocking my retreat. Whirling around, I noticed that mirrors were being held out of every cell so the inmates could watch

the ruckus. I didn't realize the mirrors were plastic and became terrified that the inmates would start smashing them to cut me up.

10 The ordinary din of the cell house had turned into a deafening roar. For the length of the tier, arms stretched into the walkway, making grabbing motions. Some of the inmates swung brooms about. "Let's get out of here—now!" Hill barked. Wheeling the food cart between us, we made a hasty retreat.

11 Downstairs, we reported what had happened. My heart was thumping, my legs felt weak. Inside the plastic gloves, my hands were soaked with sweat. Yet the attack on us wasn't considered unusual by the other guards, especially in segregation. That was strictly routine, and we didn't even file a report.

12 What was more shocking was to be sent immediately back to the same tier to pass out medication. But as I passed the cells from which we'd been attacked, the men in them simply requested their medicine. It was as if what had happened minutes before was already ancient history. From another cell, however, an inmate began raging at us. "Get my medication," he said. "Get it now, or I'm going to kill you." I was learning that whatever you're handing out, everybody wants it, and those who don't get it frequently respond by threatening to kill or maim you. Another fact of prison life.

13 Passing cell no. 632, I saw that a prisoner I had helped take to the hospital before dinner was back in his cell. When we took him out, he had been disabled by mace and was very wobbly. Hill and I had been extremely gentle, handcuffing him carefully, then practically carrying him down the stairs. As we went by his cell this time, he tossed a cup of liquid on us.

14 Back downstairs, I learned I would be going back to that tier for a third time, to finish serving dinner. This time, we planned to slip in the other side of the tier so we wouldn't have to pass the trouble cells. The plates were already prepared. "Just get in there and give them their food and get out," Hill said. I could see he was nervous, which made me even more so. "Don't stop for anything. If you get hit, just back off, 'cause if they snare you or hook you some way and get you against the bars, they'll hurt you real bad."

15 Everything went smoothly. Inmates in the three most troublesome cells were not getting dinner, so they hurled some garbage at us. But that's something else I had learned; getting no worse than garbage thrown at you is the prison equivalent of everything going smoothly.

RESPONDING TO READING

Now that you have read the essay, write a paragraph in your journal describing what kind of person you think William Recktenwald is and how he responds to his new job. In another paragraph, note any ways in which he is different from your expectations of a typical prison guard.

GAINING WORD POWER

Since our brief definitions are sometimes cursory, look up in a college-size dictionary the following words from the Terms to Recognize list. Then write a sentence using each word.

orientation	apprehensive	cursory
contraband	equivalent	

CONSIDERING CONTENT

1. What was the only piece of advice the author got during his orientation as a rookie guard?
2. What can you tell about race relations in Pontiac prison? Are the guards white and the prisoners black?
3. What qualities would a prison guard need in order to work in the segregation unit?
4. Why is the width of the galleries an important detail?
5. What kind of weapons does a guard carry?
6. At the end of the essay, why are the guards not much bothered by having food thrown at them?

CONSIDERING METHOD

1. Why does the writer quote directly the "welcoming" advice given by the veteran police officer?
2. The descriptive details in paragraph 2 have no vivid appeal to the senses. Why, then, are they included?
3. Look at the action verbs in paragraph 8: *shot, ducked, snared, lunged,* and *jerked.* Why are they more effective than *came, moved, took, stepped forward,* and *pulled?* What added feeling do you get from Recktenwald's verbs? Find five more examples of verbs that you think add color and emotion.
4. In paragraphs 7 and 9, the writer uses short-short sentences: "He wasn't" and "Now I was scared." Why do you think he uses such short statements?
5. How is the essay organized?
6. What sentence serves as a thesis statement to tell readers what the essay is about?
7. What is the author's purpose?
8. Do you think the conclusion is effective? Can you explain why?

WRITING STEP BY STEP

Did you ever hold a difficult job? If so, write an essay describing a bad day. Use specific details that will let your readers share your experience. Follow the form of "A Guard's First Night on the Job." If you don't have a job, describe a bad day at home or at school.

A. Begin by giving your readers a brief orientation—that is, explaining where you work, what you do, maybe what time you begin if you have a shift job.
B. At the end of your introduction, let your readers know that the essay will be about a bad day on the job. Your purpose may be to entertain, if you have some humorous incidents to relate, or you may write simply to inform your readers about how trying your job is, as William Recktenwald does. Or maybe you want to make the point that your line of work is pitifully underpaid for the stress it causes you.
C. Use the informal "you" to speak directly to your readers.
D. Think of one particularly troublesome detail of your job that bothers you repeatedly (like Recktenwald's problems with the cookies) and focus on that. Or jot down a number of incidents that drive you crazy and relate those in the order in which they typically occur.
E. Include specific details (like the cigarettes, pen, plastic gloves, plastic mirrors) as you describe incidents.
F. Use action verbs whenever possible (like Recktenwald's *shot, snared, lunged, grabbed, jerked, barked, hooked,* and *hurled*) to give your readers a picture of what happened.
G. If you include people, briefly identify them ("Bill Hill, a soft-spoken, six-year veteran").
H. Write a closing that reinforces how bad your day can be without actually saying so, as Recktenwald does when he concludes that merely having garbage thrown at him means his shift is "going smoothly."

OTHER WRITING IDEAS

1. Describe the most stressful vacation you ever took (or the most stressful picnic or wedding or trip to the zoo or any activity that you would expect to be pleasurable but wasn't).
2. Collaborative Writing. Get together with classmates or friends and talk about memorable "firsts," like your first day of school, your first date, your first rock concert, or your first family reunion. See how many "firsts" the group can come up with. Then, with the help

of the group, choose the "first" from your own life that will make the most interesting narrative. Decide what specific details to include, and write about the experience.

3. USING THE INTERNET. Go to **www.umkc.edu/imc/blackfir.htm** and read about a number of "African American Firsts." Were you surprised by how many of these notable accomplishments you had never heard of? Write a narrative telling about the first time you became aware of racial prejudice—directed either toward you or toward someone else.
4. WRITING ABOUT READING. We use **concrete** descriptions and examples to make abstract feelings easier to understand. What feelings does Recktenwald convey with his use of descriptive details and specific events? Identify five or six and evaluate their effectiveness. What feelings did they suggest to you?

EDITING SKILLS: COMMAS AFTER DEPENDENT ELEMENTS

Copy these sentences from the reading exactly. As you write, look for a pattern they all follow.

As he ducked away, the inmate snared the cookie box.

When we took him out, he had been disabled by mace and was very wobbly.

As we went by his cell this time, he tossed a cup of liquid on us.

Each of these sentences has two parts. Notice that the second part, after the comma, can stand alone as a sentence; it's called an **independent clause**. If you read the first part alone, it does not sound complete. The first part is called a **dependent clause** for this reason. When you write a sentence in which the *independent* part comes after a *dependent* part, you need to put in a comma to separate the two.

EXERCISE

Put commas in the following sentences:

While planning is important do not overschedule your time.

Since the best courses fill up early plan at least a term or two in advance.

If you cannot write well you will be at a disadvantage.

Now fill in the blanks to make complete sentences below:

Although studying is the major task of college life, ________________

____.

Because grades are important to some employers, ________________

____.

____________________________, Henry went to the basketball game.

Next, write three sentences that follow this dependent-comma-independent pattern.

Finally, check your essay to be sure that you have included the commas in sentences like these.

WEB SITE

www.prisonexp.org/

The Stanford Prison Experiment gives details about the controversial simulation study of the psychology of imprisonment that was conducted at Stanford University in 1971.

Preparing to Read

Did you ever think of houses as having tales to tell about the lives of the people who live in them? In your journal write the story that a house you are familiar with might tell—if it could.

More Room

Judith Ortiz Cofer

Born in Puerto Rico, Judith Ortiz Cofer moved with her family to the United States in 1955 and settled in Paterson, New Jersey. Besides essays, she writes poetry and fiction. The following piece appeared in Silent Dancing: A Partial Remembrance of a Puerto Rican Childhood, *published in 1990.*

TERMS TO RECOGNIZE

geneology *(para. 2)*	family history
acrid *(para. 4)*	harsh, bitter
malingering *(para. 4)*	pretending to be ill
inviolate *(para. 5)*	pure, virginal (never entered)
obligatory *(para. 6)*	required
purgatives *(para. 6)*	medicines to cleanse the body
animosity *(para. 9)*	hostility, resentment
coup *(para. 9)*	brilliant, surprising tactic that overcomes an opponent
vortex *(para. 9)*	whirlpool
fecund *(para. 9)*	fruitful in offspring
acceded *(para. 10)*	gave in to, went along with
emanate *(para. 12)*	emit, give off

My grandmother's house is like a chambered nautilus; it has many rooms, yet it is not a mansion. Its proportions are small and its design simple. It is a house that has grown organically, according to the needs of its inhabitants. To all of us in the family it is known as *la casa de Mamá.* It is the place of our origin; the stage for our memories and dreams of Island life. 1

I remember how in my childhood it sat on stilts; this was before it had a downstairs—it rested on its perch like a great blue bird—not a flying sort 2 of bird, more like a nesting hen, but with spread wings. Grandfather had built it soon after their marriage. He was a painter and housebuilder by trade—a poet and meditative man by nature. As each of their eight children were born, new rooms were added. After a few years, the paint didn't exactly match, nor the materials, so that there was a chronology to it, like the rings of a tree, and Mamá could tell you the history of each room in her *casa,* and thus the geneology of the family along with it.

3 Her own room is the heart of the house. Though I have seen it recently—and both woman and room have diminished in size, changed by the new perspective of my eyes, now capable of looking over countertops and tall beds—it is not this picture I carry in my memory of Mamá's *casa.* Instead, I see her room as a queen's chamber where a small woman loomed large, a throne room with a massive four-poster bed in its center, which stood taller than a child's head. It was on this bed, where her own children had been born, that the smallest grandchildren were allowed to take naps in the afternoons; here too was where Mamá secluded herself to dispense private advice to her daughters, sitting on the edge of the bed, looking down at whoever sat on the rocker where generations of babies had been sung to sleep. To me she looked like a wise empress right out of the fairy tales I was addicted to reading.

4 Though the room was dominated by the mahogany four-poster, it also contained all of Mamá's symbols of power. On her dresser there were not cosmetics but jars filled with herbs: *yerba* we were all subjected to during childhood crises. She had a steaming cup for anyone who could not, or would not, get up to face life on any given day. If the acrid aftertaste of her cures for malingering did not get you out of bed, then it was time to call *el doctor.*

5 And there was the monstrous chifforobe she kept locked with a little golden key she did not hide. This was a test of her dominion over us; though my cousins and I wanted a look inside that massive wardrobe more than anything, we never reached for that little key lying on top of her Bible on the dresser. This was also where she placed her earrings and rosary when she took them off at night. God's word was her security system. This chifforobe was the place where I imagined she kept jewels, satin slippers, and elegant silk, sequined gowns of heartbreaking fineness. I lusted after those imaginary costumes. I had heard that Mamá had been a great beauty in her youth, and the belle of many balls. My cousins had ideas as to what she kept in that wooden vault: its secret could be money (Mamá did not hand cash to strangers, banks were out of the question, so there were stories that her mattress was stuffed with dollar bills, and that she buried coins in jars in her garden under rose-bushes, or kept them in her inviolate chifforobe); there might be that legendary gun salvaged from the Spanish-American conflict over the Island. We went wild over suspected treasures that we made up simply because children have to fill locked trunks with something wonderful.

On the wall above the bed hung a heavy silver crucifix. Christ's agonized head hung directly over Mamá's pillow. I avoided looking at this weapon suspended over where her head would have lain; and on the rare occasions when I was allowed to sleep on that bed, I scooted down to the safe middle of the mattress, where her body's impression took me in like mother's lap. Having taken care of the obligatory religious decoration with the crucifix, Mamá covered the other walls with objects sent to her over the years by her children in the States. *Los Nueva Yores* was represented by, among other things, a postcard of Niagara Falls from her son Heman, postmarked, Buffalo, N.Y. In a conspicuous gold frame hung a large color photograph of her daughter Nena, her husband and their five children at the entrance to Disneyland in California. From us she had gotten a black lace fan. Father had brought it to her from a tour of duty with the Navy in Europe. (On Sundays she would remove it from its hook on the wall to fan herself at Sunday mass.) Each year more items were added as the family grew and dispersed, and every object in the room had a story attached to it, a *cuento,* which Mamá would bestow on anyone who received the privilege of a day alone with her. It was almost worth pretending to be sick, though the bitter herb purgatives of the body were a big price to pay for the spirit revivals of her storytelling. 6

Except for the times when a sick grandchild warranted the privilege, or when a heartbroken daughter came home in need of more than herbal teas, Mamá slept alone on her large bed. 7

In the family there is a story about how this came to be. 8

When one of the daughters, my mother or one of her sisters, tells the *cuento* of how Mamá came to own her nights, it is usually preceded by the qualification that Papá's exile from his wife's room was not a result of animosity between the couple. But the act had been Mamá's famous bloodless coup for her personal freedom. Papá was the benevolent dictator of her body and her life who had had to be banished from her bed so that Mamá could better serve her family. Before the telling, we had to agree that the old man—whom we all recognize in the family as an *alma de Dios,* a saintly, soft-spoken presence whose main pleasures in life, such as writing poetry and reading the Spanish large-type editions of *Reader's Digest,* always took place outside the vortex of Mamá's crowded realm—was not to blame. It was not his fault, after all, that every year or so he planted a baby-seed in Mamá's fertile body, keeping her from leading the active life she needed and desired. He loved her and the babies. He would compose odes and lyrics to celebrate births and anniversaries, and hired musicians to accompany him in singing them to his family and friends at extravagant pig-roasts he threw yearly. Mamá and the oldest girls worked for days preparing the food. Papá sat for hours in his painter's shed, also his study 9 and library, composing the songs. At these celebrations he was also known to give long speeches in praise of God, his fecund wife, and his beloved Island. As a middle child, my mother remembers these occasions as a time when the women sat in the kitchen and lamented their burdens while the men feasted out in the patio, their rum-thickened voices rising in song and praise of each other, *companeros* all.

10 It was after the birth of her eighth child, after she had lost three at birth or infancy, that Mamá made her decision. They say that Mamá had had a special way of letting her husband know that they were expecting, one that had begun when, at the beginning of their marriage, he had built her a house too confining for her taste. So, when she discovered her first pregnancy, she supposedly drew plans for another room, which he dutifully executed. Every time a child was due, she would demand, *More space, more space.* Papá acceded to her wishes, child after child, since he had learned early that Mamá's renowned temper was a thing that grew like a monster along with a new belly. In this way Mamá got the house that she wanted, but with each child she lost in health and energy. She had knowledge of her body and perceived that if she had any more children, her dreams and her plans would have to be permanently forgotten, because she would be a chronically ill woman, like Flora with her twelve children: asthma, no teeth, in bed more than on her feet.

11 And so after my youngest uncle was born, she asked Papá to build a large room at the back of the house. He did so in joyful anticipation. Mamá had asked him for special things this time: shelves on the walls, a private entrance. He thought that she meant this room to be a nursery where several children could sleep. He thought it was a wonderful idea. He painted it his favorite color—sky blue—and made large windows looking out over a green hill and the church spires beyond. But nothing happened. Mamá's belly did not grow, yet she seemed in a frenzy of activity over the house. Finally, an anxious Papá approached his wife to tell her that the new room was finished and ready to be occupied. And Mamá, they say, replied: "Good, it's for you."

12 And so it was that Mamá discovered the only means of birth control available to a Catholic woman of her time: sacrifice. She gave up the comfort of Papá's sexual love for something she deemed greater: the right to own and control her body, so that she might live to meet her grandchildren, me among them, so that she could give more of herself to the ones already there, so that she could be more than a channel for other lives, so that even now that time has robbed her of the elasticity of her body and of her amazing reservoir of energy, she can still emanate the calm joy that can only be achieved by living according to the dictates of one's own heart.

....

RESPONDING TO READING

What do you think of Mamá's chosen method of birth control? Why do you think Papá agreed to it without resentment?

GAINING WORD POWER

Cofer uses lots of unusual descriptive terms. Here is a list of nine of them.

massive (para. 3)	agonized (para. 6)
steaming (para. 4)	conspicuous (para. 6)
monstrous (para. 5)	benevolent (para. 9)
elegant (para. 5)	frenzy (para. 11)
heartbreaking (para. 5)	

Use each of these words in a sentence of your own. If the term is new to you, look it up in your dictionary before writing the sentence, and add it to your vocabulary list after you finish.

CONSIDERING CONTENT

1. What is a chambered nautilus and how is one formed? What is a chifforobe (para. 5)? Why are the children fascinated by Mamá's? Why is it kept locked, even though the key is in plain sight?
2. How are Mamá's bedroom walls decorated (para. 6)? What do these decorations tell you about her character? Why is the detail about the crucifix especially significant?
3. What is a "benevolent dictator" (para. 9)? What sort of man is Papá? What details let you know what he is like?
4. What is a "bloodless coup" (para. 9)? Explain how Mamá pulls hers off. Why is the nature of Papá's character important in understanding her success?
5. What meaning did you derive from Cofer's narrative?

CONSIDERING METHOD

1. Besides being an **analogy** (see Glossary), the description in the opening sentence of Mamá's house as "like a chambered nautilus" is also a **simile** (see Glossary). Can you find other similes in paragraphs 2 and 3? Explain why they are effective.
2. What do you call the figure of speech in the first sentence of paragraph 3: "Her own room is the heart of the house"? What does that description tell you, in a word, about Mamá?

3. Papá is referred to as a "benevolent dictator," but Cofer lets us know that Mamá also wields power in the family. What details convey this information?
4. What is the function of the single short sentence punctuated as a paragraph (para. 8)? Why does Cofer divide the essay with a space break following that sentence?
5. How does the concluding paragraph clarify the meaning of the whole piece?

Combining Strategies

Cofer employs several writing strategies in narrating the story of Mamá and the addition of her last new room. She combines descriptions, illustrations, and examples (see Chap. 4) with cause-and-effect analysis (see Chap. 9). Choose one description and one example, and explain how each appeals to your emotions and to your understanding of the situation Mamá faced.

WRITING STEP BY STEP

Write an essay in which you explain how you brought off some "bloodless coup" of your very own. In other words, tell how you managed to get your way in some matter (large or small) by outmaneuvering a person who had authority over you—a parent, teacher, older sibling, boss, coach, or law enforcement officer. Or maybe in an equal relationship with your spouse or roommate, you managed to slither out of some obligation that you felt justified in dodging, as Mamá does in Cofer's essay.

A. Tell your story, as Cofer does, in the **first person**, using *I* and *me, we, us,* and *our.*

B. Begin by setting the scene—the when and where. If your story happened in the past, let your readers know in a phrase how long ago, like Cofer's "I remember how in my childhood . . ." (para. 2). Describe, with plenty of visual details (the way Cofer presents Mamá's room), where you were—at home, in a car, a classroom, an office, on a football field—when you encountered the problem, disagreement, or conflict that you eventually resolved to your satisfaction.

C. Then, briefly explain the situation that gave rise to the problem, including only enough background details to let your readers understand how the difficulty arose.

D. Next, describe the personality of the other person involved. Use concrete details, as Cofer does in telling us the sort of man her grandfather was: "a saintly, soft-spoken presence," who enjoyed "the Spanish large-type editions of *Reader's Digest*," who loved his wife and children, who wrote poetry and sang songs, who threw "extravagant pig-roasts" with hired musicians, and who gave "long speeches in praise of God, his fecund wife, and his beloved Island" (para. 9).

E. Compose a short transitional sentence (like Cofer's single sentence in para. 8) saying that you are now going to let your readers know the way you resolved your difficulty. Make the sentence a complete paragraph.

F. Finally, present the clever strategy you used in pulling off your "bloodless coup." If you achieved some resulting benefit (as Mamá does in Cofer's final paragraph), you could mention that in your conclusion.

OTHER WRITING IDEAS

1. Tell the story of an episode that allowed you to see another side of a person you thought you knew quite well. Begin by describing the person's character as you first observed it. Then narrate the incident that changed your perception. Conclude by describing briefly how you saw the person's character afterward.
2. COLLABORATIVE WRITING. Discuss with a group of friends or classmates whether the familiar saying "Sports build character" is true or not. After you decide how you feel about the matter, think of an incident that illustrates your belief, and use it to make your point in a narrative.
3. USING THE INTERNET. There are numerous sites for writers on the Internet that offer ideas for a narrative. Visit, for instance, **www.storytellerschallenge.com/default.asp**, which gives topics in the form of "challenges." On the "Truer Than Fiction" page you will find suggestions for writing nonfiction, as well as sample essays from other writers. Another useful site is **www.storyteller.net**. The "Story of the Week" page might also spark some ideas.
4. WRITING ABOUT READING. In Cofer's essay, Mamá regains "the right to own and control her body" (para 12). What do you think of this idea? Does the author convince you that Mamá was justified in her actions? Write an essay in which you defend or challenge Mamá for what she did.

EDITING SKILLS: USING COORDINATION

Experienced writers vary their sentence types and lengths to make their writing more interesting. Even Ernest Hemingway, who was noted for his lean, simple sentences, often put two or more together to make longer ones, as Cofer does in the following example:

> In this way Mamá got the house that she wanted, but with each child she lost in health and energy.

These are the sentences that are put together:

> In this way Mamá got the house that she wanted.
> With each child she lost in health and energy.

To make short sentences into one longer sentence, you splice them together with coordinating conjunctions. There are only seven of them: *and, but, for, or, nor, yet, so.* Put a comma after each short sentence you are combining, but put a period at the very end, as in these examples:

> Mamá's belly did not grow, yet she seemed in a frenzy of activity over the house.
>
> [The house] has many rooms, yet it is not a mansion.

If you are dissatisfied with your writing because it sounds choppy, this method will help you achieve longer sentences.

EXERCISE

Using a comma and a suitable coordinating conjunction, combine the following pairs of sentences into compound sentences.

1. Laser videodiscs have superior picture quality.
   Videotape remains the mass-market leader.
2. Laser technology provides a much sharper image.
   It can pick up greater detail and more contrast from the original print.
3. Manufacturers must keep up consumer demand.
   They will never claim the market from videotapes.
4. Panasonic continues to lead the field.
   Sony has just come out with a new laserdisc player.

5. Consumers want more sophisticated equipment.
   Some companies are building compact disc players into their videodisc machines.

Now go back to your essay and combine two sentences into one long one using a coordinating conjunction. Read the new sentence out loud. If you like the effect, leave it in your essay.

WEB SITE

www.georgiaencyclopedia.org/nge/Article.jsp?id=h-488

The New Georgia Encyclopedia provides a biography of Judith Cofer along with links to several online articles and readings by her.

Student Essay Using Description and Narration

Domestic Abuse

Kelly Berlin

It was the summer of my freshman year, and I had been baby-sitting my two-year-old niece, Briana, at my sister's house every weekday for the past couple of months. It was late in the afternoon on a Friday, so I couldn't wait for my sister Kim to come home from work. Since she was already a half an hour late, I decided to take a shower so I wouldn't be late for my date later that evening. I was in the bathroom blow drying my hair when my sister and her boyfriend Scott came home. 1

My sister and Scott had been dating a couple of years, despite the disapproval of my family. Scott, even though he had a child, was not a good "father figure" for my niece. Scott was a regular drinker and smoker, and his appearance did nothing for him. He was over six-feet tall with big, bulging muscles; he wore tight clothes; and he had long, shaggy hair. My sister, on the other hand, standing only five-feet-five-inches tall, was slender with long, curly brown hair and a beautiful white smile. 2

I finished blow drying my hair and was going to ask my sister to take me home, but I heard her and Scott fighting in the bedroom. I went into the living room to watch television with my niece and to keep her mind off the shouting. I started to become worried because the yelling became more intense. All of a sudden, I heard a loud noise—not a boom, but more like a crack, a board breaking. The shouting stopped. 3

"Briana, stay here!" I left my niece in the living room and ran into the bedroom. "Scott, what in the hell did you do to her? Get away from her!" My sister was lying limp on her bed with her face down. 4

"It's none of your business. Get out of here," he yelled at me with a fierce look in his eyes. 5

I was shaking but yelled back at him with just as much determination and strength. "Bullshit, if it's not my business. She's my sister. Don't think I'm just going 6 to sit here while you push her around!" I turned, looked at my sister, and asked gently, "Kim, are you all right?"

7 "Yeah," Kim told me in a shaky voice. "I'll be all right. But my legs hurt because I hit the foot board when he pushed me down."

8 "Scott, leave NOW!" I said with all the authority I could manage.

9 "Shut up, Kelly. This isn't your house, and I'll leave whenever I please."

10 My sister pleaded with him. "Scott, please leave me alone. Just leave me alone."

11 He stormed out of the bedroom and out of the house like a raging bull. I helped my sister sit up and looked at her legs. She had two long, wide bruises forming across her upper thighs from where they hit the foot board. I told her to stay on her bed while I checked on Briana and called my mom for help.

12 My mom arrived about ten minutes later, and I told her what had happened. She was furious and was determined to do something about it. We called the police to report the incident and file charges. When the police arrived at the house, I described the scene and signed my name to the papers to file charges. Then, they took photographs of my sister's bruises and recorded her statement. But she wouldn't sign the papers. She had decided not to press charges. Well, I was confused, but I thought I could still press charges against him. I was wrong. I hadn't actually seen Scott push my sister. I had only heard them fighting and had found her lying on the bed. We tried to convince my sister to press charges against Scott, but she refused. She believed that he'd just gotten carried away, that he wouldn't do it again. We tried to make her realize that he could very well do it again, but she wouldn't believe us.

13 As it turned out, my sister was mistaken. Scott continued to beat her and control her life, and she continued to refuse to press charges. Finally, she realized he was not going to stop, so she left Scott. At one time, she put a restraining order on him, which kept him away temporarily. And fortunately, he lost his

license due to DUIs, so he had no legal transportation to get to her house. Now, the only time he harasses her is when he's high on drugs. But as time and experience have taught my sister, she now calls the police herself.

CONSIDERING CONTENT AND METHOD

1. Do you think the topic is well chosen? Is the focus narrow enough?
2. What descriptive details does the writer include? Do you think they are effective? If so, why? If not, why not?
3. Do you like Kelly's use of direct conversation in telling the story? What does it add to the account?
4. What is the point of this narrative essay? Does the writer state it directly or merely imply it? If it is stated directly, tell where.
5. What suggestions would you make to the author to improve this essay?

Chapter 4

Strategies for Making a Point: *Example* and *Illustration*

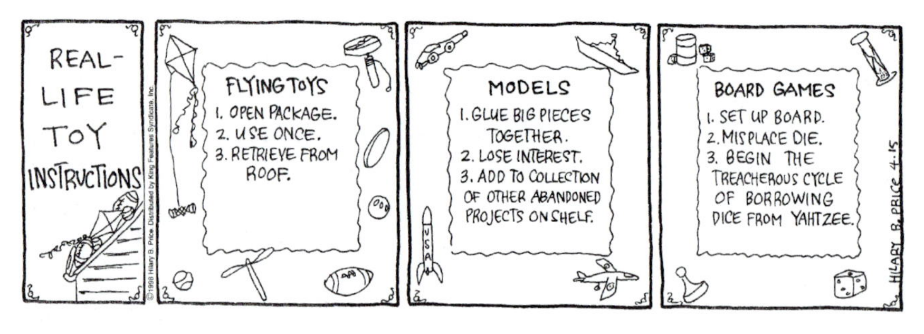

Source: © 1998 by Hillary Price. Reprinted with special permission of King Features Syndicate, Inc.

There's an old saying that a picture is worth a thousand words. It may be true. But we can't always communicate with pictures. Most of the time we have to convey our ideas with words—often with written words. You will find that your skills in describing and narrating can be put to good use in providing examples and illustrations to make a point.

Did you ever read a passage that seemed hard to get the meaning of—that remained fuzzy in your mind no matter how many times you plowed through it? Here's an example of the kind of writing we mean:

> A democratic plan of education includes more than the mere transmission of the social heritage and an attempt to reproduce existing institu-

> tions in a static form. The democratic school is also required to indoctrinate individuals with the democratic tradition which, in turn, is based on the agitative liberties of the individual and the needs of society.

If a person spoke those words to you, you could say, "What was that again?" or "Could you give me an example, please?" or "What do you mean by *agitative liberties*?" But you can't question the written page, so the meaning of whatever that writer had in mind is lost.

The difference between an **illustration** and an **example** is not clear cut. Some people use the terms to mean the same thing; some people use illustration to mean several short examples or a fairly long example, such as a brief narrative used within an essay. We don't think it makes a whole lot of difference what you call them—just be sure to use them.

THE POINT OF EXAMPLE AND ILLUSTRATION

Good writers use examples or illustrations to make their writing clear and to make it convincing. As a bonus, concrete examples make writing interesting.

Using Examples to Explain and Clarify

The paragraph about democratic education shows how vague writing is that uses only general statements. Here's another illustration to let you see how examples help in explaining ideas. We have deliberately taken the examples out of the following paragraph. See how much you can get out of it:

> You should define what you mean when any abstract, ambiguous, or controversial terms figure importantly in your writing. Serious miscommunications can occur when audience and writer do not share the same idea about what a word or phrase means, either *connotatively* or *denotatively*.

That's not too clear, is it? But read it now with the examples that were included in the original:

> You should define what you mean when any abstract, ambiguous, or controversial terms figure importantly in your writing. Serious miscommunications can occur when audience and writer do not share the same idea about what a word or phrase means, either *connotatively* (by its associations) or *denotatively* (by its direct meaning). Consider, for instance, the connotations of these words: *daddy, father, old man*. All denote *male parent,* but their understood meanings are quite different. Also, the phrase *good*

> *writing* seems clear, doesn't it? Yet three English teachers can argue endlessly about what constitutes good writing if teacher A thinks that good writing is honest, direct, and completely clear; if teacher B thinks that good writing is serious, formal, and absolutely correct; and if teacher C thinks that good writing is flashy, spirited, and highly entertaining.

A couple of the examples in that paragraph are definitions; the others explain the need for definitions. All the examples add clarity and meaning to the passage.

Using Examples and Illustrations to Convince

Consider this letter to the editor, published a few years ago in an urban newspaper:

> Liquor is something we can get along without to a very good advantage. The problem of jazz music is a very grave one in this city, also, as it produces an attitude of irresponsibility in the listener.
>
> Let's keep Kansas attractive to God-fearing people. This is the type industry is interested in hiring and this is the type needed in government and the armed forces.

Are you persuaded? Not likely, unless you agreed with the opinions before reading the letter. The writer offers nothing but unsupported personal opinions.

Examples and illustrations are essential in making a point—that is, as evidence to convince your readers that what you say is right. If you want to convince readers that they should run out and rent *Batman Forever,* you have to provide examples to explain why. You will need to discuss the thrill-packed plot, the wonderful gizmos and toys (including a Batmobile that climbs walls), the deadpan jokes, the uproarious physical humor of Jim Carrey, and Chris O'Donnell's spirited performance as Robin. The more illustrations of this sort you can provide, the more persuasive your essay will be.

THE PRINCIPLES OF EXAMPLE AND ILLUSTRATION

The success of a piece of writing often depends on how well you choose your supporting evidence.

Select Appropriate Examples

You must be sure, first of all, that the examples actually do illustrate the point you want to make. If, for instance, you are explaining how you feel about people who borrow a book and then write their own comments in the margins, be sure to focus on your feelings—of interest, outrage, violation, loss, or whatever you felt. Do not slide off the subject to discuss the interesting philosophy course you bought the book for and the time you accidentally left the book on a lunchroom counter and were quite sure you had lost it forever only to have the guy who sat behind you in class return it, saying he found it when he happened to stop in for a late lunch in the same greasy spoon that afternoon. When people do that sort of free associating in conversation, we tend to suffer through it, even as our eyes glaze over. But it will not do in writing.

Give Plenty of Examples

Keep in mind, though, that you need to supply enough examples to make your ideas clear and convincing. Say you want to persuade your readers that becoming a vegetarian is the key to a long life in a healthy body. If you offer only the single illustration of your uncle Seymour who never ate meat, never had a cold, always felt frisky, and ran in the Boston Marathon to celebrate his seventy-ninth birthday, you are not likely to sway many readers. They'll just think, "Well, wasn't he lucky?" You need either to dig up more examples—perhaps even some statistics about low-fat diets and heart disease—or else change your thesis to focus on Uncle Seymour's personal recipe for keeping fit. There's nothing wrong with using one long illustration, if that single illustration really does prove your point.

Include Specific Information

Finally, you need to develop your examples and illustrations with plenty of specific, graphic **details.** If you say that riding motorcycles is dangerous, you need to follow up with examples more specific than "Every year many people are injured in motorcycle accidents" and "The person on the motorcycle can't always tell what motorists are going to do." Instead, describe what happens when an automobile unexpectedly turns left in front of a motorcyclist traveling forty miles an hour. Mention the crushed noses, the dislocated limbs, the fractured femurs, the broken teeth, and the shattered skulls that such accidents cause. As a general rule, if you use an **abstract word,** like *dangerous,* follow it soon with a specific example, like "His splintered kneecap never did heal properly."

THE PITFALLS OF EXAMPLE AND ILLUSTRATION

If you are writing an essay developed almost entirely through the use of examples, you need to make sure those examples are connected smoothly when you furnish several in a row. Notice how the italicized **transitions** introduce the examples in this paragraph defining a psychological term:

> People who use reaction formation avoid facing an unpleasant truth by acting exactly opposite from the way they truly feel. *For example,* you may have known somebody who acts like the life of the party, always laughing and making jokes, but who you suspect is trying to fool everybody—including herself—into missing the fact that she is sad and lonely. *Another example* of reaction formation involves the person who goes overboard to be open-minded, insisting, "I'm not prejudiced! Why some of my best friends are __________!"
>
> —Ronald Adler and Neil Towne, "Defense Mechanisms"

Here are some other transitional expressions that you may find useful:

such as	that is	in the following way
namely	in this case	as an illustration
for instance	in addition	at the same time

It's quite possible to use too many transitions. Ask the friend or classmate who helps you edit your first draft to let you know if you've put in more than you need.

As you prepare to revise your essay, ask yourself (and your editorial helper) the following questions:

1. Does each of my examples really illustrate the point I'm trying to make?
2. Have I included enough examples to be convincing?
3. If I'm using a single illustration in some paragraphs, is that convincing?
4. Do any of my illustrations begin to prove the point and then stray from it?
5. Are any of my examples too short or my illustrations too long?
6. Have I used enough specific details?

WHAT TO LOOK FOR IN EXAMPLE AND ILLUSTRATION

As you study the essays in this chapter, focus on the way examples and illustrations are used.

1. In the paragraphs that have a **topic sentence** (the sentence that tells what the paragraph is about), look at the examples or illustrations and decide how convincing they are—that is, how well they explain, support, or enlarge on that idea.

2. Look for concrete, specific, sometimes visual details in the examples and illustrations themselves. Ask yourself what would be lost if these were omitted.
3. Underline the transitional terms used to introduce the examples and illustrations, and keep a list of them in your journal.

IMAGES AND IDEAS

Food Guide Pyramid
A Guide to Daily Food Choices

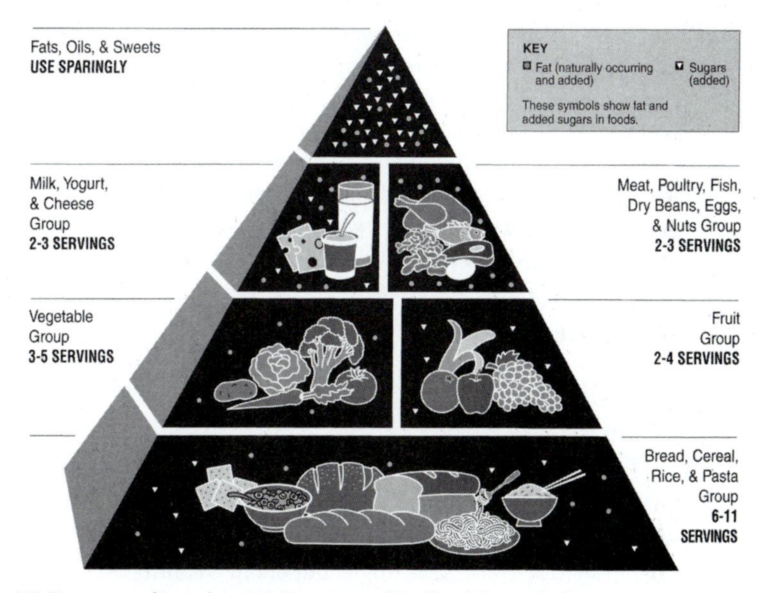

Source: U.S. Department of Agriculture/U.S. Department of Health and Human Services.

For Discussion and Writing

The U. S. Department of Agriculture food pyramid illustrates the types of food and the proportions we should eat every day for good health. Look at the examples given in the chart: Are they clear? Are there enough of them?

Most of us don't eat this way. Write about yourself or someone you know whose diet disregards the food pyramid recommendations. Give examples to show how the person's diet is out of balance according to the USDA pyramid.

Preparing to Read

List three people, living or dead, that you consider heroes, and write a sentence explaining why you think each is a hero.

My Heroes

Elizabeth Berg

Elizabeth Berg worked as a registered nurse for ten years before becoming a full-time writer, though she has written since she was a child. (She received her first rejection letter at age nine.) She has published a book a year since her first novel, Durable Goods, *in 1993, and many have become bestsellers.* Range of Motion *(1995) was made into a cable-TV movie with Rebecca deMornay in 2000. Berg's themes include the emotional effects of physical illness and "the power and salvation of female friendship." In 1999, Berg published a book of advice on the art and business of writing,* Escaping into the Open: The Art of Writing True.

TERMS TO RECOGNIZE

straight-laced *(para. 10)*	excessively strict
bowed *(para. 11)*	stooped or yielded in defeat
sophisticated *(para. 12)*	refined, worldly-wise
intent *(para. 12)*	firmly focused
luxuriating *(para. 13)*	basking, enjoying
relinquish *(para. 15)*	give up, let out of one's control
precipitated *(para. 16)*	caused

My eight-year-old daughter, Jenny, was given a school assignment not too long ago to write about a hero. "So who did you pick?" I asked her. I was imagining some possibilities: Rosa Parks, Christa McAuliffe, Sara Lee. But Jenny answered, "Laura." 1

"Who?" I asked. 2

"Laura," she said again. 3

"You mean your friend from across the street?" I asked. 4

"Yeah!" she said. 5

I was a little mystified. "How come you picked her?" I asked. "Because," Jenny answered in the ultrapatient voice of Instructor to the Hopeless, "she is my hero." 6

"Oh," I said. "I see." 7

I must confess that at first I was disappointed. I thought that if her hero was only her friend, she had failed to appreciate the magnificent contributions that real heroes have made to the world. I thought I'd better go out that afternoon and buy some books about famous scientists, artists, athletes, world leaders. That would wise her up. Also, I'd have a look at what they were teaching in her school—didn't she have an appreciation for Martin Luther King? 8

But then I thought about who I would say my heroes are, and I realized that if I told the truth, they wouldn't be famous, either. For although it is undeniable that there have been outstanding people in history who have set glorious examples and inspired me mightily, the people who inspire me most tend to be those who touch me personally, and in quiet ways. 9

For example, I had an eighth-grade English teacher named Mrs. Zinz. She was demanding and rather straight-laced, I thought, but she taught us certain critical skills for reading and writing. She was concerned not only with what we learned but also with what we were becoming: She put a premium on honesty and tried to get us to understand why it was important. She insisted that we always do our best, and in her class, we almost always did. She told me that I was a terrific creative writer and encouraged my every effort. 10

As payment for all her good work, I, along with my evil best friend, tortured her. We laughed at her in class, tried to pit our other teachers against her by telling them how unfairly she graded, and once, in a moment of extreme obnoxiousness, called her on the telephone over and over, only to hang up when she answered. Mrs. Zinz was no dummy. She knew who was calling her. In turn she called my mother, who insisted that I call Mrs. Zinz and apologize. With my face aflame, and between clenched teeth, I muttered a grossly insincere "Sorry." She accepted my apology warmly and with a style so graceful that I was infuriated all the more. And though I sulked every day in her class for the rest of the year, she never bowed by reacting to it. I got an A for the term. I moved soon after that and lost track of Mrs. Zinz. I never did apologize in any legitimate way to her. Ironically, she is still an inspiration to me, a lesson in how not to lower yourself to someone else's level, even when that person is doing everything she can to make you crazy. She is, in that way, a hero. 11

My grandfather, known to me as "Papa," was also a hero of mine. For one thing, he made all of us grandchildren laugh all of the time. He told riddles that were viewed by us as the essence of sophisticated humor. When he greeted us, he shook our hands enthusiastically and at great length, shouting, "How do! How do!" We used to line up to sit on his lap and watch him pop his dentures in and out of his mouth, an astounding feat that thrilled and terrified us—especially before we realized that the teeth were false. He was unfailingly warm and kind and knew how to make a friend 12 out of a stranger; he loved people. I saw him as a man who felt light inside, happy; and feeling that way is no small task in a world that often seems intent on taking back two for every one you get.

13 Then there is my mother-in-law, Sylvia, who at the age of retirement went back to school to pursue a lifetime dream: getting a college diploma. She bumped her bifocals into microscopes, suffered verbal abuse at the hands of an insensitive computer instructor, got used to being the last one to finish every exam, and worried about homework on weekends when she could have been luxuriating in the fact that she had nothing to do. She says that she learned an awful lot, but if you ask me, it's she who did the teaching. I am honored that our family has her love of learning to inspire us.

14 Beyond that, there are people who are heroes to me because of what they do: mail carriers, who, on days when I stay inside hiding from the cold or heat, subject themselves to hours of it; nurses, who care for those who can't care for themselves every second of every day. I admire stay-at-home mothers for their patience and their creativity in the face of almost no thanks or recognition, and working mothers for the way they juggle an awesome load of responsibilities.

15 There are people with chronic illness, for whom getting through each day is heroic. There are people who have been married for sixty years, who have lessons to teach us all. There are those who are strong enough in heart and in spirit to speak up when something feels wrong to them, to go against the majority, and oftentimes to risk themselves for the sake of others. And then there are those whom I admire most of all: people who seem to have found the secret of calm and can relinquish the race for the pleasure of seeing what's around them, right here, right now.

16 I was thinking about all this when I saw Jenny and Laura come into the house. I wanted to know a little more about what had precipitated Jenny's calling Laura a hero. Was it her sharing her brand-new toys? Being there to listen, to soothe, to make better a bad situation in the way that only good friends can? Well, as it happens, no. Jenny told me that Laura was her hero because Laura had saved her from drowning in a creek. "*What?*" I yelled.

17 Laura rolled her eyes. "Jenny, the water was only about an inch deep." Jenny shrugged and said, "So? You still saved me."

18 Laura and I let pass a certain look between us. Then she and Jenny went outside again to play.

19 If you're smart, I thought, you gratefully take your heroes where you find them. As it happens, they are everywhere. So what if the water was only an inch deep? Someone was there, caring about Jenny and showing that she did, a safe hand stretched out to another who was in trouble. This seemed heroic indeed, and later that night when I was tucking her in, I told Jenny that I thought her choice was perfect. "I know," she yawned. "Good night."

RESPONDING TO READING

How did you define "hero" before reading Berg's essay? (For help, look at the examples you gave in Preparing to Read.) Did the Berg essay change your idea of what a hero is? Why or why not?

GAINING WORD POWER

Sometimes little words can mean a lot. What, for instance, does the word *wise* mean in this sentence?

That would wise her up.

Get together with several of your classmates and see how many other meanings of the word *wise* the group can come up with. Then look it up in a dictionary. Did you think of most of them? What are the differences among the words *wise, intelligent, smart,* and *brainy*?

Now, think about the word *set*. See how many sentences the group can write using *set* with a different meaning in each one. Do at least five. Then check the dictionary to see how many meanings there are. Surprising, isn't it, how much meaning can be packed into such a simple little word?

CONSIDERING CONTENT

1. Identify the three names Berg imagines as possible heroes for her daughter in paragraph 1. Why are these women heroes? Why does Berg include Sara Lee?
2. Berg's essay revolves around the understanding of the word *hero* by the daughter, which eventually expands the mother's thinking about the concept. What is the mother considering a *hero* at the beginning of the story? How does she expand on the concept by the end?
3. What kind of mother is Elizabeth Berg? List the details that back up your answer.
4. Why did the author and her friend torture Mrs. Zinz when they were in eighth grade?
5. "If you're smart, I thought, you gratefully take your heroes where you find them" (para. 19). What does this mean? Why is taking your heroes where you find them *smart*?

CONSIDERING METHOD

1. What are the three specific examples of heroes given by Berg? What do they have in common? What groups of people does Berg consider heroic?

2. This essay first appeared in *Parents* magazine in 1992. Name several of the statements and ideas that would appeal to the audience of this magazine.
3. Find instances where Berg uses direct quotations and dialog. Think about how these sections could be rephrased without direct quotations. What effect does using exact spoken words have?
4. What is Jenny's response to her mother's compliment in the end? What parent-child dynamics are underlined? Why do you think Berg closed this way, instead of with a restatement of her concept of heroism?
5. Would you recommend this essay to a good friend? Explain why or why not.

WRITING STEP BY STEP

A. Think of two to four people in your life whom you could consider *heroes,* using Berg's concept. These could be people from your school, your home town, your ethnic subculture, your family, or your social group.

B. Introduce your essay by naming some famous or highly recognized people that you consider heroes. You can include people who are highly recognized mostly within their own field—for instance, a chess player or computer programmer or fashion designer. These people should not be personal acquaintances of yours.

C. Write a transition paragraph that highlights the difference between famous heroes and everyday heroes. Explain what makes someone an everyday hero for you.

D. Develop a paragraph or two about each of the people who are your examples of everyday heroes. Be sure to provide vivid details. Leave no doubt about why you admire each one.

E. In this development, include at least one brief narrative, as Berg does for Mrs. Zinz, Papa, and Sylvia. Consider using direct quotations.

F. Close your essay with an incident or an episode rather than a summary or restatement.

OTHER WRITING IDEAS

1. Berg writes about getting "a lesson in how not to lower yourself to someone else's level" (para. 11). Write an essay on this topic, including examples of situations in which people either did lower them-

selves to someone else's level, or refused to do so. You might decide to give examples of both giving in and resisting.

2. COLLABORATIVE WRITING. Write a collaborative essay on the previous topic, with a team of three or four. Each person should contribute an example. Work as a group to put the examples into an essay, organizing them to work smoothly together and adding an introduction and a conclusion.
3. USING THE INTERNET. In paragraph 10, Berg writes about an English teacher who "insisted that we always do our best, and in her class, we almost always did." Berg endorses the idea that people, even children, tend to rise to the expectations that others hold. Write an essay giving examples (or one long illustration) of this tendency. For inspiration, read the material on *self-fulfilling prophecy* at **http://members.fortunecity.com/nadabs/prophecy.html.**
4. WRITING ABOUT READING. What do you think of the author's concept of a hero? Do you agree with it? Are the examples convincing? Write an essay in which you analyze and evaluate the effectiveness of "My Heroes."

EDITING SKILLS: SEMICOLONS BETWEEN SENTENCES

If you tend to write mainly short, simple sentences, you may need to add some variety. If you have two sentences *that are closely related in meaning,* you can put them together—separated by a semicolon. For example,

> He was unfailingly warm and kind and knew how to make a friend out of a stranger; he loved people.

Remember that the semicolon indicates a close relationship in meaning; it can't just go between any two sentences. Also, each of the sentences must be complete in most standard English writing for a semicolon to work between them. When you read older material, like novels by Jane Austen or Charles Dickens, you may see a fragment on one side of a semicolon and a complete sentence on the other, but that is not usually done these days.

EXERCISE

Dave Barry, a popular humorist, once described the new Jolly Green Giant this way: "He no longer looks like the 'Ho, Ho, Ho' guy; he now looks like Paul McCartney on steroids." Why do you think Barry chose a semicolon between his two sentences?

Go through the essay you just wrote to see whether you have any short, closely related sentences that come one right after the other. If so, try taking out the period, putting in a semicolon, and making the second sentence's capital letter lowercase.

A word of caution: Be sure to use a *semicolon* in joining sentences. A comma definitely will not do in most cases.

WEB SITE

www.modestyarbor.com/elizabethberg.html

Writers on Writing: the June 2002 interview with Elizabeth Berg.

PREPARING TO READ

Do you feel safe when you go out alone at night? Are there certain sections of town that you would refuse to enter alone after dark? Does the fear, or lack of fear, in any way relate to your gender, your age, or the color of your skin?

"Just Walk On By": A Black Man Ponders His Power to Alter Public Space

BRENT STAPLES

Born in 1951 in Chester, Pennsylvania, Brent Staples is a journalist who also holds a Ph.D. in psychology from the University of Chicago. His memoir, Parallel Time: Growing Up in Black and White *(1994), tells the story of his childhood in Chester, a mixed-race, economically declining town. He is currently on the editorial board of the* New York Times. *The selection reprinted here was first published in* Ms. *magazine in September 1986.*

TERMS TO RECOGNIZE

uninflammatory *(para. 1)*	not likely to cause violence or excitement
unwieldy *(para. 2)*	hard to manage or to deal with
indistinguishable *(para. 2)*	not clearly different from
elicit *(para. 3)*	draw forth
warrenlike *(para. 5)*	narrow and crowded like a rabbit hutch
bandolier *(para. 5)*	a belt holding bullets, draped across the chest
lethality *(para. 6)*	being lethal or deadly
bravado *(para. 6)*	pretended courage or false confidence
ad hoc *(para. 7)*	for this case only
labyrinthine *(para. 7)*	like the winding, confusing passages in a maze
berth *(para. 9)*	a safe distance
skittish *(para. 9)*	jumpy, easily frightened
constitutionals *(para. 10)*	walks to improve one's health

My first victim was a woman—white, well dressed, probably in her early 20s. I came upon her late one evening on a deserted street in Hyde Park, a relatively affluent neighborhood in an otherwise mean, impoverished section of Chicago. As I swung onto the avenue behind her, there seemed to be a discreet, uninflammatory distance between us. Not so. She cast back a worried glance. To her, the youngish black man—a broad six feet two inches with a beard and billowing hair, both hands shoved into the pockets of a bulky military jacket—seemed menacingly close. She picked up her pace and was soon running in earnest. Within seconds she disappeared into a cross street. 1

That was more than a decade ago. I was 22 years old, a graduate student newly arrived at the University of Chicago. It was in the echo of that terrified woman's footfalls that I first began to know the unwieldy inheritance I'd come into—the ability to alter public space in ugly ways. It was clear that she thought of herself as the quarry of a mugger, a rapist, or worse. Suffering a bout of insomnia, however, I was stalking sleep, not defenseless wayfarers. As a softy who is scarcely able to take a knife to a raw chicken—let alone hold one to a person's throat—I was surprised, embarrassed, and dismayed all at once. Her flight made me feel like an accomplice in tyranny. It also made it clear that I was indistinguishable from the muggers who occasionally seeped into the area from the surrounding ghetto. I soon gathered that being perceived as dangerous is a hazard in itself: Where fear and weapons meet—as they often do in urban America—there is always the possibility of death. 2

In that first year, my first away from my hometown, I was to become thoroughly familiar with the language of fear. At dark, shadowy intersections, I could cross in front of a car stopped at a traffic light and elicit the *thunk, thunk, thunk, thunk* of the driver—black, white, male, female—hammering down the door locks. On less traveled streets after dark, I grew accustomed to but never comfortable with people crossing to the other side of the street rather than pass me. Then there were the standard unpleasantries with policemen, doormen, bouncers, cabdrivers, and others whose business it is to screen out troublesome individuals *before* there is any nastiness. 3

I moved to New York nearly two years ago and I have remained an avid night walker. In central Manhattan, the near-constant crowd covers the tense one-on-one street encounters. Elsewhere, things can get very taut indeed. 4

After dark, on the warrenlike streets of Brooklyn where I live, I often see women who fear the worst from me. They seem to have set their faces on neutral, and with their purse straps strung across their chests bandolier-style, they forge ahead as though bracing themselves against being tackled. I understand, of course, that the danger they perceive is not a hallucination. 5

Women are particularly vulnerable to street violence, and young black males are drastically overrepresented among the perpetrators of that violence. Yet these truths are no solace against the alienation that comes of being ever the suspect, an entity with whom pedestrians avoid making eye contact.

6 It is not altogether clear to me how I reached the ripe old age of 22 without being conscious of the lethality nighttime pedestrians attributed to me. Perhaps it was because in Chester, Pa., the small, angry industrial town where I came of age in the 1960s, I was scarcely noticeable against a backdrop of gang warfare, street knifings, and murders. I grew up one of the good boys, had perhaps a half-dozen fistfights. In retrospect, my shyness of combat has clear sources. As a boy, I saw countless tough guys locked away; I have since buried several, too. They were babies, really—a teenage cousin, a brother of 22, a childhood friend in his mid-20s—all gone down in episodes of bravado played out in the streets. I chose, perhaps unconsciously, to remain a shadow—timid, but a survivor.

7 The fearsomeness mistakenly attributed to me in public places often has a perilous flavor. The most frightening of these confusions occurred in the late 1970s and early 1980s, when I worked as a journalist in Chicago. One day, rushing into the office of a magazine I was writing for with a deadline story in hand, I was mistaken for a burglar. The office manager called security, and with the speed of an ad hoc posse, pursued me through the labyrinthine halls, nearly to my editor's door. I had no way of proving who I was. I could only move briskly toward the company of someone who knew me.

8 Relatively speaking, however, I never fared as badly as another black male journalist. He went to nearby Waukegan, Ill., a couple of summers ago to work on a story about a murderer who was born there. Mistaking the reporter for the killer, police officers hauled him from his car at gunpoint and but for his press credentials would probably have tried to book him. Such episodes are not uncommon. Black men trade tales like this all the time.

9 Over the years, I learned to smother the rage I felt at so often being mistaken for a criminal. Not to do so would surely have led to madness. I now take precautions to make myself less threatening. I move about with care, particularly late in the evening. I give a wide berth to nervous people on subway platforms during the wee hours. If I happen to be entering a building behind some people who appear skittish, I may walk by, letting them clear the lobby before I return, so as not to seem to be following them. I have been calm and extremely congenial on those rare occasions when I've been pulled over by the police.

10 And on late-evening constitutionals I employ what has proved to be an excellent tension-reducing measure: I whistle melodies from Beethoven and Vivaldi and the more popular classical composers. Even steely New

Yorkers hunching toward nighttime destinations seem to relax, and occasionally they even join in the tune. Virtually everybody seems to sense that a mugger wouldn't be warbling bright, sunny selections from Vivaldi's "Four Seasons." It is my equivalent of the cowbell that hikers wear when they are in bear country.

....

RESPONDING TO READING

Why don't we expect muggers to be whistling melodies from Beethoven and Vivaldi? In your journal write a brief explanation of the possible reasons.

GAINING WORD POWER

The following words appear in the reading but are not included in the Terms to Recognize. Look up each one in your dictionary and use it in a sentence of your own.

1. affluent (para. 1)
2. menacingly (para. 1)
3. dismayed (para. 2)
4. taut (para. 4)
5. vulnerable (para. 5)
6. solace (para. 5)
7. entity (para. 5)
8. attributed (para. 7)
9. precautions (para. 9)
10. congenial (para. 9)

CONSIDERING CONTENT

1. How does the author describe his physical appearance in the opening paragraph? What categories of readers would be drawn in by the opening paragraph?
2. Why does the woman take him for "a mugger, a rapist, or worse"?
3. What does the writer mean when he says "that being perceived as dangerous is a hazard in itself"? What illustrations does he offer to prove his point?
4. What kind of hometown background did Brent Staples have? What kind of person did he turn out to be?
5. At the end of paragraph 6, he speaks of three young men he was close to—"all gone down in episodes of bravado played out in the streets."

Although he doesn't tell us how any of them died, can you guess? Give examples of the kind of "episodes of bravado" that may have cost them their lives.

6. How did Staples learn to deal with the problem of being a large, young African American man in the city?
7. Why would hikers in bear country wear cowbells? Explain how that wilderness situation is similar to Staples's urban situation.

CONSIDERING METHOD

1. Explain how the brief narrative in the opening paragraph catches our interest.
2. The thesis of this selection is implied, not directly stated. Write out in your own words a statement of the author's main point.
3. This essay first appeared in *Ms.* magazine, a publication that focuses on women's issues. What did the author probably assume about his audience? How did it affect his choice of examples?
4. How does description help to make Staples's illustrations interesting and convincing in paragraphs 1 and 5?
5. Using words that sound like the noise they describe is called **onomatopoeia** (for example, the *thunk, thunk, thunk, thunk* of the car door locks in para. 3). Explain why that word choice is effective. Can you think of other examples of words that sound like what they mean?
6. This essay is developed through example and illustration, yet Staples does not tell us how the three young men died (para. 6.) Why not?
7. Explain what makes the conclusion particularly satisfying.

WRITING STEP BY STEP

Stereotypes are oversimplified groupings of people by race, gender, politics, athletic ability, ethnic origin, and so on. Staples was stereotyped because he was a young black male, and as Staples says, "young black males are drastically overrepresented among the perpetrators of . . . violence." Although there is usually some grain of truth behind stereotypes, they tend to be negative and unfair. Women, for instance, are stereotyped as weak, passive, fickle, scatterbrained, and indecisive. Men, on the other hand, are supposed to be strong-minded and assertive, but dense and unfeeling. Stereotypes are unfair because they lump lots of people into a category whether or not the characteristics fit every individual.

A. Think of a stereotype that includes you. Choose one that you think is unfair to you; your essay will explain how you are different.
B. Begin, as Staples does, with a brief narrative, a story that illustrates how you *seem* to fit the type although you actually do not.
C. In the next paragraph, define the stereotype by giving examples of several characteristics people expect you to have—or not to have. For instance, if you are a male football player, people may take you for a clumsy hulk who can barely read and write.
D. Next, explain why people would tend to place you in this stereotype.
E. Then, explain why you don't fit the stereotype, and tell about some influence while you were growing up that helped you to avoid the typical pattern of behavior. Provide concrete examples, as Staples does in paragraph 6 when he tells about his childhood among the gangs in Chester, Pennsylvania.
F. In conclusion, explain how you felt about being stereotyped and how you have learned to cope with the mistaken views of people who took you for a different kind of person than you truly are.

OTHER WRITING IDEAS

1. Write the paper outlined in the previous section but instead illustrate how you are the perfect example of a stereotype. You may want to think of a positive stereotype—or else make your essay humorous.
2. COLLABORATIVE WRITING. With a small group of classmates discuss phobias, those irrational fears that most of us have—fears of spiders, of snakes, of high places, of flying, of closed spaces. Which ones do you have? Choose your worst or your most embarrassing phobia and tell in an essay how it limits your activities, how it makes you feel, how you think you got it, and what you do to control it. Or, choose one phobia from your group's collection and develop an essay of example together.
3. USING THE INTERNET. Use Internet sources to find out about treatments for phobias. Write an essay explaining two or three different ways that people can overcome these irrational fears.
4. WRITING ABOUT READING. In paragraph 5, the author says, "Women are particularly vulnerable to street violence, and young black men are drastically overrepresented among the perpetrators of that violence." Do you agree with this comment? Write a response to this idea.

EDITING SKILLS: COMMAS AROUND INTERRUPTERS

An **interrupter** is just what it sounds like—a word or group of words that interrupts or breaks into the flow of a sentence, like the italicized words do here:

> I understand, *of course,* that the danger they perceive is not a hallucination.

You need a comma before and after the interrupter as a signal to your readers that the interrupter is an addition that can be removed without changing the meaning of the sentence. The first comma signals the start of the interruption; the second comma signals the end of the interruption. It would be quite misleading in that sentence, with its flow interrupted, to use only one comma. But if you move the *of course* to the beginning or to the end of the sentence so that it no longer interrupts the flow, then a single comma is fine:

> *Of course,* I understand that the danger they perceive is not a hallucination.
> I understand that the danger they perceive is not a hallucination, *of course.*

The principle remains the same, even when the interrupter is longer:

> One day, *rushing into the office of a magazine I was writing for with a deadline story in hand,* I was mistaken for a burglar.

Remember: put commas *around* interrupters—one before and one after.

EXERCISE

We've omitted the commas from around the interrupters in the following sentences. Figure out where they belong, and put them back in.

1. Suffering a bout of insomnia however I was stalking sleep, not defenseless wayfarers.
2. After dark on the warrenlike streets of Brooklyn where I live I often see women who fear the worst from me.
3. I chose perhaps unconsciously to remain a shadow—timid, but a survivor.
4. The office manager called security and with the speed of an ad hoc posse pursued me through the labyrinthine halls, nearly to my editor's door.

5. Relatively speaking however I never fared as badly as another black male journalist.

Now check the essay you've just written to be sure that you have punctuated interrupters correctly.

WEB SITE

www.pbs.org/blackpress/modern_journalist/staples.html

Video and audio clips of Brent Staples from the Public Broadcasting System.

Preparing to Read

Do men or women make better schoolteachers? Or does gender matter? Does the age of the students make any difference?

One Man's Kids

Daniel R. Meier

Daniel Meier received a master's degree from the Harvard Graduate School of Education in 1984. He taught first grade at schools in Brookline and Boston, Massachusetts, before getting his Ph.D. at the University of California at Berkeley. He now teaches early childhood education at San Francisco State University. His articles about teaching and his reviews of children's books have appeared in a number of educational journals. The essay reprinted here appeared in 1987 in the "About Men" series of the New York Times Magazine.

Terms to Recognize

complying *(para. 4)*	agreeing to someone else's request or command
singular *(para. 5)*	exceptional, unusual, distinguished by superiority
consoling *(para. 6)*	offering comfort and advice
intellectual *(para. 7)*	guided chiefly by knowledge or reason rather than by emotion or experience
hilarity *(para. 7)*	spirited merriment, cheerfulness
complimentary *(para. 12)*	given free as a courtesy or a favor

I teach first graders. I live in a world of skinned knees, double-knotted shoelaces, riddles that I've heard a dozen times, stale birthday cakes, hurt feelings, wandering stories, and one lost shoe ("and if you don't find it my mother'll kill me"). My work is dominated by 6-year-olds. 1

It's 10:45, the middle of snack, and I'm helping Emily open her milk carton. She has already tried the other end without success, and now there's so much paint and ink on the carton from her fingers that I'm not sure she should drink it at all. But I open it. Then I turn to help Scott clean up some milk he has just spilled onto Rebecca's whale crossword puzzle. 2

While I wipe my milk- and paint-covered hands, Jenny wants to know if I've seen that funny book about penguins that I read in class. As I hunt 3 for it in a messy pile of books, Jason wants to know if there is a new seating arrangement for lunch tables. I find the book, turn to answer Jason, then face Maya, who is fast approaching with a new knock-knock joke. After what seems like the 10th "Who's there?" I laugh and Maya is pleased.

4 Then Andrew wants to know how to spell "flukes" for his crossword. As I get to "u," I give a hand signal for Sarah to take away the snack. But just as Sarah is almost out the door, two children complain that "we haven't even had ours yet." I stop the snack mid-flight, complying with their request for graham crackers. I then return to Andrew, noticing that he has put "flu" for 9 Down, rather than 9 Across. It's now 10:50.

5 My work is not traditional male work. It's not a singular pursuit. There is not a large pile of paper to get through or one deal to transact. I don't have one area of expertise or knowledge. I don't have the singular power over language of a lawyer, the physical force of a construction worker, the command over fellow workers of a surgeon, the wheeling and dealing transactions of a businessman. My energy is not spent in pursuing, climbing, achieving, conquering, or cornering some goal or object.

6 My energy is spent in encouraging, supporting, consoling, and praising my children. In teaching, the inner rewards come from without. On any given day, quite apart from teaching reading and spelling, I bandage a cut, dry a tear, erase a frown, tape a torn doll, and locate a long-lost boot. The day is really won through matters of the heart. As my students groan, laugh, shudder, cry, exult, and wonder, I do too. I have to be soft around the edges.

7 A few years ago, when I was interviewing for an elementary-school teaching position, every principal told me with confidence that, as a male, I had an advantage over female applicants because of the lack of male teachers. But in the next breath, they asked with a hint of suspicion why I chose to work with young children. I told them that I wanted to observe and contribute to the intellectual growth of a maturing mind. What I really felt like saying, but didn't, was that I loved helping a child learn to write her name for the first time, finding someone a new friend, or sharing in the hilarity of reading about Winnie the Pooh getting so stuck in a hole that only his head and rear show.

8 I gave that answer to those principals, who were mostly male, because I thought they wanted a "male" response. This meant talking about intellectual matters. If I had taken a different course and talked about my interest in helping children in their emotional development, it would have been seen as closer to a "female" answer. I even altered my language, not once mentioning the word "love" to describe what I do indeed love about teaching. My answer worked; every principal nodded approvingly.

9 Some of the principals also asked what I saw myself doing later in my career. They wanted to know if I eventually wanted to go into educational administration. Becoming a dean of students or a principal has never been

one of my goals, but they seemed to expect me, as a male, to want to climb higher on the career stepladder. So I mentioned that, at some point, I would be interested in working with teachers as a curriculum coordinator. Again, they nodded approvingly.

If those principals had been female instead of male, I wonder whether their questions, and my answers, would have been different. My guess is that they would have been. 10

At other times, when I'm at a party or a dinner and tell someone that I teach young children, I've found that men and women respond differently. Most men ask about the subjects I teach and the courses I took in my training. Then, unless they bring up an issue such as merit pay, the conversation stops. Most women, on the other hand, begin the conversation on a more immediate and personal level. They say things like "those kids must love having a male teacher" or "that age is just wonderful, you must love it." Then, more often than not, they'll talk about their own kids or ask me specific questions about what I do. We're then off and talking shop. 11

Possibly, men would have more to say to me, and I to them, if my job had more of the trappings and benefits of more traditional male jobs. But my job has no bonuses or promotions. No complimentary box seats at the ball park. No cab fare home. No drinking buddies after work. No briefcase. No suit. (Ties get stuck in paint jars.) No power lunches. (I eat peanut butter and jelly, chips, milk, and cookies with the kids.) No taking clients out for cocktails. The only place I take my kids is to the playground. 12

Although I could have pursued a career in law or business, as several of my friends did, I chose teaching instead. My job has benefits all its own. I'm able to bake cookies without getting them stuck together as they cool, buy cheap sewing materials, take out splinters, and search just the right trash cans for useful odds and ends. I'm sometimes called "Daddy" and even "Mommy" by my students, and if there's ever a lull in the conversation at a dinner party, I can always ask those assembled if they've heard the latest riddle about why the turkey crossed the road. (He thought he was a chicken.) 13

RESPONDING TO READING

What do you think about Meier's choice of career? Do you think it's reasonable and appropriate? Why or why not?

GAINING WORD POWER

In paragraph 12 Meier uses the word *complimentary.* There is a word that's pronounced the same but has a different spelling and a different meaning—*complementary.* Do you know what each word means? English has many of these sound-alike words, and it's important to know the differences among them. They won't cause you any trouble in speaking, but they will change the meaning of your writing if you choose the wrong one.

Here is a list of words from Meier's essay. Using your dictionary, find a sound-alike word for each item in the list and write down its meaning. Then use the word you found in a sentence.

fare	right	male
new	won	course
principal	seen	whether
through	two	hole

CONSIDERING CONTENT

1. Why did Meier write this essay? What point do you think he wanted to make?
2. Meier says, "My work is not traditional male work" (para. 5)? What does he mean by that statement? Do you agree?
3. What is a "singular pursuit"? Why does the author use that phrase to describe "male work"?
4. Why did the principals who interviewed Meier have "a hint of suspicion" about him (para. 7)?
5. What kind of answers did Meier give in his job interviews (para. 8)? What kind of answers did he avoid? Why didn't he mention the word "love"?
6. The author says that men and women respond differently to him when he talks about his job (para. 11). What are the differences?
7. How does Meier feel about his job? Do you think he is being defensive or apologetic about it?

CONSIDERING METHOD

1. In the first paragraph the author says, "My work is dominated by 6-year-olds." What examples does he give to explain and support this general statement?
2. Why does Meier open his essay the way he does? Why doesn't he state his thesis until paragraph 5?
3. Find the series of words that Meier uses to describe "male work" (end of para. 5) and the words he uses to describe what he does (beginning of para. 6). What is he saying about the difference between his work and "male work"?

4. What does the author mean when he says, "In teaching, the inner rewards come from without" (para. 6)? What examples does he give to make his meaning clear?
5. What is the effect of the series of phrases that begin with "No" in paragraph 12? Why are the positive comments in parentheses? Why is the last sentence in the paragraph not in parentheses?

WRITING STEP BY STEP

Think of a workplace that you know well—a place where you have a job now or had one in the past. If your primary work is being a student, then school is your workplace.

A. First identify and briefly explain your role at the workplace ("I am a cashier, salesperson, and general trouble shooter at Posh Pups, a pet grooming and pet supply store").
B. Then name three or four personal qualities that make for success in that job. ("To work at Posh Pups, you need to be loyal and good at math and to like people as much as you like animals.")
C. Think about the order in which you want to present these qualities. You might start with a less important quality and build up to the most important one. Or you may see that two of them are related and need to be placed in back-to-back paragraphs. Make a scratch outline to help you decide how to arrange your main points and examples.
D. For each quality you name, give at least one example of how it is important in the job. Think of a specific time when each quality was needed. Tell the story of how you or your co-workers showed a particular quality (or, unfortunately, showed a lack of it).
E. If an example is long and detailed, give it its own paragraph, as Meier does in paragraphs 2, 3, and 4 of his essay. If some of your examples are only a sentence or two, try to expand them with more details that will give your readers the sights, sounds, and feel of your work.
F. Close with a summarizing statement of how you feel about this job. Or offer a recommendation to anyone who might consider going into this line of work. Try to reinforce your thesis idea without simply repeating it.
G. When you revise, look at the transitions between your paragraphs. Try to fill in any gaps between your main points. Also check to see that you used transitions to lead into your examples. Ask your instructor or your classmates to help you improve the flow of your ideas.

OTHER WRITING IDEAS

1. Select one of the following general statements, or compose one of your own. Make it the central idea of an essay full of examples and illustrations. Draw examples from your reading, your conversation, your observations, and your own experience.
   a. Action heroes in the movies today are pretty much alike.
   b. Being a good parent is probably the hardest job there is.
   c. The stereotypical female (or male) is not easy to find in our society anymore.
   d. Being a teenager can be difficult (or easy or perplexing or a lot of fun).
   e. Jealousy is a destructive emotion.
2. COLLABORATIVE WRITING. Discuss with your classmates some superstitions that you or members of your family or community have held. Frequently, these superstitions have to do with success or bad luck in sports, performances, weather, or work. Do they have any validity? How did they develop? Write an essay of example about the role that superstition plays in your life or in the life of someone you know.
3. USING THE INTERNET. Search the Web for information about the workforce of grade school teachers today. Are there more men than there used to be? Can you find information about what discourages or attracts males to grade school teaching? Integrate information you find with information from Meier to write an essay giving examples of pluses and minuses for men considering the teaching field.
4. WRITING ABOUT READING. Write an essay in which you examine the *tone* (see Glossary) of "One Man's Kids." How does the tone of the writing reflect the author's attitude toward his job?

EDITING SKILLS: USING SUBORDINATION

Writers often combine two or more ideas in a sentence by using **subordination.** When one idea is subordinate to or dependent on another, it is less important. Take a look at these sentences from Meier's essay to see how the subordinate ideas are introduced by words that make them sound less important:

> *Although* I could have pursued a career in law or business, I chose teaching instead.
>
> Then, *unless* they bring up an issue such as merit pay, the conversation stops.

If I had taken a different course and talked about my interest in helping children in their emotional development, it would have been seen as close to a "female" answer.

As you can see, each of these sentences has two parts. Notice that the second part, after the comma, can stand alone as a sentence: that part is called **independent.** If you read the first part alone, it does not sound complete. The first part is called **dependent** for this reason. The opening words—*Although, unless,* and *If*—make the first statement of each sentence dependent. These words are called *subordinating conjunctions;* they indicate that the first idea is not as important as the rest of the sentence. (You will also notice that when the dependent part comes first, a comma separates it from the independent part.)

Subordinating conjunctions are familiar words; we use them a lot. Here are some of the most common ones: *since, because, if, even if, unless, although, even though, though, as long as, after, before, when, whenever, while, until,* and *wherever.* Skillful writers use subordination to give variety to their sentences and to keep readers' attention focused on the main ideas.

EXERCISE

Imitate the following sentences. Each one begins with a dependent statement followed by an independent one. You don't have to imitate each sentence exactly; just follow the dependent-independent pattern and use the same subordinating conjunction as the model sentence. Put in the commas, too.

Model: If we want clean air, then we will have to drive more fuel-efficient cars.

Imitation: If you like lasagna, then you should try the new Italian restaurant on Division Street.

1. While I was eating my lunch, a friend walked in.
2. Wherever I go in this city, I run into old friends.
3. Before you gather up your books, be sure your notes are complete.
4. Although Selma is a good athlete, she sometimes swears at the umpire.
5. Unless she sees the error of her ways, Selma may get tossed off the team.

Write at least one more imitation of each of the preceding sentences.

Go back to your example essay to see how many sentences like these you've written. What subordinating conjunctions did you use? Did you include the commas? Now combine some more of your sentences by using subordinating conjunctions and the dependent-independent pattern. If they sound sensible, keep them in your essay.

WEB SITES

http://userwww.sfsu.edu/~dmeier/welcome.htm

On Daniel Meier's home page you will find more information about his life, his research interests, and his teaching.

www.umaine.edu/eceol

The Early Childhood Education Online site provides a wealth of information about teaching young children.

√ Preparing to Read

What income do you think marks the federal poverty line for a couple? For a family of four? What is the minimum wage in your state these days? Write down your estimates before reading the next essay.

The Working Poor

Tim Jones

Tim Jones is a writer for the Chicago Tribune. *He has covered media topics including Oprah Winfrey's Texas trial for disparaging beef and Hormel Foods's efforts to stop the use of* Spam *as a word for unwanted e-mail. He often writes stories about areas where national economics and public media interconnect.*

Terms to Recognize

tapping *(para. 2)*	taking advantage or making use of
think tank *(para. 3)*	a group organized for the purpose of researching a problem
sparsely *(para. 6)*	thinly, lightly
severance *(para. 7)*	termination of employment
vise *(para. 14)*	clamping tool that holds material between two jaws
emphysema *(para. 14)*	a serious lung disease that makes breathing difficult

McArthur, Ohio. The food line begins to form during the sunrise chill, more than two hours before the metal gates to the Care United Methodist Outreach pantry open. Hundreds of people like Theresa Ware arrive early because they fear the boxes of food stacked in neat rows will be gone by the time they push their rusty grocery carts to the head of the hours-long line. Ware keeps an eye on her watch because she can't afford to be late for work, not even if the reason is to pick up food. "This is a have-to case for us. It's humiliating," said Ware, 49, who makes $7.50 an hour working the afternoon shift at a nursing home. This recent visit was one of two food pantry stops she and her unemployed husband, Rocky, make every month. "We shouldn't have to do this," she said. 1

Theresa and Rocky Ware toil in the ranks of the working poor, a growing category of millions of Americans who play by the rules of the working world and still can't make ends meet. After tapping friends and family, maxing out their credit cards, and sufficiently swallowing their pride, at 2 least 23 million Americans stood in food lines last year—many of them the working poor, according to America's Second Harvest, the Chicago-based hunger relief organization. The surge in food demand is fueled by several forces—job losses, expired unemployment benefits, soaring health-care and housing costs, and the inability of many people to find jobs that match the income and benefits of the jobs they lost.

3 The Center on Budget and Policy Priorities, a Washington think tank, reported recently that 43 million people are living in low-income working families with children. Other government data show the number of people living below the official poverty line grew by more than 3.5 million from 2000 to 2002, to 34.6 million. And the U.S. Department of Agriculture reported that the number of Americans who don't know where their next meal will come from—categorized as "food insecure"—jumped from 31 million to 35 million between 1999 and 2002. "The reach of the economic slowdown has really pulled in a lot of folks who never expected to be poor," said Stacy Dean, director of food stamp policy for the Center on Budget and Policy Priorities. "What you see now is families turning to private relief for what often is a very small amount of help."

4 "This is not just a function of unemployment. A larger percentage of Americans are working poor, and the numbers have been growing for nine years," said Robert Forney, CEO of America's Second Harvest. "This could be the low-water mark for the economy, but for a whole lot of Americans—40 million of them—the option of [earning] a living wage and benefits? Forget it."

Exploding Demand

5 Food pantry operators across the nation—urban, suburban and rural—tell similar stories of exploding food demand from families, senior citizens, and the fastest-growing segment: the working poor. In southern Ohio, where President Lyndon Johnson declared war on poverty 40 years ago, cars will line roadsides for a half-mile or more waiting for the boxed monthly buffet of dry milk, rice, cereal, canned fruit and vegetables, instant mashed potatoes and, on good days, canned meat and chicken. "We're quickly seeing that communities that thought they were immune are now affected, whether urban or suburban," said Lisa Hamler-Podolski, executive director of the Ohio Association of Second Harvest Foodbanks. Rev. Walt Goble, who runs the Care United Methodist Outreach in McArthur, Ohio, a small and long-ago thriving village about 60 miles south of Columbus, said, "We're here from 11:30 to 4:00, or until we run out of food. Usually we run out."

6 Theresa and Rocky Ware have reluctantly joined the lines at food banks. Last year, in the sparsely populated, nine-county region of southeastern Ohio, 9.1 million pounds of food were handed out—that's up from 3.9 million pounds in 2000. In the past three years, the number of households

served by food banks has more than tripled, according to Second Harvest of Southeast Ohio.

7 Although the national economy shows fitful statistical signs of recovery, the data do not take into account that declining numbers of employers offer health insurance and many new jobs pay the minimum wage, $5.15 an hour.

- Danny Palmer, who lives in the Ohio River village of Cheshire, lost his $20-an-hour welding job and now works at Wal-Mart for $5.95 an hour. Insurance coverage he got as part of a severance package from his former employer runs out next month. He has no health coverage with Wal-Mart.
- Melissa Barringer holds three part-time jobs to augment the income she and her husband, Brian, a laborer, earn to support themselves and their three teenage children. Last year, their combined income was $18,000. "We can't keep up," Melissa Barringer said as her children ate at a soup kitchen in Coolville, Ohio.
- Oscar Sanchez shows up every Thursday for bags of canned and dry goods at the Catholic Charities' Latin American Youth Center in Chicago's Pilsen neighborhood. Sanchez, 52, is a self-employed painter who lost his construction job three years ago. His hourly wage is $7. He has no health insurance.
- And in the St. Louis suburb of Ferguson, Mary Williams works as a temp and drives her 1983 Mercury Marquis to jobs that pay $7 to $8

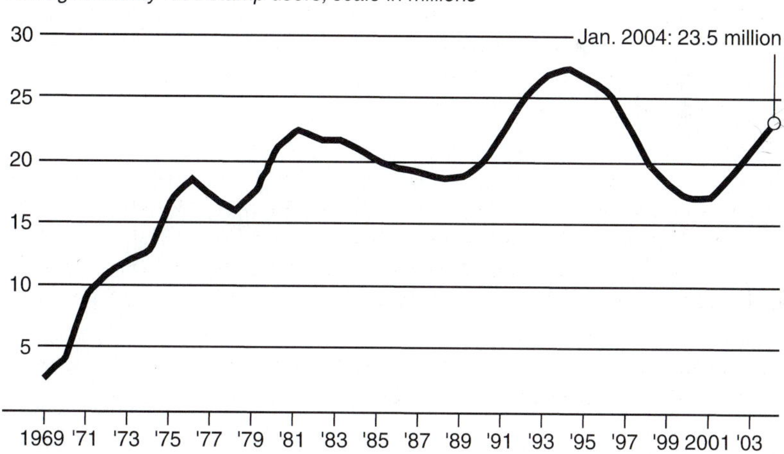

an hour. The work is not steady. Neither she nor her son has health insurance.

All of them tell different stories but have one thing in common: They have jobs and are regulars at food pantries.

What the Numbers Miss

8 The plight of people like the Wares is not reflected in monthly unemployment figures. These Americans fly uneasily beneath the radar of the government's officially recognized economic distress status, the federal poverty line of $12,490 for a couple and $18,850 for a family of four. "The official economic numbers often lag behind the real story," said Gregory Acs, a senior research associate at the Urban Institute, a Washington-based research organization.

9 "Poverty numbers usually tell only part of the story. You can be technically not poor, but you still skip bills and go once or twice a week to the food pantry," Acs said, noting that upward adjustments in poverty levels do not account for all the financial pressures facing families. For example, the share of Americans under age 65 with employer-sponsored health insurance dropped from 70 percent in 1999 to 68 percent in 2002, according to the Urban Institute. Among low-income workers, the percentage dropped from 40 percent to 35 percent.

10 Some of those who are struggling visit Felipe Ayala, who has been running Chicago's Latin American Youth Center for 32 years. "We used to run out of food at the end of the month. Now we can run out at the end of the first week," Ayala said. The increasing demand comes from people who are either out of work or, in the case of Oscar Sanchez, don't earn enough to provide a living for themselves. "It has gotten worse," Ayala said. "All the jobs are in the food industry or else they are in the suburbs. The lines here are always increasing."

11 Sanchez, who is divorced, said he goes every other Thursday to the basement of the old school. People begin to file in at 9 A.M. and move from one metal folding chair to another before they can pick up white plastic food bags on which "Thank You" is printed in red letters. They wait patiently for their turn to get the food bags. "This is what keeps me going," said Sanchez, who has used the pantry for two years. As the weather warms up, Sanchez said, he is hopeful that there will be more work and steady work. He does not expect, however, to break his pattern of regularly visiting the pantry on 17th Street.

12 Nor is Mary Williams particularly hopeful. With three years of college behind her, the 42-year-old former factory worker has not had health insurance since last summer. Williams makes about $400 a month and regularly visits a Salvation Army food pantry in St. Louis. Williams's battered car is starting to break down, and sometimes she has to drive up to an hour,

one way, to get to work. "Most of the time it's a struggle," said Williams, a single parent. "And things aren't looking any better."

13 The pressure on working families has increased in the past few years as many states have cut back Medicaid coverage in the face of their own budget troubles. Two recent reports show that as many as 1.6 million low-income people have lost publicly funded health coverage because of state budget cuts. In response, they cobble together an existence by skipping meals, borrowing from friends, moving in with family and cutting prescription pills in half or bypassing dosages altogether. Some ignore bills, which is one reason Ohio, for instance, a state hit hard by the recession, has the highest rate of loan foreclosures in the nation, according to the Mortgage Bankers Association.

14 Many are trapped in a bureaucratic vise: They make too much money to qualify for Medicaid, the government's low-income health insurance program, yet too little to buy health coverage. Because they can't afford the property taxes, the Wares have turned their home over to their adult children, who pay the tax bill. Medical bills from Rocky's treatments for emphysema are mounting. "I get bills and I say to hell with it. They can come and throw me in jail," Theresa Ware said as she loaded food into the back seat of her truck.

....

√RESPONDING TO READING

How accurate were your estimates of the poverty line and minimum wage? Did any of your ideas change from reading the facts and examples in the essay?

GAINING WORD POWER

One elderly citizen referred to himself as "temporarily embarrassed" when he was broke. Brainstorm with classmates about some other euphemisms (mild terms substituted for harsh ones) for *poverty* and *poor.*

CONSIDERING CONTENT

1. Who are the working poor? What is the main point that Jones's article makes about them?
2. Give a few reasons for an "exploding food demand" at food banks in this country. Can you think of reasons that are not covered in the article?
3. Why does the researcher Gregory Acs say that "poverty numbers usually tell only part of the story" (para. 9)? What is the rest of the story?
4. How are health insurance and home ownership related to the topic of the working poor?

5. This article was published in 2004. Has the situation of the working poor changed since its publication? If you don't know, how could you find out?
6. Paragraph 3 mentions "private relief." What is that? What other kinds of relief for the poor exist?

CONSIDERING METHOD

1. What is Jones's attitude toward the working poor? How do you know? Choose a sentence and suggest how it would be rephrased by someone with the opposite attitude.
2. From many possible examples, the author chose four to present in paragraph 7. Why do you think he chose these four? What similarities and differences do you see among these four people?
3. The main topic of the article is put into words in paragraph 2. What is the purpose of paragraph 1?
4. Notice the lavish use of descriptive statistics in the article. Did you attempt to understand the numbers? Why or why not? Were they persuasive to you?
5. Did you spend any time looking at the graph from the USDA? Look at it again. What does it show? What conclusions can you draw from it? Take a survey among your classmates about who and who did not inspect this graph when reading the article. What conclusions can you draw from your survey?
6. If this article appeared in your local newspaper, would you read it? Why or why not? The actual article is twice as long as the part we reprint here. Would you read the whole thing if it were in front of you? Why or why not?
7. Jones uses two subheadings to break up the text and guide the reader's focus. Find two more places where subheadings would be suitable, and write them. Your instructor may have you work with two or three classmates.

Combining Strategies

"The Working Poor" offers many examples of people who are working yet poor, the problems they encounter, and the solutions they find. The essay also discusses many of the causes and effects of the people's economic situation. List some of the causes and some of the effects mentioned in the article. List some other causes and effects that are not discussed in the article.

WRITING STEP BY STEP

A. Brainstorm with classmates about some way to get in trouble financially other than having a low-paying job. Some possibilities are using credit cards too much, buying a house that is too expensive, taking on expenses for others (like family members), having a high-priced hobby, making bad investments, gambling, leasing a vehicle with large payments, and sending children to costly schools.

B. Think of someone you know who has one of the problems you brainstormed. Freewrite for five minutes about the person and the problem. Choose three or four good examples of practices that have put them into financial trouble.

C. Reread Jones's first paragraph. Write a paragraph to open your essay in which you present a verbal picture that illustrates the person and problem you will be writing about.

D. Develop a description of each example, devoting one or two paragraphs to each one. Be sure to use exact details. Use numbers like prices and percentages if possible.

E. Close with a look at the future. You can choose to suggest how the person might change in the future, or you can close more negatively and tell what will happen if the present course continues. (This second type of closing is what you see in the Jones essay.)

OTHER WRITING IDEAS

1. COLLABORATIVE WRITING. The Writing Step by Step essay focused on one person with a financial problem. Working with several other students, draft an article for your school newspaper which discusses one type of financial problem that college students might have. Provide several different examples, as Jones did in "The Working Poor." Solicit examples from other classmates and friends.
2. Instead of focusing on money difficulties, write an essay of example or illustration about another life problem that many people face.
3. USING THE INTERNET. Find information on one of the institutions mentioned in the Jones article: for instance, Second Harvest, the Center on Budget and Policy Priorities, the Urban Institute, Medicaid, the Mortgage Bankers Association, or the U. S. Department of Agriculture. Write an essay using examples to explain what the institution does, or focus on an interesting part of what it does.
4. WRITING ABOUT READING. Write an essay explaining *either* that Jones's article changed your view about the working poor *or* that Jones's article had little influence on your thinking.

EDITING SKILLS: USING COMMAS IN SERIES

Notice where the commas are placed in the following examples containing items in series.

> The surge in food demand is fueled by several forces—job losses, expired unemployment benefits, soaring health-care and housing costs, and the inability of many people to find jobs that match the income and benefits of the jobs they lost.

> In southern Ohio, where President Lyndon Johnson declared war on poverty 40 years ago, cars will line roadsides for a half-mile or more waiting for the boxed monthly buffet of dry milk, rice, cereal, canned fruit and vegetables, instant mashed potatoes and, on good days, canned meat and chicken.

> In response, they cobble together an existence by skipping meals, borrowing from friends, moving in with family and cutting prescription pills in half or bypassing dosages altogether.

The commas are necessary to let you know when one item ends and another begins. Try making sense of those words without the commas to see what an uphill task reading would be without commas:

> The surge in food demand is fueled by several forces—job losses expired unemployment benefits soaring health-care and housing costs and the inability of many people to find jobs that match the income and benefits of the jobs they lost.

You can see that you have to read twice to figure out which groups of words go together. You might first read "job losses expired" and "unemployment benefits soaring," quite the opposite of the intended meaning.

Look at this sentence which includes a series:

> After tapping friends and family, maxing out their credit cards and sufficiently swallowing their pride, at least 23 million Americans stood in food lines last year.

Jones doesn't use a comma before the *and* connecting the last item in a series (which is "sufficiently swallowing their pride"). This comma is optional in standard English usage. Jones could have written, correctly:

> After tapping friends and family, maxing out their credit cards, and sufficiently swallowing their pride, at least 23 million Americans stood in food lines last year.

We like the comma before the conjunction *and*, ourselves. We think it makes the sentence easier to read on the first try. However, you will see educated writers like Jones leaving it out.

EXERCISE

Write a sentence describing a scene you know well or specially observe for this exercise. It could be your own room or a scene from nature or an event on the street. Try to string the descriptive details in a series using a model from Jones:

> Cars will line roadsides for a half-mile or more waiting for the boxed monthly buffet of dry milk, rice, cereal, canned fruit and vegetables, instant mashed potatoes and, on good days, canned meat and chicken.

WEB SITES

www.epinet.org/content.cfm/issueguides_minwage_minwage

The Economic Policy Institute's guide to information on the minimum wage includes facts, frequently asked questions, publications, tables, and charts.

www.dol.gov/esa/minwage/america.htm

On the Department of Labor's Web site on minimum wage you can find out the minimum wage in different states.

Student Essay Using Examples

My Key Chain

David C. Lair

During my four years of army service, I led a very transient life. I moved from Illinois to Missouri to California to Texas to Massachusetts to Germany and back to Illinois again, never staying in one place for very long. Consequently, I had to live a very sparse lifestyle with few possessions, and those that I did try to keep fared poorly through all my relocation. (In the army it is said that two or three moves have the same effect on one's belongings as does a house fire). Therefore, when trying to think of a possession that has been significant to me personally, my choices are narrowed to only those items that I have been able to carry on my person. Of these items, I believe that my key chain says more about myself and my life than anything else does. 1

Upon entering the service I soon learned that my eventual duty station, after I completed training at various posts in the states, would be somewhere in Germany. At this time I bought a key chain decorated with an Imperial Eagle and the inscription "Deutschland" (which means Germany); it also came equipped with a handy bottle opener. This key chain was important because it came at a time when I was looking forward to being stationed in Germany. My training was long and mentally arduous, and the key chain served as a reminder of my goal. Whenever I felt discouraged, I pulled out my key chain and thought of traveling around Germany, learning the language, meeting the people, and drinking the beer, which I had heard great things about. I already had big plans for the bottle opener. 2

When I finally arrived in Germany, my outlook on life changed, and so did the "function" of my key chain. Now I lived life for the present. I traveled, learned the language, absorbed the culture, and sampled as many brands of beer as I could find. Along with this transition, the duties of my key chain became more based in the present. Now the keys on the chain represented my 3 "home" in the city of Fulda as I wandered around the continent; now the inscription "Deutschland" made sense to me linguistically; and now the bottle opener was my most important tool. I will always look back on this time as a very happy period in my life.

4 My tour in Germany ended last December, and I returned to the states to begin my new life as a student. But the transition wasn't easy. I found that my mind often dwelled on my former lifestyle; I also found myself missing Germany. Once again, my key chain mirrored my state of mind. By this time the metal around the opener was rusted, the "Deutschland" insignia was scratched, and cracks had begun to form along the entire length of the chain. I frequently looked at my beloved belonging and remembered the fun I had had. As I sprung another cap from an imported beer, I realized that I was now living in the past.

5 Recently, while opening a beer, my key chain broke in half. The stress of opening all the beers finally drove the cracks completely through the key chain. This development caused me to reflect on my present situation. I decided that it was time to stop living in the past, and to start looking toward my future once again.

6 Over Christmas break my girlfriend and I will travel to Brazil to visit her parents, who live in Rio de Janiero. I am very excited about the trip. I am looking forward to exploring a new country once again and to meeting new people. When she heard that my key chain had broken, my girlfriend gave me a small present: a key chain with the Brazilian flag and the inscription "Brazil." Sure, there's no bottle opener, but we have "twist-tops" in the United States anyway. Now I can once again take my key chain out of my pocket and anticipate the future, while at the same time enjoying the present.

CONSIDERING CONTENT AND METHOD

1. How important is the title? Is it too simple and direct? Would you suggest a more intriguing title?
2. What general claim is the author defending and developing in this essay? Does he support his claim adequately?

3. How does the key chain function as an example throughout the essay? What changes does the key chain undergo? What do these changes exemplify?
4. What details are most effective? Are there any points you would like to hear more about?
5. What do you think of the ending? How well does it sum up the whole essay? Could it be concluded in a different way?

Chapter 5

Strategies for Clarifying Meaning: *Definition* and *Explanation*

"That's an excellent prescreened question, but before I give you my stock answer I'd like to try to disarm everyone with a carefully rehearsed joke."

"PowerTalk gives you a single mailbox icon for all incoming and outgoing mail—including fax, voice, electronic mail, and documents. Communication from online services and electronic mail from various sources are routed to your desktop mailbox when you install mail gateways supplied by the vendor," the manual for our new computer operating system cheerfully brags. Sounds great, but we have only the vaguest idea what PowerTalk *is!* How many of the terms in those two sentences would *you* need to have defined or explained? Moreover, how many terms would need definition and explanation for a reader in the 1970s?

THE POINT OF DEFINITION AND EXPLANATION

As suggested by the quotation from the computer manual and by the cartoon preceding this chapter introduction, the whole point of **definition** and explanation is to clarify things for people. Words are worse than unhelpful when they don't convey meaning but instead create confusion.

The special vocabulary of any group needs explanation when used to communicate with people outside that group. Some groups develop languages that are incredibly mysterious to outsiders. If you've ever been the outsider listening to a bunch of bridge players, computer gamers, or aerobic class addicts, you know the feeling of bafflement that grows quickly into boredom because you have no idea what they're discussing so enthusiastically. The same thing can happen when you read or write essays that don't define or explain as much material as necessary.

Some words can be defined briefly in parentheses, but complex ideas and terms need more than a phrase or sentence of explanation. In this chapter, you will read whole essays whose main purpose is to define and explain difficult or controversial concepts.

THE PRINCIPLES OF DEFINITION AND EXPLANATION

Definitions and explanations make use of a few basic techniques that can be found in almost all writing.

Descriptive Details

Can you imagine explaining anything without using descriptive details? "Why are you afraid you'll become a slave to your new silk shirt?" a friend might ask you. To explain, you would naturally give the **details** of caring for it: you must wash it in cold water by hand; then, instead of wringing it, you should wrap it in a clean towel to blot excess water; then it needs to hang dry on a padded hanger away from sun or electric light; when dry, it will be crinkled and must be ironed with a very cool iron; finally, on the day of wearing, it should be steamed while you shower and lightly re-ironed. In other words, you think you will be a servant to it.

Examples

Concepts are frequently defined or explained through examples. The classic Type A personality can be defined through examples of behavior: If Jake is Type A, he is extremely impatient in a long grocery line, he cannot find the time to go to a film starring his favorite actor, and he considers getting a promotion at work a life-or-death issue. He is likely to slam doors and throw books when angry.

In an essay in this chapter, Isaac Asimov uses himself and his garage mechanic as examples to show how subjective people's definitions of intelligence are.

Narration

Aesop's fables, which you probably remember from childhood, make use of narrative, or story, to explain a basic truth about humans. A "sour grapes" attitude is explained by the story of the fox who sees a bunch of delicious-looking grapes but can't reach them; the fox, therefore, comforts itself by deciding that they were probably sour anyway. A writer might define an abstract quality, such as heroism or courage, by telling a story with a hero who acts out the virtue.

In his essay in this chapter, Wayson Choy uses brief narratives to illustrate the "ugly, unjust" prejudice endured by his parents when they immigrated to British Columbia from China in 1918.

Comparison

Sometimes we define a word by emphasizing its likeness to something else. A **fable** is a kind of story, for example. Writers use imaginative comparison to clarify meaning, too: "A true friend is a port in a storm" and "A true friend is as comfortable as an old shoe."

A comparison may be used for surprising effect. When the phrase "inner child" first became popular, it struck us as a remarkably apt description of the juvenile traits that lurk within every adult: fun-loving, self-centered, impulsive, dependent. These parts of ourselves are easily comparable to children, and the phrase captures the idea well.

Contrast

Sometimes the best way to define something is to contrast it with something different. This technique can be quite simple: *dearth* is the opposite of *plenty*. Or the definition can consist of a comparison spiked with a contrast.

As our friend heard her five-year-old explain to her four-year-old, "Death is like going to Omaha, only longer."

In some cases, contrast is needed to correct a common misconception about a term. *Schizophrenia,* contrary to popular belief, does not involve having more than one personality. And being *educated,* according to Asimov's garage mechanic, doesn't mean being "smart."

THE PITFALLS OF DEFINITION AND EXPLANATION

When you write a definition or explanation, you risk making certain characteristic mistakes.

Missing Your Audience

Would you explain what a control key is to a computer whiz? We hope not. But if you were writing directions for a beginning word processing course, you'd be foolish *not* to define what a control key is. Analyzing your intended **audience** is important in all writing. If you misjudge what your audience needs to have defined or explained, you will either insult or confuse them.

Going in Circles

Some definitions are called circular because they don't go anywhere. They restate rather than explain. "A *smooth operator* is a person who functions without roughness," for example, tells you nothing. The second part of that definition only rewords the first part. The definition must lead somewhere, like this: "A *smooth operator* is a person who takes advantage of others by using charm and persuasion."

Abstraction

The previous circular definition also demonstrates the flaw of abstraction. It has no **concrete words** to hang onto. A high-flown sentence like "Our ideal leader is the hope of the future" provides no helpful terms about qualities we can actually identify.

Leaving Information Out

To say merely that a fable is a story doesn't get across the whole idea; it conveys only part of it: a fable has special features, like talking animals and a moral lesson. Incompleteness is a pitfall especially of short definitions, such as "Love is never having to say you're sorry." In this chapter, the essay "Mommy, What Does 'Nigger' Mean?" shows that a sentence like "*Nigger* is a degrading label for an African American" is not a full definition.

Of course, an explanation is incomplete when your reader needs to ask for clarification. Sometimes it's hard to see the holes in what you have written yourself because you already know what you're trying to say. You need to enlist a good peer editor to point out whether you have omitted important information.

WHAT TO LOOK FOR IN DEFINITIONS AND EXPLANATIONS

Here are some guidelines to follow as you study the selections in this chapter.

1. *Focus on which term or concept is being defined or explained.* This focus will help you evaluate the essay's effectiveness.
2. *Identify the ways in which the writer develops the idea,* especially through the use of details, narration, example, comparison, and contrast.
3. *Figure out who the intended audience is.* Then think about how the essay would be different if written for a different audience. What definitions and explanations would be added or deleted?
4. *Ask yourself whether the definition or explanation is complete.* If it is not complete, does the writer tell you why?

IMAGES AND IDEAS

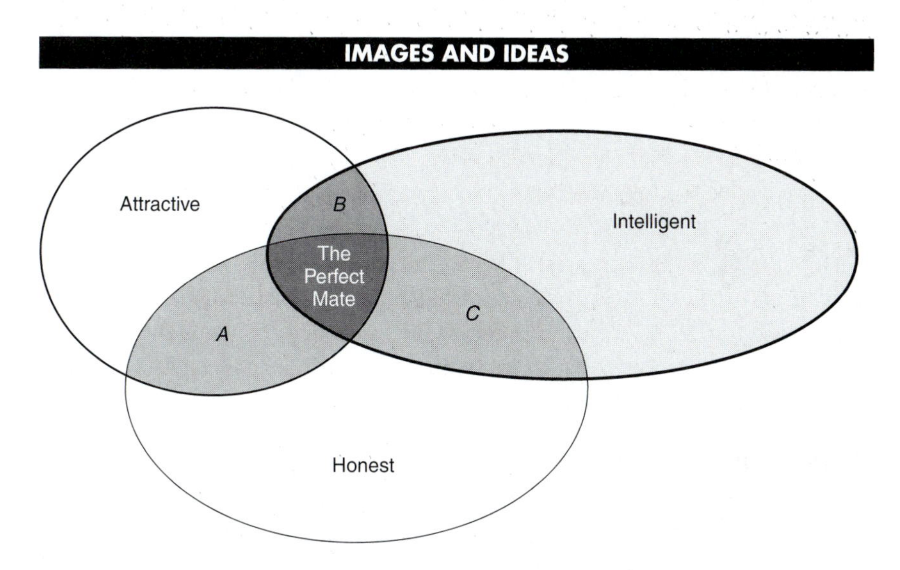

For Discussion and Writing

The diagram above shows the intersection of three qualities that together (in the central overlapping section) might define the perfect mate. Do you know people who fit into the other three overlapping sections (A, B, and C) but not into the central one? Why would they probably not be perfect mates?

You may not agree with the choice of qualities. Or, you may not agree with the sizes of the ovals, which were made in proportion to the qualities' importance. Make up your own circle diagram showing three or four qualities which, if you could find them all together, would define your perfect mate. Write a short explanation of why you chose the qualities and proportions you did.

Alternatively, you could make a circle diagram of qualities that define "Mates to Avoid at All Costs." That one could also include "Attractive."

Preparing to Read

What word could be used to label you? Consider even labels that you do not think accurate. Might someone call you a jock? a nerd? an egghead? a bimbo? a tramp? a hero? a male chauvinist? a rabid feminist? a spoiled brat? a bully? a heartbreaker?

"Mommy, What Does 'Nigger' Mean?"

Gloria Naylor

Gloria Naylor came from a rural, working-class southern background and did not enter college until she was twenty-five, after working as a missionary and a telephone operator. Naylor graduated from Brooklyn College in 1981, and in 1983 she won the American Book Award for best first novel for The Women of Brewster Place. *In this excerpt from an article first published in the* New York Times *on February 20, 1986, Naylor discusses the ways a word's meaning changes according to the context in which it is used.*

TERMS TO RECOGNIZE

necrophiliac *(para. 1)*	someone sexually attracted to corpses
verified *(para. 1)*	proven true
gravitated *(para. 2)*	moved toward
mecca *(para. 2)*	a place regarded as the center of interest or activity
inflections *(para. 3)*	tones of voice
trifling *(para. 8)*	shallow, unimportant
connotation *(para. 9)*	the idea suggested by a word or phrase, in addition to its surface meaning
stratum *(para. 12)*	level
internalization *(para. 12)*	making other people's attitudes a part of your own way of thinking

I remember the first time I heard the word "nigger." In my third-grade class, our math tests were being passed down the rows, and as I handed the papers to a little boy in back of me, I remarked that once again he 1 had received a much lower mark than I did. He snatched his test from me and spit out that word. Had he called me a nymphomaniac or a necrophiliac, I couldn't have been more puzzled. I didn't know what a nigger was, but I knew that whatever it meant, it was something he shouldn't have called me. This was verified when I raised my hand, and in a loud voice repeated what he had said and watched the teacher scold him for using a "bad" word. I was later to go home and ask the inevitable question that every black parent must face—"Mommy, what does 'nigger' mean?"

2 And what exactly did it mean? Thinking back, I realize that this could not have been the first time the word was used in my presence. I was part of a large extended family that had migrated from the rural South after World War II and formed a close-knit network that gravitated around my maternal grandparents. Their ground-floor apartment in one of the buildings they owned in Harlem was a weekend mecca for my immediate family, along with countless aunts, uncles, and cousins who brought along assorted friends. It was a bustling and open house with assorted neighbors and tenants popping in and out to exchange bits of gossip, pick up an old quarrel, or referee the ongoing checkers game in which my grandmother cheated shamelessly. They were all there to let down their hair and put up their feet after a week of labor in the factories, laundries, and shipyards of New York.

3 Amid the clamor, which could reach deafening proportions—two or three conversations going on simultaneously, punctuated by the sound of a baby's crying somewhere in the back rooms or out on the street—there was still a rigid set of rules about what was said and how. Older children were sent out of the living room when it was time to get into the juicy details about "you-know-who" up on the third floor who had gone and gotten herself "p-r-e-g-n-a-n-t!" But my parents, knowing that I could spell well beyond my years, always demanded that I follow the others out to play. Beyond sexual misconduct and death, everything else was considered harmless for our young ears. And so among the anecdotes of the triumphs and disappointments in the various workings of their lives, the word "nigger" was used in my presence, but it was set within contexts and inflections that caused it to register in my mind as something else.

4 In the singular, the word was always applied to a man who had distinguished himself in some situation that brought their approval for his strength, intelligence, or drive:

5 "Did Johnny really do that?"

6 "I'm telling you, that nigger pulled in $6,000 of overtime last year. Said he got enough for a down payment on a house."

7 When used with a possessive adjective by a woman—"my nigger"—it became a term of endearment for husband or boyfriend. But it could be more than just a term applied to a man. In their mouths it became the pure essence of manhood—a disembodied force that channeled their past history

of struggle and present survival against the odds into a victorious statement of being: "Yeah, that old foreman found out quick enough—you don't mess with a nigger."

8 In the plural, it became a description of some group within the community that had overstepped the bounds of decency as my family defined it. Parents who neglected their children, a drunken couple who fought in public, people who simply refused to look for work, those with excessively dirty mouths or unkempt households were all "trifling niggers." This particular circle could forgive hard times, unemployment, the occasional bout of depression—they had gone through all of that themselves—but the unforgivable sin was lack of self-respect.

9 A woman could never be a "nigger" in the singular, with its connotation of confirming worth. The noun "girl" was its closest equivalent in that sense, but only when used in direct address and regardless of the gender doing the addressing. "Girl" was a token of respect for a woman. The one-syllable word was drawn out to sound like three in recognition of the extra ounce of wit, nerve, or daring that the woman had shown in the situation under discussion.

10 "G-i-r-l, stop. You mean you said that to his face?"

11 But if the word was used in a third-person reference or shortened so that it almost snapped out of the mouth, it always involved some element of communal disapproval. And age became an important factor in these exchanges. It was only between individuals of the same generation, or from an older person to a younger (but never the other way around), that "girl" would be considered a compliment.

12 I don't agree with the argument that use of the word "nigger" at this social stratum of the black community was an internalization of racism. The dynamics were the exact opposite: the people in my grandmother's living room took a word that whites used to signify worthlessness or degradation and rendered it impotent. Gathering there together, they transformed "nigger" to signify the varied and complex human beings they knew themselves to be. If the word was to disappear totally from the mouths of even the most liberal of white society, no one in that room was naïve enough to believe it would disappear from white minds. Meeting the word head-on, they proved it had absolutely nothing to do with the way they were determined to live their lives.

13 So there must have been dozens of times that the word "nigger" was spoken in front of me before I reached the third grade. But I didn't "hear" it until it was said by a small pair of lips that had already learned it could be a way to humiliate me. That was the word I went home and asked my mother about. And since she knew that I had to grow up in America, she took me in her lap and explained.

....

RESPONDING TO READING

Do you agree that Naylor's relatives used the word *nigger* in a nonracist or counterracist way? Use your own reactions to the word to explain your answer.

GAINING WORD POWER

The following words appear in Gloria Naylor's essay. They have an ending, *-tion,* in common. This ending occurs frequently in our language, making nouns out of verbs: for example, the noun *conversation* comes from the verb *converse,* which means "to talk." Use your dictionary to find the verb behind each of the nouns in the following list, and give a brief definition of the verb. Begin each brief definition with the word *to.* The first one is done.

	Verb	Definition
connotation	connote	to imply or suggest
degradation		
description		
generation		
inflection		
internalization		
recognition		
situation		

Now see how it works the other way around. Here is a list of verbs that are used in Naylor's essay. Find the *-tion* noun that comes from each verb (some are obvious, but for others in the list you will need the dictionary). Be sure to copy the spelling of the noun exactly: sometimes the first part of the word will change. Then write a brief definition of the noun you have written. We did the first one for you.

	Noun	Definition
apply	application	a form to be filled out
consider		
determine		
explain		
gravitate		
humiliate		
migrate		

	Noun	**Definition**
punctuate		
realize		
receive		
repeat		
transform		
verify		

CONSIDERING CONTENT

1. What are Naylor's background and social class? How do you know? Why are they significant to the main point of the essay?
2. Why was the child puzzled when the boy called her *nigger,* even though she had heard the word before?
3. List the definitions of the word *nigger* that were used in the author's grandparents' apartment.
4. Reread paragraph 12. How did the African American community's uses of *nigger* make it not racist, according to Naylor?
5. What does the last sentence suggest about the author's mother? What does it suggest about America?

CONSIDERING METHOD

1. What kind of a little third-grader was Gloria Naylor? Why do you think she presents her childhood self as no angel? What effect does this early presentation of self have on the reader?
2. Reread the description of the author's grandparents' apartment. List at least seven words or phrases that appeal to your senses.
3. How is paragraph 12 different from the rest of the essay? How does the appearance of the page prepare you for this difference?
4. The central part of the essay includes many direct quotations. Why do you think Naylor used people's exact words so often?
5. Does Naylor envision her **audience** as primarily black or primarily white or mixed? Explain.

WRITING STEP BY STEP

Following Gloria Naylor's essay as an example, write a paper discussing a term that has various—even contradictory—meanings, depending on context: who says it, when it is used, and whether it is applied to men or women, young or old, individual or group, for example. Like Naylor, you might think of a term, such as *wife, macho, liberal, success, feminist, jock,* or *marriage,* that holds more meaning than you once thought.

A. Before you start your essay, brainstorm and jot down the various ways you have heard your term used. Ask friends and classmates for help if you need it. Remember that your readers probably have their own definitions of the term.
B. See whether your definitions fall into groups. For example, can you separate the negative meanings from the positive ones? Or do different meanings belong to different social, racial, or ethnic groups? Try at least to decide on a reasonable order in which to present your definitions. For example, the most widely used meanings might come first, with rarer and rarer meanings following it until the last one is the rarest.
C. Begin your essay with an **anecdote,** as Naylor does.
D. Launch into the body of your essay by looking back into your past. Write about how you used to think of the term, perhaps as a child.
E. Develop each meaning of the term with an explanation and examples of direct quotations using it. Keep looking at how Naylor develops her meanings.
F. Close the body of your essay with a speculation about why the term has such a variety of meanings. Look at Naylor's paragraph 12 for ideas about how to present your thoughts.
G. Use the last paragraph to reflect on the anecdote you used in the opening of your essay. This is how Naylor concludes her piece. Your discussion of the term has, by now, added a new dimension to the anecdote.

OTHER WRITING IDEAS

1. What is the first memory you have of someone saying something purposely to hurt you? Or a memory of your saying something purposely to hurt someone else? Write an essay about your own and others' response to the harsh words.
2. Collaborative Writing. Think of a phrase or word that seems to be plain but is actually used to mean many different things; for example, "Just a minute," "Well," "I'm ready," and "I'll call you," are more slippery than they seem. A group of students in your class can have some fun discussing such sayings and developing ideas for a single report or individual essays.
3. Using the Internet. Go to **www.poynter.org**; click on Diversity on the topic list; from the Diversity page, click on Diversity Tip Sheets. You will find a list of current articles about race. Locate one

or two that concern the language used to describe race and write an essay about the complexities of labeling racial groups.

4. WRITING ABOUT READING. Naylor rejects the idea that use of the word *nigger* by African Americans indicates an "internalization of racism." Look at her essay again, and think about the ways her definition of the word *nigger* supports or weakens her position. Then write your own analysis of the issue, drawing on your experiences, observations, and reading.

EDITING SKILLS: HYPHENS

Consider the use of the hyphen—the short dash—in these phrases from Gloria Naylor's essay:

third-grade class	one-syllable word
close-knit network	third-person reference
ground-floor apartment	

The hyphenated adjectives, which modify the nouns that come after them, are called *temporary compounds.* They are compound because they consist of two words; they are temporary because the two words usually exist separately. They are hyphenated because they make up a unit that seems more like one adjective than two. Sometimes you need to hyphenate in order to make your meaning clear. Think about these examples:

an Italian art specialist	an Italian-art specialist
a small auto dealer	a small-auto dealer
a comic book approach	a comic-book approach

In the second list, the hyphen shows which words go together—the specialist is not Italian, the dealer is not small, and the approach is not comic.

Sometimes, it is difficult to decide whether to hyphenate two words that frequently occur together (back-seat driver), whether to run them together as a compound word (backseat driver), or whether to leave them separate (back seat driver). There is not wide agreement even among professional writers about some of these blends. You will see "Thank you" and "Thank-you" about equally often. Your dictionary should be your guide to making the decision.

EXERCISE

Look up the following combinations to see whether words are hyphenated, compound, or separate. Write out the form your dictionary endorses.

back seat driver
happily married couple
girl friend problems
back door business deals
child like expression
an easy pick up

open and shut case
least restrictive environment
part way finished
part time worker
better fitting word
half baked plan

Look back on the essay you have just written. Did you use any compound words that needed hyphens? Did you leave out any hyphens? Check a dictionary or with your instructor before making any corrections.

WEB SITE

http://voices.cla.umn.edu/newsite/authors/NAYLORgloria.htm

Voices from the Gap—a site maintained by the University of Minnesota—provides a biography and other information on Gloria Naylor.

PREPARING TO READ

If you took an intelligence test and scored low, how would you feel? What might you say that the test failed to measure about you?

What Is Intelligence, Anyway?

ISAAC ASIMOV

Isaac Asimov (1923–1992) was an American, born in Russia, who wrote more than 200 books, including children's stories, popular science, science fiction, fantasy, and scholarly science. With so many intellectual accomplishments, he was well qualified to wonder just what intelligence is—and is not.

TERMS TO RECOGNIZE

KP *(para. 1)*	kitchen patrol (working in the kitchen)
complacent *(para. 2)*	self-satisfied
bents *(para. 2)*	interests, tendencies
oracles *(para. 3)*	divine communications
devised *(para. 4)*	made up
foist *(para. 4)*	impose
arbiter *(para. 4)*	judge
indulgently *(para. 6)*	as if doing a favor
raucously *(para. 6)*	loudly, in a disorderly way
smugly *(para. 6)*	in a self-satisfied way

What is intelligence, anyway? When I was in the Army, I received a kind of aptitude test that all soldiers took and, against a normal of 100, scored 160. No one at the base had ever seen a figure like that, and for two hours they made a big fuss over me. (It didn't mean anything. The next day I was still a buck private with KP as my highest duty.) 1

All my life I've been registering scores like that, so that I have the complacent feeling that I'm highly intelligent, and I expect other people to think so, too. Actually, though, don't such scores simply mean that I am very good at answering the type of academic questions that are considered worthy of answers by the people who make up the intelligence tests—people with intellectual bents similar to mine? 2

For instance, I had an auto repairman once, who, on these intelligence tests, could not possibly have scored more than 80, by my estimate. I always took it for granted that I was far more intelligent than he was. Yet, when anything went wrong with my car, I hastened to him with it, watched him anxiously as he explored its vitals, and listened to his pronouncements as though they were divine oracles—and he always fixed my car. 3

Well then, suppose my auto repairman devised questions for an intelligence test. Or suppose a carpenter did, or a farmer, or, indeed, almost anyone but an academician. By every one of those tests, I'd prove myself a moron. And I'd *be* a moron, too. In a world where I could not use my academic training and my verbal talents but had to do something intricate or hard, working with my hands, I would do poorly. My intelligence, then, is not absolute but is a function of the society I live in and of the fact that a small subsection of that society has managed to foist itself on the rest as an arbiter of such matters. 4

Consider my auto repairman, again. He had a habit of telling me jokes whenever he saw me. One time he raised his head from under the automobile hood to say, "Doc, a deaf-and-dumb guy went into a hardware store to ask for some nails. He put two fingers together on the counter and made hammering motions with the other hand. The clerk brought him a hammer. He shook his head and pointed to the two fingers he was hammering. The clerk brought him nails. He picked out the sizes he wanted, and left. Well, doc, the next guy who came in was a blind man. He wanted scissors. How do you suppose he asked for them?" 5

Indulgently, I lifted my right hand and made scissoring motions with my first two fingers. Whereupon my auto repairman laughed raucously and said, "Why, you dumb jerk, he used his *voice* and asked for them." Then he said, smugly, "I've been trying that on all my customers today." "Did you catch many?" I asked. "Quite a few," he said, "but I knew for sure I'd catch *you*." "Why is that?" I asked. "Because you're so goddamned educated, doc, I *knew* you couldn't be very smart." 6

And I have an uneasy feeling he had something there. 7

RESPONDING TO READING

Do you think that people who are academically intelligent are often poor at nonacademic things? Why or why not?

GAINING WORD POWER

Complete these sentences in a reasonable way, showing that you understand the vocabulary words included.

1. Troy indulgently promised to take Rachel ____________________
   ____.
2. Because he had a mechanical bent, Mark specialized in __________
   ____.
3. ______________________________, the crowd cheered raucously.
4. Dr. Morse foisted his ideas on us when he ____________________
   ____.
5. By setting himself up as the arbiter of good taste in clothes, Sheldon
   ____.

CONSIDERING CONTENT

1. Why do you think that Asimov's high intelligence score "didn't mean anything" in the army?
2. What is the basic conflict discussed in this essay?
3. Why was the repairman sure he would catch "Doc" Asimov with the joke? Was the repairman stereotyping college professors?
4. What might be on an intelligence test written by auto mechanics? carpenters? farmers? parents of preschoolers? portrait painters? What point is Asimov making with this type of suggestion?
5. Which "small subsection" (para. 4) of society has set itself up as the definers of intelligence?
6. What do you think about Asimov's "uneasy feeling"? Do you think the mechanic "had something there"? Or do you think the nature of the joke would fool most people, intelligent or not?

CONSIDERING METHOD

1. Asimov's essay contains two anecdotes (look up the word **anecdote** if you have not already done so). How do the anecdotes help you understand the conflict the writer is discussing?
2. In a fitting reflection of the content of the essay, Asimov uses both highly academic words (*oracles, bents, arbiter*) and common, informal words. Point out some of the informal words.
3. Count the number of words in each of the six sentences that make up paragraph 4. Notice the variety in sentence length. This variation

is a factor that makes writing lively instead of plodding. When you think your own writing sounds plodding, check to see whether you have a good range of sentence lengths. If they are too much alike, try to combine some of the shorter sentences into longer ones or divide long ones into shorter ones.

4. The last paragraph consists of only one sentence. What difference do you see between placing it in this position and simply adding it as a closing sentence to the preceding paragraph?
5. How do you think academics would respond to this essay? Was Asimov writing to them?

Combining Strategies

In this seven-paragraph essay exploring a definition, how many paragraphs are devoted to developing an example?

WRITING STEP BY STEP

Write an essay that investigates the meaning of an **abstract** term—an idea that cannot be directly observed, such as intelligence. You can use some of Asimov's techniques. We also suggest some other methods for developing an extended definition. Use any combination of techniques to develop your definition. To get your thinking started, consider these abstractions:

common sense	humility	educated
optimism	professionalism	cool
courage	resourcefulness	simplicity
generosity	street smart	beauty
laziness	foolish	stubbornness

Choose your own term, but be sure your choice is an abstraction (not something **concrete,** such as *submarine, pizza,* or *tennis*).

A. Begin with the question "What is ---------, anyway?"
B. If you have an anecdote that will serve as an extended investigation of the term, include the story, as Asimov did.
C. Use examples of people you know who demonstrate the abstraction you are defining. Show how they behave or think, with specific details.

D. Use description. For example, an abstract feeling may have a counterpart in a certain type of landscape or weather that you could describe, such as "fluffy white clouds" to suggest peace and contentment or "loud thunder and hard-driving rain" to indicate anger.

E. Point out differences. A good definition will make clear the difference between the word you are explaining and other words with similar meanings. You might tell how feeling peaceful is different from feeling happy or calm.

F. Provide a contrast. We often explain words by clarifying what they are *not.* For example, you could point out that laziness does not merely include lying around doing nothing, which could be depression instead.

G. In your closing, let the reader know to what extent the abstraction is important in your own life.

OTHER WRITING IDEAS

1. Write an essay similar to Asimov's on questioning the usual definition of a term. Challenge the term's usefulness or its usual meaning or everyday misuse.
2. USING THE INTERNET. Make up a word for a concept or item that doesn't have a name, as far as you know. These new words, called *sniglets,* are sometimes collected in humorous books. Here are some examples:

   BEAVO (n.): a pencil with teeth marks all over it.

   FICTATE (v.): to inform a television or screen character of impending danger under the assumption that he or she can hear you.

   OPUP (v.): to push one's glasses back up on the nose.

   At **www.almavijai.sphosting.com/Sniglets/sniglets.html**, you can view cartoons made up for sniglets. You may need to make up a related group of words and explain how you came up with them in order to write a whole essay. For example, you could make up several sniglets that describe your pet's behaviors.
3. COLLABORATIVE WRITING. Instead of presenting a unified definition of a concept, try to write an essay presenting all the different, even contradictory, meanings a certain term could have. For example, if you asked a group of people to explain what *sex appeal* means to them, you would probably get some widely varying answers. Work with a group of classmates to come up with meanings and to draft a group definition.
4. WRITING ABOUT READING. What's your response to Asimov's opening question: "What is intelligence, anyway?" Write an essay in which you define what intelligence means to you, both in yourself

and in the people you know. Use the various methods of developing a definition that are discussed in the introduction to this chapter.

EDITING SKILLS: USING THE RIGHT TENSE

One of the trickiest skills we learn as we grow up is using the correct tense for verbs. If you have ever learned a foreign language, you are fully aware of the complications of verb tense. In your native language, as you speak, you almost always choose the right tense, but in writing you can get in a snarl.

Asimov's essay shows a combination of tenses. For things that occur in the present or are true in the present, he uses plain old present tense: "I *have* the complacent feeling that I'*m* highly intelligent, and I *expect* other people to think so, too" (para. 2, italics added). For the story about the mechanic, which happened in the past, Asimov uses past tense: "Indulgently, I *lifted* my right hand and *made* scissoring motions with my first two fingers" (para. 6).

In telling stories, speakers and writers sometimes get confused as they go along, and they switch from past to present and back within the same anecdote. Haven't you heard someone tell a story this way:

> You know, if you wait too long, you have to go in person to deal with the licensing bureau. Well, yesterday I went to get my driver's license renewed. So I have to wait in line for an hour. I finally get to the front, and the guy tells me I should have brought my form that I got in the mail. So now I had to go home, and it took my whole lunch hour. When I get back, I tell the guy that if I starve to death, it's his fault. I hate these heartless bureaucracies.

The speaker gives you a sense of being there by switching to present tense, but in academic writing you usually stick to the past tense for events that took place in the past. If you're writing about what happened in a film or literary work, however, you can summarize plot in the present tense.

EXERCISE

The following paragraph mixes present and past tenses. Rewrite the paragraph, and put all the verbs into the present tense:

> *Spenser: For Hire* is based on the detective novels of Robert Parker and is several steps above the average television crime series. It is well acted and well worth watching. Robert Ulrich proved that he is an excellent actor in the part of the tough, well-read private investigator Spenser. He is believable when he fought the bad guys, and he was believable when he quotes Shakespeare. But when the script called for him to be cute and kit-

tenish with his girlfriend, Susan, he is not believable. A cute Spenser was simply embarrassing. In the books, Spenser and Susan tease and traded playful insults, but they are not kittenish. The TV show needed to fix that part of their relationship.

Now look at the verb tenses in the essay you wrote in response to the Asimov article. Which tenses did you use? Are they consistent and logical? Check over your writing, and make any needed corrections in the verb tenses.

WEB SITE

www.thomasarmstrong.com/multiple_intelligences.htm

Describes alternate ways of thinking about intelligence.

Preparing to Read

Are you aware of having an ethnic heritage? Are you proud of your cultural roots? If so, why? If you don't know where your ancestors came from, do you feel somehow left out in today's multiethnic society? If so, how do you compensate for the lack?

I'm a Banana and Proud of It

Wayson Choy

Born in 1939 and raised in British Columbia, Wayson Choy is best known for his award-winning novel The Jade Peony, *in which he focuses on his cultural heritage in Vancouver's Chinatown during the first half of this century. A winner of the prestigious Governor General's Literary Award for Non-Fiction, he presently teaches English at Humber College in Toronto. This essay first appeared in the Facts and Arguments column of the* Toronto Globe and Mail.

TERMS TO RECOGNIZE

alien *(para. 2)*	a foreigner, an outsider
concubine *(para. 3)*	a woman belonging sexually to a man not her husband
Taoist *(para. 7)*	pertaining to a Chinese religion that teaches simplicity and selflessness
assimilation *(para. 12)*	cultural absorption of a minority into the main society
paradox *(para. 13)*	an apparent contradiction that turns out to be true

Because both my parents came from China, I took Chinese. But I cannot read or write Chinese and barely speak it. I love my North American citizenship. I don't mind being called a "banana," yellow on the outside and white inside. I'm proud I'm a banana. After all, in Canada and the United States, native Indians are "apples" (red outside, white inside); blacks are "Oreo cookies" (black and white); and Chinese are "bananas." These metaphors assume, both rightly and wrongly, that the culture here has been primarily anglo-white. Cultural history made me a banana. 1

History: My father and mother arrived separately to the British Columbia coast in the early part of the century. They came as unwanted "aliens." 2

Better to be an alien here than to be dead of starvation in China. But after the Chinese Exclusion laws were passed in North America (late 1800s, early 1900s), no Chinese immigrants were granted citizenship in either Canada or the United States.

3 Like those Old China village men from *Toi San* who, in the 1850s, laid down cliff-edge train tracks through the Rockies and the Sierras, or like those first women who came as mail-order wives or concubines and who as bond-slaves were turned into cheaper laborers or even prostitutes—like many of those men and women, my father and mother survived ugly, unjust times. In 1918, two hours after he got off the boat from Hong Kong, my father was called "chink" and told to go back to China. "Chink" is a hateful racist term, stereotyping the shape of Asian eyes: "a chink in the armor," an undesirable slit. For the Elders, the past was humiliating. Eventually, the Second World War changed hostile attitudes toward the Chinese.

4 During the war, Chinese men volunteered and lost their lives as members of the American and Canadian military. When hostilities ended, many more were proudly in uniform waiting to go overseas. Record Chinatown dollars were raised to buy War Bonds. After 1945, challenged by such money and ultimate sacrifices, the Exclusion laws in both Canada and the United States were revoked. Chinatown residents claimed their citizenship and sent for their families. By 1949, after the Communists took over China, those of us who arrived here as young children, or were born here, stayed. No longer "aliens," we became legal citizens of North America. Many of us also became "bananas."

5 Historically, "banana" is not a racist term. Although it clumsily stereotypes many of the children and grandchildren of the Old Chinatowns, the term actually follows the old Chinese tendency to assign endearing nicknames to replace formal names, semicomic names to keep one humble. Thus, "banana" describes the generations who assimilated so well into North American life. In fact, our families encouraged members of my generation in the 1950s and sixties to "get ahead," to get an English education, to get a job with good pay and prestige. "Don't work like me," Chinatown parents said. "Work in an office!" The *lao wahkiu* (the Chinatown old-timers) also warned, "Never forget—you still be Chinese!"

6 None of us ever forgot. The mirror never lied.

7 Many Chinatown teenagers felt we didn't quite belong in any one world. We looked Chinese, but thought and behaved North American. Impatient Chinatown parents wanted the best of both worlds for us, but they bluntly labeled their children and grandchildren "*juk-sing*" or even "*mo no.*" Not that we were totally "shallow bamboo butt-ends" or entirely "no brain," but we had less and less understanding of Old China traditions, and less and less interest in their village histories. Father used to say we lacked Taoist ritual, Taoist manners. We were, he said, "*mo li.*"

This was true. Chinatown's younger brains, like everyone else's of whatever race, were being colonized by "white bread" U.S. family television programs. We began to feel Chinese home life was inferior. We co-operated with English-language magazines that showed us how to act and what to buy. Seductive Hollywood movies made some of us secretly weep that we did not have movie-star faces. American music made Chinese music sound like noise. By the 1970s and eighties, many of us had consciously or unconsciously distanced ourselves from our Chinatown histories. We became bananas. 8

Finally, for me, in my 40s or 50s, with the death first of my mother, then my father, I realized I did not belong anywhere unless I could understand the past. I needed to find the foundation of my Chinese-ness. I needed roots. 9

I spent my college holidays researching the past. I read Chinatown oral histories, located documents, searched out early articles. Those early citizens came back to life for me. Their long toil and blood sacrifices, the proud record of their patient, legal challenges, gave us all our present rights as citizens. Canadian and American Chinatowns set aside their family tongue differences and encouraged each other to fight injustice. There were no borders. "After all," they affirmed, *"Daaih ga tohng yahn. . . .* We are all Chinese!" 10

In my book, *The Jade Peony,* I tried to recreate this past, to explore the beginnings of the conflicts trapped within myself, the struggle between being Chinese and being North American. I discovered a truth: These "between world" struggles are universal. In every human being, there is "the Other"—something that makes each of us feel how different we are from everyone else, even family members. Yet, ironically, we are all the same, wanting the same security and happiness. I know this now. 11

I think the early Chinese pioneers actually started "going bananas" from the moment they first settled upon the West Coast. They had no choice. They adapted. They initiated assimilation. If they had not, they and their family would have starved to death. I might even suggest that all surviving Chinatown citizens eventually became bananas. Only some, of course, were more ripe than others. 12

That's why I'm proudly a banana: I accept the paradox of being both Chinese and not Chinese. Now at last, whenever I look in the mirror or hear ghost voices shouting, "You still Chinese!", I smile. I know another truth: In immigrant North America, we are all Chinese. 13

RESPONDING TO READING

Do you understand why Choy is proud to be a banana? In the last sentence, he says "In immigrant North America, we are all Chinese." How do you interpret this comment? Do you think it applies to you?

GAINING WORD POWER

Wayson Choy uses a number of *-ly* adverbs in his essay: *barely, primarily, separately, eventually, proudly, clumsily, actually, bluntly, totally, entirely, secretly, consciously or unconsciously, ironically, actually, eventually, proudly.* If you look carefully at how these words are used, you will notice that they occur in a number of different places in the sentence. Adverbs are often movable; you can put them at the beginning of the sentence or at the end, before the verb or after—in any number of spots.

Go back to Choy's essay and find six of the *-ly* words just listed above. Rewrite two or three of these sentences to illustrate how the *-ly* adverb can be used in several different places. Here's an example:

Choy's sentence:	I might even suggest that all surviving Chinatown citizens *eventually* became bananas.
Rewrites:	I might even suggest that *eventually* all surviving Chinatown citizens became bananas.
	I might even suggest that all surviving Chinatown citizens became bananas *eventually.*

Now use five of the *-ly* adverbs in sentences of your own. Experiment with several different versions to see where the adverb can be placed.

CONSIDERING CONTENT

1. What does Choy mean when he says he is a "banana"?
2. Why does he think "banana" is not a racist term? Do you agree? What about "apples" and "Oreos" and "white bread"?
3. What were the Chinese Exclusion laws and why were they eventually repealed?
4. What is the origin of the offensive term "chink"?
5. Why did Chinese parents want their children to get "English" educations?
6. What do you think is the main idea of this essay? Where is the thesis stated?

CONSIDERING METHOD

1. How do the first two introductory paragraphs prepare the readers for the rest of the essay?

2. In paragraph 1, Choy uses a **paradox** ("both rightly and wrongly") to explain the metaphors underlying the terms *bananas, apples,* and *Oreos.* Explain how these words can be both right and wrong at the same time.
3. Choy uses a number of short-short sentences: "I'm proud I'm a banana" (para. 1), "They came as unwanted 'aliens'" (para. 2), "None of us ever forgot. The mirror never lied" (para. 6), "We became bananas" (para. 8), "I needed roots" (para. 9), "I know this now" (para. 11), "They had no choice. They adapted. They initiated assimilation" (para. 12). What does he achieve with this rhetorical technique?
4. Why is the phrase "going bananas" in quotation marks in paragraph 12?
5. How do the last three sentences serve to reinforce the theme and also to unify the essay?
6. Do you think you are part of the intended **audience** for this essay? Why or why not?

WRITING STEP BY STEP

In his essay, Wayson Choy focuses his essay on defining a word. For your essay, think of some words or expressions that need defining because they are misleading, unclear, or too indirect—and define them accurately. You might choose words used in TV commercials, personal ads, real estate descriptions, fast-food restaurants, or menu language. You can focus on a single expression (like "family values") or a group of related terms (like the names for sandwiches or sizes of soft drinks or the names of real estate subdivisions or the language of the funeral business).

A. Begin by telling how the word or name came to your attention.
B. Offer the accurate or dictionary meaning.
C. Then explain what the advertiser, salespeople, or promoters of the terminology want it to mean.
D. As you explain the true meaning, explain why the contrived expression was chosen. If the word or name caught on with the public, such as "lite" or "share" (for "tell" or "discuss") or "senior citizen," speculate as to why it became so popular.
E. Conclude by calling for action from your readers—perhaps urging them to protest this obvious attempt to manipulate the public through the misuse of language.

OTHER WRITING IDEAS

1. Choy says the word "banana" is to him an "endearing nickname." Does your family have an affectionate nickname for you that captures your personality? Perhaps the family pets have nicknames that characterize their behavior. Or maybe you have a friend or relative who has an apt nickname. Write an essay explaining how the nicknames define people or pets.
2. COLLABORATIVE WRITING. With a group of three or four classmates, interview several international students on your campus, and ask for examples of expressions (such as "cramp my style" or "hit the hay") that surprised or confused them. How did they find out what the expressions really meant? Write up the results of your interviews.
3. USING THE INTERNET. At **www.crede.ucsc.edu**, you will find the Web site for the Center for Research on Education, Diversity, and Excellence. This organization explores how schools can work best in our multiethnic society. Click on "Five Standards" and you will see how CREDE defines effective pedagogy. Write an essay, based on these definitions, about how your own schooling has succeeded and failed in effectiveness.
4. WRITING ABOUT READING. Do you agree with Choy that "In every human being, there is 'the Other'—something that makes each of us feel how different we are from everyone else, even family members" (para. 11)? For Choy, that "something" was being Chinese in a "white bread" culture. What in your experience makes you feel like "the Other"?

EDITING SKILLS: CAPITALIZATION

Choy's essay illustrates many of the conventions for using capital letters in English. The author, of course, capitalizes the first word in every sentence. He also capitalizes the pronoun *I* every time he uses it. Here are some of the other capitalization rules he follows:

1. *Capitalize the names of nationalities, races, tribes, and languages:* Chinese, North American, Canadian, English, American.
2. *Capitalize the names of political, ethnic, and religious groups:* Communists, Old China traditions, Taoist.
3. *Capitalize brand names of products:* Oreos.
4. *Capitalize names of geographical locations such as cities, states, countries, and regions:* China, B.C., North America, Chinatown, U.S., West Coast.

5. *Capitalize the first and last words and all other words in titles, except articles, prepositions, and coordinating conjunctions* (a, an, the, in, into, at, to, by, and, but, or, nor, for, so, etc.): *The Jade Peony.*
6. *Capitalize the names of important government documents:* Chinese Exclusion laws, War Bonds.

Look over the essay you have just written, and check your use of capital letters. A college dictionary will give you a list of capitalization rules along with examples. Entries for specific words will also tell you when to use capitals.

EXERCISE

Supply capital letters where needed in the following sentences:

1. For years women lobbied in washington, d. c., for passage of the equal rights amendment.
2. At present, mexican americans are the fastest-growing minority in the united states.
3. The asian american population is growing rapidly, too, especially in california.
4. People of the muslim faith, adherents of islam, are also a fast-growing group in large cities like chicago and new york.
5. Our book club is reading ann tyler's dinner at the homesick restaurant.

WEB SITE

http://www.asiancanadian.net/waysonchoy_interview.html

A 2002 interview with Wayson Choy includes interesting chat about his writing process.

PREPARING TO READ

In a theater, do you watch all the credits at the end of a movie? Why or why not?

Silent Responsibility

MICHAEL AGGER

Michael Agger is an editor and writer for the "Goings on about Town" section of the New Yorker *magazine, which features very short articles about topics of interest to people who live in New York City (and sometimes to people who live elsewhere). Agger also reviews films for Microsoft Network (**www.msn.com**). The following article appeared in the October 20, 2003, edition of the* New Yorker.

TERMS TO RECOGNIZE

designation *(para. 1)*	title
countenance *(para. 1)*	facial expression
posits *(para. 1)*	suggests, proposes
denigrating *(para. 2)*	belittling
gaffer *(para. 2)*	electrician in charge of lighting
snakes *(para. 3)*	drags or pulls lengthwise
sight lines *(para. 3)*	area between the actors and the camera lens
gamely *(para. 5)*	with good nature

1 Best boy. You see the credit scrolling up the screen (other people in your row are shuffling their feet impatiently) and, just for a moment, you let yourself wonder about the designation. What comes to mind will probably be wrong. The best boy is not a beardless youth with a saintly countenance, like Billy Budd. He is not the smirking winner of an on-set popularity contest. He does not have an assistant best boy and there is no such thing as a second-best boy, although the credits for the comedy "Airplane II: The Sequel" do include a "worst boy": Adolf Hitler. Even when the best boy is a woman, the title does not change. As for its origins, they're unclear. In Victorian England, many assistants were called boys, as in "Get me another tankard of ale, boy." One theory posits that the name stuck when an English foreman was hiring theatrical laborers and said, "Give me your best boy!"

Steve Comesky, who is the best boy on a remake of "The Stepford Wives," which is now shooting in Connecticut and New York, seems a little uneasy with a title that is at once immodest (best) and somehow denigrating (boy). When people ask him what he does for a living, he says, "I tell them I'm an electrician." Comesky, a tall guy in his mid-thirties with a buzz cut and the shoulders of a wide receiver, says that his job is pretty straightforward: "I do what my boss tells me to do, pal." He works as the chief assistant to the gaffer, otherwise known as the head electrician. 2

Comesky grew up on Long Island, and got into the movie business after graduating from high school. He hangs spots, snakes cable, provides power to the stars' trailers, and helps light the scenery whenever the camera is repositioned for a new shot. He stretches his legs often while on the set; the job involves a lot of standing (always outside the actors' sight lines) and what he calls "silent responsibility"—replacing burned-out bulbs, making sure no one trips over a light cord. 3

When reporters come to visit the set of a movie like "The Stepford Wives," they come to see Nicole, or Bette. They do not, as a rule, come to see the best boy, and when a reporter does, the following things will happen. Everyone in the crew will call the best boy "sir." They will give him a peanut-butter-and-jelly sandwich with "the crusts cut off just the way you like it, sir!" They will ask him to sign papers marked "Very Important Papers." They will thank him "for all the hard work yesterday . . . the orphans couldn't be more pleased with how you saved them." A production assistant will pretend to be his personal masseuse. 4

Comesky gamely endures the practical joking: "I'll be getting this all day, pal." A long stretch of silent responsibility ensues, and then the gaffer asks Comesky to light a mounted rhinoceros head. 5

RESPONDING TO READING

"Silent Responsibility" is very short for a magazine essay. Choose an idea from it that you would like to read more about. What questions would you like to ask?

GAINING WORD POWER

Best boy, gaffer, spots, and *sight lines* are all terms used in movie production. Find five to ten more words or phrases from film production **jargon** (that is, specialized language). Write definitions for each one. List the sources—print or Internet—you used.

CONSIDERING CONTENT

1. The two incorrect **images** that Agger suggests for "best boy" are Billy Budd and "the smirking winner of an on-set popularity contest." Explain each of these images. How are they alike and different?

2. What is "silent responsibility"? Why is it important enough to serve as a title for the short essay?
3. Who are Nicole and Bette (para. 4)? Why do people visit the set to see them? Why do they not come to see the best boy?
4. What did the crew do when Agger came to see Steve Comesky? Why did they behave this way?

CONSIDERING METHOD

1. Do you think the piece should have been titled "Best Boy"? Or do you prefer the title Agger chose? Why?
2. What are some characteristics of readers who would be interested in this essay?
3. What is the final visual image in the essay? Why is this image appropriate to the main point of the piece?

WRITING STEP BY STEP

Review your definition of "silent responsibility." Many jobs involve silent responsibility, and people only notice them when they are *not* done properly, or not done at all. Think of a few occupations like this: they are usually behind-the-scenes jobs, like best boy. *Housewife* comes to mind. Choose a "silent responsibility" job that you have performed or that you know well secondhand. Interview someone who has held this job, if you have not.

A. Begin your essay by giving the job a title and explaining what most people would think on hearing the title. Tell whether this thinking would be accurate, inaccurate, or incomplete. If the job title has an interesting origin, give its background. Agger suggests possible histories of "best boy."
B. Write a paragraph about how the job-holder would describe this work. Use direct quotations if you can, as Agger did. This section should reflect a person's feelings about the job.
C. Write one or more paragraphs about the duties involved in this job. Be detailed and specific. Agger does this in paragraph 3.
D. Write one or more paragraphs setting this job in the context of the other people involved (for example, Agger shows how the crew relates to the best boy in para. 4).
E. Close the essay with a short paragraph focusing on a detail from a typical day on the job, as Agger does with the mounted rhinoceros head.

OTHER WRITING IDEAS

1. USING THE INTERNET. Write an essay about an unusual job, like farrier, sommelier, concierge, celebrity personal assistant, stunt man or stunt woman, forensic accountant, body double, animal psychologist, or hotwalker.
2. Define the jargon terms of a certain job you've had or a hobby you engage in. For example, as writers we have specialized meanings for *galleys, mss., T and E, advance, f&gs, IM, permissions, ancillaries.* To get started, make a list of the terms used at the work site or by people involved in the activity. Then group the terms according to some logical principle to help you divide your paragraphs.
3. COLLABORATIVE WRITING. Every generation develops its own slang. In small groups draw up a list of slang terms currently used in your college community. Then write an essay defining the terms and explaining why they are popular.
4. WRITING ABOUT READING. How would you describe the **tone** of Agger's essay? (Look up the term in the Glossary.) What words and phrases contribute to that tone? Is the tone appropriate for the subject matter? Can you imagine the article being written in a different tone?

EDITING SKILLS: USING COLONS

Copy the following passages exactly.

The credits for the comedy "Airplane II: The Sequel" do include a "worst boy": Adolf Hitler.

Comesky says that his job is pretty straightforward: "I do what my boss tells me to do, pal."

Comesky gamely endures the practical joking: "I'll be getting this all day, pal."

If you copied correctly, you put a colon (:) in each passage. Reread what you copied, and see whether you can come up with a rule about the use of the colon.

You probably noticed that the colon comes after a complete sentence. Go back and reread the first portion of each example. The colon then introduces something that specifies or expands on the sentence before the colon. The second part doesn't have to be a complete sentence, as you see in the Adolf Hitler example. (It doesn't have to be a direct quotation, either.) If you think of the colon as a verbal equal sign (=), you get the main relationship it suggests between the two parts. Neglecting to put a full sentence before a colon, when you are introducing a list, is one of the most frequent mistakes in written English.

EXERCISE

Complete the following passages that include colons.

1. Sami has already bought her party supplies: ________________.
2. Beau's next statement gave away his secret plans: ______________.
3. The plans were not completely reasonable: ________________.
4. ________________: a paperback detective novel, a historical romance, and a Far Side cartoon.
5. ________________: Austin only cried a little while.

Now edit the essay you just wrote, looking for places where you could have used a colon instead of a period or a semicolon. After consulting with your instructor, change the punctuation mark to a colon.

WEB SITE

www.mrqe.com

The Movie Review Query engine is the home of 38,000+ movie reviews. There you can read reviews of the 2004 *Stepford Wives* (the filming of which Michael Agger visited), as well as the original 1975 *Stepford Wives,* and even the 1980 *Revenge of the Stepford Wives.*

Student Essay Using Definition

Nothing to Be Scared Of

Kerri Mauger

For most of my life, I remember my mother being in and out of hospitals. No one was sure what was wrong with her, and it took many years to finally put a label on her. My mother has been diagnosed with schizophrenia. I am sorry to think about when my friends and I have passed a mental hospital on the street and made rude comments, but we, like most people, were very ignorant about mental illness. People may laugh and make jokes about it, but it is a reality to me. I live with it every day of my life. 1

Early on the day of my eleventh birthday party, my sister and I were downstairs watching TV. My mom came down and told us to come up to my room. We followed her upstairs, where she told us to lie on the floor. She said that there was a man outside with a gun and we had to hide so he would not come and get us. I was really scared and confused, and we stayed in my room for two hours in almost complete silence. Finally I got up and looked out the window. Our neighbors were outside gardening. When my father got home, he called the hospital, and I saw my mother taken away by force. My birthday party was canceled, and I was crushed. I realize now that my mother was hallucinating, having an experience that seemed real but was only in her imagination. Hallucinations can include all of the senses, and my mother had both seen and heard things that weren't there. 2

It has taken me many years to understand what is wrong with my mother. My dad chose to be very secretive about anything that had to do with her. I think he tried to protect my sister and me, but in the long run ignorance hurt us even more. 3

Schizophrenia is a mental disorder that makes it hard for people who have it to distinguish between what is real and what is imagined. Therefore, schizophrenics also have trouble managing emotions, thinking clearly, and dealing with other people. They cannot tell what other people's 4 talk and actions mean, and they respond strangely. People who have never encountered someone with schizophrenia may feel intimidated because of bizarre and socially unacceptable behavior. My mother may sit and listen to the voices she hears all day long, ignoring everything else. She explains that they are communicating to her through telepathy. She replies to these voices and denies that there is anything wrong with her. Some schizophrenics shout angrily at people on the street, believing that they have been insulted or threatened when they have not.

5 There is no cure for schizophrenia. Medication may help some people with the symptoms, but no medicine has helped my mother yet. Psychotherapy can assist schizophrenics in controlling their thoughts. Along with drugs and therapy, family understanding can help. The person suffering from schizophrenia needs sympathy, compassion, and respect. It is best to stay calm and nonjudgmental. Getting excited or starting to argue with the person can worsen an episode. Schizophrenia is not the person's moral fault or rational choice.

6 It is sometimes easier to laugh at something we fear and do not understand. But ignorance about something like schizophrenia can be hurtful. I love my mother dearly and after many years of being afraid and uninformed, I can finally say I am not scared of her at all. She still is not a "normal" mother, but I know the facts and can deal with whatever comes my way.

CONSIDERING CONTENT AND METHOD

1. Why does the author mention her friends and herself and the rude comments they made when passing a mental hospital?
2. Did you follow the explanation of *schizophrenia?* Do you understand the term better now that you've read this essay? Are there any questions about the disease that the author does not answer?
3. Why does the author include the incident about her birthday party (para. 2)?
4. How does the conclusion echo the introduction?
5. What is the point of this essay? Point out a sentence that states that point. How is that point reinforced in the closing?

Chapter 6

Strategies for Sorting Ideas: *Classification* and *Division*

Hanging Loose *Hanging Tough* *Just Hanging In There*

Source: © The New Yorker Collection 1988 by Edward Frascino. Reprinted with permission of The Cartoon Bank from The New Yorker Collection, cartoonbank.com. All Rights Reserved.

The next time you get ready to do your laundry, look on the back of the detergent box. There you will see directions for sorting your clothes according to the temperature of the water that you should use. The directions may read something like this:

FOR BEST CLEANING RESULTS
Sort and select temperature and begin filling washer with water.

Hot	**Warm**	**Cold**
White cottons	Bright colors	Dark colors
Colorfast pastels	Permanent press	Colors that could bleed

Hot	**Warm**	**Cold**
Diapers	Knits	Delicates
Heavily soiled items		Stains like blood and chocolate

Do you follow a procedure like this when you do your laundry? Why do you suppose the detergent makers put directions such as these on the back of the box?

THE POINT OF CLASSIFICATION AND DIVISION

As the preceding laundry example shows, separating and arranging things helps us to accomplish tasks more efficiently and more effectively. This process of "sorting things out" also helps us to clarify our thinking and understand our feelings.

Many writing tasks lend themselves to grouping information into categories. For example, you might write a paper for psychology class on the various ways people cope with the death of a loved one. For a course in economics you might write about the three basic types of unemployment (*frictional, structural,* and *cyclical*). This approach—called **classification** and **division**—enables you to present a body of information in an orderly way.

Dividing and classifying also forces you to think clearly about your topic. By breaking a subject down into its distinct parts, or categories, you can look at it more closely and decide what you want to say about each part. For example, if you are writing an essay on effective teaching styles, you might begin by dividing the teachers you have had into the "good" ones and the "bad" ones. Then you could break those broad categories down further into more precise ones: teachers who held your interest, teachers who knew their subjects, teachers who made you do busy work, and so forth. As you develop each category, you have to think about the qualities that impressed you, and this thought process leads you to a better understanding of the topic you are writing about. You may end up writing only about the good teachers, but dividing and classifying the examples will help you to organize your thinking.

THE PRINCIPLES OF CLASSIFICATION AND DIVISION

Most things can be classified or divided in more than one way, depending on the reason for grouping the items. In the laundry example, for instance, you are told to sort the clothes according to the temperature of the water. Putting bright colors with knits doesn't make any sense if you don't know the reason—or basis—for the category: *items to be washed in cold water.* When you divide and classify a topic, be sure to have a sound basis for formulating your categories. The following suggestions will help you develop a useful system of classification.

Give a Purpose to Your Classification

Merely putting facts or ideas into different groups isn't necessarily meaningful. Consider these sentences that classify for no apparent reason:

1. There are five kinds of friends in most of our lives.
2. People deal with their spare money in four basic ways.

How could you give a **purpose** to these ideas? You would have to add a *reason* or declare a *point* for the categories. Here are some revisions that give purpose to these two classifications:

1. a. There are five kinds of friends in most of our lives, and each kind is important in its own way.
   b. Most of us have five kinds of friends, and each one drives us crazy in a special way.
2. a. People deal with their spare money in four ways that reflect their overall attitudes toward life.
   b. People deal with their spare money in four ways, only one of which is truly constructive.

Establish a Clear Basis for Your Classification

If you are going to classify items or ideas, you have to decide which organizing principle you want to use to form your groups. For example, if you are classifying friends, you could group them according to how close you are to them, as Judith Viorst does in her essay in this chapter. Or you could group them according to other principles: how long you have known them, what you have in common with them, or how much time you spend with the people in each group. The important thing is to choose a workable principle and stick with it.

Make Your Groups Parallel and Equal

In a classification essay, you usually announce the groups or classes in the introduction: several ways to handle criticism, four kinds of friends, three types of stress, three levels of intelligence, and so forth. You then devote one section to each group of your classification. A section can be one paragraph or several, but the sections for the major categories should be about equal in length. In a classification essay about friendship, for instance, if you cover "childhood pals" in 150 words, then you should use approximately the same number of words for each of the other kinds of friends.

Experienced writers also present each section in a similar way. For example, in this chapter psychologist David Elkind classifies the three basic types of stress that young people experience. He labels each one with a letter, describes it, and gives **examples.** He follows the same order—label, description, examples—in covering each type. He also mentions the same two conditions when explaining each type: whether it's foreseeable and whether it's avoidable. This parallel development helps the reader to recognize the similarities and identify the distinctions among the types.

THE PITFALLS OF CLASSIFICATION AND DIVISION

The following advice will help you to avoid some of the problems that writers sometimes encounter in developing an essay of classification.

1. *Know the difference between useful and useless ways of classifying.* Sorting your clothes by brand name probably won't help you get the best cleaning results when doing the laundry. And dividing teachers into those who wear glasses and those who don't is not a very useful way to organize a paper on effective teaching styles. But classifying teachers into those who lecture, those who use a question-discussion format, and those who run small-group workshops might be significant, mainly because such groupings would allow you to discuss the teachers' philosophies, their attitudes toward students, and their effectiveness in the classroom.

2. *Be sure your classification covers everything you claim it covers.* If, for instance, you know some teachers who sometimes lecture and sometimes use small groups, you can't pretend these people don't exist just to make your classification tidy. At least mention exceptions, even if you don't give them as much space as the major categories.

3. *Don't let the basis of division shift.* If you can see a problem with the following classification system, you already understand this warning:

TYPES OF TEACHERS

A. Teachers who lecture
B. Teachers who lead discussions
C. Teachers who have a sense of humor
D. Teachers who run workshops
E. Teachers who never hold office hours

Notice that three types of teachers (a, b, and d) are grouped according to the way they run their classes, but two types (c and e) are defined by some other standard. You can see the confusion these shifting groups cause: Can teachers who lecture have a sense of humor? Don't those who use workshops also hold office hours?

4. *Be sure your groups are parallel or equal in rank.* The following classification illustrates a problem in rank:

KINDS OF POPULAR MUSIC

A. Easy listening
B. Country and western
C. Rock 'n' roll
D. Ice-T

Although Ice-T does represent a type of popular music distinct from easy listening, country, and rock, the category is not parallel with the others—it is far too small. It should be "rap" or "hip hop," with Ice-T used as an example.

5. *Avoid stereotypes.* When you write about types of behavior or put people into groups, you run the risk of oversimplifying the material. The best way to avoid this problem is to use plenty of specific examples. You can also point out exceptions and describe variations; such honesty shows that you have been thinking carefully about the topic.

WHAT TO LOOK FOR IN A CLASSIFICATION

As you read the essays in this chapter, pay attention to these points:

1. *Figure out what the author is classifying.* Then identify the basis for making up the groups and the purpose of the classification.
2. *Look for the specific groups or classes* into which the author has sorted the material. Jot down a brief list of the major categories just to see if you can keep track of them.
3. *Ask yourself if the groups are clearly defined.* Do they shift? Do they cover what they claim to cover? Are they parallel?
4. *Be alert for stereotypes.* How does the author handle exceptions and variations?
5. *Identify the audience.* How do you know who the intended readers are?

IMAGES AND IDEAS

Source: PhotoEdit.

For Discussion and Writing

Look at the students in this class photo. With your classmates, predict what will become of each one. Can you group the kids in the photo into types according to what kind of future they'll probably have? If you looked at your own class photo from high school, could you group the people into types? Did the types remain consistent as they got older, or did some people end up differently from what you expected? Using division and classification, write an essay about the class photo we provide here, or about your real or imagined class photo from your own school days.

Preparing to Read

List five or ten of the people you call "friends." Are they all friends to an equal degree? Do you have a flexible meaning for the word *friend*?

Friends, Good Friends—and Such Good Friends

Judith Viorst

A popular humorist who writes essays and light verse for many well-known magazines, Judith Viorst was born in Newark, New Jersey, in 1931 and attended Rutgers University. She has written numerous children's books, including the enduring favorite Alexander and the Terrible, Horrible, No Good, Very Bad Day *(1972), as well as several books for adults, including* Imperfect Control *(1998) and* Suddenly 60: And Other Shocks of Late-Life *(2000). The following selection appeared in her regular column in* Redbook *magazine in 1977.*

TERMS TO RECOGNIZE

ardor *(para. 1)*	enthusiasm, intensity
nonchalant *(para. 2)*	indifferent, offhand
Tuesday-doubles *(para. 7)*	tennis played by four people, in pairs, on Tuesdays
sibling rivalry *(para. 11)*	competition among children for parental favor
dormant *(para. 13)*	sleeping, inactive
revived *(para. 13)*	brought back to life
calibrated *(para. 21)*	adjusted, determined

Women are friends, I once would have said, when they totally love and support and trust each other, and bare to each other the secrets of their souls, and run—no questions asked—to help each other, and tell harsh truths to each other (no, you can't wear that dress unless you lose ten pounds first) when harsh truths must be told. Women are friends, I once would have said, when they share the same affection for Ingmar Bergman, plus train rides, cats, warm rain, charades, Camus, and hate with equal ardor Newark and Brussels sprouts and Lawrence Welk and camping. 1

In other words, I once would have said that a friend is a friend all the way, but now I believe that's a narrow point of view. For the friendships I have and the friendships I see are conducted at many levels of intensity, serve many different functions, meet different needs and range from those as all-the-way as the friendship of the soul sisters mentioned above to that of the most nonchalant and casual playmates. 2

Consider these varieties of friendship: 3

1. Convenience friends. These are the women with whom, if our paths weren't crossing all the time, we'd have no particular reason to be friends: a next-door neighbor, a woman in our car pool, the mother of one of our children's closest friends or maybe some mommy with whom we serve juice and cookies each week at the Glenwood Co-op Nursery. 4

Convenience friends are convenient indeed. They'll lend us their cups and silverware for a party. They'll drive our kids to soccer when we're sick. They'll take us to pick up our car when we need a lift to the garage. They'll even take our cats when we go on vacation. As we will for them. But we don't, with convenience friends, ever come too close or tell too much; we maintain our public face and emotional distance. "Which means," says Elaine, "that I'll talk about being overweight but not about being depressed. Which means I'll admit being mad but not blind with rage. Which means that I might say that we're pinched this month but never that I'm worried sick over money." But which doesn't mean that there isn't sufficient value to be found in these friendships of mutual aid, in convenience friends. 5

2. Special-interest friends. These friendships aren't intimate, and they needn't involve kids or silverware or cats. Their value lies in some interest jointly shared. And so we may have an office friend or a yoga friend or a tennis friend or a friend from the Women's Democratic Club. 6

"I've got one woman friend," says Joyce, "who likes, as I do, to take psychology courses. Which makes it nice for me—and nice for her. It's fun to go with someone you know and it's fun to discuss what you've learned, driving back from the classes." And for the most part, she says, that's all they discuss. "I'd say that what we're doing is *doing* together, not being together," Suzanne says of her Tuesday-doubles friends. "It's mainly a tennis relationship, but we play together well. And I guess we all need to have a couple of playmates." I agree. 7

My playmate is a shopping friend, a woman of marvelous taste, a woman who knows exactly *where* to buy *what,* and furthermore is a woman who always knows beyond a doubt what one ought to be buying. I don't have the time to keep up with what's new in eyeshadow, hemlines and shoes and whether the smock look is in or finished already. But since (oh, shame!) I care a lot about eyeshadow, hemlines and shoes, and since I don't *want* to wear smocks if the smock look is finished, I'm very glad to have a shopping friend. 8

3. Historical friends. We all have a friend who knew us when . . . maybe way back in Miss Meltzer's second grade, when our family lived in that three-room flat in Brooklyn, when our dad was out of work for seven months, when our brother Allie got in that fight where they had to call the police, when our sister married the endodontist from Yonkers, and when, the morning after we lost our virginity, she was the first, the only, friend we told. 9

The years have gone by and we've gone separate ways and we've little in common now, but we're still an intimate part of each other's past. And so whenever we go to Detroit we always go to visit this friend of our girlhood. Who knows how we looked before our teeth were straightened. Who knows how we talked before our voice got unBrooklyned. Who knows what we ate before we learned about artichokes. And who, by her presence, puts us in touch with an earlier part of ourself, a part of ourself it's important never to lose. 10

"What this friend means to me and what I mean to her," says Grace, "is having a sister without sibling rivalry. We know the texture of each other's lives. She remembers my grandmother's cabbage soup. I remember the way her uncle played the piano. There's simply no other friend who remembers those things." 11

4. Crossroads friends. Like historical friends, our crossroads friends are important for *what was*—for the friendship we shared at a crucial, now past, time of life. A time, perhaps, when we roomed in college together; or worked as eager young singles in the Big City together; or went together, as my friend Elizabeth and I did, through pregnancy, birth and that scary first year of new motherhood. Crossroads friends forge powerful links, links strong enough to endure with not much more contact than once-a-year letters at Christmas. And out of respect for those crossroads years, for those dramas and dreams we once shared, we will always be friends. 12

5. Cross-generational friends. Historical friends and crossroads friends seem to maintain a special kind of intimacy—dormant but always ready to be revived—and though we may rarely meet, whenever we do connect, it's personal and intense. Another kind of intimacy exists in the friendships that form across generations in what one woman calls her daughter-mother and her mother-daughter relationships. Evelyn's friend is her mother's age—"but I share so much more than I ever could with my mother"—a woman she talks to of music, of books, and of life. "What I get from her is the benefit of her experience. What she gets—and enjoys—from me is a youthful perspective. It's a pleasure for both of us." 13

I have in my own life a precious friend, a woman of 65 who has lived very hard, who is wise, who listens well; who has been where I am and can help me understand it; and who represents not only an ultimate ideal mother to me but also the person I'd like to be when I grow up. 14

In our daughter role we tend to do more than our share of self-revelation; in our mother role we tend to receive what's revealed. It's another kind of pleasure—playing a wise mother to a questing younger person. It's another very lovely kind of friendship. 15

6. Part-of-a-couple friends. Some of the women we call our friends we never see alone—we see them as part of a couple at couples' parties. And though we share interests in many things and respect each other's views, we aren't moved to deepen the relationship. Whatever the reason, a lack of time or—and this is more likely—a lack of chemistry, our friendship remains in the context of a group. But the fact that our feeling on seeing each other is always, "I'm *so* glad she's here" and the fact that we spend half the evening talking together says that this too, in its way, counts as a friendship. 16

(Other part-of-a-couple friends are the friends that came with the marriage, and some of these are friends we could live without. But sometimes, alas, she married our husband's best friend; and sometimes, alas, she *is* our husband's best friend. And so we find ourself dealing with her, somewhat against our will, in a spirit of what I'll call *reluctant* friendship.) 17

7. Men who are friends. I wanted to write just of women friends, but the women I've talked to won't let me—they say I must mention man-woman friendships too. For these friendships can be just as close and as dear as those that we form with women. Listen to Lucy's description of one such friendship: "We've found we have things to talk about that are different from what he talks about with my husband and different from what I talk about with his wife. So sometimes we call on the phone or meet for lunch. There are similar intellectual interests—we always pass on to each other the books that we love—but there's also something tender and caring too." 18

In a couple of crises, Lucy says, "he offered himself, for talking and for helping. And when someone died in his family he wanted me there. The sexual, flirty part of our friendship is very small, but *some*—just enough to make it fun and different." She thinks—and I agree—that the sexual part, though small, is always *some,* is always there when a man and a woman are friends. 19

It's only in the past few years that I've made friends with men, in the sense of a friendship that's *mine,* not just part of two couples. And achieving with them the ease and the trust I've found with women friends has value indeed. Under the dryer at home last week, putting on mascara and rouge, I comfortably sat and talked with a fellow named Peter. Peter, I finally decided, could handle the shock of me minus mascara under the dryer. Because we care for each other. Because we're friends. 20

8. There are medium friends, and pretty good friends, and very good friends indeed, and these friendships are defined by their level of intimacy. And what we'll reveal at each of these levels of intimacy is calibrated with care. We might tell a medium friend, for example, that yesterday we had a 21

fight with our husband. And we might tell a pretty good friend that this fight with our husband made us so mad that we slept on the couch. And we might tell a very good friend that the reason we got so mad in that fight that we slept on the couch had something to do with that girl who works in his office. But it's only to our very best friends that we're willing to tell all, to tell what's going on with that girl in his office.

22 The best of friends, I still believe, totally love and support and trust each other, and bare to each other the secrets of their souls, and run—no questions asked—to help each other, and tell harsh truths to each other when they must be told. But we needn't agree about everything (only 12-year-old girl friends agree about *everything*) to tolerate each other's point of view. To accept without judgment. To give and to take without ever keeping score. And to *be* there, as I am for them and as they are for me, to comfort our sorrows, to celebrate our joys.

RESPONDING TO READING

How many different kinds of friends do you have? How would you categorize the friends you listed in Preparing to Read? Would Viorst's categories work for you? Would you have to create any new ones?

GAINING WORD POWER

The following partial sentences include slightly different forms of the preceding Terms to Recognize. Complete each sentence in a reasonable way that shows you understand the word used. Use your dictionary to be sure you grasp the meaning.

1. Elena is an ardent hockey fan, but ______________________________
   ____.
2. Stanley did not understand the fine calibrations of politeness: he often
   ____.
3. After ______________________, Jean's social life went into a period of dormancy.
4. The writer's nonchalance about spelling was clear when ________
   ____.

CONSIDERING CONTENT

1. What ideas is Viorst giving up, according to the beginning of the essay? Why does she begin with what she no longer believes?

2. Make a list of the different kinds of friends discussed in the essay. Which kind is the exception from the others?
3. What kind of person is the author of this essay? What kind of life does she lead? Look at paragraphs 4, 5, and 11 for evidence. How might the writer's **point of view** limit her audience? Did you feel included?
4. What are the functions friends perform, according to the essay? Are there underlying similarities among most of the types?
5. Why do you think the women Viorst interviewed insisted that she discuss man-woman friendships?

CONSIDERING METHOD

1. How do you know when Viorst is beginning a section about a new type of friend? What other method might she use to signal such a shift?
2. What is the relationship between the opening of the essay and the closing?
3. What do the direct quotations do for the essay? Are they from famous people? Think of at least two reasons why they are included.
4. In paragraph 7, you can see the repetition of similar sentence openings. What words are repeated? Find another example of repetition of phrases or structures. What purpose does this repetition serve?

Combining Strategies

Look at the ways Judith Viorst develops her explanations for each type of friend. She uses narration, definition, example, comparison and contrast, and cause-and-effect reasoning. Identify instances of at least three of these methods.

WRITING STEP BY STEP

Using Viorst's essay for inspiration, write an essay that classifies several different types of romantic relationships.

A. Brainstorm, with help from friends or classmates if you want it, for ideas about the various couples and kinds of romance you see around you. Jot down everything mentioned, even if it sounds silly or overlapping—you can sort it out later.

B. Choose three to six types to discuss in your essay. When the types seem closely related, you might combine several different ones from your brainstorming notes to make up one type.

C. Make up a label for each type, as Viorst did in "Friends, Good Friends—and such Good Friends."

D. Develop a section explaining each type. Use **examples** of people who show each type of romantic attachment if you can—either people you know personally or public figures most of your readers will be able to identify. Interview your friends, as Viorst did, and use direct quotations from your interviews.

E. Be sure to show how the types are similar and different from each other by using comparisons and contrasts as you go along. Look at paragraphs 13 and 21 of Viorst's essay for samples of how to do this.

F. Use signals to show when you are shifting from one type to the next. You could use numbers and/or labels. Or you could use transitional words, like "The second type or style of romance is. . . ."

G. Write an introduction that discusses a misperception (of your own or of others) about romantic relationships. See Viorst's introduction for a model.

H. Close your essay with a statement that draws all the kinds of romance together, telling what they have in common. Alternatively, you may want to close, as Viorst did, by commenting on what you wrote in the introduction.

OTHER WRITING IDEAS

1. There are probably as many kinds of people you dislike as people you like. Write an essay categorizing the types that drive you crazy.
2. COLLABORATIVE WRITING. Judith Viorst wrote another essay in which she classifies the lies we tell. For example, she discusses lies that spare others' feelings, lies that save our own pride, and so on. Get together with a group of classmates to think of your own categories and examples of lies and liars. Have each person in the group list as many types and illustrations as he or she can; compare and discuss your lists. Then write your own essay on this topic.
3. USING THE INTERNET. Classify the variations of a single emotion: anxiety, nervousness, anger, pleasure, embarrassment, confidence, excitement, or pride, for example. Use the search term "emotional literacy" to find Web sites that might help you explore the topic. Remember to acknowledge your sources in your essay.

4. WRITING ABOUT READING. Viorst says that "the sexual part, though small, is always *some,* is always there when a man and a woman are friends." Do you agree? What do you think about friends of the other sex? Are cross-sex friendships different in quality from same-sex friendships? Would you try to restrict your sweetheart in his or her opposite-sex friendships?

EDITING SKILLS: USING PRONOUNS CONSISTENTLY

When Judith Viorst writes about her own experiences and ideas, she uses the pronouns *I* or *we* to identify herself:

> The friendships *I* have and the friendships *I* see are conducted at many levels of intensity.

> *We* all have a friend who knew *us* when. . . .

When she writes as someone addressing an audience directly, she uses the pronoun *you:*

> [N]o, *you* can't wear that dress unless *you* lose ten pounds first.

And when she writes about a person or persons, she identifies them first and then uses *he, she,* or *they* to refer to them:

> Convenience friends are convenient indeed. *They*'ll lend us *their* cups and silverware for a party.

> Evelyn's friend is *her* mother's age, . . . a woman *she* talks to of music, of books, and of life.

As you see, pronouns refer to three different "persons." Each person expresses a different point of view:

Person	**Singular**	**Plural**
First person (the person writing):	I, me, my, mine	we, us, our, ours
Second person (the person written to):	you, your, yours	you, your, yours
Third person (the person written about):	he, him, his, she her, hers, it, its	they, them, their, theirs

Writers often shift from one person to another, especially between first and third, as Judith Viorst does in her essay. But these shifts must be clear and logical. An unnecessary shift can distract or confuse a reader:

Shift: If a student wants to succeed in graduate school, you have to know the rules of the game.

The sentence is about *a student,* so it's in third person; *you* is a second person pronoun.

No shift: If a student wants to succeed in graduate school, he or she has to know the rules of the game.

No shift: If students want to succeed in graduate school, they have to know the rules of the game.

EXERCISE

Rewrite the following paragraph, making the point of view consistently *third person* (people/they). You may have to change some verbs to fit with the pronouns you changed.

Most people are interested in music, either as a spectator or as performers. You can enjoy music by watching MTV or attending concerts. We can also enjoy playing CDs and listening to music on the radio. Other people want to make their own music. If we are really serious about playing an instrument or singing, we can take lessons and join a band or choir. You might prefer, however, to play for your own enjoyment or to entertain your friends and family at parties. A person has many chances to express his love of music.

WEB SITE

www.annonline.com/interviews/980112/index.html

This site offers an interview with and information about Judith Viorst.

Preparing to Read

For the next twenty-four hours, make a note every time you hear someone describe another person's behavior (or idea or talk) as weird or strange or crazy. Also record your ideas about why the behavior was pointed out as different from normal.

I'm OK; You're a Bit Odd

Paul Chance

Paul Chance, a psychologist and teacher at Salisbury State University in Maryland, sits on the advisory board for the Cambridge Center for Behavioral Studies. He writes frequently about human behavior for various professional and educational journals. In this article, first published in 1988 in Psychology Today *magazine, Chance investigates the ways we decide whether other people are mentally healthy.*

Terms to Recognize

quirk *(para. 1)*	minor oddity
Platonic ideal *(para. 2)*	the perfect form, according to the Greek philosopher Plato
emulate *(para. 2)*	imitate
psychopath *(para. 8)*	seriously disturbed person who has aggressive, criminal tendencies
flimflam *(para. 8)*	swindle
paragon *(para. 13)*	ideal example
phobias *(para. 13)*	unreasonable and bothersome fears
benchmark *(para. 14)*	a standard level for judging quality or quantity

The new groom was happy with his bride, and everything, he explained, was fine. There was just this one peculiarity his wife had. During lovemaking she insisted that he wear his motorcycle helmet. He found it uncomfortable, and he felt just a tad foolish. Is it normal to want someone to wear a helmet during amorous activities? Does a quirk of this sort keep one off the rolls of the mentally fit? The answer depends on how you define mental fitness. There are several ways of going about it. 1

One model calls to mind the Platonic ideal. Somewhere in the heavens there exists a person who is the perfect specimen of psychological health. (Or maybe there are two of them: the perfect man may be different from the perfect woman. At least, one would hope so.) We all fall short of this ideal, of course, but it provides a model that we can emulate. Unfortunately, the Platonic answer merely begs the question, since somebody has to describe what the ideal is like. And how do we do that? 2

The everyday way of defining mental health is more subjective: if I do it, it's healthy; if I don't do it, it's sick. Is it crazy to spend Saturdays jumping out of airplanes or canoeing down rapids? Not to skydivers and white-water canoers. Is it sick to hear voices when no one is there? Not if you're the one who hears the voices—and you welcome their company. 3

This commonsense way of defining mental health sets ourselves up as the standard against which to make comparisons. There's nothing wrong with this, except that it's just possible that some of us—not me, you understand—are a bit odd ourselves. And you can't measure accurately with a bent ruler. 4

The psychodynamic model of mental health suggests that psychological fitness is a kind of balancing act. There are, according to this view, impulses in all of us that society cannot tolerate. The healthy person is not the one who always keeps these impulses under lock and key but the one who lets them out once in a while when nobody's looking. If you run around the house smashing delicate things with a hammer, for example, someone's apt to object. But if you hammer a nail into a board, and seem to have a good excuse for it, nobody minds. So the healthy person with violent impulses builds a deck behind the house. 5

The chief problem with the psychodynamic model is that it doesn't define the standard by which balance is to be measured. Building a deck may be an acceptable outlet for violent impulses, but what if every time a person feels like slugging someone he adds on to the deck until his entire backyard is covered in redwood? He's directing his impulses constructively, but his family might find him easier to live with if he just broke something once in a while. 6

Behaviorists offer a different solution. They focus on behavior, naturally, and decide whether behavior is healthy on the basis of its consequences. If the results are good, the behavior is good. In this view, there is nothing nuts about building a two-acre redwood deck so long as the person enjoys it and it doesn't get him or her into trouble. Normal behavior, then, is whatever works. 7

The behavioral approach appears to offer an objective and rational way of defining mental health. Alas, appearances are deceiving. We may agree that a person enjoys an activity, but is that enough? A sadist and a masochist may work out a mutually rewarding relationship, but does that make them healthy? A psychopath may flimflam oldsters out of their life savings and 8 do it with such charm that they love him for it, but should the rest of us emulate the psychopath?

An alternative is to let society decide. What is healthy then becomes what society finds acceptable; what is unhealthy is whatever society dislikes. Thus, aggression is abnormal among the gentle Tasaday of the Philippines but normal among the fierce Yanomamo of Venezuela. 9

The societal model has a lot of appeal, but it troubles some mental-health workers. There is something about fixing mental health to a mailing address that they find unsettling. They think that there ought to be some sort of universal standard toward which we might all strive. Besides, does it really make sense to say that murder and cannibalism are OK just because some society has approved them? And if it is, then why not apply the same standards to communities within a society? Murder is a popular activity among Baltimore youth. Shall we say that, in that city, murder is healthy? 10

A similar problem exists with the statistical model of mental health. In this case, being mentally healthy means falling close to average. Take the frequency of sexual activity among married couples, for example. Let's say that, on average, married people your age have intercourse about twice a week. That's the norm. If you indulge more or less often than that, you're abnormal—with or without a helmet. 11

There's some logic to this view. The further people deviate from the average, the more likely they are to seem strange. You may think, for instance, that limiting sex to two times a week is a bit prudish, but almost everyone is likely to think that once an hour is excessive. Again, however, there are problems. 12

Does it really make sense to hold up the average person as the paragon of mental health? This logic would have everyone cultivate a few phobias just because they happen to be commonplace, even the best students would strive to earn Cs, and all couples in the country would be frustrated by their inability to have exactly 1.8 children. 13

We can all agree that there are a lot of weirdos around, but there seems no way for us to agree about who's weird. And so there's no way for us to agree about what mental health is. That's unfortunate, because it gives us no clear goal toward which to strive and no stable benchmark against which to gauge our progress. Even so, I'm damned if I'm gonna wear a helmet to bed. 14

RESPONDING TO READING

Return to your notes from Preparing to Read. What explanations for branding something as weird are covered in Chance's essay? Does the essay change your perspective on the incidents you noted?

GAINING WORD POWER

Weird is an overused and imprecise word. Some **synonyms** from the thesaurus are *uncanny, puzzling, inexplicable, bizarre, nightmarish, irrational,* and *freakish.* Look up each of these synonyms in the dictionary. Write one sentence for each of the seven words. Make your sentences reflect the differences in meaning among the terms.

CONSIDERING CONTENT

1. Who is the **audience** for this essay? How do you know? Who is *not* the audience for the piece? Why?
2. How many approaches to defining mental health or normality does Chance cover? List them.
3. Give **examples** of your own that fit at least two of Chance's models of mental health.
4. Which model of mental health, in your opinion, would label you and your friends most mentally abnormal? Consider your family: What approach would label your family as strange?
5. In paragraph 2, Chance suggests that the ideal of psychological health would be different for men and women. Do you agree? Why or why not?
6. Where does the title of the essay come from? If you don't know, look up the first two words of the title in your library's online or card catalog. How is the essay related to the original source of the title?

CONSIDERING METHOD

1. Under each approach to defining mental health, how does Chance organize his discussion? As you were reading, when did this organization become clear to you?
2. Does each approach include a specific example? Why are the examples important? Would you have liked another example of one of the points?
3. Count how many questions Chance asks in the essay. Thinking of your other reading, is this a weird number of questions? What purpose do you think they serve?
4. What is the **tone** of the piece? For example, what tone of voice would someone use to read it out loud? Give several words that describe the tone. What clues you in to the tone?
5. Reread the last sentence. What is your reaction to it? What technique does it employ to indicate that the essay is finished?

WRITING STEP BY STEP

When we are faced with one of life's problems, we often find ourselves brainstorming to discover approaches to the problem and then considering the strengths and weaknesses of each approach. We ponder, for example, the problem of how to make housemates or family members share cleaning duties, the problem of how to make scarce money go further, or the problem of how to break unwelcome news to someone. Choose a problem that could be considered in several ways, and write an essay using Chance's discussion of approaches to mental health as a model.

A. Identify three or four approaches to the problem. On a piece of scratch paper or on your computer screen, brainstorm for a few minutes, jotting down ideas about each approach. If you need help, ask your classmates for ideas about your topic.

B. Decide on a reasonable order to present your approaches—from weakest to strongest (or vice versa), from your first hunch to your last solution, from the most ordinary to the most bizarre, or along any range that seems reasonable.

C. Choose a tone that suits your **purpose** and your **audience.** In choosing a tone, you might also consider the severity of the problem—a serious discussion of approaches to bad hair days could provide a humorous mismatch, for example. Paul Chance, in a similar way, chose a light tone to discuss a serious topic. However, a match between the topic and the tone is the usual expectation, and you should feel free to align them if you want: a humorous tone for a light topic, and a serious tone for a weighty topic.

D. Begin the essay with a little story that highlights the problem, as Chance does. Lead into a general statement of the problem.

E. Write a one- or two-paragraph discussion of your first approach. Begin with a description of the approach.

F. Give the strengths of the approach and provide an example of how it is (or would be) used.

G. Follow the strengths by discussing the weaknesses of the approach along with an example of its failings. You can, as Chance does, use the same example for both strengths and weaknesses, or if you prefer, you can use different examples to show positive and negative elements.

H. Follow steps E, F, and G for the next two or three approaches you identified in the brainstorming stage.

I. In your closing paragraph, you might follow Chance's model and write about why no approach is perfect. On the other hand, you might explain which approach you think is best, if there is one.

J. Write a closing sentence that relates back to the little story in your introduction.

OTHER WRITING IDEAS

1. Choose either a school setting or a workplace you know well, and write about the different approaches people take toward their work. Consider using specific people you know or have read about as examples of the approaches. Be sure to give enough examples and details for your reader to get a clear picture.
2. COLLABORATIVE WRITING. Working with a group of classmates, think of a quality that many people have but that they use in different ways—for example, power, beauty, charm, intelligence, or wealth. Develop an essay that explains three or four approaches to possessing this quality. Choose only one quality and explain three or four displays of it: "Some people use their power for good causes, some for control over others, and some to amass even more power." Consider evaluating each approach or explaining why people differ along these lines.
3. USING THE INTERNET. Write an essay about plagiarism. Are there different kinds? Different motives and causes of it? For help on this topic, do a search for Web sites that discuss plagiarism and how to avoid it.
4. WRITING ABOUT READING. The author offers six models or approaches for defining mental health or normalcy. Describe how each one works and explain what problem or limitation each one has. Which ones make the most sense to you?

EDITING SKILLS: CHOOSING *THERE, THEIR,* OR *THEY'RE*

Words that sound alike, called **homophones,** can be treacherous. In speech, we don't think about which one to select, but in writing we have to make a choice. Because homophones are often common words, such as *there, their,* and *they're,* we can't avoid using them. And because the meanings are quite different, we really do have to get them right or our readers will be confused.

First, look at the use of the word *there* in these sentences from "I'm OK; You're a Bit Odd":

> *There* was just this one peculiarity his wife had.
>
> *There* are, according to this view, impulses in all of us that society cannot tolerate.

In these sentences *there* is just a fill-in word; it doesn't have much meaning. We use *there* to begin sentences like these when we want to say that something or someone exists. The same spelling goes for the use of the word to indicate a place:

> Is it sick to hear voices when no one is *there*?

Now look at the use of *their* in these passages from the same essay:

> Not if you're the one who hears the voices—and you welcome *their* company.
>
> A psychopath may flimflam oldsters out of *their* life savings.

The word *their* is a possessive, showing ownership or belonging. In the first example it refers to voices; in the second example, it refers to "oldsters."

Finally, take a look at this sentence:

> Behaviorists would say that if people enjoy an action and it doesn't get them in trouble, *they're* normal.

In this example, *they're* sounds just like the others, but you can substitute the words *they are* for it. *They're* is a contraction, just like *don't* for *do not* and *she's* for *she is.* If you can substitute *they are,* you know you have the right word. If you can't, you need *there* or *their.*

All right, let's review:

> *There* are numerous sound-alike words in English.
>
> You have to know the differences in *their* meanings if you want to use them correctly.
>
> *They're* often common words that we use a lot.

EXERCISE

Fill the blanks in the following sentences with *there, their,* or *they're.*

1. ________________ are no solutions to this problem.
2. Your cousins have arrived, and ________________ going to stay for dinner.
3. When George Eliot and Jane Austen wrote ________________ great novels, ________________ were few creative opportunities for women besides writing.

4. The team members should have known that ________________ luck wouldn't hold. But ________________ definitely looking forward to next season.
5. Although it wasn't ________________ fault, the twins are sure to get blamed, and ________________ not happy about it.

Look over the essay you have just written to see if you used *there, their,* or *they're.* Did you choose the right one? Go over some past papers, too, to see how accurate you are in using these three homophones. Copy three to five sentences in which you use one, and justify your choice in each case. (For more about homophones, see Chapter 4, pp. 95–96.)

WEB SITE

www.behavior.org/columns

Read more articles and commentaries by Paul Chance at the Cambridge Center for Behavioral Studies.

Preparing to Read

How can you tell when someone has good self-esteem? How do people with good self-esteem handle life's problems?

Types of Stress for Young People

David Elkind

David Elkind is a psychologist who specializes in child development and has written influential books on the subject. In this reading from his book All Grown Up and No Place to Go *(1988), he describes three kinds of situations that we all face but that are especially stressful for teenagers.*

Terms to Recognize

expenditure *(para. 3)*	cost
potential *(para. 4, 12)*	possible
foreseeable *(para. 4)*	seen ahead of time
transient *(para. 7)*	temporary, passing away with time
jeopardize *(para. 7)*	threaten
incorporates *(para. 10)*	combines
inevitable *(para. 12)*	certain to happen
introverted *(para. 14)*	withdrawn, not sociable
extroverted *(para. 14)*	outgoing
bouts *(para. 15)*	short periods

Most situations that produce psychological stress involve some sort of conflict between self and society. So long as we satisfy a social demand at the expense of a personal need, or vice versa, the social or personal demand for action is a psychological stress. If, for example, we stay home from work because of a personal problem, we create a new demand (for an explanation, for made-up time) at our place of work. On the other hand, if we devote too much time to the demands of work, we create new demands on the part of family. If we don't manage our energy budgets well, we create more stress than is necessary. 1

The major task of psychological stress management is to find ways to balance and coordinate the demands that come from within with those that come from without. This is where a healthy sense of self and identity comes in. An 2 integrated sense of identity . . . means bringing together into a working whole a set of attitudes, values, and habits that can serve both self and society. The attainment of such a sense of identity is accompanied by a feeling of self-esteem, of liking and respecting oneself and being liked and respected by others.

3 More than anything else, the attainment of a healthy sense of identity and a feeling of self-esteem gives young people a perspective, a way of looking at themselves and others, which enables them to manage the majority of stress situations. Young people with high self-esteem look at situations from a single perspective that includes both themselves and others. They look at situations from the standpoint of what it means to their self-respect and to the respect others have for them. This integrated perspective enables them to manage the major types of stress efficiently and with a minimum expenditure of energy and personal distress.

The Three Stress Situations

4 There are three major types of stress situations that all of us encounter. One of these occurs when the potential stress is both foreseeable and avoidable. This is a *Type A* stress situation. If we are thinking about going on a roller coaster or seeing a horror movie, the stress is both foreseeable and avoidable. We may choose to expose ourselves to the stress if we find such controlled danger situations exciting or stimulating. Likewise if we know that a particular neighborhood or park is dangerous at night, the danger is both foreseeable and avoidable, and we do avoid it, unless we are looking for trouble.

5 The situation becomes more complicated when the foreseeable and avoidable danger is one for which there is much social approval and support, even though it entails much personal risk. Becoming a soldier in times of war is an example of this more complicated Type A danger. The young person who enlists wins social approval at the risk of personal harm. On the other hand, the young person who refuses to become a soldier protects himself or herself from danger at the cost of social disapproval.

6 Teenagers are often caught in this more difficult type of situation. If the peer group uses alcohol or drugs, for example, there is considerable pressure on the young person to participate. But such participation often puts teenagers at risk with parents and teachers, and also with respect to themselves. They may not like the image of themselves as drinkers or drug abusers. It is at this point that a sense of identity and a positive feeling of self-esteem stand the teenager in good stead.

7 A young person with a healthy sense of identity will weigh the danger to his or her hard-won feeling of self-esteem against the feelings associated with the loss of peer approval. When the teenager looks at the situation from this perspective, the choice is easy to make. By weighing the laboriously arrived-at feeling of self-esteem against the momentary approval of

a transient peer group, the teenager with an integrated sense of self is able to avoid potentially stressful situations. It should be said, too, that the young person's ability to foresee and avoid is both an intellectual and an emotional achievement. The teenager must be able to foresee events . . . but also to place sufficient value upon his or her self-esteem and self-respect to avoid situations that would jeopardize these feelings.

8 A second type of stress situation involves those demands which are neither foreseeable nor avoidable. These are *Type B* stress situations. Accidents are of this type, as when a youngster is hit by a baseball while watching a game, or when a teenager who happens to be at a place in school when a fight breaks out gets hurt even though he was not involved. The sudden, unexpected death of a loved one is another example of a stress that is both unforeseeable and unavoidable. Divorce of parents is unthinkable for many teenagers and therefore also unforeseeable and unavoidable.

9 Type B stress situations make the greatest demands upon young people. . . . With this type of stress teenagers have to deal with the attitudes of their friends and teachers at the same time that they are struggling with their own feelings. Such stress situations put demands upon young people both from within and from without. A youngster who has been handicapped by an accident, like the teenager who has to deal with divorce, has to adjust to new ways of relating to others as well as new ways of thinking about himself or herself.

10 Again, the young person with a strong sense of identity and a feeling of self-esteem has the best chance of managing these stress situations as well as they can be managed. In the case of divorce, for example, the teenager who incorporates other people's perspectives with his or her own is able to deal with the situation better than other teenagers who lack this perspective. For example, one young man, who went on to win honors at an Ivy League school, told his father when he and the mother divorced, "You are entitled to live your own life and to find happiness too."

11 This integrated perspective also helps young people deal with the death of a loved one. If it was an elderly grandparent who had been suffering great pain, the young person can see that from the perspective of the grandparent, dying may have been preferable to living a life of agony with no hope of recovery. As one teenager told me with regard to his grandfather who had just died, "He was in such pain, he was so doped up he couldn't really recognize me. I loved him so much I just couldn't stand to see him that way." By enabling the young person to see death from the perspective of others, including that of the person who is dying, the young person is able to mourn the loss but also to get on with life.

12 The third type of stress situation is one in which the potential stress is foreseeable but not avoidable. This is a *Type C* stress situation. A teenager who has stayed out later than he or she was supposed to foresees an

unavoidable storm at home. Likewise, exams are foreseeable but unavoidable stress situations. Being required to spend time with relatives one does not like is another stress situation that the teenager can foresee but not avoid. These are but a few examples of situations the teenager might wish to avoid but must learn to accept as inevitable.

13 To young people who have attained a solid sense of self and identity, foreseeable and unavoidable stress situations are manageable, again, because of self-esteem and the integrated perspective. They look at the situation from the perspective of themselves as well as that of the other people involved and try to prepare accordingly. They may decide, as one young man of my acquaintance did, that "with my folks, honesty is the best policy. I get into less trouble if I tell the truth than if I make up stories." In the case of visiting relatives they do not like, integrated teenagers see it from the perspective of what it means to others, such as their parents. And with respect to stress situations like exams, because they want to maintain their self-esteem, they prepare for the exam so that they will make a good showing for themselves as well as for others.

14 It is important to say, too, that integrated teenagers come in any and all personality types. Some are introverted and shy, others are extroverted and fun-loving. Some are preoccupied with intellectual concerns, others primarily with matters of the heart. Despite this diversity, they all share the prime characteristics of the integrated teenager: a set of attitudes, values, and habits that enable the young person to serve self and society, and a strong sense of self-esteem.

15 To be sure, life is complex and varied. Even the most integrated teenager, of whatever personality type, may occasionally be so overwhelmed by stress that he or she loses the integrated perspective and suffers bouts of low self-esteem. We need to remember that teenagers are new at the game of stress management and have just acquired the skills they need for this purpose. Nonetheless, the general principle holds true. The more integrated the teenager is with respect to self and identity, the better prepared he or she is to manage the basic stress situations.

RESPONDING TO READING

Identify a source of stress in your own life and see whether you can classify it as Type A, Type B, or Type C. Write in your journal about the stress and whether self-esteem contributes to how you handle it.

GAINING WORD POWER

Elkind's essay on stress uses several words and their opposites, their **antonyms,** such as *extroverted* and *introverted, inevitable* and *avoidable, foreseeable* and *unforeseeable.* Use your dictionary or a dictionary of **synonyms** and antonyms to find the opposites of the following words from the reading. Be sure to choose opposites that are the same part of speech; in the three preceding examples, all the words can be used as adjectives. You may need to use two or three words instead of one. Your dictionary will help you determine parts of speech.

transient	jeopardize
expenditure	potential
incorporate	

CONSIDERING CONTENT

1. According to Elkind, what overall conflict causes most situations of stress?
2. Which type of stress does Elkind think is most difficult for teens? Why is this type the worst?
3. Elkind concentrates on how a teenager with good self-esteem would deal with stressful situations. How would a teen with low self-esteem deal with some of these situations?
4. What is an "integrated perspective" (paras. 3, 11, 13, and 15)? Give an example. Can you think of a synonym for the term?
5. In paragraph 7, Elkind writes, the "young person's ability to foresee and avoid is both an intellectual and an emotional achievement." Explain this statement.

CONSIDERING METHOD

1. How many times does Elkind repeat his main point? Why might a writer repeat the main point several times?
2. Divide the fifteen paragraphs into five main sections. What label could you put on each section?
3. How does Elkind let you know what to expect when you are reading his essay?
4. How do you know when a new paragraph takes up a new type of stress situation?
5. How does Elkind clarify what each type of stress situation is like?
6. Is Elkind writing *to* teenagers or *about* them? How do you know?

WRITING STEP BY STEP

Using Elkind's classification of types of stress as a model, write an essay in which you investigate types of social pressure in your world. Think about the influences that cause you to think and behave the way you do. Consider which influences you resist and which you accept.

A. Identify two to four types of social pressure you see at work in your environment.

B. Think of some general strength that people could develop that would help them deal with these types of social pressure. Discuss what you mean by this strength in the introduction to the essay.
C. Write a paragraph briefly presenting the types of social pressure you intend to discuss.
D. Explain each type of social pressure in its own section of the paper. Use examples and details to clarify what you mean.
E. For each type, include an explanation of how the basic strength you discussed early in the essay would apply to this type of social pressure.
F. Put in some kind of transitional material or signal when you switch from one type to another. See the editing section at the end of this selection for help.
G. Point out some circumstances in which the basic strength might fail to help a person deal with social pressure.
H. In your closing (one or two sentences), reaffirm the general principle you have developed.

OTHER WRITING IDEAS

1. Classify and explain the types of stress that distinguish a certain period of life other than the teenage years. For example, think about the first year of college, marriage, parenthood, or retirement.
2. USING THE INTERNET. Write an article for incoming freshmen alerting them to the types of stress typically experienced by college students. Offer your readers ideas and resources for handling these pressures. You can find information at Web sites dealing with stress management and reduction, such as **http://stress.about.com/mbody.htm**
3. COLLABORATIVE WRITING. Consider the types of stresses that seem to be assigned by sex. What are the types of stress usually felt most by women? What types of stress are felt most by men? Get together in a mixed-sex group to brainstorm for ideas. Write on either sex but not both.
4. WRITING ABOUT READING. Give examples of each type of stress from your own or your friends' lives. Explain what role self-esteem plays in handling these instances of stress.

EDITING SKILLS: TRANSITIONS

Good writers use **transitions**—words and phrases that help connect ideas—to show how one sentence is related, logically, to the sentence before it. Look at these examples from the Elkind essay.

> On the other hand, the young person who refuses to become a soldier protects himself or herself from danger at the cost of social disapproval. [*On the other hand* is a transitional phrase indicating that a contrast is about to be made.]
>
> It should be said, too, that the young person's ability to foresee and avoid is both an intellectual and an emotional achievement. [The word *too* indicates that this sentence adds material consistent with the sentence before it.]
>
> Again, the young person with a strong sense of identity and a feeling of self-esteem has the best chance of managing these stress situations as well as they can be managed. [The word *again* lets you know that this sentence repeats something said earlier.]

Notice that the transitional words are separated from the main sentence with commas.

EXERCISE

Find at least two other sentences in the Elkind essay that include transitional words, and copy them exactly.

Now check your own essay to see whether you can provide more transitional "glue" to hold your thoughts together. Transitions can show several types of movement:

Addition: also, too, moreover, next, furthermore, again
Exemplification: for example, that is, for instance
Emphasis: especially, in fact, primarily, most importantly
Contrast: but, on the other hand, nevertheless, however
Comparison: likewise, similarly, also, too
Qualification: admittedly, of course, granted that
Causation: so, consequently, therefore, as a result
Conclusion: finally, in conclusion, in short, at last

Add at least one transitional word or phrase to your own essay.

WEB SITE

www.cio.com/archive/092203/elkind.html

An article by David Elkind about stress and technology, entitled "The Reality of Virtual Stress."

PREPARING TO READ

Have you ever seen *The George Lopez Show* on television? How about *Greetings from Tucson, The Brothers Garcia, An American Family,* or *Taina*? If so, what do you think of these shows? If not, why do you suppose you haven't seen them?

The Latino Show

JULEYKA LANTIGUA

A journalist, writer, editor, teacher, lecturer, and translator, Juleyka Lantigua earned a B.A. in Spanish Literature and Government from Skidmore College and an M.A. in journalism from Boston University. She has taken a variety of roles in magazine and book publishing, including the managing editorship of Urbana Latino *from 2000 to 2002. She is also a faculty mentor for Radio Ondas, National Public Radio's training program for young Latino journalists. She says she most relishes being a nationally syndicated columnist with* The Progressive *magazine's Media Project. The following article appeared in* The Progressive *in February of 2003.*

TERMS TO RECOGNIZE

virtual *(para. 6)*	existing in effect only, not recognized or acknowledged
patriarch *(para. 7)*	male leader of a family, clan, or tribe
formulaic *(para. 12)*	based on an established model or approach
caliber *(para. 13)*	a level of excellence or importance, quality
amalgamation *(para. 14)*	mixture of various elements
ethnic *(paras. 14, 18)*	relating to people who share common and distinctive racial, national, and cultural heritage
generic *(para. 15)*	standing for a whole group, general, common
discernible *(para. 17)*	able to be seen or recognized, observable
suffice *(para. 18)*	meet the needs, be enough

Lately, Latinos are everywhere on television. We're behind the anchor desk of the local news. We appear on prime time shows as guests or supporting cast members. Some of us have even landed lead roles on network sitcoms and dramas. And more of us are behind the cameras, at the writing desks, and in production meetings. 1

We're more visible not because TV executives value us as viewers but because they want to mine our purchasing power. Without offering an actual definition of this group, Nielsen Media Research estimates that there are almost nine million "Hispanic American TV households," a number that has grown by 19 percent in the last five years. 2

When we talk about the Latino television audience, we're talking about one of two broad groups of viewers: a) those whose ancestors are from Latin America, Spain, the Spanish-speaking Caribbean, Mexico, or native Mexicans in the Southwest and West Coast territories that were gobbled up by the United States; and b) those who themselves arrived from one of these countries in the last two decades (like me). This second group probably comprises the 4.5 million Latino households Nielsen qualifies as Spanish-dominant, meaning that only or mostly Spanish is spoken at home. 3

What networks and advertisers don't realize yet is that those in group "a" differ considerably from those in group "b." Latinos whose families have been here for generations, like millions of Mexican Americans, make up a type of mainstream Latino America, rooted in their heritage but living according to contemporary U.S. culture. Comedian and sitcom star George Lopez put it succinctly: "I'm just an American guy who happens to be of Mexican descent." Lopez, whose show was the most-watched program in its time slot with twelve million viewers through October 30, echoes the sentiment of millions of Latinos whose identities are American first, of Latino descent second. Many in this category are English-dominant or monolingual, having grown up in households where everyone spoke English. 4

Those in group "b" tend to have a split focus between their new lives in the States and their former or ongoing lives back in their home countries. They are also still experiencing the challenges of becoming fully integrated into American society. Group "b" has not attained the consumer power of group "a." 5

One obvious result is that the leading Spanish networks, Univision and Telemundo, have a tight hold on Spanish-dominant viewers. Another is the virtual Mexican Americanization of sitcoms, dramas, and commercials on the English networks. And no wonder: Mexican Americans and recently arrived Mexicans constitute a majority of the U.S. Latino population, 58.5 percent, or 20.6 million, according to the Census. 6

The proof is in the programming. Flipping channels during prime time might go something like this: First, there's *The George Lopez Show* (ABC), a comedy about a Mexican American airplane parts factory manager in Los Angeles. He is married to a Cuban American, and they have two children. Then, you get to *Greetings from Tucson* (WB), a comedy about a Mexican American married to an Irish American—a household navigating three cultures. On PBS, you can find *An American Family,* a drama about a Mexican American patriarch trying to hold his family together. And finally, *The Brothers Garcia* (Nickelodeon's number-one live-action series), a sitcom about a 7

Mexican American family trying to survive the pitfalls of raising their three kids through childhood and adolescence.

8 Admittedly, I have enjoyed watching some of these shows, and the writing can be inspired, funny, and original. But I have not connected with *any* enough to bring me back week after week.

9 What's missing from the shows I mentioned is a sense of belonging here *and* somewhere else, having a place to yearn for *and* memories that propel you toward your new life. That sort of immigrant motif could fuel a dramatic series about a different kind of Latino experience in the United States. But a show based on the immigrant experience may never happen, since Latinos are so determined to prove that we belong here as much as the next American.

10 George Lopez told hispaniconline.com, "They wanted to make me an immigrant of some sort. I'd rather get out of the business completely if that's what I have to do." I don't think Lopez, or any other Latino who feels like him, has an obligation to play up his immigrant heritage. He is simply demanding that he be placed in the rightful category of American of Latino descent. I respect that, but this category is not the same as Latino immigrant living in America. That's a political distinction Latino leaders are afraid to make, but it's real. And it ought to be reflected on TV.

11 Here's an idea: Create a Latino version of *Friends* where each of the amigos represents one of the larger groups of Latinos. The casting call for the twenty- and thirty-something roles might read something like this: Seeking a fifth-generation professional Chicana from L.A. who struggles to be independent from, and loyal to, her protective family; a Colombian college student activist who immigrated here as a teen and wants the U.S. to stop funding his country's civil war; an artistic Nuyorican from the Bronx (à la Jennifer Lopez) who dreams of making it big while staying true to her roots; a ninth-generation Tejana technology wiz navigating the social mores of dating outside her culture; and a Dominican law student from Washington Heights who immigrated with her parents as an infant and dreams of returning to become her country's first woman president.

12 This approach may be formulaic, but it addresses two key concepts that programmers and advertisers have not fully grasped.

13 First, the Latino audience, much like the general audience, is broken into age segments; no single show will do. There are the kids who watch *The Brothers Garcia* and *Taina,* Nickelodeon's quirky sitcom about a Puerto Rican high school student who dreams of becoming an entertainer. Their middle-age parents might cozy up to *The George Lopez Show* and *American Family.* And increasingly, there are the twenty-five-to-forty-year-olds who have the lion's share of the spending. We are the folks who wrote letters to Showtime when they threatened to cancel *Resurrection Boulevard.* We're the ones who get hooked on shows like *Frasier, Seinfeld,* and *Friends.* And that is the caliber of show we expect from Latino programs.

14 Second, we need a show to counter "the generically Latino" character. Defined by an absolute lack of ethnic identity, these supporting characters have brief exchanges with the white or black lead cast members, rarely finding themselves the protagonists of an episode or even of a storyline. *That '70s Show* is the biggest offender in this category. Barely out of his teens, the generically Latino, heavily accented, and over-the-top post-pubescent character "Fez" is a slimy amalgamation of every woman's nightmarish Latin lover. He orbits around the white cast and waits to deliver, or be the recipient of, a punch line.

15 On shows like *Law & Order* and *The Practice,* the generic Latino is usually a male criminal of varying ages and degrees of guilt or his helpless mother/sister/wife/girlfriend who is shocked by the news of his arrest and who stands by his side no matter what. To the producers' credit, these characters don't always have accents, but they are still instantly recognizable by their surnames.

16 There are exceptions to the forgettable generic Latino. Esai Morales's character on *NYPD Blue* does have storylines written for him, and he often appears in meaningful scenes. Or take nurse Carla Espinosa on the surprise hit *Scrubs.* Played by Judy Reyes, nurse Espinosa has worked for everything she's ever accomplished. She has a preoccupation with her family—namely, her needy mother. And she's in a committed relationship for which she has also worked very hard. Nurse Espinosa is the sole accent-free Latina I tune in to every week, even when it's a rerun. Of all the characters on all the shows, she is most like me and my Latina friends.

17 The other end of the spectrum has the character of Lucia Rojas-Klein on *Good Morning, Miami,* a sitcom about the team that produces a morning TV show. The drama queen Cuban American co-host who wears daringly unprofessional outfits to work and whose occasional line is barely discernible because of her belabored accent is exactly the kind of walking stereotype non-Latinos learn to love on fluffy television programs. Another show, *CSI: Miami,* has no clearly identifiable Latino characters despite being set in one of the most Latino cities in the United States.

18 For the record, I don't believe Latino actors should be limited to Latino roles. Benjamin Bratt's portrayal of Detective Curtis on *Law & Order* and countless other performers have proven that over and over. But I am saying that Latino characters and programs need to be more developed so that a mere Spanish surname does not suffice to fill the ethnic quota.

RESPONDING TO READING

Before reading this essay, had you thought about whether or not Latinos are fairly and adequately represented on TV? Did Lantigua convince you that they are not? Do you think you would be interested in watching any of the shows she describes (if you don't already)?

GAINING WORD POWER

In paragraph 4, Lantigua uses the word *monolingual,* which is formed by adding a familiar **prefix** to the word *lingual.* The prefix *mono-* means one, single, or alone; and *lingual* refers to languages. So *monolingual* means "speaking or using only one language." We use this prefix often, as in *monotone* (one tone of voice), *monopoly* (a commercial activity controlled by one company), and *monosyllable* (a word with just one syllable). Can you think of other terms that use this prefix? See if you can figure out the meaning of the following words: monologue, monochromatic, monogamy, monogram, monomania, monorail, monotheism. Then use a dictionary to check your definitions.

CONSIDERING CONTENT

1. What major division does Lantigua find in the Latino television audience (para. 3)? How do the two groups differ from each other?
2. Why are the English networks dominated by Mexican American shows? What is missing from these shows, in the author's opinion?
3. Lantigua says that "a show based on the immigrant experience may never happen" (para. 9). Why not?
4. What is wrong with "the generically Latino" character (para. 14)?
5. Why does the author like nurse Espinosa on *Scrubs*? Why does she not like the character played by Lucia Roja-Klein on *Good Morning, Miami*?

CONSIDERING METHOD

1. What evidence is there that Lantigua is writing to Latino readers? What other **audiences** does she also hope to reach?
2. The author first divides the Latino audience in paragraph 3. What further division does she make later in the article? What is the point of this second division?
3. How many examples does the author use in her classification? What purpose do they serve?
4. Find three uses of statistics in the article. What points do they make?
5. What is the purpose of describing a Latino version of *Friends* (para. 11)? Is it an effective technique?

WRITING STEP BY STEP

Write an essay that classifies the way that a specific group of people are presented on TV. You can choose an ethnic or racial group, as Juleyka Lantigua does, or you can look at a different classification: children, athletes, married couples, doctors, gay people, police, teachers, and so forth. You might want to narrow your topic by focusing on a particular type of show—for example, how parents are portrayed in situation comedies, or how politicians are presented in dramatic shows.

A. Begin by making a list of titles, characters, and other details for the group you have chosen. Ask friends and classmates to help you brainstorm examples; consult *TV Guide* or some other publications about television to gather specific material.
B. Divide the examples into at least three categories that illustrate the way TV shows draw on popular images and **stereotypes.**
C. Reread Lantigua's opening paragraphs. Write an opening for your own essay in which you explain a misconception or distinction that television producers and writers don't seem to recognize. Explain why they are wrong. Or write an opening that gives your overall evaluation of the way your target group is portrayed.
D. Introduce your categories by identifying the basis for your division and classification. Then name and briefly summarize your categories.
E. Choose an especially useful **example** to explain each category. In each case, examine the image or stereotype in some detail. You will probably need to explain your examples; your readers may not immediately see the distinctions you are making and may not be familiar with the shows.
F. Write a closing that refers to the points you made in your introduction (step C). Like Lantigua, you might suggest what changes could be made in the way this group is presented.

OTHER WRITING IDEAS

1. Using the directions in the Writing Step by Step assignment, write an essay that classifies stereotypes—ethnic or otherwise—that are used in popular magazine advertisements.
2. Collaborative Writing. Get together with a group of classmates to discuss how each of you views your roots. Then write an essay that classifies people according to their attitudes toward their own backgrounds—ethnic, religious, cultural, racial, or class roots. For example, in what ways do people who come from a particular economic class relate to their past? What stances do people raised in a specific faith take toward religion in later life?
3. Using the Internet. Visit several Web sites aimed at a specific ethnic group, such as the Latino Issues Forum (**www.lif.org/index.html**) and Hispanic Online (**www.hispaniconline.com/**),

or the Asian American Net (**www.asianamerican.net**) and Asia-Nation:The Landscape of Asian America (**www.asian-nation.org**). Using a classification and division approach, write a report on the topics that seem to be important to this group.

4. WRITING ABOUT READING. Throughout her article, the author uses the word *Latino* to refer to the people under discussion. What does that term mean? How does it differ, if it does, from *Hispanic* (para. 2)? Lantigua also uses the words *Chicana, Colombian, Nuyorican, Dominican,* and *Tejana*. Write an essay about the use of ethnic language in "The Latino Show." Define the terms and explain why the author uses them as she does.

EDITING SKILLS: USING APOSTROPHES

Using apostrophes makes most writers at least a bit nervous. To get a grip on this slippery punctuation mark, think about these basic rules:

1. Use an apostrophe to indicate that two words have been pushed together to form a contraction. Put the apostrophe where you leave the letters out.

   would not = wouldn't   is not = isn't
   you are = you're   he is = he's
   they will = they'll   we have = we've

2. Use an apostrophe plus *s* with singular nouns to show possession.

   the heat of the sun = the sun's heat
   the rent for one year = one year's rent
   the wig belonging to Judy = Judy's wig
   cards from UNICEF = UNICEF's cards

3. Use an apostrophe after the *s* with plurals to make them possessive.

   papers from three students = three students' papers
   alibis of four criminals = four criminals' alibis

   If the plural does not already end in *s*, add an *s* after the apostrophe.

   games for children = children's games
   poetry by men = men's poetry

EXERCISE

In the Lantigua essay, there are seventeen separate words that use apostrophes. Copy the seventeen words exactly. Don't count repeats. Next to each one, write whether the apostrophe makes the word a contraction or a possessive. What does the apostrophe in the title *That '70s Show* indicate?

Now return to your classification essay, and make sure that you have used apostrophes correctly.

WEB SITE

www.laprensa-sandiego.org/archieve/april09-04/study.htm

Report of a study entitled "Looking for Latino Regulars on Prime-Time Television: The Fall 2003 Season," conducted by UCLA's Chicano Studies Research Center.

Student Essay Using Division and Classification

Contemplating Homicide at the Mall

Bobby Lincoln

1 I think I started to dislike shopping when I started having to pay for all of the things I bought and, consequently, had to start paying attention to every little detail. All of a sudden, shopping wasn't just a question of whether or not I liked something or wanted it; it now became a question of whether or not I *needed* it, and if so, how much? I had to consider the quality of the merchandise, the store's reputation, the limits of my bank account, and so on. But nothing has proved more of a challenge and an annoyance to my shopping life than the salespeople. Oh, they're everywhere, and they're more treacherous than ever. Like the flu, they have evolved into many strains in an attempt to wear down our defenses. In an effort to understand their intentions, you need to be aware of the different tactics they employ to make your shopping experience a walk in the park—a dark, scary, annoying park that tries to steal your money.

2 First you have the Chatty Cathy variety, who may or may not have gulped down four cups of coffee before starting her shift—but most likely did. Consider my last visit to Build-A-Bear, the children's store that gives kids the opportunity to choose, stuff, dress, and name their own teddy bear (or a variety of animals). No sooner did my nephew and I walk in the door than Ms. Chat—perky, bubbly, and way too excited to be awake that early on her summer vacation—met us head on. "Hi, and welcome to Build-A-Bear," she said with a force that amazed me, because I didn't think anyone could smile ear-to-ear and still talk. "We have a new stuffed animal, a golden retriever, and it's *so cute,* and have you been here before, because if you haven't then I'll tell you what to do, because it's really *fun* and easy but mostly just fun and if you have any questions at all you just let me know and I'll help you with whatever you need, and—." We had already walked away. Cathy got the point.

The caffeine buzzers contrast with their opposite, the Mighty Uninterested. These folks mainly show up at work to do as little as possible to earn their hourly wages. I had the privilege of working with several when I was a cashier for a local discount retail store. (Lesson learned: never, *ever* work at a discount retail store.) One sticks out in particular. His name was Darryl, and it's a wonder he's still alive without having someone to eat and breathe for him; that's how unwilling he was to do anything. There were several nights when he would crouch like a Hobbit on the lowest shelf of his counter, completely hiding from any looming customer who might want to—*gasp!*—pay for some merchandise. On nights when he wasn't discovered by a manager, he'd spend an hour or two in this position. And when they did find him out, he acted like *he* was the injured party. 3

Then, of course, you have the ones who are downright rude: the Snobs. My most obvious run-in with a Snob came in a small jewelry store in Paris, France (and in no way am I trying to suggest the French are rude; this is mere coincidence). It was the end of a backpacking trip through Europe, and I wanted to buy something nice and inexpensive for my sister. I did my best to speak the three or four French words I knew, in hopes the saleslady would cut me off and let me speak English. Apparently, however, she didn't speak English, so I had to resort to sign language, using gestures not unlike the ones that a trained gorilla would use when he wants to eat a banana or play peek-a-boo. On top of this embarrassment, I kept mixing French with the other languages I had been speaking during my trip, so *Oui* became *Sí* and *Merci* was *Danke.* And Madame Snob just stood there, watching my desperate attempts to make myself understood, not offering one bit of help. I finally gave up and waited until I got back home to buy my sister's gift. 4

As I said, these salespeople are everywhere, and they come in all forms. They are frequently evil, but they can be stopped. If they're rude to you, smile sweetly, belch loudly, and walk away. If they seem uninterested, dangle the promise of extra commission in 5 front of their noses and watch them spring to life. But perhaps most importantly, down a few double espressos of your own before you enter the store. Being a little ahead of their game might make all the difference.

CONSIDERING CONTENT AND METHOD

1. How would you describe the **tone** of this essay? Is it appropriate? Does it work?
2. The author uses just one example for each type. Is that enough? Are the examples sufficiently developed to fulfill their purpose?
3. Does the author have a serious point to make? If so, what is it?
4. Lincoln makes use of comparisons in his descriptions. Point out some that you think are effective.
5. What do you think of the conclusion? How successful is it?

Strategies for Examining Two Subjects: *Comparison* and *Contrast*

Source: © Michael Jantze. Reprinted with special permission of King Features Syndicate.

Every day we use **comparison-and-contrast** thinking to make decisions in our personal lives. Should I wear the pink shirt or the teal? Should I take my lunch to work or eat out? Will it be Taco Bell or McDonald's? Should I write a check or use my credit card? Often the factors influencing an everyday decision are weighed so quickly in our minds that we may not be aware of having examined both sides. But when faced with a major decision, we think longer and may even make a list of pros and cons: "Reasons for Getting a New Car" on the left side of the sheet; "Reasons for Keeping Old Blue" on the right side. Intelligent decision making helps keep our lives from lapsing into chaos. Comparing and contrasting serve as useful tools in the struggle.

THE POINT OF COMPARISON AND CONTRAST

Technically, we *compare* things that are similar and *contrast* things that are different. But in order to contrast things, we have to compare them. And a comparison will often point out the differences of items being compared. You can see how fuzzy the distinction gets, so don't worry about it.

Using Comparisons to Explain

Using comparisons is a dandy writing strategy for explaining something your readers don't know much about by comparing it to something they do know about. For instance, geologists like to explain the levels in the earth's crust by comparing them to the layers in an onion. Or if you want to explain something complicated, like the way our eyes work, you could compare the human eye to a camera, which is simpler and more easily understood.

Using Comparisons to Persuade

Comparisons can be an effective technique in persuasive writing to help clarify and convince. People who believe that illegal drugs should be decriminalized often compare the current drug-related gang warfare to the alcohol-related crime wave during Prohibition in the 1930s. When alcohol was made legal again, the gangsters were out of business and the violence subsided. The same thing would happen today, the argument goes, if drugs were legalized. Such a comparison, called an **analogy,** can be quite convincing if you can think of one that is sensible and clearly parallel.

Using Contrast to Decide

You can set up a contrast to help clarify differences—if, for instance, you are faced with a choice between products or people or pets or proposals (or whatever). You might want to make this investigation for your own benefit or to convince someone else that one thing is preferable to another. Say, your boss asks you to find out which photocopying machine would be the best buy for the office. Because the machines are all fairly similar, your report would focus on the differences in order to determine which one would give the best results for the least money. Or if you are trying to determine whether to borrow money or get a part-time job to help pay next year's tuition, you should consider carefully the advantages and disadvantages of each choice.

THE PRINCIPLES OF COMPARISON AND CONTRAST

Since the purpose of comparing and contrasting is to show similarities or differences, good organization is crucial to success. There are two standard ways of composing this kind of writing: the **block-by-block pattern** and the

point-by-point pattern. These patterns can, of course, be expanded to include consideration of more than two elements.

Using the Block-by-Block Plan

Particularly useful for responding to comparison-or-contrast essay examination topics is this simple method of organization:

1. State your purpose.
2. Present key features of the first part of your comparison.
3. Make a transition.
4. Present corresponding features of the second part of your comparison.
5. Draw your conclusions.

This simple plan serves perfectly for showing how something has changed or developed: your earliest views about AIDS compared with the way you think now; Madonna's first album compared with her latest one; Picasso's early work compared with his later paintings; hair styles ten years ago compared with hair styles today. You will see how effectively this plan works when you read the first two essays in this chapter.

Similarities and Differences

A variation on the block-by-block plan is to cluster the points of comparison and contrast according to similarities and differences. Consider, for instance, the proverbial apples and oranges comparison. An advertising executive for the apple industry might organize a fact sheet promoting apples by first clustering the similarities between apples and oranges (their nutritional value and availability, for example) and then emphasizing their differences (more varied uses for apples and their lower cost). The **purpose,** to sell apples, is best served by posing the comparison but emphasizing the favorable contrasts.

Using the Point-by-Point Plan

When you have time to carefully organize your ideas, you may want to present the material in a way that highlights individual differences. In other words, you choose the points of comparison that best illustrate the similarities or differences and then organize the material to contrast those features.

Each major section of the comparison covers one point in terms of both (or all) the things being compared or contrasted. For example, a nutritionist might organize the fact sheet on apples and oranges by drawing conclusions one section at a time about the relative nutritional value, availability, and use, covering both apples and oranges under each of these points. For consistency, either apples or oranges would be discussed first in each section, like this:

Apples versus Oranges

1. Relative cost
   a. Orange prices remain fairly constant throughout the year.
   b. Apples are least expensive during the fall harvest season.
2. Nutritional value
   a. Oranges contain lots of vitamins A and C and have 62 calories.
   b. Apples offer fiber and have 81 calories.
3. Availability
   a. Oranges are generally available year round.
   b. Apples are best during the fall.
4. Use
   a. Oranges can be used as snacks, salads, main dishes, and desserts.
   b. Apples can be used as snacks, salads, main dishes, and desserts.

You could use exactly the same material in writing a block-style essay as you would with this point-by-point pattern. The differences lie in the way you arrange your material.

THE PITFALLS OF COMPARISON AND CONTRAST

If you are using a block pattern, be sure to follow the same order in making your comparison. If you are writing about the differences between waterbeds and airbeds, you could start by explaining the main features of the waterbed—something like this:

1. fill it with a hose
2. adjusting for comfort tricky
3. needs a heater
4. heavy when full of water
5. reasonable in cost

Then, after switching to your discussion of airbeds, take up the same features in the same order:

1. inflate it with a button
2. adjusting for comfort easy
3. no need for a heater

4. light when full of air
5. expensive to buy

Keep the same order in your point-by-point organization, too. After grouping all of the preceding information under headings something like these:

1. comfort
2. convenience
3. cost

be careful to discuss first waterbeds, then airbeds; waterbeds, then airbeds; waterbeds, then airbeds—in presenting the material in each category. As you read Mark Twain's "Two Views of the Mississippi," notice that in his second viewing, he describes the same details in the same order as he did in his first viewing.

Avoid Using Too Many Transitional Words

True, comparison-and-contrast writing involves making lots of **transitions,** but you won't necessarily need to signal each one with a transitional word. In the block pattern, you will need an obvious transition to let your readers know that you're going to shift to the second item or idea. You'll write a sentence such as, "Airbeds, *on the other hand,* are an excellent buy for several reasons" or "*On the contrary,* waterbeds can sometimes prove troublesome to maintain."

But when you follow the point-by-point pattern, you will probably not want to signal every shift back and forth with a transitional term. (Did you notice that the term *but* in the previous sentence provided a transition between paragraphs?) With the point-by-point method of organization, you will be making contrasts under each point. If you try to signal each one, your prose may get clunky. Once you have established the shifts back and forth, let your readers be guided by the pattern—and make sure you stick to it.

Avoid Repetition in Concluding

In a brief comparison or contrast essay, you need not summarize all your points at the end. Your readers can remember what you've said. Instead, draw a meaningful conclusion about the material. Maybe you'll want to assert that two apparently different religions are basically the same. Or you might stress that the differences are so crucial that conflict will always exist. If you are comparing products, come out in favor of one over the other—or else declare that neither amounts to a hill of beans—or perhaps invite your readers to judge for themselves. Just be sure to make a point of some sort at the end.

WHAT TO LOOK FOR IN COMPARISON AND CONTRAST

As you study the essays in this chapter, pay particular attention to the organization and the *continuity* (what makes it flow).

1. Look for a pattern (or maybe a combination of patterns) that each writer uses in presenting the material. Decide whether you think the method of organization is effective or whether there might be a better way.
2. Underline the transitional terms (such as *on the other hand, but, still, yet, on the contrary, nevertheless, however, contrary to, conversely, consequently, then, in other words, therefore, hence, thus, granted that, after all*). Notice what sort of meaningful links are provided between paragraphs when no purely transitional terms appear.
3. Look at the introductions and conclusions to see what strategies are used to set up the comparison or contrast at the beginning and to reinforce the writer's point at the end.

IMAGES AND IDEAS

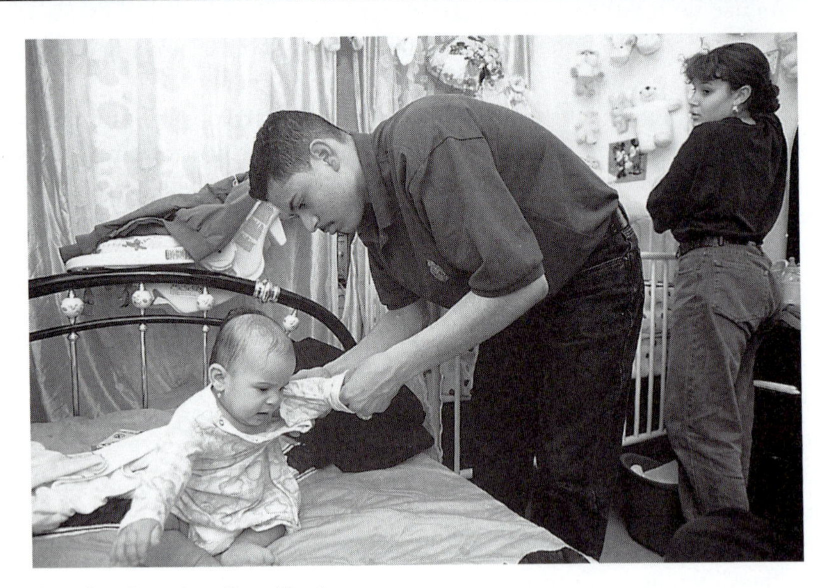

Source: Getty Images Inc.—Stone Allstock

For Discussion and Writing

Look closely at the young couple in this photo. How do you think having a baby has changed their lives? How do you think your life would change if you had a child? Get together with a group of friends or classmates, and discuss specific ways in which your lives would be (or perhaps already have been) altered by becoming a parent. Consider the trials as well as the joys. Besides the love, the burbling smiles, and the tottering first steps, mention the sleepless nights, the loss of leisure time, and the ongoing financial expenses. During this discussion, compile two lists—one pro and one con. Consider the lists carefully. If you were going to write on this issue, would you recommend becoming a parent or remaining child free?

Preparing to Read

Do you think it's possible for two people to look at the same thing and see it quite differently? Consider how a loving owner might view an overweight, spindly legged dog and how a neighbor might view the same animal. Can you think of other examples?

Two Views of the Mississippi

Mark Twain

Before becoming Mark Twain, America's most beloved humorist, Samuel Clemens (1830–1910) was a riverboat pilot, a journalist, and an unsuccessful gold miner. He said late in his life that his days on the river were the happiest he ever spent. In Life on the Mississippi *he explains in detail how he became a pilot—how he learned to "read" the river. In the following slightly edited passage, Twain tells of one drawback to that otherwise rewarding experience.*

Terms to Recognize

trifling *(para. 1)*	small, a tiny detail
acquisition *(para. 1)*	something gained or acquired
solitary *(para. 2)*	all alone
conspicuous *(para. 2)*	obvious
opal *(para. 2)*	a gemstone showing many shades of pink, blue, lavender, and gold
radiating *(para. 2)*	spreading
somber *(para. 2)*	dark
unobstructed *(para. 2)*	not hidden or blocked from view
rapture *(para. 3)*	joy, delight
wrought *(para. 3)*	brought about, caused
bluff reef *(para. 3)*	steep ridge of sand just beneath the surface of the water
shoaling *(para. 3)*	building up mud and sand
compassing *(para. 4)*	guiding (as with a compass)

1 Now when I had mastered the language of this water, and had come to know every trifling feature that bordered the great river as familiarly as I knew the letters of the alphabet, I had made a valuable acquisition. But I had lost something, too. I had lost something which could

never be restored to me while I lived. All the grace, the beauty, the poetry had gone out of the majestic river!

2 I still keep in mind a certain wonderful sunset which I witnessed when steamboating was new to me. A broad expanse of the river was turned to blood; in the middle distance the red hue brightened into gold, through which a solitary log came floating black and conspicuous; in one place a long, slanting mark lay sparkling upon the water; in another the surface was broken by boiling, tumbling rings, that were as many-tinted as an opal; where the ruddy flush was faintest was a smooth spot that was covered with graceful circles and radiating lines, ever so delicately traced; the shore on our left was densely wooded, and the somber shadow that fell from this forest was broken in one place by a long, ruffled trail that shone like silver; and high above the forest wall a clean-stemmed dead tree waved a single leafy bough that glowed like a flame in the unobstructed splendor that was flowing in the sun. There were graceful curves, reflected images, woody heights, soft distances; and over the whole scene, far and near, the dissolving lights drifted steadily, enriching it every passing moment with new marvels of coloring.

3 I stood like one bewitched. I drank it in, in a speechless rapture. The world was new to me, and I had never seen anything like this at home. But as I have said, a day came when I began to cease from noting the glories and charms which the moon and sun and the twilight wrought upon the river's face; another day came when I ceased altogether to note them. Then, if that sunset scene had been repeated, I should have looked upon it without rapture, and should have commented upon it, inwardly, after this fashion: "This sun means that we are going to have wind to-morrow; that floating log means that the river is rising, small thanks to it; that slanting mark on the water refers to a bluff reef which is going to kill somebody's steamboat one of these nights, if it keeps on stretching out like that; those tumbling 'boils' show a dissolving bar and a changing channel there; the lines and circles in the slick water over yonder are a warning that that troublesome place is shoaling up dangerously; that silver streak in the shadow of the forest is the 'break' from a new snag, and he has located himself in the very best place he could have found to fish for steamboats; that tall dead tree, with a single living branch, is not going to last long, and then how is a body ever going to get through this blind place at night without the friendly old landmark?"

4 No, the romance and beauty were all gone from the river. All the value any feature of it had for me now was the amount of usefulness it could furnish toward compassing the safe piloting of a steamboat.

RESPONDING TO READING

Do you agree with Twain that gaining knowledge takes away the appreciation of beauty? Have you ever had the opposite experience, when learning increased your appreciation of a work of art?

GAINING WORD POWER

Twain says the sunset put him in a state of "speechless rapture." The **suffix** *-less* means "without," "lacking," or "not able to." So, speechless means "without speech" or "unable to speak."

For the following list of words, write out brief definitions that show what the **root** word means with the suffix added. If you don't know the meaning of the root word, look it up in your trusty dictionary.

1. clueless
2. ruthless
3. mirthless
4. dauntless
5. meaningless
6. peerless

Now, use each word in a sentence that conveys the meaning—something beyond "I am clueless" or "She is clueless" or "Clyde is clueless."

CONSIDERING CONTENT

1. What two modes of writing does Twain mainly use in this selection?
2. In the opening sentence, what does he mean by the "language" of the river?
3. What did he lose by learning to be a riverboat pilot? What did he gain?
4. What contrast does he present in paragraphs 2 and 3?
5. At the beginning of his description in the second paragraph, he says, "A broad expanse of the river was turned to blood. . . ." What does that mean? Do you know what that figure of speech is called?
6. Can you find a **thesis** statement? If so, where?

CONSIDERING METHOD

1. Point out two **similes** (comparisons using *like* or *as*) in paragraph 2.
2. The second sentence in paragraph 2 is extremely long. Reread this complicated sentence and carefully decide where you would put in periods to create several shorter sentences.
3. What major transitions does Twain use—first in setting up the contrast, then in shifting from the before to the after?
4. In paragraph 3, Twain personifies the river—that is, he gives it human characteristics. Point out three examples of this technique of **personification.** Why do you think he used this figure of speech?
5. Does the contrast follow a point-by-point or a block pattern of organization? Make a brief outline of the way the comparison is arranged.

WRITING STEP BY STEP

Many times we feel nostalgic (warm and sentimental) about the past—about favorite former recording groups, TV programs, friends, cars, houses, articles of clothing, and so forth. Get together with several friends or classmates and talk about what you liked best in "the good old days." Take notes when meaningful memories from your past come to mind.

Choose something long gone that you still feel a fondness for—something now replaced by something newer, but not necessarily better. Then, using a block pattern, write an essay contrasting your lost prize with today's version.

A. Begin by letting your readers know that you have a fine, new whatever—fancy bicycle, designer jeans, sports car, neighborhood bakery, favorite restaurant, TV series, or rock group—but that the new version can't measure up to the one you enjoyed before.

B. Then, describe your former favorite. Use plenty of precise descriptive details, as Mark Twain does in his second paragraph. Show your reader what makes you feel the way you do about this treasure from the past.

C. Next, make a transition similar to Twain's: "But as I have said, . . ." Or write your own: "Then, after old Shep died, I bought an unbelievably dumb purebred Russian wolfhound" or "Finally, I sent my faithful Ford to the junkyard and spent a fortune on a classic MG that gives me nothing but trouble."

D. Now describe your replacement, focusing on its shortcomings—the many ways in which it doesn't measure up to the one you loved previously. Follow the same order here in presenting the new failings that you followed in describing the old virtues. (Notice that Twain uses exactly the same order in telling the two ways he saw the river. In his third paragraph, he begins with the sunset, then mentions the log, then the slanting mark on the water, then the "tumbling 'boils,' " then the lines and circles, then the streak in the shadow, and finally the dead tree—just as he did in his second paragraph.)

E. In your final paragraph, express again your regret at having to make do with this unsatisfactory replacement and your longing to have the old something back.

OTHER WRITING IDEAS

1. Write an essay following the preceding step-by-step instructions, but choose instead a thing that you replaced with something new that you think is a great deal better.
2. USING THE INTERNET. Use a search engine such as AlltheWeb or Google to locate information about piloting a steamboat today. What

technological advances have made the job much easier and safer? Write an essay explaining how riverboat piloting today differs from piloting in Twain's day.

3. COLLABORATIVE WRITING. People of different ages sometimes see things differently. Write about an attitude (toward work, religion, education, sports, television, the law, or something else) that you think is different for people who are older or younger than you are. To gather material, consult with your peers as well as people from the other generation to find out what their understanding of this difference is. Incorporate any useful ideas into your essay.
4. WRITING ABOUT READING. Do you agree with Twain that gaining knowledge can take some of the "romance" out of life? Have you ever had a similar experience—perhaps learning something that robbed you of a childish illusion or discovering something that caused you to question your former understanding of an event? Explain in writing how the change occurred.

EDITING SKILLS: QUOTATION MARKS INSIDE QUOTATIONS

Once in a while, when you have enclosed conversation or quoted material in quotation marks, you may need quotation marks around a word or phrase within the passage. You can't use regular quotation marks inside quotation marks because your reader would never be able to tell what ended where. So inside regular double quotation marks, you use single quotation marks around any words that also need quotation marks.

Notice that in his third paragraph, Twain puts a passage in quotation marks because, as he describes his second way of looking at the river, he is giving us a "conversation" with himself. Within that pretended speech, the word "boils" also needs quotation marks because it has an unusual meaning (as did the word "conversation" in the previous sentence). So, he puts "boils" in single quotation marks because it is already inside double marks. Here is another example:

> Mr. Blackwell observed, "Baggy clothes are definitely 'in' this season."

On your keyboard, you may have an opening single quotation mark. Check the keys beside the numbers at the top. The closing mark is one you use all the time: the apostrophe. It can also serve as the left-hand quotation mark if you don't have a special key for an opening mark.

Here are some other uses of quotation marks, including some instances where single marks are used inside double ones:

1. To enclose words used as words:

   It's all right to begin a sentence with "and" as long as you don't do it too often.

   (You can use italics or underlining to indicate words used as words, if you prefer.)

2. To enclose titles of short works—short stories, poems, essays, chapters of books, or song titles:

   Edgar Allan Poe's story "The Pit and the Pendulum"
   Robert Frost's poem "The Road Not Taken"
   Richard Selzer's essay "The Discus Thrower"
   Bob Dylan's song "Mr. Tambourine Man"

   But underline or put in italics the titles of longer, separately published works:

   Alice Walker's novel The Color Purple
   Arthur Miller's play Death of a Salesman
   Steven Spielberg's movie *Jurassic Park*
   Georges Bizet's opera *Carmen*

3. To enclose quoted material within a quotation:

   "What did your instructor do when you didn't turn in your term paper?"

   "I went into her office with my paper in my hand, and she said, 'I hope you just got out of the hospital; otherwise, you just failed the course.'"

EXERCISE

In the following sentences, insert regular quotation marks and single quotation marks where needed. You may use italics in some cases if you prefer.

1. Proofread carefully to be sure you have not confused its and it's.
2. Reggie said to me, You be there on time or I'm leaving without you, Squirt.
3. Then I said to Reggie, You call me Squirt again and I'm not going anywhere with you ever again!
4. Serena wrote an analysis of Sandra Cisneros's short story Woman Hollering Creek.
5. I liked this poem, Kesha observed, but when Frost says The woods are lovely, dark, and deep, what does deep mean?

Now go through your writing for this class (your essays and your journal entries), and see whether you've been using quotation marks correctly. Then make up a sentence of your own using single quotation marks inside regular quotation marks.

WEB SITE

www.greatriver.com/

The Mississippi River Home Page provides information about riverboats today, as well as in the late 19th century when Twain was piloting.

Preparing to Read

Are you the kind of person who gets fidgety if the socks aren't tidily arranged by color in your drawer? Or are you a casual type who seldom manages to get the socks from the laundry basket into the drawer? Have you tried to change, or do you not feel any need to?

Neat People vs. Sloppy People

Suzanne Britt

Suzanne Britt is a freelance journalist who teaches English at Meredith College in Raleigh, North Carolina. She has written for the New York Times, *the* Baltimore Sun, *and* Newsday. *In her collection of witty essays* Skinny People Are Dull and Crunchy like Carrots *(1982), Britt clearly favors fat folks. In the following essay, which comes from her book* Show and Tell *(1983), she sides solidly with slobs, demonstrating that taking an unusual stand—even a not very reasonable stand—can be an effective strategy for humorous writing.*

Terms to Recognize

vs. *(title)*	abbreviation of *versus,* meaning in contrast with or against
rectitude *(para. 2)*	uprightness, correctness
metier *(para. 3)*	a person's area of strength or expertise
excavation *(para. 5)*	digging out, uncovering and removing
meticulously *(para. 5)*	giving great attention to details
scrupulously *(para. 5)*	carefully doing the right thing
cavalier *(para. 6)*	free and easy
vicious *(para. 9)*	hateful, spiteful
salvaging *(para. 9)*	saving from destruction, rescuing
swath *(para. 12)*	a long strip
organic *(para. 12)*	having to do with living things

I've finally figured out the difference between neat people and sloppy people. The distinction is, as always, moral. Neat people are lazier and meaner than sloppy people. 1

Sloppy people, you see, are not really sloppy. Their sloppiness is merely the unfortunate consequence of their extreme moral rectitude. Sloppy people carry in their mind's eye a heavenly vision, a precise plan, that is so stupendous, so perfect, it can't be achieved in this world or the next. 2

Sloppy people live in Never-Never-Land. Someday is their metier. Someday they are planning to alphabetize all their books and set up home catalogues. Someday they will go through their wardrobes and mark certain items for tentative mending and certain items for passing on to relatives of similar shape and size. Someday sloppy people will make family scrapbooks into which they will put newspaper clippings, postcards, locks of hair, and the dried corsage from their senior prom. Someday they will file everything on the surface of their desks, including the cash receipts from coffee purchases at the snack shop. Someday they will sit down and read all the back issues of *The New Yorker.* 3

For all these noble reasons and more, sloppy people never get neat. They aim too high and wide. They save everything, planning someday to file, order, and straighten out the world. But while these ambitious plans take clearer and clearer shape in their heads, the books spill from the shelves onto the floor, the clothes pile up in the hamper and closet, the family mementos accumulate in every drawer, the surface of the desk is buried under mounds of paper, and the unread magazines threaten to reach the ceiling. 4

Sloppy people can't bear to part with anything. They give loving attention to every detail. When sloppy people say they're going to tackle the surface of the desk, they really mean it. Not a paper will go unturned; not a rubber band will go unboxed. Four hours or two weeks into the excavation, the desk looks exactly the same, primarily because the sloppy person is meticulously creating new piles of papers with new headings and scrupulously stopping to read all the old book catalogues before he throws them away. A neat person would just bulldoze the desk. 5

Neat people are bums and clods at heart. They have cavalier attitudes toward possessions, including family heirlooms. Everything is just another dust-catcher to them. If anything collects dust, it's got to go and that's that. Neat people will toy with the idea of throwing the children out of the house just to cut down the clutter. 6

Neat people don't care about process. They like results. What they want to do is get the whole thing over with so they can sit down and watch the rasslin' on TV. Neat people operate on two unvarying principles: Never handle any item twice, and throw everything away. 7

The only thing messy in a neat person's house is the trash can. The minute something comes to a neat person's hand, he will look at it, try to decide if it has immediate use and, finding none, throw it in the trash. 8

Neat people are especially vicious with mail. They never go through their mail unless they are standing directly over a trash can. If the trash 9 can is beside the mailbox, even better. All ads, catalogues, pleas for charitable contributions, church bulletins, and money-saving coupons go straight into the trash can without being opened. All letters from home, postcards from Europe, bills and paychecks are opened, immediately responded to, then dropped in the trash can. Neat people keep their receipts only for tax purposes. That's it. No sentimental salvaging of birthday cards or the last letter a dying relative ever wrote. Into the trash it goes.

Neat people place neatness above everything, even economics. They are incredibly wasteful. Neat people throw away several toys every time they walk through the den. I knew a neat person once who threw away a perfectly good dish drainer because it had mold on it. The drainer was too much trouble to wash. And neat people sell their furniture when they move. They will sell a La-Z-Boy recliner while you are reclining in it. 10

Neat people are no good to borrow from. Neat people buy everything in expensive little single portions. They get their flour and sugar in two-pound bags. They wouldn't consider clipping a coupon, saving a leftover, reusing plastic non-dairy whipped cream containers, or rinsing off tin foil and draping it over the unmoldy dish drainer. You can never borrow a neat person's newspaper to see what's playing at the movies. Neat people have the paper all wadded up and in the trash by 7:05 A.M. 11

Neat people cut a clean swath through the organic as well as the inorganic world. People, animals, and things are all one to them. They are so insensitive. After they've finished with the pantry, the medicine cabinet, and the attic, they will throw out the red geranium (too many leaves), sell the dog (too many fleas), and send the children off to boarding school (too many scuff marks on the hardwood floors). 12

RESPONDING TO READING

Do you think that Britt is right in preferring sloppy people? In your journal, make a list of some advantages of being neat; then make a list of some disadvantages of being sloppy.

GAINING WORD POWER

In her last paragraph Britt writes of the "*organic* as well as the *inorganic* world." You know from the Terms to Recognize that *organic* means "having to do with living things." So what does the same word mean when you add the **prefix** *in-*? That prefix commonly means "no," "not," or "without." Thus, in her sentence *inorganic* means "not organic"—having to do with things that are *not* living.

This prefix is quite common, partly because it has several different meanings. Besides making a word negative, *in-* often means "in," "into," "within," or "toward," as in a baseball *infield,* an *inboard* motor, or to *instill* values.

Knowing these meanings of this prefix can help you make sense of new words. For each of the following terms, write out the meaning of the prefix *in-* , followed by the meaning of the **root** word, as in this example using *indefinite:*

in- = not *definite* = certain, precise, clear

Get help from your dictionary if you need it.

inhale	inhuman	insomnia
injustice	inland	inability
inroad	invisible	indigestion
indirect	inlay	insane

CONSIDERING CONTENT

1. In the first paragraph, what is the moral difference that Britt finds between neat and sloppy people? Is this the difference you would give? Is she serious about this moral difference? How can you tell?
2. Does the essay have a **thesis** statement? If so, where does it appear?
3. What is "Never-Never Land," mentioned at the beginning of paragraph 3? Do you know where the term comes from?
4. What kinds of things does Britt say messy people are always planning to do? Point out her exact details.
5. What faults does she find with neat people? Again, point out details from every paragraph.
6. Is she being fair? If not, can you explain why not?

CONSIDERING METHOD

1. How does Britt let you know in her opening paragraphs that she isn't entirely serious?
2. Which pattern of organization does she use in presenting her contrast?
3. Can you find a transitional sentence that leads smoothly into the second part of the contrast?

4. Point out several words that Britt uses for humorous exaggeration. Point out several humorous examples.
5. You probably noticed that this essay has no **conclusion.** Would it be more effective if it had one, do you think? Try writing a brief concluding sentence and see whether it adds or detracts.

Combining Strategies

Britt employs a number of writing strategies in developing her essay. Her organization is a block-by-block comparison/contrast, and the entire piece involves definition—neat people and sloppy people. She also employs lots of examples, illustrations, explanations, and descriptions. Although the writing is humorous, its tongue-in-cheek purpose is persuasion. Reexamine the essay and identify an example of each of these strategies.

WRITING STEP BY STEP

Get together with a few friends or classmates to discuss various kinds of people who can be classified into types the way Britt does with neat versus sloppy (such as *plump vs. thin, fun-loving vs. serious, perky vs. droopy,* or *exercise nuts vs. couch potatoes*). In this brainstorming session, jot down any details that might be useful to include in an essay comparing the two types you choose. You can make your contrast either humorous or serious. If you decide to be humorous, you will probably also decide to defend the less positive group—praising fat folks instead of thin people, for instance. Or you may decide to make fun of both sides by contrasting two negative types, such as *eggheads vs. airheads.*

A. Begin with a two- or three-sentence introduction letting your readers know that you'll be contrasting two types of people and favoring one group (as Britt puts it, "Neat people are lazier and meaner than sloppy people").

B. Organize your contrast in a block pattern, first discussing the important characteristics of one type, then the same or similar characteristics of the other type.

C. Be sure to include plenty of examples that will show your readers the behavior you're explaining. Take another look at Britt's essay, and notice the kinds of details she uses—how many of them and how specific: not just *a corsage* but *the dried corsage from their senior prom,* not just *a recliner* but a *La-Z-Boy recliner,* not just *the geranium* but *the red geranium.*

D. In the body of your essay, begin every paragraph (as Britt does) with the name of your type—*plump people,* for instance. Vary this system of deliberate repetition once or twice by adding a transition, like the one Britt adds at the start of her fourth paragraph: "*For all these noble reasons and more,* sloppy people never get neat." (Leave out the *noble* part unless you're being funny.) Consider transitions like these:

Besides all these features, plump people . . .

In addition to these troubles, plump people . . .

Furthermore, plump people . . .

As a matter of fact, plump people . . .

Without question, plump people . . .

E. When you finish discussing your first type, include a transitional sentence. Britt shifts smoothly from praising sloppy people to criticizing neat ones with this sentence: "A neat person would just bulldoze the desk" (end of para. 5).

If you can't come up with a similar sentence that supplies a bridge from one type to the next, it's just fine to begin the second half of the contrast with a transitional term like one of these:

Thin people, *on the other hand,* . . .

Contrary to popular belief, thin people . . .

Conversely, thin people . . .

Thin people, *however,* . . .

On the contrary, thin people . . .

By contrast, thin people . . .

F. If you can think of some insight concerning your contrast, offer it as a conclusion. But don't simply tack on something obvious. That's worse than no conclusion at all. Be sure to end your last paragraph with an impressive sentence—either a short, forceful one or a nicely balanced one, like Britt's:

> After they've finished with the pantry, the medicine cabinet, and the attic, they will throw out the red geranium (too many leaves), sell the dog (too many fleas), and send the children off to boarding school (too many scuff marks on the hardwood floors).

Work on that final sentence. Make it one that leaves your readers feeling satisfied.

OTHER WRITING IDEAS

1. Using a block-by-block pattern, contrast two types of players in a sport or game such as basketball, tennis, poker, chess, or a video game. In your conclusion explain what you think causes the differences between the two types.
2. USING THE INTERNET. Britt humorously presents neat people as being overly particular, but excessively neat people sometimes can have serious problems. Visit the Website of the Obsessive Compulsive Foundation (**www.ocfoundation.org/ocf1010a.htm**) and read about this psychological disorder. Write an essay contrasting an obsessively neat person with someone who is just normally neat. You could use Adrian Monk, the obsessively neat detective on the popular TV series *Monk,* as a prime example; if you're not a *Monk* watcher, substitute someone you know who exhibits obsessive/compulsive traits. You might also compare the obsessive form of another trait—such as punctuality, caution, or perfectionism—with its normal form.
3. COLLABORATIVE WRITING. Get together with friends or classmates, and discuss major life changes that you and they have experienced. Talk about what your lives were like before and after the changes occurred. Consider situations such as life before marriage and life after; life in your folks' home and life in the dorm or your own apartment; life at your previous job and life at your new one. Then write an essay about one of your life changes, telling first what your life was like before, then how things were after. Focus on one factor of your experience that changed greatly, like the amount of freedom or the amount of responsibility you had.
4. WRITING ABOUT READING. Starting with the lists you made in responding to Britt's essay, write one of your own organized like hers (but probably with a more serious tone), showing how neatness makes life easier and sloppiness leads to problems. Include plenty of details and examples to show the advantages of being tidy and the folly of being perpetually scattered.

EDITING SKILLS: USING APOSTROPHES

Apostrophes probably cause more problems than other marks of punctuation. Even experienced writers feel shaky about using them. If you study the following rules about using apostrophes, you should be able to use them more confidently.

1. *Use an apostrophe to indicate that a noun is possessive.* Possessive nouns usually indicate ownership, as in *Miguel's hat* or *the lawyer's briefcase.* But sometimes the ownership is only loosely suggested, as in *the rope's length* or *a week's wages.* If you are not sure whether a noun is possessive, try turning it into an *of* phrase: *the length of the rope, the wages of a week.*
   a. If the noun does not end in *s,* add *'s.*

      Rita climbed into the driver's seat.

      The women's lounge is being redecorated.

   b. If the noun is singular and ends in *s,* add *'s.*

      The boss's car is still in the parking lot.

      Have you met Lois's sister?

   c. If the noun is plural and ends in *s,* add only an apostrophe.

      The workers' lockers have been moved.

      A good doctor always listens to patients' complaints.

2. *Use an apostrophe with contractions.* Contractions are two-word combinations formed by omitting certain letters. The apostrophe goes where the letters are left out, not where the two words are joined.

does not = doesn't	he is or he has = he's
would not = wouldn't	let us = let's
you are = you're	I am = I'm

3. *Do not use an apostrophe to form the plural of a noun.* The letter *s* gets pressed into service in a number of ways; its most common use is to show that a noun is plural (more than one of whatever the noun names). No apostrophe is needed with a simple plural:

   Two *members* of the starting team are suspended for the next three *games* for repeated curfew *violations.*

EXERCISE

1. Examine this sentence from paragraph 11 in Suzanne Britt's essay: "You can never borrow a neat person's newspaper to see what's playing at the movies." Two words in this sentence end in *'s;* one is possessive and one is a contraction. Do you see the difference? Explain how you can tell.
2. Find two other possessive nouns in Britt's essay, and explain what they mean by turning each into an *of* phrase.
3. Find two other contractions that end in *'s* in Britt's essay. What do these contractions stand for?

4. Find six other contractions in Britt's essay, and explain their meaning.
5. Find several examples of plural nouns that end in *s* (without an apostrophe).

Now go back over the essay you have just written, and check your use of apostrophes. Have you left an apostrophe out of a possessive? Have you put an unneeded apostrophe in a plural noun? Have you misplaced any apostrophes in contractions?

WEB SITE

http://owl.english.purdue.edu/handouts/grammar/g_apost.html

The Purdue University Online Writing Lab's handout on using the apostrophe. You can print it out and use it when writing and editing your papers.

PREPARING TO READ

Have you ever noticed that girls and boys play differently? Think about preschool, playground, and athletic activities. Why do you think those differences occur?

A Whole New Ballgame

BRENDAN O'SHAUGHNESSY

A former teacher and coach at St. Ignatius College Prep in Chicago, Brendan O'Shaughnessy now lives with his family in Indianapolis, where he writes for the Times of Northwest Indiana *as the paper's state bureau chief. He has also written freelance for the* Chicago Tribune, *which is where the article reprinted below appeared on December 1, 2002.*

TERMS TO RECOGNIZE

baiting *(para. 2)*	teasing, needling
insinuation *(para. 2)*	an implied reference
incredulous *(para. 4)*	unbelieving, doubting, skeptical
gaffe *(para. 4)*	a social blunder or mistake
demeanor *(para. 10)*	outward behavior
intimidated *(para. 11)*	made timid or fearful

The first day of freshman basketball tryouts, I learned that coaching girls is different. I was demonstrating the correct way to set a cross screen. I positioned my legs shoulder-width apart and crossed my hands—fists clenched—over my groin to protect myself from the injury that all men fear. I paused, confused, understanding from the girls' bewildered looks that something was wrong. The other coach, a 15-year veteran of coaching girls, recognized my rookie mistake and bailed me out. He raised his arms and covered his chest, and I knew that I had entered alien territory. 1

I had coached boys basketball for six years before circumstances in the athletic department forced me to switch to "the other side." I looked forward to the challenge in the same way that I had anticipated the move from teaching at an all-boys' school to a coed institution five years before. At the very least, I figured, I would be more likely to get cookies at Christmas and 2 a gift at the awards banquet. Baiting a feminist friend, I told her that I was excited about the change because I could be more relaxed, less intense, and besides, I wouldn't get any technicals. I just assumed girls didn't take their basketball as seriously as boys. The insinuation hit its mark. She scolded me, saying that girls were just as eager to win and play well as boys. She also suggested I read Madeleine Blais's *In These Girls, Hope Is a Muscle,* a book about a girls team's basketball season.

3 From the book and from my teaching experience, I began the season with certain expectations about coaching girls. I would need to be more encouraging, less critical. Most boys need a little tearing down before they can be rebuilt on a more solid fundamentals base. Boys want to be Allen Iverson and inherently assume they know more than their old-school coach, who watched "Hoosiers" one too many times. Girls, whose experience of playground games and watching the all-stars is often limited, do not start with as many bad habits. I expected they would be more coachable. They wouldn't need their inflated athletic egos broken down, but rather built from the ground up.

4 Smugly thinking I was prepared, I got a rude awakening with my screen-setting gaffe that first day. Imagine my incredulous stare when a girl trying out, in an attempt to explain why she had thrown up and had to sit out of wind sprints, told me she hadn't run since gym class—the year before. I was also surprised—and relieved—that we did not have to cut, since only 20 girls stuck out the trials for the two teams. With boys, two or three times as many students usually came out for the teams as could be taken.

5 I immediately noted differences in the early practices. Girls' attention to directions was far superior to the boys, most of whom found it physically impossible not to be distracted by any movement anywhere in the gym. Whereas the boys generally either went deadpan or shot me the evil "how dare you" death stare when I corrected their play, the girls often sincerely apologized for any mistake. My stereotypically gawky center, when told not to leave her feet on defense, said, "I know. I'm sorry. I'm terrible." Embarrassed, I tripped out a halting reassurance. I tried to build up her confidence by calling her "the rebound machine," but she just thought I was goofy.

6 Strangest of all, they actually wanted to talk to me and the other coach, something teenage boys found equivalent to having their nose hairs, if they had any, individually plucked out in front of an audience of teenage girls. The girls came running up before practice to tell us about their classes, about who said what at lunch, about who had spilled perfume on her uniform. Uncomfortable after years of boys slinking away into corners, I usually responded, "Stretch out."

7 Before the first game, I realized that some of their silliness was simply due to their age, not their sex. In the pregame huddle, the other coach said we needed to play hard or go home with a big L. One of the girls asked if

everyone would have to take the "L" home instead of the bus if we lost. During the game, one player attempted to high-five a referee after making a shot. But it was more than their tender age. While I was giving a post-game speech, one player interrupted and said, "Those are the coolest sweat-pants. They zip all the way down." When my grandfather died, the whole team signed a condolence card with individual attempts to comfort me. Another time, returning from a late game, when the bus broke down on the highway in 15-degree weather, one player cut the tension with, "Coach, want a chocolate-chip cookie? I made them."

I began to observe that the team split into two groups: the hard, aggressive players and the softer, nice players. One side had girls who would steal a ball from their teammates in order to shoot. The other side had girls who apologized to their defenders if they scored. Some would crash the boards and clear out space with vicious elbows, and others would avoid any chance of injury or even breaking a sweat. The aggressive group rolled their eyes at the limp-wristed run of one girl they called "the dancer" or "Basketball Barbie." The timid girls rolled their eyes and called our best shooter a "ball hog." 8

After six wins and a growing gulf between the cliques, we experienced our first loss. Actually, we got blown out by 35 points. We could barely get the ball down the court. A coach learns all he needs to know about his team by how they react to a loss. My team began to motivate each other in practice. They started to pull for each other. Best of all, the gap between the groups of player types began to slowly close. In time, we were a single unit again. 9

And I was swept up in the intensity of their effort. I don't know exactly when it began, but it was cemented when I was called for a technical foul in a Christmas tournament game. Whereas boys' freshman coaches tend to be overly passionate, like myself, sporting buzz cuts and angry demeanors, girls' coaches usually were more welcoming. One informed us that her name was Poppy, offered our team bagels and Gatorade, and said, "We're all about fun here." It was all I could do to refrain from saying, "We're all about kicking your butt." 10

Fast forward to the conference championship game, where we faced the same team that had blown us out by 35 earlier in the season. Since then, this powerhouse had won every game, none by fewer than 20 points. Not intimidated this time, our girls played them even for a quarter. When the opponent went on a second-quarter run, I impolitely objected to an over-the-back foul and was hit with another technical. Shocked, I realized that I had been given more technicals in a single season of coaching girls than I ever had as a boys' coach. My feminist friend would be proud. The team responded. The collective jaw of the bench dropped to the hardwood when "Basketball Barbie" hit a shot, slapped the floor and yelled, "C'mon, 11 girls, let's play some defense." I couldn't have been more pleased if it had been my own daughter.

No, it didn't lead to a win, but we never gave up either, clawing to a nine-point loss and the bittersweet distinction of holding that team to their narrowest margin of victory all season. Even in defeat, the girls had come a long way in their separate challenges. Some had overcome a natural timidity by learning to play aggressively, and others had learned to trust their team. My lesson? New depths to the same game I've always loved. 12

RESPONDING TO READING

Have you watched both girls and boys play basketball? Did you observe any differences between the teams in the way they played? Do you think, as O'Shaughnessy does, that the girls are just as dedicated to winning as the boys?

GAINING WORD POWER

Decide whether the italicized words in the following list are used correctly or not. Then, write *yes* or *no* beside each sentence. After you finish, look back at the Terms to Recognize definitions to check your work.

____ 1. Billy Bob's *demeanor* made him appear lazy and disinterested.
____ 2. Secretly he studied hard and made *incredulous* improvement.
____ 3. Then a jealous classmate *intimidated* that he had cheated.
____ 4. I thought that such an *insinuation* was outrageous and spiteful.
____ 5. So, I bought him a double latte *gaffe* as a reward for his success.

CONSIDERING CONTENT

1. What are the rewards O'Shaughnessy anticipates or initially expects as a result of switching from coaching boys to coaching girls?
2. In paragraph 2, what are *technicals*? Why did he think he wouldn't get any technicals coaching girls?
3. What expectations did he have about how he would need to coach girls differently than boys? What differences does he actually find?
4. In paragraphs 4, 5, and 6, he says that in early practices he notes differences between the girls' and boys' behavior. What are some of those differences?
5. The girls refer to one of their teammates with a "limp-wristed run" as "Basketball Barbie." What sort of player do you envision from that name?
6. How does being called for a technical foul let the coach know his team is now "a unit" and "cemented"? Why would his feminist friend be proud that he received more technicals coaching girls than boys?

CONSIDERING METHOD

1. How does O'Shaughnessy capture your interest in his opening paragraph?
2. His essay is organized using a point-by-point contrast. Why do you think he chose this pattern rather than using the simple block-by-block method?
3. What other patterns of development can you identify in the essay?
4. Why do you think he mentions his feminist friend near the beginning and again near the end?
5. What does his use of basketball jargon (*cross screen, wind sprints, blown out, over the back foul*) tell you about his intended audience?
6. What is the author's thesis? Is it stated or implied?

WRITING STEP BY STEP

Think of a situation, place, object, or person *then* and *now.* Choose something that has changed a lot: wedding receptions years ago and today, your desk before and after cleaning, your sluggish first computer and your smart new computer, your childhood tennis shoes and your new athletic shoes, your granny then and now.

Write a comparison focusing mainly on the *now,* but mentioning the *then* occasionally to show the contrast. If you can get your friends to help, brainstorm with them to think of good points you can use in drawing the comparison. You need qualities that fit both *then* and *now,* but think of a lot more details about the *now.*

A. Begin by giving the subject of your comparison in a sentence or two. Don't state it directly but imply it: "I had no idea that wedding receptions were not always elaborate productions until my mom told me about her wedding thirty years ago," or "I was slow to appreciate the benefits of my new computer because it took so much effort to learn to use it. My old computer was easy by comparison."
B. As you discuss the *now,* remind your readers, whenever you take up a new point, how it was *then.* Use transitional words when you need them: *by comparison, on the other hand, on the contrary, but, still, after all, like, nevertheless, contrary to, however, granted that.*
C. In your last paragraph, tell how you feel about the change. Would you rather have the *then* instead of the *now*—or do you find the *now* a great improvement?
D. If you're stuck for an ending, try concluding with a question and then answering it: "If I had known then how much easier my com-

puter would make writing, would I have complained so loudly? Maybe not," or "If I could go back to planning a reception that cost less than $10,000, would I? You bet I would!"

OTHER WRITING IDEAS

1. Think of two possible views on one of these aspects of life: the value of work, the role of family, or the importance of sports. Write a paper contrasting two people who represent the two views.
2. COLLABORATIVE WRITING. In a small group discuss gender stereotypes—the way society expects women and men to behave. From the notes you take, choose three or four categories—such as manner of speaking, walking, dressing, and showing emotion; or typical careers, leisure activities, and tastes in movies. Organize your essay using the point-by-point method, first doing the women's role, then the men's (or the reverse, but be consistent from section to section). At the end, draw a conclusion about how society treats people who do not conform to these roles and expectations.
3. USING THE INTERNET. Visit the Gender Equity in Sports site at **http://bailiwick.lib.uiowa.edu/ge/GEREDESIGN.html**, click on the "About Title IX" link, and read the material. Then, with this information in mind, speculate in writing about the role Title IX may have played in bringing about the changes that O'Shaughnessy observed.
4. WRITING ABOUT READING. Consider how our society's expectations for girls are reflected in O'Shaughnessy's article. Does it also reveal society's expectations for boys? Discuss the ways in which the author shows how gender roles in organized sports have changed and whether you see these changes as positive or negative.

EDITING SKILLS: CHOOSING *ITS* OR *IT'S*

Look at this sentence from O'Shaughnessy's article:

The insinuation hit its mark.

Why does he use *its* and not *it's* in that sentence?

Now look at the uses of *it's* and *its* in the following sentences:

It's been a hard day's night.

It's raining on my parade.

This knife has lost *its* edge.

That horse just won *its* last race.

Notice that in the first two sentences, the words *it has* or *it is* can be substituted for *it's.* Not so in the next two sentences. Those two uses of *its*—without the apostrophe—are possessive, as in "Each piece has *its* own place in the puzzle."

You can use the substitution test to see whether you have used the right form. If you have written *it's,* you should be able to read the sentence with *it is* or *it has* instead. That substitution won't work with the possessive *its,* which carries a sense of ownership or belonging.

The plan failed because of *its* flaws.

I like that sitcom in spite of *its* dysfunctional characters.

Many writers are tempted to put an apostrophe in the possessive *its* because so many other possessives require apostrophes: the *plan's* flaws, the *sitcom's* characters. No wonder it's confusing. Try putting the possessive *its* on a mental list with the other possessives—*his, hers, ours, theirs*—which also have no apostrophes. This mental grouping may help you choose the correct form.

EXERCISE

Fill in the blanks in the following sentences with *its* or *it's.* Be prepared to explain your choices.

1. Your proposal has much in ______ favor, but ______ unlikely that the committee will vote for it.
2. ______ an old car, but ______ paint job is new.
3. The dog bit ______ own tail; ______ not an exceptionally smart dog.
4. ______ important that a company give ______ employees job security.
5. ______ a whole new ballgame.

WEB SITE

www.detnews.com/2003/editorial/0302/08/d07-79750.htm

On these pages you will find a lively debate about men, women, and sports conducted through the "Letters to the Editor" columns of the *Detroit News.*

PREPARING TO READ

Before looking at the following essay, consider whether you think children are born smart or whether they get smart by studying hard in school. Jot down your thoughts in your journal.

The Trouble with Talent: Are We Born Smart or Do We Get Smart?

KATHY SEAL

A California-based freelance journalist, Kathy Seal frequently writes about children and education in such popular magazines as Parents *and* Family Circle. *In the essay we reprint here, first published in* Lear's *magazine in July 1993, she examines an attitude that may help to explain why math scores of American children have fallen far behind those of children in Japan.*

TERMS TO RECOGNIZE

rote *(para. 4)*	routine, mechanical repetition
efficacy *(para. 12)*	ability to bring about an effect
rampant *(para. 12)*	widespread, out of control
per se *(para. 16)*	in and of itself
mammoth *(para. 19)*	huge
conviction *(para. 19)*	firmly held belief

Jim Stigler was in an awkward position. Fascinated by the fact that Asian students routinely do better than American kids at elementary math, the UCLA psychologist wanted to test whether persistence might be the key factor. So he designed and administered an experiment in which he gave the same insolvable math problem to separate small groups of Japanese and American children. 1

Sure enough, most American kids attacked the problem, struggled briefly—then gave up. The Japanese kids, however, worked on and on and 2 on. Eventually, Stigler stopped the experiment when it began to feel inhumane: If the Japanese kids were uninterrupted, they seemed willing to plow on indefinitely.

3 "The Japanese kids assumed that if they kept working, they'd eventually get it," Stigler recalls. "The Americans thought 'Either you get it or you don't.'"

4 Stigler's work, detailed in his 1992 book *The Learning Gap,* shatters our stereotypical notion that Asian education relies on rote and drill. In fact, Japanese and Chinese elementary schoolteachers believe that their chief task is to stimulate thinking. They tell their students that anyone who thinks long enough about a problem can move toward its solution.

5 Stigler concludes that the Asian belief in hard work as the key to success is one reason why Asians outperform us academically. Americans are persuaded that success in school requires inborn talent. "If you believe that achievement is mostly caused by ability," Stigler says, "at some fundamental level you don't believe in education. You believe education is sorting kids, and that kids in some categories can't learn. The Japanese believe *everybody* can master the curriculum if you give them the time."

6 Stigler and his coauthor, Harold W. Stevenson of the University of Michigan, are among a growing number of educational psychologists who argue that the American fixation on innate ability causes us to waste the potential of many of our children. He says that this national focus on the importance of natural talent is producing kids who give up easily and artful dodgers who would rather look smart than actually learn something.

7 Cross-cultural achievement tests show how wide the gap is: In a series of studies spanning a ten-year period, Stigler and Stevenson compared math-test scores at more than 75 elementary schools in Sendai, Japan; T'aipei, Taiwan; Beijing, China; Minneapolis; and Chicago. In each study, the scores of fifth graders in the best-performing American school were lower than the scores of their counterparts in the worst-performing Asian school. In other studies, Stigler and Stevenson found significant gaps in reading tests as well.

8 Respect for hard work pervades Asian culture. Many folk tales make the point that diligence can achieve any goal—for example, the poet Li Po's story of the woman who grinds a piece of iron into a needle, and Mao Tse-tung's recounting of an old man who removes a mountain with just a hoe. The accent on academic effort in Asian countries demonstrates how expectations for children are both higher and more democratic there than in America. "If learning is gradual and proceeds step by step," says Stigler, "anyone can gain knowledge."

9 To illustrate this emphasis, Stigler videotaped a Japanese teacher at work. The first image on screen is that of a young woman standing in front of a class of fifth graders. She bows quickly. "Today," she says, "we will be studying triangles." The teacher reminds the children that they already

know how to find the area of a rectangle. Then she distributes a quantity of large paper triangles—some equilateral, others right or isosceles—and asks the class to think about "the best way to find the area of a triangle." For the next 14½ minutes, 44 children cut, paste, fold, draw, and talk to each other. Eventually nine kids come to the blackboard and take turns explaining how they have arranged the triangles into shapes for which they can find the areas. Finally, the teacher helps the children to see that all nine solutions boil down to the same formula: $a = (b \times h) \div 2$ (the area equals the product of the base multiplied by the height, divided by two).

10 Stigler says that the snaillike pace of the lesson—52 minutes from start to finish—allows the brighter students enough time to understand the concept in depth, as they think through nine different ways to find the areas of the three kinds of triangles. Meanwhile, slower students—even learning-disabled students—benefit from hearing one concept explained in many different ways. Thus children of varied abilities have the same learning opportunity; and the result is that a large number of Japanese children advance relatively far in math.

11 Americans, on the other hand, group children by ability throughout their school careers. Assigning students to curricular tracks according to ability is common, but it happens even in schools where formal tracking is not practiced.

12 So kids always know who the teacher thinks is "very smart, sorta smart, and kinda dumb," says social psychologist Jeff Howard, president of the Efficacy Institute, a nonprofit consulting firm in Lexington, Massachusetts, that specializes in education issues. "The idea of genetic intellectual inferiority is rampant in [American] society, especially as applied to African-American kids."

13 A consequence is that many kids face lower expectations and a watered-down curriculum. "A student who is bright is expected just to 'get it,'" Stigler says. "Duller kids are assumed to lack the necessary ability for ever learning certain material."

14 Our national mania for positive self-esteem too often leads us to puff up kids' confidence, and we may forget to tell them that genius is 98 percent perspiration. In fact, our reverence for innate intelligence has gone so far that many Americans believe people who work hard in school must lack ability. "Our idealization of a gifted person is someone so smart they don't have to try," says Sandra Graham of UCLA's Graduate School of Education.

15 Columbia University psychologist Carol Dweck has conducted a fascinating series of studies over the past decade documenting the dangers of believing that geniuses are born rather than made. In one study, Dweck and UCLA researcher Valanne Henderson asked 229 seventh graders whether people are "born smart" or "get smart" by working hard. Then they compared the students' sixth and seventh grade achievement scores.

The scores of kids with the get-smart beliefs stayed high or improved, and those of the kids subscribing to the born-smart assumption stayed low or declined. Surprisingly, even kids who believed in working hard but who had low confidence in their abilities did very well. And the kids whose scores dropped the most were the born-smart believers with high confidence.

Dweck's conclusion: "If we want our kids to succeed, we should emphasize effort and steer away from praising or blaming intelligence per se." 16

Psychologist Ellen Leggett, a former student of Dweck's at Harvard, has found that bright girls are more likely than boys to believe that people are born smart. That finding could help to explain why many American girls stop taking high school math and science before boys do. 17

Seeing intelligence as an inborn trait also turns children into quitters, says Dweck. "Kids who believe you're born smart or not are always worried about their intelligence, so they're afraid to take risks," Dweck explains. "But kids who think you can get smart aren't threatened by a difficult task or by failures, and find it kind of exciting to figure out what went wrong and to keep at it." Or, in Jeff Howard's words, "If I know I'm too stupid to learn, why should I bang my head against the wall trying to learn?" 18

Getting Americans to give up their worship of natural ability and to replace it with the Asian belief in effort seems a mammoth undertaking. But Dweck maintains that it's possible to train kids to believe in hard work. The key to bringing kids around, says Dweck, is for the adults close to them to talk and act upon a conviction that effort is what counts. 19

The Efficacy Institute is working on exactly that. The institute's work is based on theories that Howard developed as a doctoral candidate at Harvard, as he investigated why black students weren't performing in school as well as whites and Asians. Using the slogan "Think you can; work hard; get smart," the institute conducts a seminar for teachers that weans them from the born-smart belief system. 20

"We tell teachers to talk to kids with the presumption that they can all get As in their tests," explains project specialist Kim Taylor. Most kids respond immediately to their teachers' changed expectations, Howard says. As proof, he cites achievement-test scores of 137 third grade students from six Detroit public schools who were enrolled in the Efficacy Institute program during 1989 and 1990. The students' scores rose 2.4 grade levels (from 2.8 to 5.2) in one year, compared with a control group of peers whose scores only went up by less than half a grade level. 21

Institute trainers now work in approximately 55 school districts, from Baltimore to St. Louis to Sacramento. In five cities, they're working to train every teacher and administrator in the school district. 22

While current efforts for change are modest, no less a force than the Clinton administration is weaving this new thinking into its education agenda. During a talk this past spring to the California Teachers Association, U.S. Secretary of Education Richard Riley pledged to work on setting national standards in education. "These standards," he says, "must be for all of our young people, regardless of their economic background. We must convince people that children aren't born smart. They get smart." 23

....

RESPONDING TO READING

Did reading the essay change your mind about kids being born smart or getting smart through hard work? In your journal explain how you think Americans' attitudes on the subject could be changed, especially if the government or the National Education Association decided to spend money on the effort.

GAINING WORD POWER

In paragraph 10, Kathy Seal uses the interesting word *snaillike* to describe the slow pace of the Japanese teacher's math instruction. The term *-like* is a **combining form** meaning "resembling or characteristic of." By adding *-like* to other words (many of them names for animals), you can produce useful new descriptive terms. You could write, for instance, of a child's shell-like ear, and your readers would be able to picture the delicate curve of the ear. Notice that you hyphenate *shell-like,* when there are three *l*s together, but not *snaillike,* when there are only two.

1. Add *-like* to the end of each word listed below.

war	child	bird	barn	ostrich
bell	lady	cat	flower	cow

2. Then write a definition that includes the characteristic conveyed by that new word: *slow as a snail, curved like a shell.*
3. Finally, write a sentence for each new word that makes use of the descriptive characteristic: The *snaillike* traffic on the freeway resulted from an accident.

CONSIDERING CONTENT

1. According to researchers, what happened when groups of American kids and groups of Japanese kids were given insolvable math problems to work on?

2. How do most Americans think Asian students are taught? According to researchers quoted by Seal, is this impression true?
3. How did the Japanese instructor teach about triangles in the example given in paragraph 9? What are the advantages of this method of teaching?
4. What is the typical American attitude about learning—that kids are born smart or that they get smart by working hard? What is the Japanese attitude?
5. Do you know what Seal means by "our national mania for positive self-esteem" (para. 14)? What problems do researchers think it causes?
6. How is the Efficacy Institute trying to change the American attitude and instead "train kids to believe in hard work" (para. 19)?

CONSIDERING METHOD

1. How does Seal's opening sentence help get you interested in her material?
2. Find two specific examples that help readers understand why the Japanese kids beat the American kids on achievement tests. How helpful are these examples?
3. What other method of providing evidence and explaining ideas does the author use in the second part of her article about Americans' attitudes?
4. Seal employs the block organization for her contrast. Find the sentence in which she makes the transition from explaining how Japanese kids learn to considering how American kids learn. What transitional term does she use?
5. The essay has a brief concluding section following the contrast of the two educational systems. What is the purpose of these final paragraphs (19–22)?
6. The essay concludes with a direct quotation (para. 23). What makes this ending effective?

WRITING STEP BY STEP

Think of an issue on which you and your parents—or you and your spouse—strongly disagree. With your parents, for instance, you might differ in your attitudes toward premarital sex; or they might disapprove of your taste in music, clothing, or hairstyles. With your spouse, you might disagree about household chores, child care, financial matters, or vacation plans. Choose an issue that you feel confident you are right about, but be sure that you are also quite familiar with the evidence for the opposing point of view.

A. Begin your essay by presenting the problem, as Kathy Seal does in her introduction when she states "the fact that Asian students routinely do better than American kids at elementary math" (para. 1). You can start by admitting that you have this heated disagreement in your immediate family about whatever it is.

B. Next explain how other members of your family view this matter, just as Seal explains how Japanese schoolchildren are taught math (paras. 2–10). Try to include brief specific examples as Seal does in paragraphs 8 and 9. And be fair. Give as much space to presenting this opposing view as you will give to your own viewpoint in the second part of the essay.

C. Write a transitional sentence similar to the one that Seal uses at the start of paragraph 11: "Americans, on the other hand, group children by ability throughout their school careers."

D. Now, present your side of the issue. Offer plenty of specific examples. The evidence that Seal uses in analyzing the attitudes of Americans about how children learn are mainly quotations from researchers, but these quotations are full of specifics—problems, beliefs, research studies, testing results.

E. If possible, include a final section similar to Seal's paragraphs 19–22, in which she tells what the Efficacy Institute is doing to encourage American kids to work harder in school. Try to think of a way of resolving the problem you described in your essay. It's possible that focusing on the opposing viewpoint (as you did in the first part of your essay) may reveal a middle ground. Look for a compromise. If you find a solution that stops short of involving the law and justice system, present it here.

F. If you can think of no way of resolving the problem, conclude by admitting that you and your family will simply have to agree to disagree. But consider the final quotation in Seal's essay. Modeling the deliberate repetition of a word, try to end with two forceful sentences similar to these: "We must convince people that children aren't born smart. They get smart" (para. 23).

OTHER WRITING IDEAS

1. Using a block pattern of organization, write a comparison for your classmates of two sports. Focus your attention on which one is more complicated to play or more fun to watch, and use your comparison to explain why you think so.

2. USING THE INTERNET. Write an essay, using your own experience as well as those of people you know, to contrast the pros and cons of family vacations. Use an Internet search engine to visit the sites of typical vacation destinations (the Grand Canyon, Las Vegas, Disneyland, and so forth) to gather additional ideas. In your brainstorming, consider the preferences of both the parents and the children, but write from the viewpoint of one or the other.
3. COLLABORATIVE WRITING. Seal thinks Americans should "give up their worship of natural ability" and "replace it with the Asian belief in effort." She quotes psychologist Carol Dweck's assertion "that it's possible to train kids to believe in hard work" (para. 19). Do you believe that hard work is what's required in learning? Get together with a group of classmates and discuss whether this is an entirely useful idea. What advantages did Seal include in her essay? What additional advantages or distinct disadvantages were generated in your brainstorming session? Write an essay contrasting the two viewpoints and conclude by recommending one side or the other.
4. WRITING ABOUT READING. Seal provides an interesting illustration of the way a Japanese teacher presents a concept in math instruction (para. 9). Write an essay contrasting that technique with the way you were taught some concept (like in math, grammar, or spelling) in grade school or high school. Conclude by stating which method you think is better and why.

EDITING SKILLS: USING DASHES

The dash is a handy mark of punctuation that will give **emphasis** to whatever follows it—as long as you don't use it too often. Notice how the dash works in this sentence from Seal's essay:

> Sure enough, most American kids attacked the problem, struggled briefly—and then gave up.

Seal could have used a comma after *briefly,* but she chose the dash because readers pay more attention to what follows a dash than to what follows a comma. Using the dash is unusual, unexpected—thus emphatic.

You can also use a pair of dashes to set off a few words in the middle of a sentence if you want to emphasize them, as Seal does here:

> Meanwhile, slower students—even learning-disabled students—benefit from hearing one concept explained in many different ways.

Commas would be quite correct there but dashes give emphasis to the words set off.

There's another handy use for the dash—to avoid comma clutter. Look at this sentence from Seal's essay:

> Then she distributes a quantity of large paper triangles—some equilateral, others right or isosceles—and asks the class to think about "the best way to find the area of a triangle."

Again, commas would be correct before and after that phrase, but when the group of words set off contains one or more commas, putting dashes around it makes the whole sentence easier to read.

Finally, you can use a dash instead of a colon to introduce an example, as Seal does here:

> Many folk tales make the point that diligence can achieve any goal—for example, the poet Li Po's story of . . . and Mao Tse-tung's recounting of. . . .

When typing, use two hyphens to form a dash (--), but if you are writing on a word processor, you may find that it has a separate key for the dash. In either case, do not put a space before or after the dash. And remember not to use dashes too often, or they will lose their good effect.

EXERCISE

Put in a dash or replace a comma with a dash whenever you think it would improve the sentence.

1. My friend Yolanda said she just turned twenty-nine for the third time.
2. It's time I started saving for a Florida vacation, for a cruise to the Bahamas, for a new Lexus, for my retirement.
3. All kinds of spices, even pepper, garlic, and onion, give Eddie indigestion.
4. Madonna's costume, what there was of it, shocked even broad-minded me.
5. Marvin had only one chance and a slim one at that.

Now examine the sentences in the essay you just wrote. Can you improve any of them by inserting dashes? Try to revise at least two of your sentences using dashes.

WEB SITE

www.efficacy.org/.

To find further information about ways in which "virtually all children can get smart," visit this Web site for the Efficacy Institute.

Student Essay Using Comparison and Contrast

Shopping Online

Dana Webb

1 Do you ever stand in front of your closet and think you don't have anything to wear? Then you look outside and see it's snowing and too cold to go out of the house for a trip to the mall. Maybe you live an hour from the closest mall and don't have time to make the trip. You might even hate trying on clothes in the tiny dressing rooms while being annoyed by the sales clerks bringing five more shirts in your size that you don't even like. Whatever the case, there is an available alternative. With only a computer and a credit card, you can shop online from your very own home. Two of my favorite shopping Web sites, Gap.com and Abercrombie.com, offer fun and easy ways to view and buy merchandise online, but they're set up differently and appeal to different kinds of shoppers.

2 Just as the home says much about the family, the homepage says much about the Web site. Since the homepage is the first thing you see, it's the key to whether you're going to continue shopping. The homepage should tell you what the Web site is all about, but also be interesting enough to draw you in. Gap definitely has a more user-friendly homepage than Abercrombie. Gap has all the clothing options clearly displayed so you can go to the type you're interested in straight from the homepage. It has links for men, women, kids, baby, and maternity clothing, making it easy to find what you're looking for. Abercrombie's homepage goes more for style and looks. It opens with just a picture of one of the striking models Abercrombie is famous for. You click on the model to start shopping. It also contains links to a number of other features: Lifestyle, A&FTV, music, info, ANFMail, and A&Fquarterly. There's only one link to shopping. Abercrombie's homepage is made to get viewers involved; Gap's homepage is designed to let shoppers shop.

3 You may be hesitant to shop online because you think it's too complicated. Actually, both Web sites are easy

to use. They both list and describe all types of clothes with pictures. It is easy to find sizes and colors. The process for finding the type of clothes you're looking for is similar on both sites: once you're inside the section you want (such as men's or women's), you're given a list of types of clothing, such as pants, sweaters, and so on. If you click on sweaters, you then see a list of all the sweaters to look at. On both sites, you can put everything into a shopping bag, so that all your purchases are together when you are ready to check out. Even though the process of shopping is basically the same for the two Web sites, there are some important differences.

In order for one Web site to be more successful than another, it has to have something unique. That is exactly what Abercrombie.com has. It has something called "the dressing room." Once inside the dressing room, you can click on pictures to mix and match any top you want with any bottom article of clothing to see how they look together. This is an enjoyable way to shop online, and it helps you decide what you want. Abercrombie also has a special deal running during the Christmas season called the Wishlist. You can make a list of things you would like, and e-mail this list to a family member or friend—an effective way to get the word out about Abercrombie. 4

Gap.com also has some unique links that Abercrombie does not have. The Gap homepage has an easy return policy, which gives you assurance about purchasing merchandise online. Gap.com also has links to Old Navy and Banana Republic on the homepage, which are under the same line as Gap. These are the only links Gap.com has other than for shopping. 5

By contrast, Abercrombie.com has a variety of things to do other than shop. You can apply for an Abercrombie credit card. You can sign up for an e-mail account to receive news about the latest styles; you can also e-mail Abercrombie and tell them what you think about the Web site. Since the models are such a big draw for Abercrombie, you can, of course, download pictures of models and make them desktop backgrounds and screen 6 savers. There is even information about how to apply to be a model. Abercrombie.com also has media links to MTV.com, ComedyCentral.com, Nintendo.com, and SportsIllustrated.com.

7 The options available on Abercrombie.com reflect its customer base. Abercrombie is a store with a distinctive style of merchandise marketed for teenagers, so its Web site appeals to the hip, energetic lifestyle of young people. Gap, on the other hand, has clothing for infants, children, teens, moms, and dads. Its Web site appeals to people who want to buy clothes, not party on the Internet. The two Web sites accurately represent the two different stores.

CONSIDERING CONTENT AND METHOD

1. What is the **tone** of this essay? Point out several sentences or phrases that let you identify the tone.
2. Does the introduction catch your interest? How?
3. Who is the **audience** for this essay? How can you tell?
4. What is the **purpose** of this essay? Is the author trying to persuade her readers of anything?
5. Does Dana provide adequate, appropriate evidence to support her views about each site? Does she give you a clear picture of what the sites look like and how each works?

Strategies for Explaining How Things Work: *Process* and *Directions*

"Here you go, right on page 13 of the manual: 'Never stop walking while the treadmill is on.'"

Listen to our rural relatives explaining how to get to the family reunion: "Just follow the hard road down to where the Snivelys' cow barn used to stand before the fire; then turn off on the gravel track and go a piece until you get to the top of the second big rise after the creek. Look for Rabbithash's old pickup." All eighty-six aunts, uncles, and cousins find these directions perfectly clear, but anyone from farther away than Clay City is going to have some difficulty getting there before the potato salad goes funny.

THE POINT OF WRITING ABOUT PROCESS AND DIRECTIONS

When you want to include second-cousins-once-removed and even complete strangers in your audience, you will try to write out directions that don't rely on so much in-group information (such as where the Snivelys' barn *used* to be). Communicating directions so that almost anyone can understand them is a difficult task, as you know if you have ever tried to do it. Explaining a process is quite similar: after you've performed a certain task over and over, it's hard to explain to someone else exactly how it's done. Do you remember how hard it was to shift the manual transmission while turning the corner when you first learned to drive? Could you easily explain it to a sixteen-year-old? And when a child asks, "Where does the rain come from?" wouldn't most people rather come up with a cute story than really grapple with the workings of nature?

But sometimes, frequently on the job, you must come up with an orderly, step-by-step explanation of how something is done or how it works. The explanation may be just part of a larger report or essay; for example, proposing a solution to the company's mail problems must include an account of how the current mail system works. This chapter includes models of several types of **process writing.**

THE PRINCIPLES OF PROCESS AND DIRECTIONS

The basic organizing principle behind process and direction writing involves time. You are usually concerned with a series of events, and these events may not float through your mind in the same order they should appear in your written work. Your readers will be frustrated and confused by flashbacks or detours to supply information that you should have covered earlier. Therefore, the scratch outline takes on great importance in this type of writing effort. A blank piece of unlined paper or a new document in your word processing program will help you to get started. On this page, you will list the steps or stages of the process as you first think of them—

only be sure to space the items widely apart. In the spaces you can add points you forgot the first time through: these can be major steps ("Collect the dog shampoo and old towels before you attempt to collect the dog"), substeps ("Pile up more old towels than you think you will possibly need"), or warnings ("Don't speak to your beast in a tone of panicky sweetness; he'll know you're up to something"). Once you consider your notes complete, read through them while visualizing the process to pinpoint anything you have forgotten.

"How to Wash Your Dog" doesn't represent the only angle you can take on process writing, although it is a useful one. In "Shopping Can Be a Challenge," Carol Fleischman uses an extended example to outline the strategies and pitfalls of being a shopper who also happens to be blind. Garrison Keillor's directions on "How to Write a Personal Letter" combine practical and emotional features of the process—not only showing us how to do it but also persuading us that it is worth doing. Emily Nelson's "Making Fake Flakes" reveals a trick of movie-making magic by tracing the prop department's creation of potato-flake "snow" in a variety of settings. And Dave Barry's essay begins with introductory material telling how a Japanese audience views American football before launching into his humorous explanation of the rules of the game. In your future writing projects, all these techniques for explaining a process and giving directions will be useful.

THE PITFALLS OF PROCESS AND DIRECTIONS

The problems you encounter in process writing usually have their roots in understanding your audience. For instance, several years ago when we first joined the Internet, we were assured that the printed directions we received were quite complete. They began, "Once you are connected to the CCSO Terminal, follow these steps to log on." Once we were *what?* To *what?* Obviously, the directions were written for people much more "connected" than we are. You have no doubt had similar experiences; hundreds of cartoons around Christmas time portray frantic parents trying to follow "easy" assembly instructions for their children's toys.

Reviewing Your Process

When you revise your process writing, think about the people who will be reading it. Ask yourself these questions:

1. Have I chosen the best starting point? Think about how much your audience already knows before you decide where to begin describ-

ing the process. Don't assume your readers have background knowledge that they may not have. And remember that your audience may need to be persuaded to learn what you wish to teach.

2. Have I provided enough definitions of terms? See Chapter 5 for help in writing definitions and deciding when they are needed.
3. Have I been specific enough in the details? "Dig a trench" is more specific than "Dig a hole," but how deep should the trench be? How wide? How long?

Addressing Your Audience

Another decision you will need to face in your process writing concerns not only who your audience is but also how you intend to speak to them. In this book, we address you, our readers, as "you." This straightforward, informal voice is desirable in much writing. Sometimes, you can keep the informality yet leave the "you" out, using **imperative** sentences (or commands) such as "Gather the towels before catching the dog," or "Dig a trench two feet wide." You may also choose to describe a **third person** performing or observing the process: "The experienced video game player works quickly," or "The first feature a palm reader examines is the life line, running from between the thumb and first finger in a curve down to the wrist." However you decide to deal with addressing your audience, you should be careful not to mix these approaches accidentally.

WHAT TO LOOK FOR IN PROCESS AND DIRECTIONS

The readings in this chapter differ greatly in their treatment of topics, even though they have process and directions in common. As you read, consider these questions:

1. What are the differences among the introductions? Can you account for these differences by looking at the purpose of each author?
2. How does each author signal where a new step, stage, or part begins?
3. How does each author address the readers? Is this way of addressing readers suitable?
4. Are there any points at which you would like further details or explanation? Where, what, and why?
5. What strategies other than process and directions appear within these readings—for example, narration, description, or comparison and contrast?

IMAGES AND IDEAS

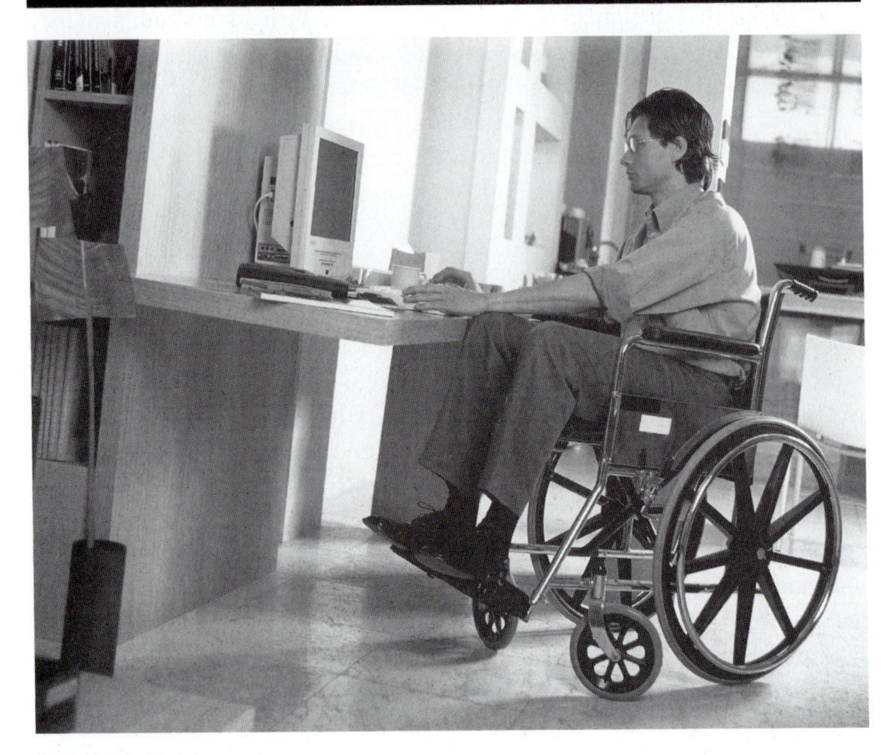

Source: © Javier Pierini

For Discussion and Writing

Computers and other technological advancements have significantly changed the lives of people with disabilities. What kinds of advantages are suggested by this photograph of a writer at work? Jot down a sketch outline of the main steps of your own current physical and mental writing rituals and practices—where and how you like to get started, for example. Next, revise your outline to offer your writing process as a model for a classmate who is wheelchair-bound and might require more flexibility than you included in your plan.

Preparing to Read

How do you respond when you encounter people with disabilities? Have you had any specific experiences that led to significant or even slight changes in your attitudes or behaviors in these situations?

Shopping Can Be a Challenge

Carol Fleischman

Carol Fleischman, who lives in Niagara Falls, New York, is a regular contributor to the Buffalo News. *Her articles cover a wide range of everyday events, from yard sales to wedding showers, but her recurring theme is life as a blind person, especially the joys and challenges of working with assistance dogs. "Shopping Can Be a Challenge" first appeared in 1998 and won the International Association of Assistance Dog Partners "Best Article of the Year Award."*

Terms to Recognize

ritual *(para. 1)*	set form or order of events
beeline *(para. 2)*	straight path, direct route
booming *(para. 5)*	deep, resonant sound
in earshot *(para. 6)*	within hearing distance
chastised*(para. 14)*	strongly criticized, scolded
preoccupied *(para. 17)*	busy, distracted
reminiscent *(para. 18)*	similar to, reminding one of
foiled *(para. 18)*	prevented, stopped

Have you ever tried to buy a dress when you can't see? I have, because I'm blind. At one time, I would shop with friends. This ritual ended after the time I happily brought home a dress a friend had helped me choose, and my husband, Don, offered a surprising observation: "The fit is great, but do you like all those huge fish?" The dress went back. Now I rely on Don and my guide dog, Misty, as my shopping partners. 1

We enter the store and make a beeline for the dress department. Don usually sees two or three salespeople scatter. The aisles empty as if a bomber had come on the scene. Then I realize I'm holding the "live wire." But I'm not judgmental—once I, too, was uneasy around large dogs. Although 2

Misty is better behaved than most children, I know a 65-pound German shepherd is imposing.

3 On one recent shopping trip, a brave saleswoman finally approached us. "Can I help you?" she said to my husband.

4 "Yes, I'm looking for a dress," I replied. (After all, I'm the one who will be wearing it.) "Maybe something in red or white."

5 "RED OR WHITE," she said, speaking very slowly and loudly even though my hearing is fine. I managed not to fall as Misty jumped back on my feet, frightened by the woman's booming voice.

6 Don was distracted too. I heard him rustling through hangers on a nearby rack. I called his name softly to get his attention, and another man answered my call. Bad luck. What were the chances of two Dons being in earshot?

7 "This is great!" Don said, holding up a treasure. I swept my hand over the dress to examine it. It had a neckline that plunged to the hemline. "Hmmm," I said. I walk three miles daily with Misty and stay current with fashion, but I'm positive this costume would look best on one of the Spice Girls. Finally, I chose three dresses to try on.

8 Another shopper distracted Misty, even though the harness sign reads: "Please do not pet. I'm working." She said, "Your dog reminds me of my Max, who I recently put to sleep," so I was sympathetic. We discussed her loss for 15 minutes. Some therapists don't spend that much time with grieving clients.

9 Don was back. He told me the route to travel to the dressing room. I commanded Misty: right, left, right, and straight ahead. We wove our way past several small voices.

10 "Mom, why is that dog in the store?"

11 "Mom, is that a dog or a wolf?"

12 And my personal favorite: "But that lady's eyes are open."

13 I trust parents to explain: "The lady is giving her guide dog commands. Her dog is a helping dog. They're partners." I questioned whether this positive message had been communicated, though, when I heard an adult say: "Oh, there's one of those blind dogs."

14 Other people, though well-intentioned, can interfere with my effective use of Misty. Guide dogs are highly trained and very dependable but occasionally make potentially dangerous mistakes. On my way through the aisles, Misty bumped me into a pointed rack, requiring my quick action. I used a firm tone to correct her, and she dived to the ground like a dying actress. Witnessing this performance, another shopper chastised me for being cruel. I was shocked. Misty's pride was hurt, but I needed to point out the error in order to avoid future mistakes. If I did not discipline her, what would prevent Misty from walking me off the curb into traffic?

15 Composing myself, I was delighted by the saleswoman's suggestion: "Can I take you to your dressing room?" I was less delighted when she

grabbed me and pushed me ahead while Misty trailed us on a leash. I wriggled out of the woman's hold. Gently pushing her ahead, I lightly held her elbow in sighted-guide technique (called so because the person who sees goes first).

"This is better. Please put my hand on the door knob. I'll take it from here," I said. 16

In the room, Misty plopped down and sighed with boredom. I sighed with relief that she was still with me. On one shopping trip, I was so preoccupied with trying on clothes that Misty sneaked out beneath the dressing room's doors. I heard her tags jingling as she left but was half-dressed and couldn't retrieve her. Fortunately, Don was outside the door and snagged her leash. 17

I modeled the dresses for Don and, feeling numb, bought all three. As we left the store, Misty's magnetism, reminiscent of the Pied Piper's, attracted a toddler who draped himself over her. She remained calm, as he tried to ride her. The boy's fun was soon foiled by his frantic mother. When we returned to our car, I gave Misty a treat and lots of praise. A good day's work deserves a good day's pay for both of us. 18

"Shop till you drop" or "retail therapy" could never be my motto. To me, "charge" means going into battle. 19

RESPONDING TO READING

Does shopping ever feel like a battle to you? Do any of your reasons for feeling that way overlap with Fleischman's? Can you come up with other imaginative comparisons for the shopping experience?

GAINING WORD POWER

In paragraph 8, Fleischman writes about a dog who was "*recently* put to sleep." In paragraph 3, we read the phrase "one *recent* shopping trip." Both words share the meaning of new or not long ago. *Recent* is an adjective, describing the noun *trip. Recently* is the adverb form of the same word; adverbs describe words other than nouns, in this case the verb *put.* Most adverbs are made by adding *-ly* to adjectives; for example, *happily* comes from *happy* (para. 1), *usually* comes from *usual* (para. 2), and *finally* comes from *final* (para. 3).

This reading includes ten other adverbs that end in *-ly.* Find at least five of them. Then write their adjective form, checking with the dictionary for spelling if necessary. Then express the shared meaning between the two forms, again using the dictionary. Here's an example to get you started:

Adverb: slowly

Adjective: slow

Shared meaning: at a low speed, not moving quickly

CONSIDERING CONTENT

1. Why did Fleischman stop shopping with her friends? What makes her husband and dog better shopping partners?
2. List the mistakes made by the people Fleischman encounters as she shops? Now make a list of proper etiquette for interacting with a person who is blind.
3. What reasons does Fleishman offer for sighted people's troublesome behaviors? Can you provide other reasons?
4. Why does Fleishman say she is "feeling numb" in paragraph 18? What other emotions does she experience during this shopping trip?

CONSIDERING METHOD

1. Identify the stages of Fleischman's typical shopping trip. Why does she choose not to use a more direct, step-by-step format for her process essay?
2. Describe Fleischman's tone. Does it change over the course of the essay?
3. How does Fleischman offer proof in paragraph 7 that she stays "current with fashion"?
4. In the second paragraph, Fleischman writes "The aisles empty as if a bomber had come on the scene." Find other examples of **metaphor** and word choice that continue this battle metaphor.

Combining Strategies

Fleischman uses an extended example, narration, and description to provide a clear understanding of her shopping process and to achieve a persuasive purpose. What are the advantages of her approach? What other strategies might be used with this topic if one were writing for a different audience, such as trainers of assistance dogs?

WRITING STEP BY STEP

You probably have your own shopping style. Write an essay called "Shopping Can Be ________________," filling in the blank with the word that best captures your feelings about this everyday activity. You will be following the form of "Shopping Can Be a Challenge."

A. Start by asking a question that will catch most readers' interests.
B. Develop your introduction with an illustration of a past shopping experience that shaped your current technique or attitude.
C. Use an extended example of a typical shopping expedition to organize your essay. Be sure to provide concrete support—names, dialogue, description of people and places.
D. Select one of the comparisons you came up with in your Responding to Reading response and use it to help convey your tone and purpose throughout the essay. For example, you might compare shopping to a treasure hunt, which would suggest an entertaining, exciting experience. Also revise your phrasing—nouns, adjectives, adverbs, and verbs—to further communicate your attitude. For example, if your image as a shopper is "The Great Hunter," here is a revision you might make:

First draft: The hunt was almost over, but I managed to buy some discount socks at the last minute.

Revision: The hunt was almost over, but I managed to snare some small game (discount socks) at the last minute.

OTHER WRITING IDEAS

1. Choose an activity that you do, and explain to your readers how they would benefit from doing it too.
2. COLLABORATIVE WRITING. Meet with your classmates to discuss other challenges that people with disabilities experience in their everyday activities. Talk about a range of disabilities and then construct a list of problems commonly encountered by people with these different types of disabilities. Consider things like conducting a conversation or performing simple actions. Write an essay based on the results of your group brainstorming and personal experiences.
3. USING THE INTERNET. Interacting with the blind is also an issue in the workplace. Visit Internet sites offering advice on this topic, such as the U.S. Business Leadership site at **www.usbln.com/bestprac/blind.html**, and then write a process essay on meeting and interacting with the blind in a work environment.
4. WRITING ABOUT READING. Fleischman's dog Misty died a few years after Fleischman wrote this article. She wrote of its devastating effects in another article in April 2001. Using a newspaper database such as Lexis-Nexis, find this article and use it and "Shopping Can Be a Challenge" to outline Fleischman's guidelines for working with an assistance dog.

EDITING SKILLS: USING PARENTHESES

When you want to include an idea that is interesting but not crucial to your discussion, put it in parentheses. Here's an example from Fleischman's third and fourth paragraphs:

> On one recent shopping trip, a brave saleswoman finally approached us. "Can I help you?" she said to my husband.
>
> "Yes, I'm looking for a dress," I replied. (After all, I'm the one who will be wearing it.)

The interaction is captured in the dialogue (speech or conversation recorded in writing), but in the final parenthetical sentence, Fleischman comments on the dialogue, and the parentheses tell the reader that the remark is reinforcement for the implied (but unspoken) point she is making. This example shows the use of parentheses to highlight a mocking tone.

Parentheses are also used in these ways:

1. To enclose brief definitions within a sentence (like the examples in the previous paragraph):

   > I tightly held her elbow in a sight-guide technique (called so because the person who sees goes first).
   >
   > Writers use transitions to improve *continuity* (the flow of their ideas).

   Notice that the period goes after the parenthesis at the end unless the whole statement begins and ends within the parentheses.
2. To enclose examples and brief explanations (like the one in item 1 and this one you are presently reading).
3. To enclose dates within a sentence:

   > John Stuart Mill (1806–1873) favored women's equality with men.

EXERCISE

Find five examples of parentheses in this chapter, and explain which of the uses discussed applies to each example.

Then look at the essay you just completed. Is there any material that should be enclosed in parentheses? Consider adding a definition, a date, a comment, or a brief explanation with parentheses that would add to the reader's understanding of your ideas.

WEB SITES

www.support-dogs.org.uk

The Web site Support Dogs Online offers advice on a variety of uses for assistance dogs, from alerting people with epilepsy about an upcoming seizure to performing everyday tasks, such as turning on lights and opening doors.

Preparing to Read

How do you feel when you receive a letter? Does anyone regularly write to you? To whom do you write, and why? Do you keep letters and read them again later?

How to Write a Personal Letter

Garrison Keillor

Garrison Keillor is a famous radio program host. His show, A Prairie Home Companion, *reaches over 4 million listeners a week and is especially popular because of Keillor's spoken essays about the Minnesota town of Lake Wobegon, a make-believe place where "all the men are strong, all the women are good-looking, and all the children are above average." Listeners are delighted by the charming quirkiness of Lake Wobegon citizens and their everyday lives. In the essay reprinted here from* We Are Still Married *(1989), Keillor's neighborly style comes through as he gives advice and support to letter writers.*

Terms to Recognize

wahoo *(para. 2)*	probably a type of *yahoo,* which is a coarse, crude person
anonymity *(para. 4)*	namelessness, being unknown
obligatory *(para. 6)*	required by custom or etiquette
sensate *(para. 6)*	filled with feelings
sensuous *(para. 8)*	pleasing to the senses
salutation *(para. 10)*	the greeting that opens a letter
declarative *(para. 10)*	making a statement
episode *(para. 13)*	an incident or event, a unit of a longer story
urinary tract *(para. 13)*	the system relating to the kidneys and bladder and their functions
means *(para. 14)*	mode or process

We shy persons need to write a letter now and then, or else we'll dry up and blow away. It's true. And I speak as one who loves to reach for the phone, dial the number, and talk. The telephone 1 is to shyness what Hawaii is to February; it's a way out of the woods. *And yet:* a letter is better.

2 Such a sweet gift—a piece of handmade writing, in an envelope that is not a bill, sitting in our friend's path when she trudges home from a long day spent among wahoos and savages, a day our words will help repair. They don't need to be immortal, just sincere. She can read them twice and again tomorrow: *You're someone I care about, Corinne, and think of often, and every time I do, you make me smile.*

3 We need to write; otherwise nobody will know who we are. They will have only a vague impression of us as A Nice Person, because, frankly, we don't shine at conversation, we lack the confidence to thrust our faces forward and say, "Hi, I'm Heather Hooten; let me tell you about my week." Mostly we say "Uh-huh" and "Oh really." People smile and look over our shoulder, looking for someone else to meet.

4 So a shy person sits down and writes a letter. To be known by another person—to meet and talk freely on the page—to be close despite distance. To escape from anonymity and be our own sweet selves and express the music of our souls.

5 Same thing that moves a giant rock star to sing his heart out in front of 123,000 people moves us to take ballpoint in hand and write a few lines to our dear Aunt Eleanor. *We want to be known.* We want her to know that we have fallen in love, that we quit our job, that we're moving to New York, and we want to say a few things that might not get said in casual conversation: *Thank you for what you've meant to me. I am very happy right now.*

6 The first step in writing letters is to get over the guilt of *not* writing. You don't "owe" anybody a letter. Letters are a gift. The burning shame you feel when you see unanswered mail makes it harder to pick up a pen and makes for a cheerless letter when you finally do. *I feel bad about not writing, but I've been so busy,* etc. Skip this. Few letters are obligatory, and they are *Thanks for the wonderful gift* and *I am terribly sorry to hear about George's death* and *Yes, you're welcome to stay with us next month.* Write these promptly if you want to keep your friends. Don't worry about the others, except love letters, of course. When your true love writes *Dear Light of My Life, Joy of My Heart, O Lovely Pulsating Core of My Sensate Life,* some response is called for.

7 Some of the best letters are tossed off in a burst of inspiration, so keep your writing stuff in one place where you can sit down for a few minutes and—*Dear Roy, I am in the middle of an essay but thought I'd drop you a line. Hi to your sweetie too*—dash off a note to a pal. Envelopes, stamps, address book, everything in a drawer so you can write fast when the pen is hot.

8 A blank white 8″ × 11″ sheet can look as big as Montana if the pen's not so hot—try a smaller page and write boldly. Get a pen that makes a sensuous line, get a comfortable typewriter, a friendly word processor—whichever feels easy to the hand.

Sit for a few minutes with the blank sheet of paper in front of you, and meditate on the person you will write to, let your friend come to mind until you can almost see her or him in the room with you. Remember the last time you saw each other and how your friend looked and what you said and what perhaps was unsaid between you, and when your friend becomes real to you, start to write. 9

Write the salutation—*Dear* You—and take a deep breath and plunge in. A simple declarative sentence will do, followed by another and another. Tell us what you're doing and tell it like you were talking to us. Don't think about grammar, don't think about style, don't try to write dramatically, just give us your news. Where did you go, who did you see, what did they say, what do you think? 10

If you don't know where to begin, start with the present: *I'm sitting at the kitchen table on a rainy Saturday morning. Everyone is gone and the house is quiet.* Let your simple description of the present moment lead to something else; let the letter drift gently along. 11

The toughest letter to crank out is one that is meant to impress, as we all know from writing job applications; if it's hard work to slip off a letter to a friend, maybe you're trying too hard to be terrific. A letter is only a report to someone who already likes you for reasons other than your brilliance. Take it easy. 12

Don't worry about form. It's not a term paper. When you come to the end of one episode, just start a new paragraph. You can go from a few lines about the sad state of pro football to the fight with your mother to your fond memories of Mexico to your cat's urinary-tract infection to a few thoughts on personal indebtedness and on to the kitchen sink and what's in it. The more you write, the easier it gets, and when you have a True True Friend to write to, a *compadre,* a soul sibling, then it's like driving a car; you just press on the gas. 13

Don't tear up the page and start over when you write a bad line—try to write your way out of it. Make mistakes and plunge on. Let the letter cook along and let yourself be bold. Outrage, confusion, love—whatever is in your mind, let it find a way to the page. Writing is a means of discovery, always, and when you come to the end and write *Yours ever* or *Hugs and Kisses,* you'll know something you didn't when you wrote *Dear Pal.* 14

Probably your friend will put your letter away, and it'll be read again a few years from now—and it will improve with age. And forty years from now, your friend's grandkids will dig it out of the attic and read it, a sweet and precious relic of the ancient Eighties that gives them a sudden clear glimpse of you and her and the world we old-timers knew. You will have then created an object of art. Your simple lines about where you went, who you saw, what they said, will speak to those children, and they will feel in their hearts the humanity of our times. 15

You can't pick up a phone and call the future and tell them about our times. You have to pick up a piece of paper. 16

RESPONDING TO READING

After reading Keillor's essay, are you encouraged to try writing a letter to a friend? Why or why not? Answer these questions in your journal.

GAINING WORD POWER

Add to the following fragments, making them into reasonable sentences. Be sure that your additions show your understanding of the terms from the reading.

1. President Weber began his after-dinner speech with an obligatory ________________.
2. To protect her anonymity, the writer ________________.
3. Preparing ________________ is more sensuous than ________________.
4. Henry sat down and wrote this salutation: ________________.
5. An office memo usually begins with a plain declarative statement, such as "________________."

CONSIDERING CONTENT

1. Why are letters preferable to face-to-face communication for shy people? What other reasons might make letters preferable to conversation, according to Keillor?
2. What are some differences between Keillor's advice and other writing instructions or rules you have heard? Why do you think these differences exist?
3. Did you ever have to write a letter or essay "meant to impress" (para. 12)? What was the experience like? How did you feel about the writing you produced?
4. Note two or three spots where Keillor deals with the emotions a letter writer might have. Why is it important to give advice about these?
5. What does Keillor mean by "an object of art" in paragraph 15? What has changed the ordinary letter into art? Do you think this claim makes sense or exaggerates?

6. What about using e-mail or blogs or listserves? Does Keillor's essay have any relevance to these electronic forms of communication? Do you think he would see them as similar to or different from postal letters?

CONSIDERING METHOD

1. Make a brief list of Keillor's pieces of advice. What is the reasoning behind the order he uses?
2. Who is the "we" Keillor refers to in the essay? Who is the "you"? What kinds of people might think the essay does not apply to them?
3. Does this writing strike you as formal or **informal?** Point out words and phrases that influenced your decision. Why do you think Keillor made this choice about the level of formality?
4. What are the lines and phrases printed in italics? Can you identify their purpose?

WRITING STEP BY STEP

In the reading, Keillor advocates writing a letter, even when a phone call is possible. Write an essay in which you promote doing something in the old-fashioned way even though new ways are available. For possible inspiration, think about writing by hand rather than on a word processor, baking bread, sewing clothes, building furniture, doing math without a calculator, reading a novel rather than watching the movie, conducting a courtship, or raising children.

A. In the beginning of your essay, suggest one or two reasons the old way might be better than the new way.
B. Use "I," "we," and "you" to refer to yourself, yourself and your readers, and your readers.
C. Use everyday language and familiar examples to get your points across to a wide audience.
D. Explain why some people feel hesitant about doing things the old way.
E. Suggest ways that the reader can overcome this reluctance. Look at paragraphs 6 through 13 for examples of how Keillor does this.
F. Include, if relevant, some "don'ts" to help your reader avoid problems, as Keillor does in paragraphs 10, 12, and 13.
G. In your closing, reinforce your main point by looking at the positive effect(s) the actions you promote could have. You might look into the future, as Keillor does.

OTHER WRITING IDEAS

1. Try writing a letter to a friend, using Keillor's advice if you want. At the same time, take notes about how you go about performing the task. Record the thoughts and feelings that you experience along the way as well as the techniques you use. Write an essay describing the experience of writing the letter. Use direct quotations from the letter in italics to illustrate your points.
2. COLLABORATIVE WRITING. The reading emphasizes writing letters as an outlet for shy people. Another form of communication, public speaking, brings out the shy side of almost everyone. Meet with a small group of your classmates to discuss your emotional responses to public speaking. Include solutions you have come up with for handling these feelings. Write an essay modeled on Keillor's in which you give emotional and practical advice to a person who reluctantly must make a speech or give a presentation.
3. USING THE INTERNET. Visit the "Rule 1: Remember the Human" link on the Netiquette Web site at **www.albion.com/netiquette/rule1.html**. Then take a look at "A Beginner's Guide to Effective E-mail" at **www.webfoot.com/advice/email.top.html**. Based on the advice given on these sites and your own experience, write a brief but usable set of instructions on writing e-mail for someone new to this form of communicating.
4. WRITING ABOUT READING. Keillor asserts that "writing is a means of discovery" (para. 14). What does this mean? Write an essay about a piece of writing you once did (or tried to do) that led you to an unexpected discovery. The discovery could be about yourself, the writing process, the subject matter, school, the intended audience, or a combination of things.

EDITING SKILLS: BUSINESS LETTERS

Although Keillor gives good advice about writing personal letters, you will also be faced with writing business letters—applications for jobs, requests for action or information from people you don't know personally, or explanations of your proposals or ideas to other people in your workplace. Here is a diagram explaining the format of a business letter:

```
7012 E. Front St.
Bloomington, IL 61701-5413
December 11, 2004
```

Your address: Provide your own address in case the letter gets separated from the envelope. Use letterhead stationery instead if you have it.

Today's date: Your reader will know exactly when you wrote the letter.

Dr. Gina Fisher
Department of Communications
Illinois Valley Community College
Oglesby, IL 61348

Inside address: Give the address where you are sending the letter. If the envelope gets ruined in the mail, the letter can still be delivered.

Dear Dr. Fisher:

Salutation: Use a colon after the greeting. If you do not know the exact name of the person you are writing to, use a title: Dear Director of Admissions.

I was your student in a Communications 211 course two years ago, and I am writing to ask you a favor.

The course was extremely interesting for me, and on the basis of some of the readings you assigned, I decided to major in psychology here at Illinois State.

Since taking your course, I have moved twice and have lost the bibliography about helping behavior that you gave us. This list of articles would be helpful to me again in my present courses. If you still have copies of it, I would greatly appreciate your sending me one.

Thank you for your trouble, and thank you, too, for starting me on a rewarding course of study. I am enclosing an SASE for your reply.

Ending: Say something positive if at all possible. "Thank you for your attention" is a good all-purpose ending.

SASE: Stands for **s**elf-**a**ddressed **s**tamped **e**nvelope. Enclosing one of these, addressed to yourself with postage already attached, will increase your chance of a quick reply.

Sincerely,

Complimentary close: *Sincerely* is fine for almost any case. Save cute closings for personal friends.

Leave four blank lines for your signature.

Susan Lee

Type your full name.

EXERCISE

After studying the form, type a brief letter from Dr. Fisher to Susan Lee telling her that the requested bibliography is being sent to her along with another list of books that Susan may find interesting.

WEB SITE

http://prairiehome.publicradio.org/features/

Take a look at this link on the Prairie Home Companion Web site. Do you see connections between the way Keillor approaches letter writing and his radio show's content and style?

Preparing to Read

How important are special effects to your enjoyment of movies and TV? When you watch TV sitcoms and dramas, do you notice when something seems wrong or fake or out of place?

Making Fake Flakes

Emily Nelson

Emily Nelson is a staff reporter for the Wall Street Journal. *Her reports are often feature articles drawn from recent business and entertainment news. "Making Fake Flakes" first appeared in the* Journal *in June 2003.*

TERMS TO RECOGNIZE

riddled *(para. 1)*	punctured with numerous holes
props *(para. 2)*	articles or objects used by actors
sloshing *(para. 2)*	splashing, floundering
formula *(para. 3)*	a set model or prescription
alien *(para. 4)*	outsider, a being from outer space
lavish *(para. 4)*	abundant, extravagant, grand
acrylic *(para. 5)*	synthetic plastic
dubbing *(para. 6)*	inserting, substituting for the original
mussed *(para. 8)*	made messy, disordered
pulverizes *(para. 11)*	crushes, breaks into pieces

A special-effects whiz in Hollywood, Howard Jensen can create fireball explosions and make cars look riddled with bullets. But his current job—to create falling snow for TV shows such as "Boston Public"—is one of the hardest, he says. So he turned to potato flakes. 1

Turn on the TV, and that gentle, romantic snow falling is, quite likely, the same stuff sold at the supermarket for making mashed potatoes. Props crews can easily load the potato flakes into rolling drums on cranes that sprinkle down flakes. Industrial fans, called "Ritters" in the industry after their inventor, keep the flakes aloft. From afar, the effect looks just like idealized snow. Up close, however, potato flakes look like . . . potato flakes. 2

They don't melt when they land on a person or on the ground. And they have a way of turning into potatoes. The cast of "ER" once found itself sloshing through mashed potatoes when it rained during filming.

3 Snow, it turns out, is a big production worry for TV studios these days. For years, much of TV was set indoors in the same style that made Archie Bunker's living room so familiar in "All in the Family." Today, viewers raised on big-budget movies expect greater realism and seasonal, outdoor shots in TV productions, a formula that's worked well for NBC's "Law and Order." Many sitcoms, including "Malcom in the Middle," regularly use outside scenes. Moreover, networks are asking shows to change seasons with the calendar, which gives viewers a sense they're watching in real time.

4 But while TV studios can create talking animals and alien neighbors, snow can be trickier. Unlike rain, snow doesn't make any noise. For a good thunderstorm, technicians simply mix in lightning and raindrop sounds on a soundtrack. Even a frosty puff of breath is hard to reproduce without expensive digital manipulation. And unlike lavish movies that travel to wintry locales and use elaborate post-production digital effects, budgets in television are often tight. Production schedules are so short that most dramatic shows must crank out an episode every eight days, leaving little money or time to add in special effects.

5 Depending on a script's demands—blizzard or flurries?—TV crews mix in a variety of materials. Besides potato flakes, firefighters' foam, acrylic flakes, soap and starch products, slivers of plastic (which can sting actors in high winds), polyester blankets (best for roofs and backdrops), and real ice are popular.

6 Snow scenes on TV call for unusual shortcuts. Characters on a sitcom standing in the snow, for example, rarely talk. That's because the snow-blowing machine makes a racket—usually replaced with background music alone, to save the expense of dubbing in dialogue. When the plot on "Dawson's Creek" called for the cast to go on a ski trip, the outdoor shots were close-ups, allowing the crew to cover just a small area with fake snow—in this case a white foam used by fire departments.

7 "We've attempted many ways, and I can't say any of them look too convincing," says Paul Stupin, the show's executive producer, who says small TV screens are more forgiving than movie screens. "Dawson's Creek," set in Massachusetts, also takes meteorological license, filming in North Carolina where it rarely snows. On "Friends" recently, the cast admired a snowfall by looking out the window—a cheap shorthand that requires attaching a tumbler above the window frame, which sprinkles white plastic flakes. Producers hoped viewers didn't notice that in the same scene, Chandler walks in without so much as a flake on his coat.

8 Jonathan Prince, executive producer and writer for NBC drama "American Dreams," wanted snow so badly for a recent shoot that he changed

the plot of an episode to afford it. For a romantic but fake snowfall on star Meg Pryor's first kiss, Mr. Prince had Meg and her family arrive late to church, thereby avoiding the need to hire congregation extras. In search of the perfect setting for a snowball fight, Mr. Prince ordered in truckloads of crushed ice. On the day before filming, as the cast played with the ice on the Los Angeles lot, Mr. Prince watched, in horror, as hair and make-up got mussed and, even worse, it looked like they were throwing white rocks.

9 "Real snowballs don't explode in powdery fluff the way we wish they do," Mr. Prince says regretfully. "They only do that on television." So the props crew brought in potato flakes. "We were throwing mashed potatoes at each other," he says. "There are no runny noses. It's perfect TV."

10 Sometimes TV shows in colder climates manage to write snow out of the script to avoid finding or making it, as when writers for "Everwood," a new TV drama set in Colorado, added an unusually warm breeze that the town celebrates with a festival in one episode. They reversed the weather in two other episodes, writing in a blizzard that keeps everyone housebound and keeps the cameras inside. A show where the cast never gets outside is called "a bottle episode."

11 Another problem with real ice is that it's expensive. Potato flakes cost about $90 per 50-pound box; last year, a winter episode of "Ally McBeal" required five boxes. North Hollywood Ice Co., charges $130 per ton of ice, usually with a 10-ton minimum, plus tax and $35 an hour each for five operators. Sold in large blocks, the ice is fed through a noisy snow blower that chips it, pulverizes it, and blows it out airborne through a hose. This is good for laying snow on the ground but not falling snow. Crushed ice doesn't float—it falls with a thud or melts. Potato flakes or plastic versions fall more realistically. "We still haven't figured out a way to get it lofting down and pretty," says Fred Rymond, manager at North Hollywood, which hauled 30 tons of ice, in blocks weighing 300 to 400 pounds each, to a set for a recent blizzard on "Boston Public."

12 Idahoan Foods sells its potato flakes in major supermarket chains such as Albertson's and Safeway, but it also ships the same stuff in 800-pound shipments to special-effects companies, says Ruth Shriver, a sales coordinator at the Lewisville, Idaho, potato processor. When Ms. Shriver spotted snow falling while watching "The West Wing" recently, she said, "Is that potato flakes? It could be my stuff."

RESPONDING TO READING

Will having learned how artificial snow is made change your experience of watching TV? Will it take away from the romantic mood the director is often aiming for in outdoor winter scenes, for example?

GAINING WORD POWER

romantic	seasonal	frosty
wintry	dramatic	powdery

The preceding words are all used as *adjectives* in the reading. That is, they describe or modify other words (nouns) to make their meaning more specific and vivid. Look back through the reading to find the adjective-plus-noun combinations. We have done the first example for you.

romantic ____snow____	wintry ______________
seasonal ______________	dramatic ______________
frosty ______________	powdery ______________

The adjective *romantic* obviously comes from the noun *romance*. Using your dictionary, find the nouns that are related to these other four adjectives:

adj.:industrial	noun:______________
adj.:expensive	noun:______________
adj.:digital	noun:______________
adj.:meteorological	noun:______________

Now list one other related word, its meaning, and its part of speech for the four adjectives. Use your dictionary for help. For example, *expensive* is related to the verb *expend,* which means "to lay out." Both have to do with spending. Choose one set of three related words and try using them in three sentences of your own.

CONSIDERING CONTENT

1. What are the most common materials used to create fake snow on TV?
2. Why do TV characters who are simply standing in snow stay silent? What other pitfalls do the special effects people run into when using artificial snow and how do they solve each problem?
3. What do time and cost have to do with the decisions TV studios make about their use of special effects?
4. What's a "Ritter"? What other production terms are introduced and defined in the essay?

CONSIDERING METHOD

1. Why does Nelson begin her introduction with images of "fireball explosions" and cars "riddled with bullets"?
2. How many different snow-making techniques are described in the essay? Can you identify a plan for how Nelson ordered her examples?
3. Find examples of Nelson's use of humor in "Making Fake Flakes." Why did she choose this tone for a process essay?
4. Why does Nelson use so many different TV shows to illustrate her points? Would a single extended example work?
5. Why does Nelson include the **anecdote** about Ruth Shriver at the end of the essay?

WRITING STEP BY STEP

Choose a topic and a process similar to the ones in "Making Fake Flakes." Your topic should be something everyone is familiar with, like snow on TV; however, the details of the process itself should offer readers a few surprises or new insights, like how difficult it is to create that snow using potato flakes and ice. You might, for example, reveal some behind the scenes secret about a job you have performed (dipping perfect softserve ice cream cones, frying perfect fries, quieting uncontrollable children) or a hobby you have perfected (staying upright while windsurfing, keeping up your stamina on a cross-country bike race, sewing a designer knock-off at half the price).

A. Direct your writing to a general audience. Assume that the majority of your readers would not be familiar with the details of the process, that most of them, for example, have never worked at a fast-food restaurant or sewn their own clothes.
B. Begin by calling your readers' attention to the difference between what they think they know about your process and the point you wish to make about it. Nelson, for example, wishes to counter the idea that complex special effects are more difficult to achieve than simple ones.
C. Because your purpose is to inform, not instruct, organize your essay as Nelson does, providing the what, when, how, and why of your process by providing several successful examples and a few failed ones.
D. Develop your essay with specific details and colorful, concrete language to keep your reader's interest and attention. Nelson, for example, describes cars "riddled with bullets," and amuses her readers with images of the cast of *ER* "sloshing through mashed potatoes when it rained during filming."

E. If possible, conclude your essay with an amusing **anecdote** as Nelson does.

OTHER WRITING IDEAS

1. Write an essay in which you use process to explain a problem and then suggest a solution. For example, show how being overweight is a problem and then offer guidelines for losing weight. Other topics are smoking, excessive alcohol consumption, lack of experience, poor study habits, and chronic lateness.
2. Collaborative Writing. Identify and explain the stages of an emotional process, such as falling in love (or falling out of love). Work with a group of your classmates to come up with emotional processes and ideas about the different stages before you write your individual essay.
3. Using the Internet. Look on the Internet, using a search engine such as Google or Yahoo, to find a complicated set of instructions, like those for programming a cell phone. Study them until you figure them out. Then write a new set of instructions that a beginner would find easy to follow.
4. Writing about Reading. "Making Fake Flakes" is about creating illusions, making what's not real seem real. Have you ever tried to create an illusion, to make people believe in something that wasn't true? For example, did you ever convince your parents you were ready to take on adult responsibilities, when, in fact, you weren't quite sure yourself that you were ready? Select such an experience and explain how you achieved your goal of creating an illusion.

EDITING SKILLS: PUNCTUATING QUOTATIONS

Quotations from experts and from your reading will often enhance your writing. However, most people find the punctuation of quotations tricky. This exercise will help you put the periods, commas, and quotations marks in the right places. (For advice on using single quotation marks inside regular double quotations marks, see pages 200–201.)

EXERCISE

Copy the following passages *exactly:*

"We've attempted many ways," says Paul Stupin, the show's executive producer, "and I can't say any of them look too convincing."

> "Real snowballs don't explode in powdery fluff the way we wish they do," Mr. Prince says regretfully. "They only do that on television." So the props crew brought in potato flakes. "We were throwing mashed potatoes at each other," he says. "There are no runny noses. It's perfect TV."
>
> When Ms. Shriver spotted snow falling while watching "The West Wing" recently, she said, "Is that potato flakes? It could be my stuff."

Notice that you can put the tag line (the part that tells who is being quoted) before the quotation, after the quotation, or in the middle of the quotation. Exchange your copies with a classmate and check each other's writing for exact, accurate placement of all the punctuation. Then look at your process essay to make sure you have punctuated quotations correctly.

WEB SITE

http://stuff.howstuffworks.com/question171.htm

At this HowStuffWorks.Com site, the complex real-time and digital processes for creating the *Star Wars* light sabers are concisely and clearly explained. Links on the site take you to explanations of other special effects techniques, as well.

Preparing to Read

Did you ever watch a sporting event (such as rugby, curling, cricket, squash, soccer, or even football, baseball, or basketball) without knowing the rules of the game? How did you figure out what was going on? Would you have enjoyed it more if you had known the rules?

There Are Rules, You Know

Dave Barry

Dave Barry, who was born in 1947 in Armonk, New York, is one of America's most popular humorists. A graduate of Haverford College, Barry began his career as a reporter for the Associated Press in Philadelphia. He is now a syndicated columnist for the Miami Herald *and the author of more than twenty humorous books. He won the Pulitzer Prize for commentary in 1988. In the following essay, which appeared on January 30, 2000, Barry focuses his usual wit and irreverent humor on the subject of one of our national obsessions—the Super Bowl.*

Terms to Recognize

blitzkrieg *(para. 1)* a fierce offensive attack
ligament *(para. 1)* a band of tissue connecting bones or cartilages
hurling *(para. 3)* an Irish game resembling field hockey
kilometer *(para. 4)* metric unit of measure, 1000 meters
extraction *(para. 6)* origin, lineage

We are coming upon the Super Bowl, which is by far the most important sporting event in the world as measured in total tons of free shrimp consumed by sportscasters. This year the Super Bowl will be broadcast to many foreign nations, which, almost by definition, contain numerous foreigners. These people are often puzzled by American football, a highly complex sport that requires a knowledge of many technical terms such as "run," "pass," "cornerbacker," "blitzkrieg," "Texas Leaguer," "ligament," and "Hank Stram." This complexity makes the game difficult for foreigners to grasp. 1

I know this because some years ago, while visiting Japan, I watched the Miami Dolphins and the Oakland Raiders play a demonstration game in a Tokyo stadium where, for a zesty snack, you could buy pieces of fried oc- 2 topus on a stick. The fans were polite, but they had no clue what was going on. The only thing that aroused their interest was the Dolphins cheerleaders. The game would stop for a time out, and the cheerleaders would start jumping around, and immediately the fans would go WILD, cheering and thrusting their octopus nuggets into the air.

I'm not being critical here. I've been on the other side of this coin. While visiting Ireland, I watched an Irish sport called "hurling" (really) in which men who are not wearing helmets basically beat each other senseless with sticks. In terms of violence, this sport makes American football look like "Pat the Bunny." I'd never seen this sport, so I relied on the fans around me to answer my questions ("Is that player dead?" "Did all that blood come out of his EAR?" etc.). So I know how hard it can be to understand a foreign sport, which is why today to help you foreign persons follow the Super Bowl, I am presenting: THE RULES OF AMERICAN PROFESSIONAL FOOTBALL. 3

Football is played on a field that is 100 yards (374 kilometers) long and is covered with lines called "hash marks" to indicate where players have lost their breakfasts. On either side of the field are the benches, where the 350 players who are not involved in the game sit and wave to their moms. Behind each bench is a big plastic jug of Gatorade. The object of the game is to be the first team to dump this on the "coach," a very angry man who hates everybody. 4

The game is divided into four 15-minute quarters, each of which lasts a little over three hours. Timeouts may be called by anybody at any time for any reason, including political unrest in Guatemala. Between the second and third quarters, there is a half-time musical extravaganza in which Neil Diamond, Toni Tennille, the Muppets, and the late Al Hirt join with every human being who has ever auditioned for "Star Search" to perform "A Tribute to Medleys." 5

The game begins when a small man of foreign extraction kicks the pigskin, or "ball," as far as possible, then wisely scuffles off the field. The referee then places the ball on an imaginary "line of scrimmage," which is visible only to the referee and his imaginary friend, Mr. Pootywinkle. On either side of this line, the two teams form "huddles," where they decide who will perform the traditional celebratory dance when the upcoming "play" is over. 6

The "play" itself happens very quickly so you foreign persons must not blink or you'll miss it. Here's what happens: 7

1. A large player called the "center" squats over the ball, and then the "quarterdeck" touches him in a way that would get them both executed in the Middle East.
2. All the players run into each other and fall down.

3. Certain players leap to their feet and perform celebratory dances, while referees add to the festivity by hurling brightly colored flags into the air.

Now comes the heart and soul of football: Watching slow-motion replays of the players falling down. You'll see this from every possible point of reference, including the Hubble telescope. You'll see so many replays that at some point you'll swear that, in the background, you can see Mr. Pootywinkle. 8

When the replays are finally over, the referee formally announces that the play does not count. Then it's time for eight commercials featuring sport utility vehicles climbing Mt. Everest, and it's back to the huddles for more non-stop action! 9

Yes, foreign persons, football is a complex sport, but you'll find that if you take the time to watch this year's Super Bowl, you will soon discover why every year so many millions of Americans are glued to their television sets. Watching rental videos. 10

....

RESPONDING TO READING

Are you a football fan? Did being or not being one influence your response to the essay? Do you think Barry is a dedicated football fan himself? What makes you think so?

GAINING WORD POWER

Barry twice uses the unusual word *celebratory,* which takes the familiar verb *celebrate* and turns it into an adjective (a word describing something) by adding the **suffix** *-ory.* When making adjectives, the suffix means *of, relating to, characterized by, serving for, producing,* or *maintaining*—as in *migratory* (the verb *migrate* plus *-ory*). This same suffix can also turn verbs into nouns (names of persons, places, or things) as in *observatory* (the verb *observe* plus *-ory*). When forming nouns, the *-ory* means *a place for* or *something that serves for.*

Identify the following words as either adjectives or nouns; then use each one in a sentence. What verbs do they come from? Consult your dictionary if you're unsure of the meaning.

1. crematory
2. sensory
3. oratory
4. laudatory
5. accusatory

6. exploratory
7. obligatory
8. mandatory
9. congratulatory
10. laboratory

CONSIDERING CONTENT

1. What's funny about Barry's list of "technical terms" in the opening paragraph?
2. Why is the Irish game of hurling (as Barry describes it) more violent than American football (para. 3)?
3. What are Barry's criticisms of football referees in paragraph 6? Do you think they are justified?
4. Why does the author describe the coach as "a very angry man who hates everybody" (para. 4)?
5. What does the last line tell you about Barry's attitude toward the Super Bowl?

CONSIDERING METHOD

1. Do you think the "foreign persons" Barry addresses would understand professional football any better after reading his explanation of the rules? Who is the real **audience** for this piece? Why does the author adopt this fictional audience?
2. How does the device of the "foreign persons" serve to unify the essay (see paras. 6 and 7)?
3. Where does Barry state his **thesis?** Why do you think he placed it there?
4. Name the figure of speech involving exaggeration that Barry employs for humorous effect (for instance, in para. 4: "the 350 players who are not involved in the game"). Find several other examples in paragraph 5.
5. Why is the last sentence written as a fragment? Would it be more effective if written as a complete sentence?

WRITING STEP BY STEP

Barry's tone is clearly humorous, as he makes fun of professional football and the hoopla that surrounds it. In your essay, think of some sport or cultural practice that involves rituals and present the "rules" that are involved. Consider, for instance, professional wrestling, NASCAR racing, formal weddings, graduation ceremonies, fraternity initiations, or rock concerts. If you have a flair for humorous writing, make fun of the procedure. If not, present the process in a neutral tone to inform someone who might want to know all about it. The effectiveness of your essay in either case will depend on your use of well-chosen, specific details to illustrate each "rule."

A. Begin your essay, as Barry does, with some background on the procedure you're going to explain. Try for an entertaining **anecdote** like the one in the Tokyo stadium involving cheerleaders and octopus on a stick (para. 2). Work in your thesis gracefully before moving on to the body of your paper.
B. Describe the place in which your chosen ceremony or entertainment event typically takes places (as Barry does in para. 4).
C. Briefly mention the purpose of the event. Barry does so in a single sentence at the end of paragraph 4.
D. Go through the procedure chronologically, providing details. If there are actual rules involved, consider numbering them.
E. Conclude with your positive or negative evaluation of this event.

OTHER WRITING IDEAS

1. Write the paper outlined in the previous section, but instead describe the steps of something you created that didn't turn out as planned. Then tell the outcome. Perhaps you devised a scheme that backfired or performed a lab experiment that flopped; maybe you botched a project at work, couldn't assemble a bicycle, or tried a weight-loss program that didn't work.
2. Collaborative Writing. Write a humorous essay in which you explain how *not* to do something: the wrong way to study for a test, the wrong way to impress a first date, the wrong way to make lasagna, and so on. Get together with a group of classmates to brainstorm topics and strategies for making the essay humorous.
3. Using the Internet. Find an Internet site that explains the rules of the sport you wrote about in your Preparing to Read response—or of any sport that you want to know more about. After studying the rules, write your own set of basic instructions for someone wanting to play the sport.
4. Writing about Reading. Barry's humorous essay suggests several serious criticisms of both football and its fans. Make a list of these, and then write an essay defending the sport and those who love to watch it. Or write an essay that suggests other problems you associate with the sport and its followers.

EDITING SKILLS: USING COMMAS

Writers put a comma after a word or group of words that comes in front of the main part of the sentence. That main part is called the **independent clause** because it can stand alone as a sentence. Look at these examples from Barry's essay (italics added):

> *While visiting in Ireland,* I watched an Irish sport called "hurling" (really) in which men who are not wearing helmets basically beat each other senseless with sticks.

> *On either side of this line,* the two teams form "huddles," where they decide who will perform the traditional celebratory dance when the upcoming "play" is over.

> *When the replays are finally over,* the referee formally announces that the play does not count.

The italicized words and groups of words that come before the commas are called **dependent elements:** they cannot stand alone as sentences and are not really necessary to the meaning of the sentences.

Writers also put commas before and after dependent elements when they come in the middle of an independent sentence. Here are some examples from Barry's essay (italics added):

> I know this because some years ago, *while visiting Japan,* I watched the Miami Dolphins and the Oakland Raiders play a demonstration game in a Tokyo stadium where, *for a zesty snack,* you could buy pieces of fried octopus on a stick.

> Yes, *foreign persons,* football is a complex sport. . . .

These dependent elements are called **interrupters:** they break up the flow of the sentence and are not really necessary to the meaning of the sentence.

EXERCISE

In Barry's essay, find two more examples of introductory words or groups of words that are separated from the independent clause by a comma. Then find two more examples of interrupters that are set off by commas.

Now copy the following sentences, putting in commas where needed:

1. The rabbit nevertheless does shed fur.

2. Like most people Ralph does not know the name of his congressional representative.
3. Speaking of travel would you like to go to Seattle next week?
4. Jazz some people believe is American's greatest contribution to the arts.
5. The result as we have seen is not a pretty one.
6. The patient said the nurse is acting strangely.

Finally, check the essay you have just written to be sure you have used commas to separate dependent elements from the rest of the sentence.

WEB SITES

http://weblog.herald.com/column/davebarry/

Join Barry and his readers in keeping his daily blog (an online collaborative journal).

Student Essay Using Process and Directions

A Graceful Stride

Ann Moroney

1 A common misconception about hurdle races is that the runner "jumps" over the hurdle. This observation is completely false. A hurdle runner does just that—runs. The hurdler takes long steps, or strides, over the hurdle and sprints on to the next one. As simple as this process may sound, running hurdles is complicated. For one thing, the hurdle itself is thirty-three inches high, which comes to about the hip. "Stepping" over something this high is really quite difficult, as I found out my freshman year in high school. I also had to master a number of other skills to run hurdles efficiently.

2 The first problem to tackle when it comes to running hurdles is the fear. At every one of my races, I sat in my blocks looking at the line of hurdles in front of me with fear in my heart. No matter how good I became or how long I'd been running, I always felt *fear:* fear that I'd trip over the hurdle and fall on the track, fear that I'd knock the hurdle over in front of everybody, and fear that I'd actually get over the first hurdle with no problem and become a hurdler for four years. Throughout my experience, I found only one way to get over the fear: stop thinking and start running.

3 The main thing to concentrate on is form. Form is the way the runner carries himself or herself over the hurdles. The better the form is, the faster the run will be. Form includes many elements. The runner must coordinate legs, arms, torso, and eyes into a fluid sprint through ten hurdles on 100 meters of track.

4 The lead leg, the one to go over the hurdle first, should be slightly bent, but more or less straight in front of the body. The toe should be pointing toward the sky and the heel should line up with the middle of the hurdle. (As my former track coach Ms. Tolefree always said, "Heel to the Gill," referring to the brand name printed directly in the middle of the hurdle.) Just as

the body comes over the hurdle, the lead leg should snap down to the ground. The goal is to bring the leg down as close to the back of the hurdle as possible without actually hitting it. The sooner the feet are on the ground, the faster the runner can continue running.

5 The runner must also concentrate on the trail leg—the one that comes second over the hurdle. The knee should be bent with the leg at the side, the thigh parallel to the hurdle. (Imagine sitting on the ground with one leg bent at the side, mimicking a frog.) As the lead leg snaps down, the trail leg will be following over the hurdle. The hip will rotate so the leg is once again perpendicular to the ground. The knee should snap up to the chest as close as possible, as this will help the runner stretch out the stride. (As Ms. Tolefree was fond of saying, "Your knee is like the scope on a rifle. It directs where your stride will go. If your knee is high, your stride will be long. If it's low, your stride will be short.") After clearing the hurdle, the runner must continue striding through the rest of the hurdles.

6 Another important part of form is arms; they not only keep balance, but they also keep the form tight. The arm opposite the lead leg is the lead arm. It should be reaching out in front of the body toward the toe of the lead leg. The other arm, the one opposite the trail leg, is the trail arm. It should come back in a biceps-flexing fashion and stay at the side until the body is clear of the hurdle. It's usually a good idea to keep the hands open; a closed fist takes up energy and tenses the muscles in the rest of the body.

7 The final elements of form have to do with small, yet significant, mechanics. Air time is a major factor when running hurdles. The goal is to remain in the air for as little time as possible; the faster the feet return to the track, the faster the runner can continue running. So the body should be as low to the hurdle as possible without hitting it. Also, the torso should lean slightly forward—but not be bent, because taking the time to straighten up will slow the runner down. A slight lean forward will get the runner over the hurdle and on to the

next one. The final point involves the eyes. Hurdlers should not look at the hurdle they are about to step over. Looking further down the track instead of directly in front of them keeps runners running.

8 Running hurdles takes time, practice, and commitment. During the split second it takes to clear a hurdle, the runner must take into account a wide variety of movements to run the race successfully. A good, hard practice schedule is the only way to acquire a productive hurdling technique. And developing a graceful stride is not as easy as it looks.

CONSIDERING CONTENT AND METHOD

1. Why does the author begin with a misconception? Is that a good way to begin?
2. What information did you find new and interesting? Was there anything you didn't understand? Did this essay increase your appreciation of hurdling?
3. Were you surprised that this essay was written by a female? Why or why not?
4. Where does the author include personal comments? Do these comments add or detract from the explanations?
5. What do you think of the comments from the coach, Ms. Tolefree? Why are they included?

Chapter 9

Strategies for Analyzing Why Things Happen: *Cause* and *Effect*

BIZARRO by Dan Piraro

Source: Reprinted with special permission of King Features Syndicate.

Human beings are naturally curious. We all want to know why. Why does the car keep stalling? Why are some people better at math than others? Why does a leaf change color? Why did Jennifer Lopez and Ben Affleck separate? This common human impulse to understand why things happen provides a powerful motive for reading and writing.

THE POINT OF CAUSE-AND-EFFECT WRITING

We study **causes** and their **effects** in order to understand events and solve problems. If we can find out why the car keeps stalling, we can fix it. If we can figure out why some people are good at math, maybe we can help those who aren't. If we know why a leaf changes color, we'll have a greater appreciation for nature and its processes. If we know why J-Lo and Ben broke up, we might get on *Oprah*. A lot of the writing done in college courses requires cause-and-effect thinking; students are frequently asked to explain things like the origins of the Russian Revolution, the roots of prejudice, the causes of volcanic eruptions, the effects of hunger on learning, the reasons for Hamlet's delay. The good news about this kind of writing is that it feeds off the natural curiosity of both reader and writer. Inquiring minds want to know why, and you get to tell them.

When you develop an essay by analyzing causes, you are explaining to your readers *why* something happened. If you go on to explore the effects, you are analyzing *what* happened—the consequences. For example, if your topic is divorce and you write "Why Teenage Marriages Fail," that's primarily a cause paper. But if you write "What Divorce Does to Young Children," that's primarily an effect paper. You will probably stick to one **purpose** in a single essay, but you might take up both causes and effects if you have the time and the assignment allows you to.

THE PRINCIPLES OF CAUSE-AND-EFFECT WRITING

Analyzing cause-effect relationships is one of the primary methods of reasoning. It requires careful thinking and planning.

Types of Causes and Effects

When you think about causes and effects, you need to realize that they can be *immediate* or *remote*. The immediate causes are usually the obvious ones; they occur just before a result appears. An immediate cause for breaking up with your boyfriend might be that he didn't call you last night to tell you he'd be two hours late. But you also know there are deeper, more important reasons for the breakup—for example, his habit of forgetting to call and his general lack of concern for your feelings. These are the remote causes, the ones further removed from the effect they produce. They are also called *underlying causes* because they are often more difficult to see.

Effects can also be immediate or remote. The immediate effect of failing to get gas is that your car stops running. But the remote effects can stretch out for quite a while: you block traffic and cause an accident; you're late for class and miss an important lecture; you do poorly on the next exam and get a lower grade in the class; your car insurance goes up because of the accident; you have to change majors because you don't have the required grade point average to be admitted into advanced courses. Remote effects are also called *long-term effects.*

Patterns of Cause and Effect

1. If you want to **focus** on causes, begin by describing a condition or problem or result (such as breaking up with your boyfriend), and then fully explain the causes or reasons (for the breakup). With this approach, you may be able to use **chronological order** (according to time) if you can trace the causes from the earliest to the most recent. More likely, though, your organization will fall into some **logical order** that reflects the relative importance of the causes: from the least significant to the most critical or from personal reasons to more general ones.

2. If you want to focus on effects, start with some condition or event and explain the consequences. For example, you might begin by describing the breakup with your girlfriend and then go on to show how it affected you. Again, you can present the effects chronologically: at first you were depressed; then you began to spend more time with your friends; you also had more time to study, so your grades improved; finally, you began to date again and found a much better girlfriend. Or you can arrange the effects according to importance: from a fairly obvious result to the most subtle, from effects on yourself to effects on other people.

3. Since causes and effects are closely related, you might find that tracing a chain or sequence of events, including both causes and effects, is the best way to approach your topic. In a causal chain the first cause produces an effect, which becomes the cause of the next effect, and so on. In the example about running out of gas, there were two chains. First, running out gas caused a blocked intersection, which led to an accident, which resulted (some time later) in higher insurance costs. Also, running out of gas caused

you to be late for class and miss an important lecture, which contributed to a lower grade, which affected your grade-point average, which prevented you from getting into the advanced program, which caused you to change majors. If you decide to describe a chain of causes and effects, be sure to outline it carefully.

THE PITFALLS OF CAUSE-AND-EFFECT WRITING

An explanation of causes and effects won't be successful if your readers find your thinking fuzzy or flawed. Here are some ways to avoid the most common faults of cause-effect reasoning:

1. Don't mix causes and effects. When talking informally about why things happen, you may shuttle back and forth between causes and their effects. But in writing, you need to follow a clear pattern: focus on causes, focus on effects, or describe an orderly chain of causes and effects.

2. Don't settle for obvious causes and immediate effects. Your explanations will be much more convincing if you look for underlying causes and long-term effects. As Jade Snow Wong shows in her essay in this chapter, her date with a fellow student provided an opportunity to stand up to her parents, but it was not the cause of her rebellion. The "real" causes ran much deeper. It is, of course, possible to go too far back in searching for causes. You'll need to exercise some judgment in deciding which reasons are still valid and relevant.

3. Don't oversimplify. Most conditions and events are complex, involving multiple causes and numerous effects. In a short essay, you may have to concentrate on the primary reasons, but be sure to let your readers know that's what you are doing.

4. Don't omit any key links in a chain of causes and effects. You don't have to spell out every single step in a sequence of events, but be certain that your readers will be able to follow and make all the right connections themselves.

5. Don't worry about absolute proof. In explaining causes and effects, you can't always prove conclusively why something happened. But offer as much evidence as you can to help the reader see the connections that you see. You always need to support your causes and effects with specific details—examples drawn from personal experience, statistics, and statements

by experts. You may want to conduct interviews and collect your own information or visit the library to find material on your topic.

WHAT TO LOOK FOR IN CAUSE-AND-EFFECT WRITING

As you read the selections in this chapter, ask yourself these questions:

1. Does the writer focus on causes or effects, or does the essay consider both? Make a list of the causes and the effects as you read.
2. Look for the point, or purpose, of the author's explanation of causes and effects.
3. Decide what kind of causes or effects the author presents—immediate, remote, or both. Does the writer follow a chain of causal relationships?
4. Does the author have a particular audience in mind? How can you tell?
5. What is the tone of the essay? Does the author sound serious, humorous, angry, irritated, sad, regretful, or something else? How does that tone relate to the author's purpose and audience?
6. Notice what writing strategies the author uses to develop the explanations. Take note of examples, descriptions, narratives, comparisons, definitions, and so on.

IMAGES AND IDEAS

:-)   :-(:-D   :-O   :´(

For Discussion and Writing

What are these symbols? Do you use them?

Write an essay about why these symbols came into use and what has resulted from their use. Some prewriting questions to consider: What purpose do they serve? What effect do they have on the writer and the receiver? Do you think they have the desired effect? Were there any unexpected effects of the creation of these symbols? Do you think they can be understood cross-culturally? Why or why not?

Preparing to Read

Do you tell your parents and your best friends everything? Do you share some things with your folks and other things only with your friends? And what do you keep to yourself? Is revealing the truth always the best policy?

The Biggest Play of His Life

Rick Reilly

Rick Reilly began writing for his hometown newspaper, the Boulder Daily Camera, *while attending the University of Colorado. After graduation in 1981, he wrote for the* Denver Post *and then the* Los Angeles Times, *before moving to* Sports Illustrated *in 1985, where his opinion column "The Life of Reilly" appears weekly. He has been named National Sportswriter of the Year seven times and has written several novels, screenplays, and sports biographies. The following selection ran as one of Reilly's columns in May of 2000.*

Terms to Recognize

sexual orientation *(para. 5)* the direction of one's sexual interest toward the same, opposite, or both sexes

fricasseed *(para. 6)* cut into pieces and stewed in gravy

yahoo *(para. 8)* a crude or vicious person (from the race of brutes in Jonathan Swift's *Gulliver's Travels*)

schmo *(para. 8)* a stupid or obnoxious person (slang)

ruckus *(para. 13)* a disturbance or commotion

visionary *(para. 13)* someone who sees things as they ought to be; an idealist

One of the captains of the high school football team had something big he wanted to tell the other players. "I was so anxious," remembers middle linebacker Corey Johnson, a senior at Masconomet High in Topsfield, Massachusetts, "I thought I was going to vomit." He took a hard gulp. "I want to let all of you guys know something about me." He tried not to let his voice quake. "I'm coming out as an openly gay student here." 1

His teammates' eyes and mouths went wide as soup plates. "I hope this won't change anything," Corey quickly went on. "I didn't come on to you 2 last year in the locker room, and I won't this year. I didn't touch you last year in the locker room, and I won't this year."

3 Awkward silence.

4 "Besides, who says you guys are good enough anyway?"

5 And you know what happened? They laughed! But that's not the best part. The best part is what happened next. Nothing. Corey's teammates had no problem with his sexual orientation. His coach had no problem with it. His mom and dad and his sister had no problem with it. His teachers, his counselor—nobody—had a problem with it.

6 O.K., somebody scrawled FOOTBALL FAG on a door at school. True, one cementhead parent asked coach Jim Pugh to have the team take a new vote on the captaincy, but Pugh told him to stuff it. And, yeah, one week the opposing team's captain kept hollering, "Get the fag!" but his coach finally benched him (and Masconomet fricasseed that team 25–0).

7 No opponent refused to play against Corey. No opposing coach said, "Boys, the Lord wants you to go out and crush that heathen!" Nobody held up a sign at a Masconomet game that read WHICH SIDE ARE YOU ON, COREY? Nope. Corey Johnson, guard-linebacker, wrestler, lacrosse player, just went out and played his senior football season, same as ever. Masconomet did well (7–4 for the season, 25–8 with Corey, a two-way starter for three years). Now Corey is getting on with his life, hopeful as ever. He'll graduate with his class next month, think about playing small-college football, and become a gay activist, a journey that began at the Millennium March on Washington for Equality.

8 Can't wait for Corey to be on a gay parade float when some beer-bellied yahoo hollers, "Hey, girls! Shoe sale next corner!" The football captain might turn the poor schmo into a smudge mark.

9 Corey can take the hits now, but hiding the truth about himself was so depressing in his sophomore and junior years that he let his grades drop, skipped practice, and even skipped school. When an adult friend started ripping homosexuals at a Super Bowl party in January 1998, Corey couldn't decide whether to punch him or cry. He knew he had to do something.

10 First, he told a guidance counselor he was gay and then a few teachers. They all supported him. A year later he told his parents. Fine. Then his best friend, Sean. Uh-oh. Big problem. Sean started crying. Corey asked him what was wrong. "I'm sorry you couldn't share this with me before," Sean said. They're still best friends.

11 Since coming out, Corey says, he has heard from "hundreds" on the Internet, including athletes who wish they had the guts to come out, too. "But," says Corey, "they always say, 'At my school? No way. It'd be impossible.'"

12 At Masconomet, a public school with an enrollment of thirteen hundred, Corey is the football captain who had even more moral courage than phys-

ical. He's admired by his teammates. In fact, nothing much changed between them, except on bus rides home after wins, when the whole team sang *YMCA* together. Well, it isn't *Hunker Down, You Hairy Bulldogs,* but it works.

13 Maybe we're actually getting somewhere in the U.S. A young man who leads young men comes out as gay, and it makes such a ruckus you can still hear the crickets chirp. In fact, last month the Boston Gay, Lesbian, and Straight Education Network handed its Visionary Award not just to Corey, but also to his teammates. Can you imagine that? A high school football team getting an award for *tolerance*?

14 When I was growing up, my best friend was a hilarious kid I'll call Danny. Along about high school, he stopped coming around. Then, in college, he showed up in the Gay Club photo in the yearbook. After that, Danny didn't take my calls. It's a lousy feeling. I guess I'm not the kind of person he could've shared that with.

RESPONDING TO READING

Did the reaction of Corey Johnson's family, friends, and teammates surprise you? Why? Would the reaction have been the same if Corey had been a female? What if he had been less popular or less talented?

GAINING WORD POWER

Look at these two sentences from Rick Reilly's article:

> True, one cementhead parent asked coach Jim Pugh to have the team take a *new* vote on the captaincy. . . .
>
> He *knew* he had to do something.

Why did Reilly use *new* in the first sentence but *knew* in the second? Words like *new* and *knew,* which are pronounced the same but have different meanings and different spellings, are called **homophones.** There are a lot of these sound-alike words in English, and they cause problems for some writers. You can't just learn how to spell these words; you have to match the spelling with the meaning. And your computer's spell-checker won't help you choose the right one. Keeping a list of the ones that give you trouble will increase your awareness and save you time when you edit.

Each of the following words from Reilly's essay has a common homophone. Identify the homophone, and then write sentences that show the differences in meaning.

1. one (para. 1, 6)
2. know (para. 1, 5)
3. week (para. 6)
4. for (para. 7)
5. wait (para. 8)
6. whether (para. 9)
7. hear (para. 13)
8. In paragraphs 10 and 14, Reilly uses the word *then,* and in paragraph 12 he uses *than.* What's the difference between *then* and *than*? Although they're not pronounced exactly the same, their sounds are similar enough to cause confusion in writing. Compose sentences that show the difference in meaning.

CONSIDERING CONTENT

1. The article suggests that the author was quite surprised by the outcome of Corey Johnson's experience. What reaction did Reilly probably expect? Why?
2. How did hiding his sexual preference affect Johnson? What steps did he take to deal with his situation?
3. What was the reaction of Corey's best friend Sean? How do you think Sean's response influenced Johnson's decision to come out to the football team?
4. Reilly says "Maybe we're actually getting somewhere in the U.S." What does he mean? Do you agree?
5. The author says Johnson had even more moral courage than physical. Do you agree?

CONSIDERING METHOD

1. Why does Reilly open with the coming out scene? How else might he have started?
2. What assumptions did Reilly probably make about his original **audience,** the readers of *Sports Illustrated*?
3. Identify several examples of **slang** or **colloquial language** in this article. What other details of style does Reilly use to make his writing sound casual and familiar?
4. What is the point of the final paragraph? Why does Reilly include this information? Is it significant that he says "I'm" (present tense) instead of "I was" (past tense)?

WRITING STEP BY STEP

Write an essay about a time when you were reluctant to reveal something to parents, teachers, classmates, or friends because you feared a negative reaction. Perhaps it was something you had done, a choice you had made, or an opinion that you felt strongly about. Maybe it involved a fear, a physical problem, a strange belief, a hidden accomplishment, or an unusual hobby.

A. Begin by narrating the scene in which you acknowledge or make public what you've been hiding, as Reilly does in the first four paragraphs of his article. If you don't want to use this approach, you could also start by describing what it is that you were afraid to reveal.
B. Give the reasons for your reluctance to reveal the truth. If other people advised you not to say anything, quote their advice directly.
C. Explain why you wanted to tell these particular people and why you expected a negative response. What effects did you expect your revelation to have? Also describe how you felt about not being able to share your secret.
D. Explain why you decided to reveal your secret, and describe what happened when you did. Was the response what you expected, or did the reaction surprise you? Use dialogue to help recreate this scene.
E. Tell how things worked out. Are you still on good terms with the people you told? Give some supporting evidence, as Reilly does in paragraph 12, to show how your relationship was affected by your revelation.
F. Conclude with some comments about how you feel now. Are you glad you revealed your secret? If you had to do it again, would you do the same thing? You might also suggest how others might benefit from knowing your story.

OTHER WRITING IDEAS

1. Think of a prejudice or a belief that offends you. Explain why people think this way and why it offends you.
2. Collaborative Writing. Get together with a small group of classmates to talk about experiences or incidents from the past that had a strong effect on you. Perhaps you learned an important lesson about yourself or changed the way you think about yourself. Ask one another questions to dredge up specific details about what happened. Then write an essay about the experience or incident and its effect on you.

3. USING THE INTERNET. "The Biggest Play of His Life" is a coming out story: an account of a gay person revealing his or her sexual orientation. Search the Internet for more examples—for instance, on a site like ComingOutStories.com. Read a number of them and write a cause-effect essay about the coming out process for readers unfamiliar with the topic. You can focus on the causes (why people decide to come out), the effects (what happens when they do), or both.
4. WRITING ABOUT READING. Consider what "The Biggest Play of His Life" might be like if Corey Johnson had written it himself. How would Johnson's version differ from Reilly's? What would Johnson emphasize that Reilly merely touched on? What does Reilly include that Johnson might not? Write an essay in which you explore the possible differences in Johnson's own account.

EDITING SKILLS: SENTENCE FRAGMENTS

Look at the following groups of words from paragraph 10 in Rick Reilly's article:

> A year later he told his parents. Fine. Then his best friend, Sean. Uh-oh. Big problem. Sean started crying.

Which of these are grammatically complete sentences? Which ones are not?

The first and last word groups are independent sentences. They both contain a subject and a verb: "he told" and "Sean started." But the other three are sentence fragments. In order to make them complete, you would have to add a subject and a verb, like this:

> Everything [subj] was [vb] fine.
> Then he told his best friend, Sean.
> Uh-oh, his reaction created a big problem.

In the case of the interjection "Uh-oh," you can just attach it to the beginning of the next sentence.

EXERCISE

There are five other sentence fragments in Reilly's article. List them and then rewrite each in order to make it a complete sentence. Why do you think Reilly used fragments in these places instead of complete sentences? How does the use of fragments suit his tone and audience?

WEB SITES

http://sportsillustrated.cnn.com/inside_game/archives/rick_reilly/

This Insider Archive contains a biography of Reilly and access to all of his columns going back to January 1999.

http://owl.english.purdue.edu/handouts/grammar/g_frag.html

Get downloadable handouts on sentence fragments, complete with exercises and answer keys, from Purdue University's Online Writing Lab.

Preparing to Read

Do you enjoy horror movies? Why do you like to watch them? How do you react when you view them? If you don't enjoy horror movies, can you explain why they do not appeal to you?

Why We Crave Horror Movies

Stephen King

You know Stephen King as the master of terror, the author of a string of best-selling horror novels. And you have probably seen some of the movies made from his novels: Carrie, Misery, Christine, Pet Sematary, *and* Firestarter *(among others). In the following essay, King offers an entertaining explanation of why people like being scared out of their wits.*

TERMS TO RECOGNIZE

grimaces *(para. 1)*	twisted facial expressions
hysterical *(para. 1)*	emotionally uncontrolled
province *(para. 3)*	proper area or sphere
depleted *(para. 3)*	used up, drained, worn out
innately *(para. 4)*	naturally, essentially
reactionary *(para. 4)*	wanting to return to an earlier time
menaced *(para. 6)*	threatened, endangered
voyeur *(para. 6)*	a peeping Tom, someone who enjoys watching something private or forbidden
penchant *(para. 7)*	a strong fondness or inclination
psychic *(para. 7)*	mental, psychological
status quo *(para. 9)*	existing condition or state of affairs
sanctions *(para. 10)*	expressions of disapproval, punishments
anarchistic *(para. 11)*	disorderly, ignoring the rules
morbidity *(para. 12)*	an interest in gruesome and horrible things
subterranean *(para. 12)*	underground

I think we're all mentally ill; those of us outside the asylums only hide it a little better—and maybe not all that much better, after all. We've all known people who talk to themselves, people who sometimes squinch 1 their faces into horrible grimaces when they believe no one is watching, people who have some hysterical fear—of snakes, the dark, the tight place, the long drop . . . and, of course, those final worms and grubs that are waiting so patiently underground.

2 When we pay our four or five bucks and seat ourselves at tenth-row center in a theater showing a horror movie, we are daring the nightmare.

3 Why? Some of the reasons are simple and obvious. To show that we can, that we are not afraid, that we can ride this roller coaster. Which is not to say that a really good horror movie may not surprise a scream out of us at some point, the way we may scream when the roller coaster twists through a complete 360 or plows through a lake at the bottom of the drop. And horror movies, like roller coasters, have always been the special province of the young; by the time one turns 40 or 50, one's appetite for double twists or 360-degree loops may be considerably depleted.

4 We also go to re-establish our feelings of essential normality; the horror movie is innately conservative, even reactionary. Freda Jackson as the horrible melting woman in *Die, Monster, Die!* confirms for us that no matter how far we may be removed from the beauty of a Robert Redford or a Diana Ross, we are still light-years from true ugliness.

5 And we go to have fun.

6 Ah, but this is where the ground starts to slope away, isn't it? Because this is a very peculiar sort of fun, indeed. The fun comes from seeing others menaced—sometimes killed. One critic has suggested that if pro football has become the voyeur's version of combat, then the horror film has become the modern version of the public lynching.

7 It is true that the mythic, "fairy-tale" horror film intends to take away the shades of gray. . . . It urges us to put away our more civilized and adult penchant for analysis and to become children again, seeing things in pure blacks and whites. It may be that horror movies provide psychic relief on this level because this invitation to lapse into simplicity, irrationality and even outright madness is extended so rarely. We are told we may allow our emotions a free rein . . . or no rein at all.

8 If we are all insane, then sanity becomes a matter of degree. If your insanity leads you to carve up women like Jack the Ripper or the Cleveland Torso Murderer, we clap you away in the funny farm (but neither of those two amateur-night surgeons was ever caught, heh-heh-heh); if, on the other hand, your insanity leads you only to talk to yourself when you're under stress or to pick your nose on your morning bus, then you are left alone to go about your business . . . though it is doubtful that you will ever be invited to the best parties.

9 The potential lyncher is in almost all of us (excluding saints, past and present; but then, most saints have been crazy in their own ways), and every now and then, he has to be let loose to scream and roll around in the grass. Our emotions and our fears form their own body, and we recognize that it

demands its own exercise to maintain proper muscle tone. Certain of these emotional muscles are accepted—even exalted—in civilized society; they are, of course, the emotions that tend to maintain the status quo of civilization itself. Love, friendship, loyalty, kindness—these are all the emotions that we applaud, emotions that have been immortalized in the couplets of Hallmark cards and in the verses (I don't dare call it poetry) of Leonard Nimoy.

When we exhibit these emotions, society showers us with positive reinforcement; we learn this even before we get out of diapers. When, as children, we hug our rotten little puke of a sister and give her a kiss, all the aunts and uncles smile and twit and cry, "Isn't he the sweetest little thing?" Such coveted treats as chocolate-covered graham crackers often follow. But if we deliberately slam the rotten little puke of a sister's fingers in the door, sanctions follow—angry remonstrance from parents, aunts and uncles; instead of a chocolate-covered graham cracker, a spanking. 10

But anticivilization emotions don't go away, and they demand periodic exercise. We have such "sick" jokes as, "What's the difference between a truckload of bowling balls and a truckload of dead babies?" (You can't unload a truckload of bowling balls with a pitchfork . . . a joke, by the way, that I heard originally from a ten-year-old.) Such a joke may surprise a laugh or a grin out of us even as we recoil, a possibility that confirms the thesis: If we share a brotherhood of man, then we also share an insanity of man. None of which is intended as a defense of either the sick joke or insanity but merely as an explanation of why the best horror films, like the best fairy tales, manage to be reactionary, anarchistic, and revolutionary all at the same time. 11

The mythic horror movie, like the sick joke, has a dirty job to do. It deliberately appeals to all that is worst in us. It is morbidity unchained, our most base instincts let free, our nastiest fantasies realized . . . and it all happens, fittingly enough, in the dark. For those reasons, good liberals often shy away from horror films. For myself, I like to see the most aggressive of them—*Dawn of the Dead*, for instance—as lifting a trap door in the civilized forebrain and throwing a basket of raw meat to the hungry alligators swimming around in that subterranean river beneath. 12

Why bother? Because it keeps them from getting out, man. It keeps them down there and me up here. It was Lennon and McCartney who said that all you need is love, and I would agree with that. 13

As long as you keep the gators fed. 14

RESPONDING TO READING

Do you agree with King that we are all mentally ill and that our "anticivilization emotions don't go away"? Respond to these ideas in your journal.

GAINING WORD POWER

What does it mean to do "a complete 360"? (King uses the expression in para. 3.) You may know that 360 refers to the number of degrees in a circle and that the phrase means to go all the way around to where you started, to travel in a complete circle. One way of expressing the opposite idea is to say you did "an about-face," which is a military command for pivoting around to face in the opposite direction.

Both of these phrases—"a complete 360" and "an about-face"—are figurative expressions; they're supposed to put a picture in our minds by referring to an object (like a circle) or an action (like a military maneuver). **Figurative language** requires interpretation; we have to figure out the references and imagine the picture that the writer wants to put in our minds. Here are some more figurative phrases that Stephen King uses in his essay. Explain what they mean and tell what you see in your mind's eye. If you're not sure of what King means, ask other people for their interpretations.

1. "those final worms and grubs that are waiting so patiently underground"
2. "tenth-row center"
3. "shades of gray"
4. "clap you away in the funny farm"
5. "the couplets of Hallmark cards"
6. "before we get out of diapers"
7. "lifting a trap door in the civilized forebrain and throwing a basket of raw meat to the hungry alligators swimming around in that subterranean river beneath"

CONSIDERING CONTENT

1. What does King mean when he says that we are all mentally ill? What is the nature of the "insanity" that we share?
2. What are the obvious reasons for enjoying horror movies? What are some of the not-so-obvious reasons?
3. King refers to our "anticivilization emotions" that occasionally need exercise. What are these emotions? Why do they need to be exercised?
4. Do you think Stephen King is biased in his defense of horror movies? Do you think you would respond differently to this essay if it were written by someone you had never heard of?

CONSIDERING METHOD

1. King begins with a startling statement about insanity. What is the effect and point of this opening?
2. How does he maintain the insanity theme throughout the essay? Do you think this is an effective strategy?
3. At what point in the essay does King reveal that he will be dealing with causes?
4. Does King deal with immediate causes or long-term causes or both? In what order does he arrange the causes he discusses?
5. Identify comments or passages that you think are humorous. What does the humor contribute to this essay?
6. What is the function of the two one-sentence paragraphs (paras. 5 and 14)?

Combining Strategies

King uses several *comparisons* to explain and support his defense of horror movies. What points does King make by comparing horror movies to the following?

a. riding a roller coaster
b. public lynching
c. mythic fairy tales
d. sick jokes.

Do you think these are good comparisons?

WRITING STEP BY STEP

Think of another form of entertainment that people seem to crave, and write an essay explaining why. You might write about soap operas, MTV, video games, action movies, reality TV, sports programs, jogging (or other exercise activities), bodybuilding, shopping, online chat rooms, or the like.

A. Begin with a startling statement, as Stephen King does.
B. Give background about the subject by explaining your opening statement.
C. Discuss the obvious or immediate causes first. Identify at least three reasons and explain them.

D. Then examine the long-term or underlying causes. Show how these causes produce the craving that you are writing about.
E. Be sure to name specific programs or games.
F. Use figurative language and analogies to help explain the causes. Even include a little humor the way King does, if you want to.
G. Write a conclusion in which you comment on the behavior you have analyzed. Make your comment an outgrowth of your discussion of underlying causes.

OTHER WRITING IDEAS

1. Write an essay about the effects of watching too much TV. Use your own observations and experiences to support your conclusions.
2. COLLABORATIVE WRITING. Working with a group of classmates, interview a variety of people to find out why they like a certain type of entertainment (TV hospital shows, arcade games, stock car races, line dancing, rock concerts, ballet, situation comedies, or the like). Ask them to give you specific reasons. Then write a group report explaining why this form of entertainment is popular. Use several direct quotations from the interviews to support your explanations.
3. USING THE INTERNET. Find out what researchers say about the effects of television violence on children. Be sure to locate sites that offer more than one point of view. Then write an essay for parents, informing them of these effects and suggesting what they can do to counter them.
4. WRITING ABOUT READING. Check out King's ideas about the cravings for horror movies by watching one and analyzing your reactions to it. Did the movie "re-establish [your] feelings of essential normality"? In what ways did it provide "psychic relief" or satisfy your "anticivilization emotions"? How did you feel before watching the movie? How did you feel afterwards?

EDITING SKILLS: CHECKING PRONOUN REFERENCE

Whenever you use pronouns (words such as *he, she, it, his, her, their,* etc., that stand in for nouns), you must be sure that each has a clear antecedent (the noun that the pronoun stands for). If the antecedents aren't clear, your readers can get lost. Look at this sentence from Stephen King's essay:

> We've all known people *who* talk to *themselves,* people *who* sometimes squinch *their* faces into horrible grimaces when *they* believe no one is watching. . . .

The pronouns *themselves, who, their,* and *they* all clearly refer to the antecedent *people.*

Several pronouns are especially troublesome when they appear without specific antecedents: *this, which,* and *these.* King uses these words eleven times. Sometimes he places them right before the nouns they refer to—"this roller coaster," "this level," "this invitation," "these emotions"—so there is no misunderstanding about what they mean. But in other instances, he uses *this, these,* and *which* as free-standing pronouns. Find these other uses (in paras. 3, 6, 9, 10, and 11), and see if you can name the thing or idea to which these pronouns refer. Do you think King has used these pronouns clearly?

EXERCISE

The following sentences are vague because they contain pronouns without clear antecedents. Rewrite each sentence to make it clear.

Example: Clyde made Gary do his homework.

Revisions: Gary did his homework because Clyde made him. Gary did Clyde's homework because Clyde made him.

1. Hampton aced the history test because he made it so easy.
2. Carlo is working on the railroad in Tennessee, which depresses him.
3. Kesha dropped out of school after they took away fall break.
4. Our cat chased the ground squirrel until he got tired.
5. Although Juan's mother is a chemist, Juan hates it.
6. Rosa tried to support herself by painting and acting. This was a mistake.
7. Mel blamed his failure on his choice of occupation, which was unfortunate.
8. Washington wore a sombrero and sequined gloves to the prom. This was a big hit.

Now look at the essay you have just written. Pay particular attention to your use of pronouns. Make sure each has a clear antecedent.

WEB SITE

www.stephenking.com/

Find out more about Stephen King's views on horror stories and other aspects of writing at the official online resource for news and information about King and his works.

Preparing to Read

Did your parents ever forbid you to do something that you went ahead and did anyway? What happened? How did you feel about it later?

Fifth Chinese Daughter

Jade Snow Wong

Jade Snow Wong grew up in San Francisco, the daughter of immigrant parents who brought with them from China an ancient and rigid set of family traditions. The following excerpt from Wong's autobiography, Fifth Chinese Daughter, *recounts the inevitable conflict between traditional parents and a young woman who is beginning to discover a different world beyond her home and neighborhood.*

TERMS TO RECOGNIZE

oblige *(para. 2)*	accommodate, please
adamant *(para. 3)*	unyielding, inflexible
edict *(para. 3)*	command, rule
incurred *(para. 4)*	acquired, taken on
nepotism *(para. 7)*	favoritism shown to relatives
incredulous *(para. 12)*	unwilling to admit or accept what is heard; unbelieving
unfilial *(para. 12)*	disrespectful to parents
revered *(para. 12)*	honored, respected
innuendos *(para. 14)*	sly suggestions, indirect hints
devastated *(para. 15)*	crushed, overwhelmed
perplexed *(para. 16)*	puzzled, confused

1 By the time I was graduating from high school, my parents had done their best to produce an intelligent, obedient daughter, who would know more than the average Chinatown girl and should do better than average at a conventional job, her earnings brought home in repayment for their years of child support. Then, they hoped, she would marry a nice Chinese boy and make him a good wife, as well as an above-average mother for his children. Chinese custom used to decree that families should "introduce" chosen partners to each other's children. The groom's family

should pay handsomely to the bride's family for rearing a well-bred daughter. They should also pay all bills for a glorious wedding banquet for several hundred guests. Their daughter belonged to the groom's family and must henceforth seek permission from all persons in his home before returning to her parents for a visit.

2 But having been set upon a new path, I did not oblige my parents with the expected conventional ending. At fifteen, I had moved away from home to work for room and board and a salary of twenty dollars per month. Having found that I could subsist independently, I thought it regrettable to terminate my education. Upon graduating from high school at the age of sixteen, I asked my parents to assist me in college expenses. I pleaded with my father, for his years of encouraging me to be above mediocrity in both Chinese and American studies had made me wish for some undefined but brighter future.

3 My father was briefly adamant. He must conserve his resources for my oldest brother's medical training. Though I desired to continue on an above-average course, his material means were insufficient to support that ambition. He added that if I had the talent, I could provide for my own college education. When he had spoken, no discussion was expected. After this edict, no daughter questioned.

4 But this matter involved my whole future—it was not simply asking for permission to go to a night church meeting (forbidden also). Though for years I had accepted the authority of the one I honored most, his decision that night embittered me as nothing ever had. My oldest brother had so many privileges, had incurred unusual expenses for luxuries which were taken for granted as his birthright, yet these were part of a system I had accepted. Now I suddenly wondered at my father's interpretation of the Christian code: was it intended to discriminate against a girl after all, or was it simply convenient for my father's economics and cultural prejudice? Did a daughter have any right to expect more than a fate of obedience, according to the old Chinese standard? As long as I could remember, I had been told that a female followed three men during her lifetime: as a girl, her father; as a wife, her husband; as an old woman, her son.

5 My indignation mounted against that tradition and I decided then that my past could not determine my future. I knew that more education would prepare me for a different expectation than my other female schoolmates, few of whom were to complete a college degree. I, too, had my father's unshakable faith in the justice of God, and I shared his unconcern with popular opinion.

6 So I decided to enter junior college, now San Francisco's City College, because the fees were lowest. I lived at home and supported myself with an after-school job which required long hours of housework and cooking but paid me twenty dollars per month, of which I saved as much as possible.

The thrills derived from reading and learning, in ways ranging from chemistry experiments to English compositions, from considering new ideas of sociology to the logic of Latin, convinced me that I had made a correct choice. I was kept in a state of perpetual mental excitement by new Western subjects and concepts and did not mind long hours of work and study. I also made new friends, which led to another painful incident with my parents, who had heretofore discouraged even girlhood friendships.

7 The college subject which had the most jolted me was sociology. The instructor fired my mind with his interpretation of family relationships. As he explained to our class, it used to be an economic asset for American farming families to be large, since children were useful to perform agricultural chores. But this situation no longer applied and children should be regarded as individuals with their own rights. Unquestioning obedience should be replaced with parental understanding. So at sixteen, discontented as I was with my parents' apparent indifference to me, those words of my sociology professor gave voice to my sentiments. How old-fashioned was the dead-end attitude of my parents! How ignorant they were of modern thought and progress! The family unit had been China's strength for centuries, but it had also been her weakness, for corruption, nepotism, and greed were all justified in the name of the family's welfare. My new ideas festered; I longed to release them.

8 One afternoon on a Saturday, which was normally occupied with my housework job, I was unexpectedly released by my employer, who was departing for a country weekend. It was a rare joy to have free time and I wanted to enjoy myself for a change. There had been a Chinese-American boy who shared some classes with me. Sometimes we had found each other walking to the same 8:00 A.M. class. He was not a special boyfriend, but I had enjoyed talking to him and had confided in him some of my problems. Impulsively, I telephoned him. I knew I must be breaking rules, and I felt shy and scared. At the same time, I was excited at this newly found forwardness, with nothing more purposeful than to suggest another walk together.

9 He understood my awkwardness and shared my anticipation. He asked me to "dress up" for my first movie date. My clothes were limited but I changed to look more graceful in silk stockings and found a bright ribbon for my long black hair. Daddy watched, catching my mood, observing the dashing preparations. He asked me where I was going without his permission and with whom.

10 I refused to answer him. I thought of my rights! I thought he surely would not try to understand. Thereupon Daddy thundered his displeasure and forbade my departure.

11 I found a new courage as I heard my voice announce calmly that I was no longer a child, and if I could work my way through college, I would choose my own friends. It was my right as a person.

My mother had heard the commotion and joined my father to face me; both appeared shocked and incredulous. Daddy at once demanded the source of this unfilial, non-Chinese theory. And when I quoted my college professor, reminding him that he had always felt teachers should be revered, my father denounced that professor as a foreigner who was disregarding the superiority of our Chinese culture, with its sound family strength. My father did not spare me; I was condemned as an ingrate for echoing dishonorable opinions which should only be temporary whims, yet nonetheless inexcusable. 12

The scene was not yet over. I completed my proclamation to my father, who had never allowed me to learn how to dance, by adding that I was attending a movie, unchaperoned, with a boy I met at college. 13

My startled father was sure that my reputation would be subject to whispered innuendos. I must be bent on disgracing the family name; I was ruining my future, for surely I would yield to temptation. My mother underscored him by saying that I hadn't any notion of the problems endured by parents of a young girl. 14

I would not give in. I reminded them that they and I were not in China, that I wasn't going out with just anybody but someone I trusted! Daddy gave a roar that no man could be trusted, but I devastated them in declaring that I wished the freedom to find my own answers. 15

Both parents were thoroughly angered, scolded me for being shameless, and predicted that I would some day tell them I was wrong. But I dimly perceived that they were conceding defeat and were perplexed at this breakdown of their training. I was too old to beat and too bold to intimidate. 16

RESPONDING TO READING

Has college affected you in the same way that it affected Jade Snow Wong? Write in your journal about the effects college has had on you; compare your reactions to Wong's.

GAINING WORD POWER

Come up with definitions of your own for the italicized words in the following sentences. Use the clues from the surrounding words and sentences to help you. Then check your definition against a dictionary definition.

1. "Chinese custom used to *decree* that families should 'introduce' chosen partners to each other's children." (para. 1)
2. "Having found that I could *subsist* independently, I thought it regrettable to *terminate* my education." (para. 2)

3. "I pleaded with my father, for his years of encouraging me to be above *mediocrity* in both Chinese and American studies had made me wish for some undefined but brighter future." (para. 2)
4. "My *indignation* mounted against that tradition and I decided then that my past could not determine my future." (para. 5)
5. "I was kept in a state of *perpetual* mental excitement by new Western subjects and concepts and did not mind long hours of work and study." (para. 6)
6. "My new ideas *festered;* I longed to release them." (para. 7)
7. "He was not a special boyfriend, but I had enjoyed talking to him and had *confided* in him some of my problems. *Impulsively,* I telephoned him. I knew I must be breaking rules, and I felt shy and scared." (para. 8)
8. "I was too old to beat and too bold to *intimidate.*" (para. 16)

CONSIDERING CONTENT

1. What did you learn about Wong's parents and their traditional beliefs in the first paragraph?
2. What caused Wong to rebel against her parents? What situations and experiences contributed to her quest for independence?
3. Why was Wong sure that more education would prepare her for a different future from her classmates? Did it?
4. Why did the sociology course affect Wong so strongly? What did she learn in that course that changed her views and fed her rebellion?
5. What action did Wong take to demonstrate her independence? Why did she choose this particular way of disobeying her parents? Do her actions seem risky and rebellious to you?
6. In what ways was Wong a lot like her father? Which of his beliefs and attitudes actually contributed to his daughter's rebellion?

CONSIDERING METHOD

1. Why is the first paragraph important? How does the information in this paragraph prepare you for Wong's actions?
2. What does Wong accomplish by mentioning her oldest brother? What is she trying to show?
3. What is the function of the questions in paragraph 4? Why does Wong express these ideas in questions instead of in statements? (See pages 363–64 for an explanation of rhetorical questions.)
4. How much time passes in this selection? How does Wong indicate the passage of time? Where does she move back in time? Why does she do so?

5. Paragraph 8 opens with these words: "One afternoon on a Saturday . . ." What do these words signal? What mode of writing begins at this point?
6. Wong doesn't use direct quotations or quoted conversation. How does she report what she said to her parents and what they said to her? What is the effect of this indirect presentation?
7. How would you describe the **tone** of this selection? Does Wong sound angry, sad, upset, calm, confident, relieved, happy, or what?

WRITING STEP BY STEP

As we grow older, we sometimes change our minds about our parents. We begin to see things more from their point of view. Think of a piece of advice, a household rule, or a parental opinion that you used to hate but now understand. If you don't want to write about your parents, think of another adult, such as a teacher or a coach, who set down rules or gave advice that you once rejected but now think appropriate. Write an essay in which you explain what caused you to change your mind.

A. Begin with a statement of your general point. You might say something like "I used to think my parents' advice was outdated and pointless, but a lot of it makes sense to me today" or "Now that I'm coaching pee-wee soccer, I follow some of the same rules I used to hate when I was a player."
B. Explain the particular rule, opinion, advice, or guideline you used to resist. Re-create specific incidents and arguments you used to have that illustrate your earlier response.
C. Analyze your change in attitude. Cite specific reasons or causes, and explain them.
D. If there was a key incident or turning point in your thinking, focus on that event and describe it in detail.
E. Conclude by revealing how you feel about this change in thinking. Does it mean that you've grown up or that you've sold out?

OTHER WRITING IDEAS

1. Write an essay about why adolescents leave home. Or write an essay about the effects of their leaving home.
2. Collaborative Writing. Explain why some jobs or professions seem to be done mainly by women (like child care, housework, nursing, elementary school teaching). As you plan your paper, talk to some

women who do these jobs or who are preparing to enter one of these professions, and ask their opinions. Also talk to some men, and ask them how they would feel about teaching grade school, being a nurse, cleaning houses, or taking care of children for a living.

3. USING THE INTERNET. Talk to someone who comes from a cultural background different from yours. If you can, spend some time with this person's friends and family. Also locate Internet sites that will give you information about this cultural group. Then, with your classmates as audience, write an essay explaining what you learned about this culture and how it affected your attitudes toward these people.
4. WRITING ABOUT READING. In the last paragraph, Wong writes that she "dimly perceived" that her parents "were conceding defeat and were perplexed at this breakdown of their training" (para. 16). What were they conceding and why were they perplexed? Write an essay that answers these questions. Include personal experiences that might help to explain or support your conclusions.

EDITING SKILLS: USING PARALLEL STRUCTURE

Look at these last two sentences from paragraph 3 in Wong's essay:

When he had spoken, no discussion was expected.

After his edict, no daughter questioned.

When you look at the sentences in this format, with one on top of the other, you can see how they match up in size and form. This correspondence is called **parallelism,** or parallel structure. Parallel structure often occurs within a sentence, as the following sentences from Wong's essay show. (The parallel parts are italicized for you.)

As long as I could remember, I had been told that a female followed three men during her lifetime: *as a girl, her father; as a wife, her husband; as an old woman, her son.*

I was *too old to beat* and *too bold to intimidate.*

Writers use parallel structure to catch the reader's attention and, as Wong does, to compare or contrast important points. Parallel structures also add **emphasis** and variety to a sentence or paragraph.

EXERCISE

Copy two of the sentences previously quoted; write two or three sentences of your own that imitate Wong's use of parallel structure. Then examine the following sentences, and write your own sentences that imitate the parallel structures.

1. "This freedom, like all freedoms, has its dangers and its responsibilities."—James Baldwin

   Imitation: This new attendance policy, like all school policies, has its supporters and its critics.

2. "We must stop talking about the American dream and start listening to the dreams of Americans."—Reubin Askew

   Imitation: We must stop asking silly questions and start questioning silly policies.

Look over your own essay and revise two or three sentences, using parallel structure. Read the new sentences aloud. If they sound clear and effective, keep them in your essay.

WEB SITE

http://sun3.lib.uci.edu/~dtsang/aas2.htm

This is the Web site for Asian American Studies Resources: an extensive collection of sites for studying Asian cultures, complied by Daniel Tsang, the Asian American Studies Bibliographer for the library at the University of California at Irvine.

PREPARING TO READ

Do you eat fast food? How much and how often? Do you think it's healthy?

Supersize Me

GREG CRITSER

Educated at Occidental College and UCLA, Greg Critser lives in Pasadena, California. He is a regular contributor to USA Today, *the* Los Angeles Times, *and* Harper's Magazine. *In 1999, his articles on obesity won a James Beard nomination for best feature writing, and he is frequently interviewed by PBS and other news media on the subject of food politics. The following selection comes from the second chapter of his book* Fat Land: How Americans Became the Fattest People in the World *(2003).*

TERMS TO RECOGNIZE

contentious *(para. 4)*	quarrelsome, heated
truculent *(para. 9)*	fierce, hostile, aggressively self-assertive
entrepreneurism *(para. 10)*	the practice of taking business risks
primal *(para. 10)*	primary, first in importance
maligned *(para. 14)*	criticized, attacked, smeared
table d'hôte *(para. 15)*	a complete meal offered at a fixed price
à la carte *(para. 15)*	a menu or list that prices each item separately
immutable *(para. 21)*	unchangeable, permanent
maw *(para. 23)*	the mouth or jaws of a hungry animal

1 David Wallerstein, a director of the McDonald's Corporation, hated the fifth deadly sin because it kept people from buying more hamburgers. Wallerstein had first waged war on the injunction against gluttony as a young executive in the theater business. At the staid Balaban Theaters chain in the early 1960s, Wallerstein had realized that the movie business was really a margin business; it wasn't the sale of low-markup movie tickets that generated profits but rather the sale of high-markup snacks like popcorn and Coke. To sell more of such items, he had, by the mid-1960s, tried about every trick in the conventional retailer's book: two-for-one specials, combo deals, matinee specials, etc. But at the end of any

given day, as he tallied up his receipts, Wallerstein inevitably came up with about the same amount of profit.

2 Thinking about it one night, he had a realization: People did not want to buy two boxes of popcorn *no matter what.* They didn't want to be seen eating two boxes of popcorn. It looked piggish. So Wallerstein flipped the equation around: Perhaps he could get more people to spend just a little more on popcorn if he made the boxes bigger and increased the price only a little. The popcorn cost a pittance anyway, and he'd already paid for the salt and the seasoning and the counter help and the popping machine. So he put up signs advertising jumbo-size popcorn. The results after the first week were astounding. Not only were individual sales of popcorn increasing; with them rose individual sales of that other high-profit item, Coca-Cola.

3 Later, at McDonald's in the mid-1970s, Wallerstein faced a similar problem: With consumers watching their pennies, restaurant customers were coming to the Golden Arches less and less frequently. Worse, when they did, they were "cherry-picking," buying only, say, a small Coke and a burger, or, worse, just a burger, which yielded razor-thin profit margins. How could he get people back to buying more fries? His popcorn experience certainly suggested one solution—sell them a jumbo-size bag of the crispy treats.

4 Yet try as he may, Wallerstein could not convince Ray Kroc, McDonald's founder, to sign on to the idea. As recounted in interviews with his associates and in John F. Love's 1985 book, *McDonald's: Behind the Arches,* the exchange between the two men could be quite contentious on the issue. "If people want more fries," Kroc would say, "they can buy two bags."

5 "But Ray," Wallerstein would say, "they don't want to eat two bags—they don't want to look like a glutton."

6 To convince Kroc, Wallerstein decided to do his own survey of customer behavior, and began observing various Chicago-area McDonald's. Sitting in one store after another, sipping his drink and watching hundreds of Chicagoans chomp their way through their little bag of fries, Wallerstein could see: People *wanted* more fries.

7 "How do you know that?" Kroc asked the next morning when Wallerstein presented his findings.

8 "Because they're eating the entire bagful, Ray," Wallerstein said. "They even scrape and pinch around at the bottom of the bag for more and eat the salt!"

9 Kroc gave in. Within months receipts were up, customer counts were up, and franchisees—the often truculent heart and soul of the McDonald's success—were happier than ever.

10 Many franchisees wanted to take the concept even further, offering large-size versions of other menu items. At this sudden burst of entrepreneurism, however, McDonald's mid-level managers hesitated. Many of them viewed large-sizing as a form of "discounting," with all the negative connotations

such a word evoked. In a business where "wholesome" and "dependable" were the primary PR watchwords, large-sizing could become a major image problem. Who knew what the franchisees, with their primal desires and shortcutting ways, would do next? No, large-sizing was something to be controlled tightly from Chicago, if it were to be considered at all.

11 Yet as McDonald's headquarters would soon find out, large-sizing was a new kind of marketing magic—a magic that could not so easily be put back into those crinkly little-size bags.

12 Max Cooper, a Birmingham franchisee, was not unfamiliar with marketing and magic; for most of his adult life he had been paid to conjure sales from little more than hot air and smoke. Brash, blunt-spoken, and witty, Cooper had acquired his talents while working as an old-fashioned public relations agent—the kind, as he liked to say, who "got you into the newspaper columns instead of trying to keep you out." In the 1950s with his partner, Al Golin, he had formed what later became Golin Harris, one of the world's more influential public relations firms. In the mid-1960s, first as a consultant and later as an executive, he had helped create many of McDonald's most successful early campaigns. He had been the prime mover in the launch of Ronald McDonald.

13 By the 1970s Cooper, tired of "selling for someone else," bought a couple of McDonald's franchises in Birmingham, moved his split-off ad agency there, and set up shop as an independent businessman. As he began expanding, he noticed what many other McDonald's operators were noticing: declining customer counts. Sitting around a table and kibitzing with a few like-minded associates one day in 1975, "we started talking about how we could build sales—how we could do it and be profitable," Cooper recalled in a recent interview. "And we realized we could do one of three things. We could cut costs, but there's a limit to that. We could cut prices, but that too has its limits. Then we could raise sales profitably—sales, after all, could be limitless when you think about it. We realized we could do that by taking the high-profit drink and fry and then packaging it with the low-profit burger. We realized that if you could get them to buy three items for what they perceived as less, you could substantially drive up the number of walk-ins. Sales would follow."

14 But trying to sell that to corporate headquarters was next to impossible. "We were maligned! Oh, were we maligned," he recalls. "A 99-cent anything was heresy to them. They would come and say 'You're just cutting prices! What are we gonna look like to everybody else?' "

15 "No no no," Cooper would shoot back. "You have to think of the analogy to a fine French restaurant. You always pay less for a *table d'hôte* meal than you pay for *à la carte*, don't you?"

16 "Yes, but—"

"Well, this is a *table d'hôte,* dammit! You're getting more people to the table spending as much as they would before—and coming more often!" 17

Finally headquarters relented, although by now it hardly mattered. Cooper had by then begun his own rogue campaign. He was selling what the industry would later call "value meals"—the origin of what we now call supersizing. Using local radio, he advertised a "Big Mac and Company," a "Fish, Fry, Drink and Pie," a "4th of July Value Combo." Sales, Cooper says, "went through the roof. Just like I told them they would." 18

★ ★ ★ ★ ★

Though it is difficult to gauge the exact impact of supersizing upon the appetite of the average consumer, there are clues about it in the now growing field of satiety—the science of understanding human satisfaction. A 2001 study by nutritional researchers at Penn State University, for example, sought to find out whether the presence of larger portions *in themselves* induced people to eat more. Men and women volunteers, all reporting the same level of hunger, were served lunch on four separate occasions. In each session, the size of the main entree was increased, from 500 to 625 to 700 and finally to 1000 grams. After four weeks, the pattern became clear: As portions increased, all participants ate increasingly larger amounts, despite their stable hunger levels. As the scholars wrote: "Subjects consumed approximately 30 percent more energy when served the largest as opposed to the smallest portion." They had documented that satiety is not satiety. Human hunger could be expanded by merely offering more and bigger options. 19

Certainly the best nutritional data suggest so as well. Between 1970 and 1994, the USDA reports, the amount of food available in the American food supply increased 15 percent—from 3300 to 3800 calories or by about 500 calories per person per day. During about the same period (1977—1995), average individual caloric intake increased by almost 200 calories, from 1876 calories a day to 2043 calories a day. One could argue which came first, the appetite or the bigger burger, but the calories—they were on the plate and in our mouths. 20

By the end of the century, supersizing—the ultimate expression of the value meal revolution—reigned. As of 1996, some 25 percent of the $97 billion spent on fast food came from items promoted on the basis of either larger size or extra portions. A serving of McDonald's french fries had ballooned from 200 calories (1960) to 320 calories (late 1970s) to 450 calories (mid-1990s) to 540 calories (late 1990s) to the present 610 calories. In fact, everything on the menu had exploded in size. What was once a 590-calorie McDonald's meal was now . . . 1550 calories. By 1999, heavy users—people who eat fast food more than twenty times a month—accounted for $66 billion of the $110 billion spent on fast food. Twenty times a month is 21 now McDonald's marketing goal for every fast-food eater. The average Joe or Jane thought nothing of buying Little Caesar's pizza "by the foot," of supersizing that lunchtime burger or supersupersizing an afternoon snack. Kids had come to see bigger everything—bigger sodas, bigger snacks, bigger candy, and even bigger doughnuts—as the norm; there was no such thing as a fixed, immutable size for anything, because anything could be made a lot bigger for just a tad more.

22 There was more to all of this than just eating more. Bigness: The concept seemed to fuel the marketing of just about everything, from cars (SUVs) to homes (mini-manses) to clothes (super-baggy) and then back again to food (as in the Del Taco Macho Meal, which weighed four pounds). The social scientists and the marketing gurus were going crazy trying to keep up with the trend. "Bigness is addictive because it is about power," commented Inna Zall, a teen marketing consultant, in a page-one story in *USA Today.* While few teenage boys can actually finish a 64-ounce Double Gulp, she added, "it's empowering to hold one in your hand."

23 The pioneers of supersize had achieved David Wallerstein's dream. They had banished the shame of gluttony and opened the maw of the American eater wider than even they had ever imagined.

RESPONDING TO READING

After reading this selection, do you think you'll try to change your eating habits? Why or why not?

GAINING WORD POWER

Explain as thoroughly as you can what the following phrases mean in Critser's essay.

1. the fifth deadly sin (para. 1)
2. high-markup snacks (para.1)
3. cost a pittance (para. 2)
4. mid-level managers (para. 10)
5. hot air and smoke (para. 12)
6. the prime mover (para. 12)
7. kibitzing with like-minded associates (para. 13)
8. went through the roof (para. 18)

CONSIDERING CONTENT

1. What did David Wallerstein figure out about moviegoers and the amount of popcorn they bought? What did he do about it?
2. Why did McDonald's managers resist the idea of offering large-size versions of menu items?

3. What scheme did Max Cooper come up with to raise sales at McDonald's?
4. What did the 2001 Penn State study reveal about the effect of portion size on eating behavior?
5. What link between supersizing and caloric intake does Critser make? Does this conclusion seem valid and reasonable?
6. According to Critser, how has the concept of "bigness" spread beyond the fast food market? Do you agree that "Bigness is addictive because it is about power"?

CONSIDERING METHOD

1. What is Critser's **point of view?** Who is his intended **audience,** and what message does he want to convey to this audience?
2. Explain the **analogy** between "value meals" and the menu in a fine French restaurant (paras. 14–18). Does this seem like a valid comparison?
3. Who are the two sources for the first part of the article (paras. 1–18)? Why does Critser rely on these people for the information in this part?
4. Look at the statistics in paragraphs 19, 20, and 21. What point do they support in each paragraph? What are their sources?
5. What effects does the closing paragraph achieve?

WRITING STEP BY STEP

Greg Critser examines the problem of increased obesity in Americans and uncovers a surprising cause: the supersizing of fast foods. In this writing assignment you will have a similar goal.

Think of a problem or condition that you know something about, and look beneath the surface to discover both the *immediate* and the *remote* (or *underlying*) causes. One place to start is on your campus: the lack of a day-care center, a parking crisis, the widespread occurrence of cheating or plagiarism, trouble getting the courses needed to graduate, the ineffectiveness of student government, or the difficulty in obtaining funds for a new student group. You could also write about a family problem, if you think it would interest your readers.

A. Brainstorming is a productive way to discover the best topic for this assignment. Don't stop at the first problem that comes to mind, but remember that you must know the topic well enough to explore it in some depth. Also review past issues of your campus newspaper or chat with classmates, teachers, and friends to identify possibilities.

B. Once you've decided what to write about, make a list of all the causes that you and others have thought of so far, even the ones you don't necessarily see as important or correct.

C. Cluster these causes on your list. Put immediate or easily observable causes in one group, indirect or less easily seen ones in another. Next, cluster around the indirect causes, trying to identify where those causes might originate. Critser's examination led him to supersizing and then to value meals, which led him to the connection between portion size and consumption, and finally to the possible link between increased calories and the growing popularity of fast food.

D. You will need to select the most important causes for the problem to allow you space to explain each one fully. Consider which points are least familiar, even surprising, to your readers and most likely to influence their thinking on the problem. A new perspective can sometimes move those with set ideas to reconsider their views.

E. A sketch outline of your chain of causes will let you know how well prepared you are to begin your draft. If the chain reveals a need for more examples or explanations, return to brainstorming and further conversations with people who know about the problem.

F. Before you begin the draft, review Critser's organization. He takes a chronological approach, starting with the first step in the process of increasing serving sizes. He could have started with the final effects—increased calories and obesity—and worked his way back to the primary causes. Your readers' familiarity with the problem and their attitudes toward it should help you decide where to start and how to organize your points.

G. In your conclusion, consider using Critser's technique of summing up his main message and driving the point home. Or you could try putting the next step in the hands of your readers by suggesting what they can do with the information and insights you have given them.

OTHER WRITING IDEAS

1. Brainstorm a list of causes for some problem you are facing, such as paying your bills, getting all your work done, dealing with a difficult friend or teacher, sticking to your diet. Organize the list into immediate and long-term causes. Write an essay analyzing the causes of your problem and concluding with your solution.
2. COLLABORATIVE WRITING. Conduct an informal survey among your friends and relatives about their favorite junk food: What do they like about it? When do they eat it? How often? Using this information, do some freewriting about why you think people love

this kind of food. Then gather your thoughts, and write an essay about the popularity of junk food.

3. USING THE INTERNET. Do an online search for information about the Slow Food Movement. Then, for a introductory class in health studies or your school newspaper, write an article about the origins and goals of this movement. Be sure to give credit to the sources of your information.
4. WRITING ABOUT READING. In the last five paragraphs of his essay, Critser examines the links between supersizing and increased food consumption in America. Review those paragraphs and write an evaluation of them. What claims does Critser make? How persuasive are his evidence and conclusions?

EDITING SKILLS: ELIMINATING WORDINESS

A good writer would never write only short, simple sentences, but a good clean sentence is better than a wordy, cluttered one. Look at these spare sentences from Critser's article:

It looked piggish.

People wanted more fries.

Kroc gave in.

A less skillful writer might have used a lot more words to say the same thing, like this:

It would make them appear to be gluttonous and piggish to people who might be watching.

The general public desired to consume additional french-fried potatoes.

Finally, after much hemming and hawing, Kroc relented and yielded in the end to Wallerstein's opinion.

There's nothing actually wrong with those sentences except what English teachers call *verbiage* and everybody else calls **wordiness.** Be careful not to say the same thing twice ("gluttonous and piggish," "relented and yielded," and "finally" and "in the end"). And avoid these common expressions:

Wordy	**Concise**
tall in height	tall
past history	history
blue in color	blue
advance forward	advance
expensive in price	expensive

consensus of opinion	consensus
continue to remain	remain
join together	join
few in number	few
positive benefits	benefits
at this point in time	at this time

EXERCISE

If you have trouble saying things succinctly, practice by streamlining the following wordy sentences. Keep the same meaning but eliminate the extra words. Here's an example:

> The masculine-gendered style used online in ListServe communications is, at this point in time, characterized by an adversarial attitude.
>
> (Revised) The male style used in communications on ListServe is adversarial.

1. It is my desire to be called Ishmael.
2. In my opinion there are many diverse elements about this problem that one probably ought to at least think about before arriving at an opinion on the matter.
3. The obnoxious children were seldom corrected or reprimanded because their baffled and adoring parents thought their objectionable behavior was normal and acceptable.
4. There came a time when, based on what I had been reading, I arrived at the feeling that the food we buy at the supermarkets to eat is sometimes, perhaps often, bad for us.
5. By and large, a stitch sewed or basted as soon as a rip is discovered may well save nine times the amount of sewing necessary if the job is put off even for a relatively short time.

Now go through your final draft one more time looking only for verbal clutter. Then, in the words of Mark Twain, "When in doubt, leave it out."

WEB SITES

www.supersizeme.com/

This site gives details about the film *Super Size Me,* the tongue-in-cheek documentary about a man who eats nothing but McDonald's food, three times a day for thirty days.

www.techcentralstation.com/supersizecon.html

This site, called the Super Size Con, attempts to expose "the truth about the film *Super Size Me*" by uncovering its inaccuracies and refuting its claims.

Student Essay Analyzing Causes

Blogging: An Emerging Addiction

Amee Bohrer

My blog career started innocently. My friend Virginia had one, and I became addicted to it, checking it a couple times a week and then every day, and then a few times a day. It was an easy way for me to keep up with her life even though she lives in Maryland. 1

Blog is a slang word for web log. It's an Internet hobby quickly gaining popularity. People blog for many reasons. Most often they use it as an online diary, to keep in touch with friends or business associates, and for a guaranteed outlet for political commentary. College students especially use blogs to sound off about their personal observations. It appeals specifically to writers, or to anyone who has something to say and wants to be published online. Any idiot can get a blog—that's the beauty of it. 2

My friend started bugging me to start one, but I shied away from it for months by claiming it was too complicated to set up. Then one day I was bored, so I figured I would give it the old college try. It was deceptively easy. Within five minutes, I was the proud owner of a newborn blog. I wasn't just a frustrated college student, but a frustrated college student who was now a published freelance writer! 3

It was slow at first, and no one ever read it. Then I told a few friends, and they started reading my blog. Next, I installed a tag board, which allows people to have a conversation similar to instant messaging on a blog, except it's strictly a message board instead of real-time communication. That addition opened up a whole new dimension, and soon I discovered that my friends were visiting just to check out the new postings. They began reacting to the content of my posts, bantering about gnat haikus and a number of wonderfully peculiar things. We were united by the tag board; it was the next best thing to talking to them on the phone or getting an e-mail every day. 4

Maintaining a blog popular with a few select friends who knew about it made me feel important—like I was doing something productive with my writing. Then one day my oldest friend said, "Oh, so you're actually going to write me a real e-mail, not just a big impersonal blog?" I was shocked to find anger underneath the sarcasm. I then considered and indeed saw the truth in her remark. 5

Before my blog became popular with my friends, I had written them e-mails on a regular basis. But the blog allowed me to make announcements and update them on my life all at once, and I noticed that now I just substituted a comment in my post or on the tag board to them. In turn, instead of calling or e-mailing me, now they just posted comments on my tag board. It was quite a change in our friendship. 6

I realized blogging had taken over my life. At times when I might have done homework, called a friend, or eaten, I was now blogging. Every time something important happened, or when boredom struck, I went straight to my blog. It had become an addiction. 7

I also began to post way more than "too much information," either about myself or other people. I even used it as a passive-aggressive way to get back at my ex-boyfriend, lambasting him because I knew his best friend read my blog and would deliver my sentiments to him. That's when I realized I had gone too far. In the beginning, I had adopted a hardcore philosophy, thinking it was my duty to blog everything. It didn't occur to me not to post something just because it might offend someone. But after a while, I came to see that a blog is not a newspaper, nor should it resemble a tell-all tabloid. 8

Now I feel I have control of my addiction, and I monitor the content I post very closely. It's still a lot of fun; but I've learned that although blogging can be a great way to exercise your writing muscles and publish your brilliant insights, it can also damage friendships and monopolize your time. Bloggers beware. 9

CONSIDERING CONTENT AND METHOD

1. What is a blog or blogging? Does the author define these terms accurately and adequately?
2. According to the author, why do people keep a blog?
3. Why did she get addicted? How did she recognize her addiction?
4. What did she finally realize about blogging? What are her current feelings about blogging?
5. Do you blog? If so, do you agree with the author's conclusions about blogging? Does her cause-effect reasoning seem sound?

Chapter 10

Strategies for Influencing Others: *Argument* and *Persuasion*

Source: Rex Babin/*The Sacramento Bee.*

You already know when to use persuasion: when you hope to get your readers to agree with you and maybe, as a result, to take some kind of action—

like giving you a refund or picketing city hall. And on essay tests and in classroom assignments you are often asked to take a stand or support a conclusion.

All the writing strategies that you studied in previous chapters can be used to persuade your readers or to argue a point. You can use comparison/contrast to demonstrate the superiority of whole wheat over white bread, or cause-effect analysis to convince your readers that rap music reinforces violent attitudes toward women, or definition to show that computer nerds are more interesting than most people think. In this chapter, you will use all the strategies for organizing your thoughts and ideas as you learn how to influence readers' opinions about controversial or unfamiliar issues.

THE POINT OF ARGUMENT AND PERSUASION

Your purpose in this type of writing is to encourage the readers to accept your point of view, solution, plan, or complaint as their own. Traditionally, the word **persuasion** refers to attempts to sway the readers' emotions, while the word **argument** refers to tactics that address the readers' logic. Most convincing writing today mixes the two types of appeal. A personal testimony from a paraplegic accident victim pleading with readers to use their seat belts persuades through emotional identification. A list of statistics concerning injury rates before and after seat belt laws went into effect argues the point through rationality. A combination of the two tactics would probably be quite effective. In everyday language, *persuasion* means influence over the audience, whether emotional or rational.

THE PRINCIPLES OF ARGUMENT AND PERSUASION

Presenting a conventional persuasive essay involves five tasks:

1. State the **issue** your essay will address, and put it in a context: Why is it controversial or problematic? Why do people care about it? Why do people disagree about it?
2. State your main point or thesis. What point of view, solution, or stance do you wish the readers to adopt? This is your **claim.**
3. Provide well-developed **evidence** on your own side of the issue. You can develop your point through facts, statistics, examples, expert testimony, and logical reasoning (cause and effect, analogy), just to name a few strategies. This is the longest part of a conventional argument, and each piece of evidence will probably take a paragraph or more to develop.
4. Respond to opposing viewpoints. This is called the **refutation** section. Especially when arguments against your own are widely known,

you need to acknowledge them and counter them, or your essay will have obvious holes in it. You might minimize their importance, demonstrate that they are not logical or factual, or offer alternative ways of thinking about them.

5. Close by reminding your reader of your main point and the strength of your evidence. Many persuasive essays include a call to action, encouraging readers to do something in support of your cause.

Writers often alter this conventional plan, especially tasks 3 and 4. There are several typical arrangements you can use to fit your topic, audience, purpose, and the nature of your evidence:

The Counterargument. You can anticipate what people on the other side of the issue would say and organize your argument as a point-by-point refutation. This approach works well for tackling controversial topics and for clearing up common misconceptions. In this chapter, for example, "Death Penalty Showdown" argues for capital punishment by countering several anti–death penalty arguments, one by one.

The Pro and Con Argument. Another way to present an argument is to look at the pros and cons of an issue. These may also be called advantages and disadvantages or strengths and weaknesses. This approach is commonly used when you are trying to make a difficult decision or settle a dispute amicably.

The Problem-Solution Argument. You can use this approach when you want to argue directly for a change in a policy or system. You need to identify the problem, demonstrate that it's relevant or serious, explain why current methods aren't working, propose a solution, and show how it will work. Sometimes you can merely call people's attention to a problem without offering a solution, but most of the time you will want to recommend a course of action to reduce or eliminate the problem. In this chapter, "The War on Drugs" presents a lengthy indictment of the current drug prosecution policies and ends with the author's proposal for a solution.

THE ELEMENTS OF GOOD ARGUMENT

Claims, evidence, and refutation are the basic building blocks of argument. You must understand them in order to write persuasively and effectively.

Claims

The "engine" that propels any argument is its claim. Claims fall into three categories:

1. *Claims of fact* assert that something is true: "Women currently do not receive equal pay for equal work."

2. *Claims of value* support or deny the worth or merit of something: "Getting a college education is more important than playing sports."
3. *Claims of policy* state that certain conditions, courses of action, or practices should be adopted: "A six-month maternity leave policy ought to be required by law."

For any argument to be effective, it must begin with a claim that's significant, reasonable, and supportable by evidence.

Evidence

Some of the most frequently used kinds of evidence are these:

- *Personal observation or experience:* "I came the closest to achieving a decent fit between income and expenses only when I worked seven days a week." This is the least widely accepted form of evidence because it comes from a narrow band of experience and may not be representative. But it can be very persuasive in communicating the human significance of an issue.
- *Facts:* "Rents usually have to be less than 30 percent of one's income to be considered 'affordable.'" Facts are noncontroversial pieces of information that can be confirmed through observation or by generally accepted sources; they are persuasive to the extent that they relate to the statement they support.
- *Relevant examples:* "One waitress shares a room in a boarding house for $250 a week; another lives with her mother; the night cook pays $170 a week for a one-person trailer; the hostess lives in a van parked behind a shopping center at night." Examples are most persuasive when they are objective and clearly relevant.
- *Testimony:* "Maggie Spade of the Economic Policy Institute explains that the low-income housing crisis exists partly because it is not reflected in the official poverty rate, which is based on the cost of food, not shelter." Testimony should come from sources that the readers will accept as credible; it carries its greatest weight when the meaning of the facts or data is not self-evident and some judgment or interpretation is required to reach a conclusion.
- *Data:* "Almost 60 percent of poor renters, amounting to a total of 4.4 million households, spend more than half of their income on shelter." This is probably the most readily accepted form of evidence because of its apparent objectivity and scope.

To be persuasive, all evidence in support of a claim must meet certain standards:

1. It must be reasonably up to date.
2. It must be sufficient in scope.
3. It must be relevant to the claim.

Thus, you wouldn't use 1995 unemployment data in support of a 2005 policy decision (recentness). Neither would you rely on unemployment data from only one month to formulate a long-term unemployment policy (scope). Nor would you use employment data from Canada to comment on the American unemployment picture, unless your argument involved an overall comparison of the two countries' economies (relevance).

Refutation

Arguments always assume that other points of view are possible; otherwise, there would be no reason to argue. You may feel so strongly about an issue that you want to attack those who disagree with you and ridicule their opinions, but that strategy can backfire, especially if you're trying to influence readers who are undecided. Your case will be strengthened if you treat opposing views with respect and understanding.

This acknowledgment of and response to the opposing views is called *refutation*. There is no best place in an argument to refute the opposition. Sometimes you will want to bring up opposing arguments early and deal with them right away. Another approach is to anticipate objections as you develop your own case point by point. Wherever you decide to include your refutation, your goal is to point out problems with the opponents' reasoning and evidence. You can refute opposing arguments by showing that they are unsound, unfair, or flawed in their logic. Frequently, you will present contrasting evidence to reveal the weakness of your opponents' views and to reinforce your own position.

When an opposing argument is so compelling that it cannot be easily countered, you should concede its strength. This approach will establish that you are knowledgeable and fair-minded. You can sometimes accept the opponents' line of thought up to a certain point—but no further. Or you can show that their strong point addresses only *one* part of a complex problem.

A SAMPLE ANNOTATED ARGUMENT

The following selection appeared in *U.S. News & World Report* in the December 30, 1996, issue. Although this article is a news report, it also contains an argument. The writer makes a claim of fact: that needle exchange programs are effective in reducing the rate of HIV infection. He also presents a claim involving policy: that the federal ban on funding these programs should be lifted. The marginal notes identify key features and strategies in building an argument.

Ignoring the Solution

JOSHUA WOLF SHENK

Statement of the problem

AIDS was once the scourge of the gay community. Soon, it will be largely a drug addict's disease. Scientists believe that 50 percent of all new HIV infections occur among intravenous drug users, with an additional 20 percent or so occurring among junkies' sex partners. The syringe is the Typhoid Mary of the 1990s. Yet what worked best in curtailing the spread of HIV among homosexuals—mass education campaigns promoting safe sex—has been ineffective with drug addicts lurking in society's shadows. What does seem to work is giving drug users clean needles.

Primary claim (thesis)

Solution

Since 1986, some 100 small needle-exchange programs have sprouted up around the country, through which used syringes are traded for new, sterile ones—no questions asked. Often run by private groups with limited funds, these experiments have been the object of intense scrutiny by major universities and federal health agencies. The conclusion? The programs work. Studies have shown up to a sevenfold reduction in all blood-borne diseases, a 33 percent projected drop in HIV infections, and 25 percent fewer cases of dangerous behavior, such as needle sharing.

Evidence (data)

Claim of fact

Further claim

Besides saving lives, these needle exchanges deliver a huge financial payoff. Consider the case of an HIV-positive addict who infects eight others in a one-year period (a very modest estimate). If each turns to Medicaid to pay his or her lifetime medical costs (at an average $119,000 plus), that's about a $1 million burden for tax payers—money that could have been saved if the one addict had been in a needle-exchange program.

Evidence (example, data)

Evidence for the effectiveness of needle exchanges is not airtight. Drug users who partici-

Concession

pate in needle exchanges may be more safety conscious and thus at less risk of contracting HIV in the first place. But studies also show that those who participate improve their own behavior over time. So evidence that needle exchanges have at least some positive effects is strong.

Refutation

Restatement of claim

Evidence (expert testimony)

On balance, the studies are persuasive enough that physician Scott Hitt, chairman of President Clinton's Advisory Council on HIV/AIDS, rebuked his own president for banning the use of federal AIDS funds for needle exchanges. Hitt is joined in the endorsement of needle exchanges—and the call for more federal involvement—by the National Academy of Sciences, the Centers for Disease Control and Prevention, and the General Accounting Office.

Implied claim of policy

Refutation of first objection

The administration worries that needle exchanges might increase drug use. It's a reasonable fear but one not borne out by research, according to the CDC. Only a handful of needle-exchange studies have tracked drug use, but their conclusions jibe with anecdotal evidence and common sense: While addicts prefer clean needles, they will eagerly opt for the abundant supply of dirty ones in the face of a monstrous drug craving.

Refutation of second objection

Some worry that needle exchanges are the classic "Band-Aid"—dealing with HIV infection but not the underlying drug addiction. But needle exchanges have actually worked as a bridge into real treatment. One program in Tacoma, Wash., made nearly 1,000 referrals to drug treatment programs in two years. Others worry that needle exchanges, cheap as they are, will siphon funds from zero-tolerance treatment efforts. But the real problem is that all anti-addiction programs are woefully underfunded.

Evidence (example)

Refutation of third objection

Conclusion

It's hard to avoid the suspicion that these concerns have less to do with science or public health than with politics; specifically, a reluctance to muddy the "just say no" message. But there's another message leaders should heed—that no one has to die needlessly. Peter Lurie, a leading University of California researcher, estimates that nearly 10,000 lives could have been saved over the

Evidence (expert testimony)

past few years by an aggressive expansion of needle-exchange programs. Wasn't the war on drugs supposed to be about saving lives?

THE PITFALLS OF ARGUMENT AND PERSUASION

If you follow the preceding suggestions, you should be able to write a convincing essay. But there are some risks to be aware of in this kind of writing.

Taking on Too Much

Narrowing your topic is always a good idea, but it's especially important in argumentation. You won't be able to write a sensible essay on "What's Wrong with the American Economy." Select a more manageable problem, such as unemployment or unbridled greed. Even those problems are probably too broad to cover in a short essay. Always consider moving the issue closer to home. For example, if someone in your family has been unemployed, you are probably equipped to write about the psychological effects of unemployment on the individual. If you have found yourself in terrifying and unnecessary credit card debt, you can probably write persuasively about uncontrolled consumerism. In these cases, you have credibility to discuss the issue. Credibility is an important part of your appeal, and you will notice that each author in this chapter establishes the right to claim knowledge about his or her subject, either explicitly or implicitly.

Mistaking the Audience

The readers you most want to reach with a controversial essay are the ones in the middle—those who are undecided and might be swayed by your ideas. People with extreme opinions on either side are likely to be unmovable. Even with persuadable audiences, you must expect some resistance. Be sure your tone does not bolster this resistance by being insulting or condescending. The voice of sweet reason and a "we're in this together" attitude invite your readers to agree with you.

Logical Fallacies

Flaws in reasoning can undermine your cause and harm your credibility. Be sure that you are not guilty of these common logical fallacies:

1. *Overgeneralization.* Recently, a national survey found that the number of unmarried women among highly educated people was much larger than among the less educated. Articles on the alarming shortage of men willing to marry educated females abounded, and solutions to

the difficulty were proposed. Actually, most of the single, educated females had chosen their unmarried status. The problem was the false generalization that all unmarried women *wanted* husbands and were seeking them. The shortage of marriage-minded men was not proven.

2. *Either-or thinking.* Be sure that you don't present only two alternatives when more exist. For example, some writers in education want us to believe that either we set national standards for mathematical achievement, or our children will continue to fall behind other countries' children in math skills. The fallacy is that national standards are not the only route to high math achievement: smaller classes, better teaching conditions, early intervention policies, and parent involvement are just a few ideas left out of the either-or reasoning. (In fact, the existence of national standards is not correlated with math achievement internationally.)
3. *False analogy.* Analogy is a compelling form of argument, but you must take care that the two cases you compare are really similar. If you argue that your college should imitate a successful general education program used at another university, you must be sure that the two institutions have similar students, faculty, goals, and organizational structures. Otherwise, adopting a plan that works for someone else could be disastrous. Recently, a U.S. congressman claimed that foreign dignitaries, while in our country, should say the Pledge of Allegiance because, similarly, we show respect at the Olympics when other countries' national anthems are played. The analogy falls apart when you consider what the Pledge of Allegiance actually *says.*
4. *Faulty claims about causation.* Remember that two things that occur closely in time or one after the other are not necessarily causally related. In our cities, ice cream sales and murder rates both increase in the summer; can we say that eating ice cream causes aggression? No, probably it's the heat that encourages both. We say "children who are hugged are more likely to be nice," and "children who are beaten are more likely to be unpleasant." But the causes and effects could be the other way around. Maybe nice children are hugged *because* they're nice, and unpleasant children get beaten *because* they're unpleasant.

WHAT TO LOOK FOR IN ARGUMENT AND PERSUASION

Here are some questions to ask yourself as you study the selections in this chapter.

1. What issue is being addressed? What stance or point of view am I being asked to adopt?

2. What claim or claims does the author make? What kind of claims are they? Are they significant and reasonable?
3. Does the author explain the issue clearly and completely? Are there any holes or gaps in the author's thinking?
4. What evidence does the author use to support the claims? Is this evidence sufficient, relevant, and up to date?
5. Does the writer refute possible objections? How effective is this refutation? Has the author ignored any important objections?
6. Is the conclusion satisfactory? Has the author persuaded me?

IMAGES AND IDEAS

Source: The Advertising Archive.

For Discussion and Writing

What point does this poster make? Is it an example of argument or persuasion? Create your own parody of a well-known advertisement. A parody imitates a serious work but exaggerates or adds details that poke fun at the original. Decide how you want to change the original to make your point.

PREPARING TO READ

Do you think our society pays too much attention to athletes and athletics? Do you play a sport? If so, what do you think your chances are of becoming a professional?

Send Your Children to the Libraries

ARTHUR ASHE

Arthur Ashe was the first African-American player to win a major men's tennis tournament: the U.S. Open in 1968 and the Wimbledon championship in 1975. Ashe survived heart surgery in 1979 and announced his retirement from competition in 1980, although he continued to serve as the nonplaying captain of the U.S. Davis Cup team. He became infected with the AIDS virus, probably through a blood transfusion that he received during his second bypass operation in 1983; he died of complications from AIDS in 1993. In this letter, published in the New York Times *in 1977, Ashe argues that the lure of professional sports is actually harmful to black athletes.*

TERMS TO RECOGNIZE

pretentious *(para. 1)*	falsely superior
expends *(para. 1)*	spends, uses up
dubious *(para. 1)*	doubtful, questionable
emulate *(para. 2)*	follow, copy
Forest Hills *(para. 3)*	location of the U.S. Open tennis tournament in 1968
attributing *(para. 4)*	assigning, crediting
viable *(para. 5)*	possible, workable
Wimbledon *(para. 12)*	suburb of London and site of the world-famous tennis tournament

Since my sophomore year at University of California, Los Angeles, I have become convinced that we blacks spend too much time on the playing fields and too little time in the libraries. Please don't think of this attitude as being pretentious just because I am a black, single, profes- 1 sional athlete. I don't have children, but I can make observations. I strongly believe the black culture expends too much time, energy and effort raising, praising and teasing our black children as to the dubious glories of professional sports.

All children need models to emulate—parents, relatives or friends. But when the child starts school, the influence of the parent is shared by teachers and classmates, by the lure of books, movies, ministers and newspapers, but most of all by television. 2

Which televised events have the greatest number of viewers? Sports—the Olympics, Super Bowl, Masters, World Series, pro basketball playoffs, Forest Hills. ABC-TV even has sports on Monday night prime time from April to December. So your child gets a massive dose of O. J. Simpson, Kareem Abdul-Jabbar, Muhammad Ali, Reggie Jackson, Dr. J. and Lee Elder and other pro athletes. And it is only natural that your child will dream of being a pro athlete himself. 3

But consider these facts: For the major professional sports of hockey, football, basketball, baseball, golf, tennis and boxing, there are roughly only 3,170 major league positions available (attributing 200 positions to golf, 200 to tennis and 100 to boxing). And the annual turnover is small. We blacks are a subculture of about 28 million. Of the 13½ million men, 5–6 million are under twenty years of age, so your son has less than one chance in a thousand of becoming a pro. Less than one in a thousand. Would you bet your son's future on something with odds of 999 to 1 against you? I wouldn't. 4

Unless a child is exceptionally gifted, you should know by the time he enters high school whether he has a future as an athlete. But what is more important is what happens if he doesn't graduate or doesn't land a college scholarship and doesn't have a viable alternative job career. Our high school dropout rate is several times the national average, which contributes to our unemployment rate of roughly twice the national average. 5

And how do you fight the figures in the newspapers every day? Ali has earned more than $30 million boxing, O. J. just signed for 2½ million, Dr. J. for almost $3 million, Reggie Jackson for $2.8 million, Nate Archibald for $400,000 a year. All that money, recognition, attention, free cars, girls, jobs in the off-season—no wonder there is Pop Warner football, Little League baseball, National Junior League tennis, hockey practice at 5 A.M. and pickup basketball games in any center city at any hour. 6

There must be some way to assure that the 999 who try but don't make it to pro sports don't wind up on the street corners or in the unemployment lines. Unfortunately, our most widely recognized role models are athletes and entertainers—"runnin'" and "jumpin'" and "singin'" and "dancin'." While we are 60 percent of the National Basketball Association, we are less than 4 percent of the doctors and lawyers. While we are about 35 percent of major league baseball, we are less than 2 percent of the 7 engineers. While we are about 40 percent of the National Football League, we are less than 11 percent of construction workers such as carpenters and bricklayers.

8 Our greatest heroes of the century have been athletes—Jack Johnson, Joe Louis and Muhammad Ali. Racial and economic discrimination forced us to channel our energies into athletics and entertainment. These were the ways out of the ghetto, the ways to that Cadillac, those alligator shoes, that cashmere sport coat. Somehow, parents must instill a desire for learning alongside the desire to be Walt Frazier. Why not start by sending black professional athletes to high schools to explain the facts of life?

9 I have often addressed high school audiences and my message is always the same. For every hour you spend on the athletic field, spend two in the library. Even if you make it as a pro athlete, your career will be over by the time you are thirty-five. So you will need that diploma. Have these pro athletes explain what happens if you break a leg, get a sore arm, have one bad year or don't make the cut for five or six tournaments. Explain to them the star system, wherein for every O. J. earning millions there are six or seven others making $15,000 or $20,000 or $30,000 a year.

10 But don't just have Walt Frazier or O. J. or Abdul-Jabbar address your class. Invite a benchwarmer or a guy who didn't make it. Ask him if he sleeps every night. Ask him whether he was graduated. Ask him what he would do if he became disabled tomorrow. Ask him where his old high school athletic buddies are.

11 We have been on the same roads—sports and entertainment—too long. We need to pull over, fill up at the library and speed away to Congress and the Supreme Court, the unions and the business world. We need more Barbara Jordans, Andrew Youngs, union cardholders, Nikki Giovannis and Earl Graveses. Don't worry: We will still be able to sing and dance and run and jump better than anybody else.

12 I'll never forget how proud my grandmother was when I graduated from UCLA in 1966. Never mind the Davis Cup in 1968, 1969, and 1970. Never mind the Wimbledon title, Forest Hills, etc. To this day, she still doesn't know what those names mean. What mattered to her was that of her more than thirty children and grandchildren, I was the first to be graduated from college, and a famous college at that. Somehow, that made up for all those floors she scrubbed all those years.

RESPONDING TO READING

Do you think the athletes at your school receive special treatment? Do they spend too much time on their sports and too little time on their studies? In your journal express your opinion about the place of athletics in colleges and universities.

GAINING WORD POWER

Explain in your own words the meaning of the following phrases. Use clues from surrounding sentences to help you.

1. the lure of books, movies (para. 2)
2. a massive dose (para. 3)
3. blacks are a subculture (para. 4)
4. a child is exceptionally gifted (para. 5)
5. pickup basketball games (para. 6)
6. economic discrimination (para. 8)
7. instill a desire for learning (para. 8)
8. make the cut (para. 9)
9. the star system (para. 9)

CONSIDERING CONTENT

1. What problem is Ashe concerned about? What solution does he propose?
2. Ashe is clearly addressing the parents of black sons. Are his opinions relevant to other races or to parents of girls? Why or why not?
3. In paragraph 8, Ashe says that "economic discrimination" has forced African Americans into sports and entertainment. What does he mean? Do you agree?
4. According to Ashe, why do African-American athletes need a diploma? Do these reasons apply to other races and to nonathletes?
5. Who are Barbara Jordan, Andrew Young, Nikki Giovanni, and Earl Graves? Why does Ashe say "We need more" of these people?
6. In paragraph 11, Ashe says that blacks have been "on the same roads—sports and entertainment—too long." Does he want blacks to avoid these careers?
7. This letter was written in 1977, almost thirty years ago. Are Ashe's views still relevant?

CONSIDERING METHOD

1. Ashe states the problem in his first sentence. Where does he restate it? Why does he restate the problem several times?
2. What criticisms is Ashe anticipating in his opening paragraph?
3. In his letter, Ashe uses a lot of statistics. How convincing are they?
4. What other kinds of evidence does Ashe use to support his main points?
5. How does Ashe make it clear that he is part of the audience he's addressing? Why does he want his readers to know that he is not an outsider?

6. Notice Ashe's frequent use of **parallel structure,** items in a series, and intentional repetition. Find several examples of each. What effect is Ashe trying to achieve with these elements?
7. Explain the **metaphor** that Ashe uses in paragraph 11.
8. Why does Ashe conclude with comments about his grandmother? Is this an effective ending?
9. What stereotypes about blacks does Ashe refer to? Why does he use the word *black* and not *African American?*

WRITING STEP BY STEP

Write a letter to a newspaper in which you encourage fellow students or fellow citizens to join you in solving some problem that affects you all. Choose a problem close to home—something on campus or in your community that you want fixed, improved, regulated, legalized, banned, or reorganized.

A. Begin, as Ashe does, with a clear and direct statement of the problem.
B. Explain your interest or involvement in the situation. Make it clear to your readers that you are not an outsider: use the pronouns *we, us, our,* and *ours.*
C. Describe the problem and, if necessary, explain why you think it needs a solution. Your readers may not realize how serious the problem is or how much it affects them.
D. Use actual examples and, if possible, give statistics to support your claims.
E. Emphasize your points and keep your readers' attention by using parallel structure, items in series, and intentional repetition.
F. State your solution, and explain how it will work.
G. If appropriate, show why other solutions won't work and why your proposal is the best one.
H. Conclude with a personal **anecdote** or comment, like Ashe's story about his grandmother's pride.

OTHER WRITING IDEAS

1. Write a complaint letter about some problem you have experienced as a consumer. Identify the problem, explain why it needs to be solved, and offer your solution. Address your complaint to the person or agency you think will be able to solve your problem: the store manager, the company owner, the product manufacturer, the Better Business Bureau, the state Consumer Protection Office, Tony Soprano.
2. COLLABORATIVE WRITING. With a group of friends or fellow students, discuss whether there is some rule change that would make a

popular sport safer, fairer, or more interesting to watch. Using a **problem-solution** approach, write a letter to a sports magazine (such as *Sports Illustrated*) or to an athletic organization (such as the NCAA) stating your proposal for changing the rules. Focus on the reasons why you think the change would be beneficial.

3. USING THE INTERNET. Visit the Web site **www.steroids.org/overview.htm**, and read the information presented there on steroid use. Then write an essay for a high school newspaper persuading young athletes not to take steroids.
4. WRITING ABOUT READING. In his letter Ashe addresses his argument to the parents of student athletes. Write a letter to this same audience in which you argue that participation in sports is good for students.

EDITING SKILLS: AVOIDING SEXIST LANGUAGE

As the opening sentence of paragraph 5, Arthur Ashe writes, "Unless a *child* is exceptionally gifted, you should know by the time *he* enters high school whether *he* has a future as an athlete" (our italics). When this letter was published in 1977, it was acceptable to use masculine pronouns to refer to virtually all living beings as if they were male. And, of course, Ashe is thinking of *sons* as becoming professional athletes—even though he later mentions that we need more "Barbara Jordans" and "Nikki Giovannis" (para. 11).

These days you need to avoid using masculine pronouns to refer to people of both sexes. If Ashe were still alive today, he probably would not write, "And it is only natural that your *child* will dream of being a pro athlete *himself*" (para. 3). More likely he would cast that sentence in the plural this way: "And it is only natural that your *children* will dream of being pro athletes *themselves.*" Both sexes are included in all plural pronouns. People don't think either male or female when they read *we, us, our, you, your, they, them, their, ourselves, yourselves, themselves.* So, if you simply write in the plural, the problem disappears.

Occasionally you can revise a sentence to eliminate the pronoun, like this:

(sexist) A tennis player must practice daily to stay at the top of *his* form.

(revised) A tennis player must practice daily to stay in top form.

Or if you find yourself once in a while needing to write a singular sentence for some good reason, it's quite all right to use both male and female pronouns, like this:

Everyone on board must wear a lifejacket for *his or her* own safety.

Just don't do it this way too often, or your writing will get annoyingly cluttered.

EXERCISE

Because your readers may be bothered by **sexist language,** you should write in the plural most of the time. For practice, rewrite the following sentences in the plural to get rid of the italicized pronouns.

1. An Olympic swimmer needs to work out daily to perform at *her* best.
2. But a championship bridge player can take a week off without ruining *his* game.
3. A professional athlete needs to watch *his* diet, as well as exercise *her* body.
4. A chess player is constantly exercising *his* mind, but *he* can eat whatever *he* pleases.
5. Even an amateur golfer gets plenty of exercise when *she* plays eighteen holes, unless *she* rides in a cart.

Now go back over the essay you just wrote and check the personal pronouns. Did you use any (such as *he, his, him* or *she, her, hers*) that unfairly or inaccurately exclude the other sex? Make any necessary changes—such as using *he or she* or rewriting in the plural to allow *they, their,* or *them.*

WEB SITE

www.chron.com/content/chronicle/sports/special/barriers/ashe.html

This site presents a comprehensive look at Arthur Ashe's life, his accomplishments, and his dedication to social causes.

PREPARING TO READ

Have you ever been arrested or known anyone who has? Do you think people are ever unfairly punished in this country? Or do you think most people get the sentences they deserve?

The War on Drugs

BILL BRYSON

Having spent two decades living in England, American writer Bill Bryson now lives in New Hampshire. His books include A Walk in the Woods: Rediscovering America on the Appalachian Trail *(1998) and* In a Sunburned Country *(2000). The following essay was written in 1996 for Bryson's weekly column in a New Hampshire newspaper and appears in his collection* I'm a Stranger Here Myself: Notes on Returning to America after Twenty Years Away *(1999).*

TERMS TO RECOGNIZE

extenuating *(para. 1)*	making less serious by providing a partial excuse
ferocity *(para. 2)*	intense fierceness
Newt Gingrich *(para. 2)*	Georgia Republican, Speaker of the House of Representatives from 1995 to 1997
disproportionate *(para. 3)*	unequal, out of proportion in size or amount
affidavit *(para. 7)*	an official statement or legal document
zealous *(para. 9)*	fanatical, passionate
vindictiveness *(para. 9)*	an attitude of revenge, spitefulness

I recently learned from an old friend in Iowa that if you are caught in possession of a single dose of LSD you face a mandatory sentence of seven years in prison without possibility of parole. Never mind that you are, say, eighteen years old and of previous good character, that this will ruin your life, that it will cost the state $25,000 a year to keep you incarcerated. Never mind that perhaps you didn't even know you had the LSD—that a friend put it in the glovebox of your car without your knowledge or maybe saw police coming through the door at a party and shoved it into your hand before you could react. Never mind any extenuating circumstances whatever. This is America in the 1990s and there are no exceptions where drugs are concerned. Sorry, but that's the way it is. Next. 1

It would be nearly impossible to exaggerate the ferocity with which the United States now prosecutes drug offenders. In fifteen states you can be sentenced to life in prison for owning a single marijuana plant. Newt Gingrich, the House Speaker, recently proposed that anyone caught bringing as little as two ounces of marijuana into the United States should be imprisoned for life without possibility of parole. Anyone caught bringing more than two ounces would be executed. 2

According to a 1990 study, 90 percent of all first-time drug offenders in federal courts were sentenced to an average of five years in prison. Violent first-time offenders, by contrast, were imprisoned less often and received on average just four years in prison. You are, in short, less likely to go to prison for kicking an old lady down the stairs than you are for being caught in possession of a single dose of any illicit drug. Call me soft, but that seems to me a trifle disproportionate. 3

Please understand it is not remotely my intention here to speak in favor of drugs. I appreciate that drugs can mess you up in a big way. I have an old school friend who made one LSD voyage too many in about 1977 and since that time has sat on a rocker on his parents' front porch examining the backs of his hands and smiling to himself. So I know what drugs can do. I just haven't reached the point where it seems to me appropriate to put someone to death for being an idiot. 4

Not many of my fellow countrymen would agree with me. It is the clear and fervent wish of most Americans to put drug users behind bars, and they are prepared to pay almost any price to achieve this. The people of Texas recently voted down a $750 million bond proposal to build new schools but overwhelmingly endorsed a $1 billion bond for new prisons mostly to house people convicted of drug offenses. 5

America's prison population has more than doubled since 1982. There are now 1,630,000 people in prison in the United States. That is more than the populations of all but the three largest cities in the country. Sixty percent of federal prisoners are serving time for nonviolent offenses, mostly to do with drugs. America's prisons are crammed with nonviolent petty criminals whose problem is a weakness for illegal substances. Because most drug offenses carry mandatory sentences and exclude the possibility of parole, other prisoners are having to be released early to make room for all the new drug offenders pouring into the system. In consequence, the average convicted murderer in the United States now serves less than six years, the average rapist just five. Moreover, once he is out, the murderer or rapist is immediately eligible for welfare, food stamps, and other federal assistance. A convicted drug user, no matter how desperate his circumstances may become, is denied these benefits for the rest of his life. 6

The persecution doesn't end there. My friend in Iowa once spent four months in a state prison for a drug offense. That was almost twenty years 7 ago. He did his time and since then has been completely clean. Recently, he applied for a temporary job with the U.S. Postal Service as a holiday relief mail sorter. Not only did he not get the job, but a week or so later he received by recorded delivery an affidavit threatening him with prosecution for failing to declare on his application that he had a felony conviction involving drugs. The Postal Service had taken the trouble, you understand, to run a background check for drug convictions on someone applying for a temporary job sorting mail. Apparently it does this as a matter of routine—but only with respect to drugs. Had he killed his grandmother and raped his sister twenty-five years ago, he would in all likelihood have gotten the job.

8 It gets more amazing. The government can seize your property if it was used in connection with a drug offense, even if you did not know it. In Connecticut, according to a recent article in the *Atlantic Monthly* magazine, a federal prosecutor named Leslie C. Ohta made a name for herself by seizing the property of almost anyone even tangentially connected with a drug offense—including a couple in their eighties whose grandson was found to be selling marijuana out of his bedroom. The couple had no idea that their grandson had marijuana in the house (let me repeat: they were in their eighties) and of course had nothing to do with it themselves. They lost the house anyway.

9 The saddest part of this zealous vindictiveness is that it simply does not work. America spends $50 billion a year fighting drugs, and yet drug use goes on and on. Confounded and frustrated, the government enacts increasingly draconian laws until we find ourselves at the ludicrous point where the Speaker of the House can seriously propose to execute people—strap them to a gurney and snuff out their lives—for possessing the botanical equivalent of two bottles of vodka, and no one anywhere seems to question it.

10 My solution to the problem would be twofold. First, I would make it a criminal offense to be Newt Gingrich. This wouldn't do anything to reduce the drug problem, but it would make me feel much better. Then I would take most of that $50 billion and spend it on rehabilitation and prevention. Some of it could be used to take busloads of youngsters to look at that school friend of mine on his Iowa porch. I am sure it would persuade most of them not to try drugs in the first place. It would certainly be less brutal and pointless than trying to lock them all up for the rest of their lives.

RESPONDING TO READING

What do you think should be done about the drug problem in this country? Do you agree with Bryson's solution? Why or why not?

GAINING WORD POWER

Look up the following words in a college dictionary. Then write a sentence of your own for each word.

1. mandatory (para. 1)
2. incarcerated (para. 1)
3. illicit (para. 3)
4. fervent (para. 5)
5. petty (para. 6)
6. tangentially (para. 8)
7. draconian (para. 9)
8. ludicrous (para. 9)

CONSIDERING CONTENT

1. What is the point of Bryson's opening example concerning the punishment for possession of LSD in Iowa?
2. Would you agree that the sentences for first-time drug offenders are "disproportionate" when compared to the sentences for violent first-time offenders?
3. In paragraph 4, Bryson says he is not speaking in favor of drugs. What then is the purpose of his essay?
4. According to Bryson, in what ways are convicted drug users treated unfairly?
5. Who is Leslie C. Ohta and what did she do? How do you feel about her actions?
6. Bryson says that punishing drug users harshly "simply does not work" (para. 9). What evidence does he give to support this claim? Do you agree?
7. What solution does the author offer? How reasonable is it?

CONSIDERING METHOD

1. Why does Bryson begin by referring to "an old friend in Iowa"? How does he use his friends in Iowa to unify his essay?
2. How does the author establish his credibility on this topic?
3. Identify four of the contrasts Bryson uses to develop his claim about the unfairness of the war on drugs. How effective are they?
4. What is the point of paragraph 4? Why does Bryson include this paragraph?
5. Find at least four uses of statistics, and discuss their effectiveness.

6. What kind of evidence is used in paragraph 7? How does it differ from the evidence used elsewhere?
7. Identify several examples of humor in this essay. How would you describe that humor? Is it effective?

WRITING STEP BY STEP

Each of us is aware of some system that doesn't (or didn't) work. Think about fads such as miracle weight-loss plans; customs such as dating and marriage; practices such as spanking or social promotion; university or school rules or requirements; local government policies such as curfews; or even national laws such as Prohibition. Surely there is somewhere in society that you perceive inadequacy in addressing a real problem. Write an essay based on Bryson's example in which you explain what doesn't work and why.

A. Be sure to establish your credibility for writing about this topic. If your credentials aren't based on personal experience, maybe they are based on some other knowledge or expertise.
B. Write a paragraph telling why the current situation exists. What is the system you're criticizing supposed to do?
C. Write one paragraph explaining that the problem is real but that the present system doesn't address it effectively.
D. Devote two or three paragraphs to evidence for your own claim that the system doesn't work. Each paragraph should develop one point on your side of the matter. Use at least two different types of evidence (facts, statistics, expert testimony, examples, logical reasoning, personal experience).
E. In your closing paragraph, suggest a better approach for efforts to deal with the problem or situation. You don't have to go into detail (Bryson doesn't), but you do need to indicate why you consider this approach more workable.

OTHER WRITING IDEAS

1. Argue for the alteration or abolition of one of our culture's rituals. Examples: elaborate wedding ceremonies, expensive funerals, grade school graduations, Christmas gift giving, proms, baby showers, presidential election campaigns, traditional dating, beauty contests, and so on. Choose as your audience people who would not automatically agree with you.
2. COLLABORATIVE WRITING. Write about a habit (smoking, procrastinating, overeating, or being late, for example) that you think is

damaging. To get started, meet with a group of classmates or friends to share stories about bad habits and how to break them. When you write your essay, present the habit as a problem, describe its seriousness, and explain your solution. Your audience will be people who have the habit and see nothing wrong with it.

3. USING THE INTERNET. Study the Web site of the Drug Reform Coordination Network at **www.drcnet.org/DARE/** where you will find an examination of the work being done by DARE (Drug Abuse Resistance Education). Then argue either that DARE is a valuable tool in the war against drugs or that DARE is ineffective and a waste of time and money.
4. WRITING ABOUT READING. Write a response to Bryson's essay. Do you agree with his claims? How would you wage the war on drugs in the United States?

EDITING SKILLS: USING COMPOUND VERBS

Look at the verbs in the following sentences from Bill Bryson's essay:

> Violent first-time offenders, by contrast, were imprisoned less often and received on average just four years in prison.
> The people of Texas recently voted down a $750 million bond proposal to build new schools but overwhelmingly endorsed a $1 billion bond for new prisons.
> Because most drug offenses carry mandatory sentences and exclude the possibility of parole, other prisoners are having to be released early.
> He did his time and since then has been completely clean.

The verbs that we have highlighted are compound verbs—two verbs that have the same subject and are connected by a coordinating conjunction. With compound verbs you can expand a simple sentence and eliminate wordiness (as we just did in this sentence). Most of the time you will use *and:*

> The Internal Revenue Service seized our assets *and* confiscated our computers.

Sometimes you can use *but, or,* and *yet:*

> The FBI questioned our neighbors *but* ignored our cleaning crew.
> They suspected us of espionage *or* wanted us for tax evasion.

You can also join more than two verbs:

The gymnast leaped up, grabbed the bar, *and* swung into her first maneuver.

EXERCISE

After studying the preceding examples, combine the following groups of simple sentences by using compound verbs.

1. *TV Guide* lists the latest shows. It also describes them briefly.
2. Many people are renting movies. They are playing them on their videocassette recorders.
3. *Frasier* uses recycled plots. This show also features brilliant dialogue.
4. Tillie bakes bread. She talks on the phone. She drinks coffee at the same time.
5. The nurse put down the chart. He grabbed the patient's wrist. He took her temperature.
6. In medieval Europe the peasants were distracted by war. They were weakened by malnutrition. They were exhausted by the struggle to make a living.

Now check the essay you have just written. See if you can combine a couple of sentences by using a compound verb.

WEB SITES

www.rwjf.org/news/featureDetail.jsp?id=37

Visit the Web site of the Robert Wood Johnson Foundation for a rundown of the remarkable way drug penalties vary from state to state.

www.usdoj.gov/dea/agency/penalties.htm

The Web site of the federal Drug Enforcement Agency provides an up-to-date listing of federal drug penalties.

PREPARING TO READ

If you were dying, slowly and painfully, from an incurable disease, what would you want the hospital staff to do? Should they let you die, or prolong your life (and your pain)?

A Crime of Compassion

BARBARA HUTTMANN

Barbara Huttmann is the associate director of nursing for Children's Hospital in San Francisco. She has written two books about the rights of patients: The Patient's Advocate *and* Code Blue: A Nurse's True-Life Story. *In the following essay, which originally appeared on the "My Turn" page of* Newsweek *magazine in 1983, Huttmann tells about her decision to let a suffering patient die.*

TERMS TO RECOGNIZE

resuscitated *(para. 3)*	revived, brought back to life
haggard *(para. 5)*	worn out
IV solutions *(para. 6)*	liquids given by injection (IV stands for intravenous—"in the vein")
irrigate *(para. 7)*	wash out, flush
lucid *(para. 10)*	aware, clear-minded
impotence *(para. 10)*	powerlessness
imperative *(para. 11)*	command, directive
riddled *(para. 13)*	pierced with numerous holes
pallor *(para. 15)*	paleness, lack of color

"Murderer," a man shouted. "God help patients who get *you* for a nurse." 1

"What gives you the right to play God?" another one asked. 2

It was the *Phil Donahue Show* where the guest is a fatted calf and the audience a 220-strong flock of vultures hungering to pick at the bones. I had told them about Mac, one of my favorite cancer patients. "We resuscitated 3 him 52 times in just one month. I refused to resuscitate him again. I simply sat there and held his hand while he died."

4 There wasn't time to explain that Mac was a young, witty, macho cop who walked into the hospital with 32 pounds of attack equipment, looking as if he could single-handedly protect the whole city, if not the entire state. "Can't get rid of this cough," he said. Otherwise, he felt great.

5 Before the day was over, tests confirmed that he had lung cancer. And before the year was over, I loved him, his wife, Maura, and their three kids as if they were my own. All the nurses loved him. And we all battled his disease for six months without ever giving death a second thought. Six months isn't such a long time in the whole scheme of things, but it was long enough to see him lose his youth, his wit, his macho, his hair, his bowel and bladder control, his sense of taste and smell, and his ability to do the slightest thing for himself. It was also long enough to watch Maura's transformation from a young woman into a haggard, beaten old lady.

6 When Mac had wasted away to a 60-pound skeleton kept alive by liquid food we poured down a tube, IV solutions we dripped into his veins, and oxygen we piped to a mask on his face, he begged us: "Mercy . . . for God's sake, please just let me go."

7 The first time he stopped breathing, the nurse pushed the button that calls a "code blue" throughout the hospital and sends a team rushing to resuscitate the patient. Each time he stopped breathing, sometimes two or three times in one day, the code team came again. The doctors and technicians worked their miracles and walked away. The nurses stayed to wipe the saliva that drooled from his mouth, irrigate the big craters of bedsores that covered his hips, suction the lung fluids that threatened to drown him, clean the feces that burned his skin like lye, pour the liquid food down the tube attached to his stomach, put pillows between his knees to ease the bone-on-bone pain, turn him every hour to keep the bedsores from getting worse, and change his gown and linen every two hours to keep him from being soaked in perspiration.

8 At night I went home and tried to scrub away the smell of decaying flesh that seemed woven into the fabric of my uniform. It was in my hair, the upholstery of my car—there was no washing it away. And every night I prayed that Mac would die, that his agonized eyes would never again plead with me to let him die.

9 Every morning I asked his doctor for a "no-code" order. Without that order, we had to resuscitate every patient who stopped breathing. His doctor was one of several who believe we must extend life as long as we have the means and knowledge to do it. To not do it is to be liable for negligence, at least in the eyes of many people, including some nurses. I thought about what it would be like to stand before a judge, accused of murder, if Mac stopped breathing and I didn't call a code.

And after the fifty-second code, when Mac was still lucid enough to beg for death again, and Maura was crumbled in my arms again, and when no amount of pain medication stilled his moaning and agony, I wondered about a spiritual judge. Was all this misery and suffering supposed to be building character or infusing us all with the sense of humility that comes from impotence? 10

Had we, the whole medical community, become so arrogant that we believed in the illusion of salvation through science? Had we become so self-righteous that we thought meddling in God's work was our duty, our moral imperative and our legal obligation? Did we really believe that we had the right to force "life" on a suffering man who had begged for the right to die? 11

Such questions haunted me more than ever early one morning when Maura went home to change her clothes and I was bathing Mac. He had been still for so long, I thought he at last had the blessed relief of coma. Then he opened his eyes and moaned, "Pain . . . no more . . . Barbara . . . do something . . . God, let me go." 12

The desperation in his eyes and voice riddled me with guilt. "I'll stop," I told him as I injected the pain medication. 13

I sat on the bed and held Mac's hands in mine. He pressed his bony fingers against my hand and muttered, "Thanks." Then there was one soft sigh and I felt his hands go cold in mine. "Mac?" I whispered, as I waited for his chest to rise and fall again. 14

A clutch of panic banded my chest, drew my finger to the code button, urged me to do something, anything . . . but sit there alone with death. I kept one finger on the button, without pressing it, as a waxen pallor slowly transformed his face from person to empty shell. Nothing I've ever done in my 47 years has taken so much effort as it took *not* to press that code button. 15

Eventually, when I was as sure as I could be that the code team would fail to bring him back, I entered the legal twilight zone and pushed the button. The team tried. And while they were trying, Maura walked into the room and shrieked, "No . . . don't let them do this to him . . . for God's sake . . . please, no more." 16

Cradling her in my arms was like cradling myself, Mac, and all those patients and nurses who had been in this place before, who do the best they can in a death-denying society. 17

So a TV audience accused me of murder. Perhaps I am guilty. If a doctor had written a no-code order, which is the only *legal* alternative, would he have felt any less guilty? Until there is legislation making it a criminal act to code a patient who has requested the right to die, we will all of us risk the same fate as Mac. For whatever reason, we developed the means to prolong life, and now we are forced to use it. We do not have the right to die. 18

RESPONDING TO READING

Huttmann says "We do not have the right to die." Should we have this right? In your journal write down your thoughts and feelings about "the right to die."

GAINING WORD POWER

Writers sometimes make passing references to familiar or significant people, places, objects, or events from history, the Bible, and literature. These references are called **allusions;** they help a writer to set the tone or heighten the meaning without going into a long explanation. Explain the following phrases from Barbara Huttmann's essay. Use a dictionary or other reference works to help you.

1. The "fatted calf" (para. 3) refers to the biblical parable of the Prodigal Son. If you don't remember what happens to the fatted calf, look up the parable (Luke 15:11-32), and explain why Huttmann makes this reference.
2. The "twilight zone" (para. 16) was the name of an old TV program. Do you know the show? What quality of that show is Huttmann calling on in this reference? Can you see how it fits her point?
3. "Code blue" is a medical term. You can probably figure out its general meaning from the essay. But why is it a "code," and why is the code "blue" (instead of some other color)?
4. What is the legal definition of "liable" (para. 9)? Do you think Huttmann used the word because of its legal associations?
5. The title of the essay is a variation of the phrase "crime of passion." Do you know what a crime of passion is? How does the meaning of this phrase relate to Huttmann's title?

CONSIDERING CONTENT

1. How does this selection qualify as a **problem-solution** essay? What problem is Huttmann presenting? What is her solution?
2. Explain what a "no-code" order is. Why wouldn't Mac's doctor issue one?
3. Huttmann says that Mac's doctor believes "we must extend life as long as we have the means and knowledge to do it." Is that what you believe? What does Huttmann believe?
4. What does the author mean by "a spiritual judge"?
5. Explain the question that ends paragraph 10.

6. Huttmann was accused of "playing God" for letting Mac die. How does she turn the accusation around in paragraph 11? According to Huttmann, who is playing God?
7. Did Huttmann commit a crime? Explain your answer.
8. Do we live in a "death-denying society," as the author claims (para. 17)? What do you think she means by this phrase?
9. Do you think Huttmann feels guilty? Why did she publicly reveal what happened?

CONSIDERING METHOD

1. Why does Huttmann begin by quoting audience members of the *Phil Donahue Show?* Why does she return to the TV audience again in her last paragraph?
2. Why does Huttmann describe Mac on the day he entered the hospital (para. 4)? How does she use a contrast to that description in the next paragraph?
3. What contrast between doctors and nurses does the writer present in paragraph 7? How does this contrast relate to the difference of opinion about the "no-code" order (para. 9)?
4. Why does Huttmann go into so much detail about what the nurses did for Mac (paras. 6 and 7)?
5. How does Huttmann attempt to enlist sympathy for Mac's situation and for hers? Identify specific details that appeal to the readers' emotions.
6. Paragraph 11 is made up entirely of questions. Why does the author use this method of presenting these ideas?
7. What purpose do the quotations from Mac and Maura serve (paras. 6, 12, 14, and 16)?
8. Explain the **irony** in Huttmann's next-to-last sentence. (Irony is the use of language to express an unexpected outcome or to suggest that something is not what it seems to be.)

Combining Strategies

Huttmann employs a number of strategies to produce her compelling and persuasive essay. She uses narration, description, example, comparison/contrast, and process. Identify at least three instances of these methods.

WRITING STEP BY STEP

Write an essay about a time that you had to make a difficult choice, one that involved a conflict of values. Perhaps a good friend asked you to give her financial and moral support for getting an abortion, and you're opposed to abortion. Or maybe your parents got divorced, and you had to choose which one to live with. Perhaps the decision involved having a beloved pet put to death, standing up for an unpopular opinion, or revealing the dishonesty of someone you liked and admired.

A. Write the essay as a first-person account.
B. Begin with a narrative of the events leading up to the moment of decision. Make the narrative come to life with quotations and concrete details.
C. Tell about making the decision. How did you solve your dilemma? Did you have to compromise?
D. Narrate in detail the sequence of events that followed your decision. Explain how you felt at the time.
E. Discuss the consequences of your decision. How did it affect you? How did it affect others? Did people support you?
F. In your conclusion, express your current feelings about the decision and its consequences. Did things turn out all right? Do you regret your decision? If you had to do it again, would you make the same choice?

OTHER WRITING IDEAS

1. Defend some type of TV show that often receives negative criticism (sitcoms, soap operas, sports coverage, reality shows, music TV, shopping networks, and so on). Address an audience that views your choice of shows unfavorably.
2. COLLABORATIVE WRITING. Write an essay about the problems of being in the minority: of being a smoker in an antismoking society, of being left-handed in a world made for right-handers, of being a bicyclist on roads full of automobiles, of being overweight in a society where thin is in—or something similar. Talk to other members of the minority to get more information for arguing for your rights.
3. USING THE INTERNET. Look online for information about right to die laws, which Huttmann argues that our society needs. Check out the Public Agenda Web site (**publicagenda.org/**), and find out how

laws legalizing assisted suicide are working out in the state of Oregon and in the Netherlands. Then write an essay arguing either for or against legalizing a suffering terminal patient's right to die.

4. WRITING ABOUT READING. Huttmann's essay appeared in *Newsweek* as a "My Turn" piece. Write a letter to the editor of *Newsweek* supporting her viewpoint or arguing against it, using examples from your own experience or from your reading.

EDITING SKILLS: USING SHORT SENTENCES FOR EMPHASIS

Experienced writers vary their sentences, both in structure and in length. If you look at Barbara Huttmann's sentences, you will see that most of them are at least ten words long and many are over twenty. But sometimes she throws in a very short sentence for effect:

> There wasn't time to explain that Mac was a young, witty, macho cop who walked into the hospital with 32 pounds of attack equipment, looking as if he could single-handedly protect the whole city, if not the entire state [40 words]. "Can't get rid of this cough," he said [8 words]. Otherwise, he felt great [4 words].

Huttmann uses another four-word sentence in the last paragraph—"Perhaps I am guilty"—and a six-word sentence in the third paragraph: "I refused to resuscitate him again." These sentences express very important points; they grab our attention by being noticeably different from the longer sentences around them. The other unusually short sentence in Huttmann's essay is the last one. It contains eight one-syllable words, which drive the final point home: "We do not have the right to die." As you can see, Huttmann doesn't use short sentences very often, but when she does, she creates a strong effect.

EXERCISE

Look through Bill Bryson's essay in this chapter (p. 325), and find the five sentences that contain fewer than eight words. Note where he places them, and describe their effect.

Examine the sentences in your essay. Can you find a place to use a short sentence for **emphasis**? Try to end a paragraph with a short statement. Also take a look at your conclusion; that's another good place to sum up the main point in a short sentence. If you can't think of a new sentence, try shortening one that you've already written.

WEB SITE

http://www.chevychase.com/dr.death.html

For information about our society's most famous advocate of assisted suicide, read the unofficial home page of Dr. Jack Kevorkian, dubbed "Dr. Death."

DEBATE: EXAMINING THE DEATH PENALTY

In 1972 the U.S. Supreme Court declared executions unconstitutional. Four years later, the Court approved their resumption. Between 1976 and 2004, 944 people were put to death in the United States. As of July 2004, there were more than 3,697 inmates on death row in this country.

The use of death as a punishment is one of America's most divisive issues. It raises difficult questions: Does it deter violent crime? Is it fair to those who died to let their murderers go on living? Can innocent persons be executed by mistake? Can the death penalty be administered fairly? Here we present two articles that examine these and other questions.

Preparing to Read

Does your state sanction capital punishment? If you had to vote in a referendum to legalize or outlaw the death penalty in your state, how would you vote? What reasons would you give for your decision?

Death Penalty Showdown

DAVID LEIBOWITZ

David Leibowitz is a graduate of Florida State University and holds master's degrees from Temple and New York universities. He worked as a reporter for several newspapers in New Jersey before becoming a regular columnist for the Arizona Republic *in Phoenix. The following article appeared in the* Arizona Republic *in May of 1999.*

TERMS TO RECOGNIZE

illusory *(para. 4)* based on an erroneous belief or perception, unreal
systemic *(para. 4)* relating to the entire system
dispensed *(para. 9)* administered or handed down
frivolous *(para. 10)* trivial, unworthy of serious attention
charade *(para. 13)* a pretense that is easily perceived
curtailed *(para. 16)* cut short, stopped

Always, I recall the dead woman's hands. I saw them in a photograph once, not long before I watched her killer die from a dose of poison injected by the state of Arizona. The woman's name was Amelia Schoville, and in that picture her hands resembled claws, swollen with blood, thumbs bound by a shoelace, fingers straining up from the dirty mattress where she died. 1

I remembered those hands on the night [in April 1998] when Amelia's killer, Jose Roberto Villafuerte, was put to death, 5,540 days after he killed her. I recalled Amelia's hands again May 5, 1999, when Robert Wayne Vickers finally got a deadly needle after twenty-one years on Arizona's Death Row. 2

Imagine those hands, ever-empty, ever-reaching. Now, multiply them by every Death Row convict on every Death Row in each of the thirty-eight states that uses the death penalty. Think of all those victims, all those empty hands. All that justice denied. 3

The Death Penalty Is Just and Legal

Make no mistake: Although foes of capital punishment will try to cloud this issue by injecting religion or by scrambling after some illusory moral high ground, the death penalty, when imposed and carried out, is justice. Execution represents a proportional, measured response to mankind's most barbarous act. It has precedent, it has been ruled legal countless times by countless courts, and it is supported by an overwhelming majority of Americans. In a nation where justice is often represented by a set of scales, execution as punishment for a depraved murder marks the ultimate—and only—systemic balance. 4

Joe Maziarz has spent the past dozen years prosecuting death penalty appeals for the Arizona Attorney General's Office. "If society is not willing to exact justice, what purpose do we serve for the citizens?" he asks. "If someone had killed my wife, I can't take the law into my own hands. I rely on society to do that. For some murders, the only justice, from a societal perspective, is the death penalty." 5

Case in point: "Bonzai Bob" Vickers. A one-man crime wave who developed an animal's lust for blood, Vickers first earned a spot on Arizona's Death Row in 1978, for the jailhouse murder of his cellmate, Frank Ponciano. Feeling wronged because Ponciano had taken his Kool-Aid and failed to awaken him for lunch, Vickers stabbed his victim to death, then carved "Bonzai" into Ponciano's back. Vickers' lone regret, as told to a prison psychologist? He didn't have time to add a swastika beside the misspelled Japanese war cry. 6

Four years later, already a resident of Cellblock Six, Vickers somehow managed to top himself. After fellow killer Buster Holsinger made a suggestive remark about a photo of Vickers' eleven-year-old niece, Vickers fash- 7 ioned a firebomb using hair gel. He burned Holsinger to death, then tossed in a second firebomb for good measure. "Did I do a good job?" Vickers asked investigators later. "I told them they should have gassed me in December, when they had a chance." Instead, his execution took an additional seventeen years. That was time enough for Vickers to complete his legend: 158 major violations while behind prison walls, including a dozen assaults on corrections officers, twenty attacks on inmates and forty charges of making weapons.

8 Foes of the death penalty often argue that execution has no deterrent effect. That may well be true for society at-large—especially given the decades between a crime and an execution—but one thing is certain: What happened May 5 in the Florence prison permanently deterred "Bonzai Bob" Vickers, who badly needed deterring.

Strengthen the Death Penalty

9 Another favorite anti-death penalty argument holds at its crux the notion that our justice system is "broken." Death sentences are inconsistently dispensed, say the abolitionists; the decades of appeals represent "cruel and unusual punishment"; the system is not cost-effective. Thus, they say, capital punishment should be abandoned. Wrong. It should be fixed.

10 The Anti-Terrorism and Effective Death Penalty Act, approved in 1996 after nearly twenty years of congressional debate on the subject of federal appeals, was a small step in that direction. The act, with its time limits on filings and judges' decisions, and its limitations on successive federal appeals, has significantly streamlined the judicial process, mostly by cutting down on capital defense lawyers' favorite trick: frivolous claims of mental incompetency.

11 "Usually, it's limited only by the imagination of the defense attorney," Maziarz says. "They shop around for these shrinks that will basically say, based upon these brain scans . . . and looking at this person's background, that they have a 'brain disorder.' . . . It's like trying to grab ahold of Jell-O. There's no way to prove or disprove anything."

The Incompetency Charade

12 Find a sympathetic federal judge, and the process stops cold, even when guilt or innocence is no longer up for contention. As proof, look no further than the case of Michael Poland, convicted in 1979, along with his brother, in the killing of two armored-car guards while stealing $288,000. In October 1998, a federal judge in Hawaii stayed Poland's execution only two hours before its enactment. At the center of his ruling: A defense-team psychologist who claimed the "stress" Poland endured on Death Row left him incompetent for execution.

The amazing thing: Just sixteen days before that psychologist ruled him insane, Michael Poland was sane enough to bribe a prison investigator in a brazen try to escape Death Row. Assistant Attorney General Paul McMurdie has spent years on the Poland case. I spoke with him hours after the stay. "(Mental) incompetence like this isn't all of a sudden just going to happen," he told me. "This is nothing but a charade to make sure the execution didn't take place." And it worked, at least temporarily—Poland comes up for execution again on June 16th, 1999. [He was executed that day.] 13

Two obvious outcomes of ploys like this: The killer's legal bill soars, while the execution abolitionists rant. 14

Improve the Appeals Process

"To me, it's like a self-fulfilling prophecy," says McMurdie's colleague, Maziarz. "The anti-death penalty forces try to make it as expensive as possible. They unfortunately convince the courts to spare no expense and no time in allowing all this to keep mushrooming. Then, when they're successful in doing that, they point to that and say, 'Ah ha, look how expensive it is . . . to get someone executed.' " Again, the answer would appear to lie with repairing the process, with making the death penalty a consistent, certain answer to the worst kinds of murder. 15

Of course, saying the appeals process needs to be fixed isn't saying it needs to be curtailed: In Arizona, there is no time limitation on convicted murderers making claims of actual innocence, nor should there be. Newly discovered evidence and new facts can be brought forth at any time, even if repeated claims of sudden insanity have been limited. "The number one thing is making sure that the guilty are convicted and the innocent are not," Maziarz says. "We don't want to execute innocent people." 16

What we do want to do, instead, is consistent with the aim of the American system of justice: To adequately, proportionally punish those who would violate the code that governs society. The very worst of us, those like "Bonzai Bob" Vickers and Jose Villafuerte, deserve as much as a consequence. And those like Rick Schoville, a man who waited fifteen years to get justice in the murder of his mother, deserve nothing less. 17

"It brought a closure to it," Schoville says thirteen months after the fact, "and for that I'm really, really happy . . . . Now that I see more of these people getting executed, I personally think it's about time that's actually taking place. I know how I felt, what a relief it was to finally bring that part of it to an end. I can only imagine how victims of these other people must be feeling. These guys are finally getting what they deserve." 18

As he speaks, I imagine his mother's hands—now untied, now set free, finally getting to touch what had always been just out of grasp. 19

....

When Justice Lets Us Down

JIM DWYER, PETER NEUFELD, AND BARRY SCHECK

A two-time Pulitzer Prize winner in journalism, Jim Dwyer has worked at several newspapers in New Jersey and New York. He is currently a columnist for the New York Daily News. *While covering a trial for his column, Dwyer met attorneys Peter Neufeld and Barry Scheck. As public defenders in the South Bronx, Neufeld and Scheck worked on a case in which they concluded that the defendant's innocence would have been established by DNA testing, but the evidence was ignored. This experience led them to found the Innocence Project in 1992, which has since used DNA evidence to exonerate thirty-seven people convicted of crimes they did not commit. The two attorneys teamed with Dwyer to write a book detailing the work of the Innocence Project. The following selection, published in* Newsweek *on February 14, 2002, is an excerpt from that book,* Actual Innocence.

TERMS TO RECOGNIZE

stay *(para. 2)*	a court-ordered delay or halt
litany *(para. 3)*	a repeated phrase, a rundown or list
inept *(para. 4)*	incompetent, unskilled
moratorium *(para.5)*	a suspension of action
coerced *(para. 7)*	forced, pressured by threats
fabricated *(para. 7)*	made-up, fictitious
statutory *(para. 8)*	authorized by law
exonerations *(para. 8)*	freedom from blame, acquittals

The warden was seated at the head of a long table when Ron Williamson was led into the office and told to sit down. Once, Williamson had been a professional baseball player, the hero jock who married the beauty queen in his small Oklahoma town. Now, on an August morning in 1994, at 41, the athlete had passed directly from his prime to a state beyond age. His hair had gone stringy and white; his face had shrunk to a skeletal mask wrapped in pasty, toneless skin. 1

The warden said he had a duty to carry out, and he read from a piece of paper: You have been sentenced to die by lethal injection, and such sentence will be carried out at 12:30 A.M., the 24th of September, 1994. The prison had received no stay of execution, the warden said, so Williamson would be brought to the holding area next to the death chamber until the 24th. 2

He was led back to a cell, screaming from that moment on, night and day, even after they moved him into another unit with double doors to muffle the noise. On Sept. 17, Williamson was shifted into the special cell for prisoners with less than a week to live. By then, the screaming had torn his throat to ribbons, but everyone knew his raspy, desperate litany: "I did not rape or kill Debra Sue Carter! I am an innocent man!" In Norman, Okla., a public defender named Janet Chesley frantically scrambled a team to move the case into federal court, assembling a mass of papers and fresh arguments. With just five days to go, they won a stay. 3

One year later, U.S. district court Judge Frank Seay would rule that Williamson's trial had been a constitutional shambles. His conviction and death sentence—ratified by every state court in Oklahoma—had been plagued by unreliable informants, prosecutorial misconduct, an inept defense lawyer, bogus scientific evidence, and a witness who himself was a likely suspect, the federal judge said. He vacated the conviction. 4

Last April, before Williamson could be tried again, DNA tests arranged by the Innocence Project at the Benjamin N. Cardozo Law School in New York proved that he was actually innocent. Across the country, the Ron Williamson story has happened over and over. Since 1976, Illinois has executed 12 people—and freed 13 from death row as innocent. Last week Illinois Governor George Ryan declared a moratorium on executions. The next day a California judge threw out the convictions of nine people after prosecutors said they were among at least 32 framed by a group of corrupt police officers within the Los Angeles Police Department. New innocence cases, based on DNA tests, are on the horizon in California, Texas, Florida and Louisiana. 5

A rare moment of enlightenment is at hand. In the last decade, DNA tests have provided stone-cold proof that 69 people were sent to prison and death row in North America for crimes they did not commit. The number has been rising at a rate of more than one a month. What matters most is not how the wrongly convicted got out of jail, but how they got into it. "How do you prevent another innocent man or woman from paying the ultimate penalty for a crime he or she did not commit?" asked Governor Ryan, a Republican and death-penalty supporter. "Today I cannot answer that question." 6

In 22 states, the fabric of false guilt has been laid bare, and the same vivid threads bind a wealthy Oklahoma businessman and a Maryland fisherman; a Marine corporal in California and a boiler repairman in Virginia; a Chicago drifter and a Louisiana construction worker; a Missouri schoolteacher and the Oklahoma ballplayer. Sometimes, it turns out, eyewitnesses make mistakes. Snitches tell lies. Confessions are coerced or fabricated. Racism trumps the truth. Lab tests are rigged. Defense lawyers sleep. Cops lie. 7

DNA testing can't solve these problems but it reveals their existence. Many can be fixed with simple reforms. To simply ignore them means that more criminals go free, and the innocent will suffer. Clyde Charles learned the crushing weight of the status quo: he was freed just before Christmas, after 19 years of his life were squandered in Louisiana's Angola prison. The last nine years were spent fighting for the DNA lab work that cleared him in a matter of hours. In 48 states, prisoners don't have statutory rights to tests that could prove their innocence, and too often authorities stubbornly resist. In more than half its exonerations, the Innocence Project was forced into fierce litigation simply to get the DNA tests. 8

"Our procedure," wrote Justice Learned Hand in 1923, "has always been haunted by the ghost of the innocent man convicted. It is an unreal dream." Nearly 75 years later, Judge Seay in Oklahoma wrote an epilogue to the writ of habeas corpus for Ron Williamson: "God help us, if ever in this great country we turn our heads while people who have not had fair trials are executed. That almost happened in this case." For all the gigabytes of crime statistics kept in the United States, no account is taken of the innocent person, wrongly convicted, ultimately exonerated. No one has the job of figuring out what went wrong, or who did wrong. The moment has come to do so. 9

GAINING WORD POWER

Because these essays argue for a certain point of view, they use language to undercut opposing ideas. For instance, by saying that the foes of capital punishment are "*scrambling* after some *illusory* moral high ground," David Leibowitz suggests that opponents of the death penalty are struggling ("scrambling") to find nonexistent ("illusory") support for their position. If, on the other hand, Leibowitz admired the opponents' arguments, he might say they were "*aspiring* to a higher moral *standard.*" Find four words or phrases in each essay that are used to discredit or otherwise undercut the opposition. Briefly explain the negative quality that these words or phrases convey.

CONSIDERING CONTENT

1. Which essay did you find most persuasive, and why?
2. How does Leibowitz refute the claim that execution has no deterrent effect? How effective is his refutation?
3. According to Leibowitz, how should the death penalty process be fixed?
4. The authors of "When Justice Lets Us Down" claim that "A rare moment of enlightenment is at hand" (para. 6). Who is being enlightened and about what?

5. According to Dwyer, Neufeld, and Scheck, what problems with the justice system can DNA testing reveal?
6. If DNA tests can quickly provide clear evidence of innocence or guilt, why do you think they're so difficult to obtain?

CONSIDERING THE METHOD

1. How do the authors of each essay establish their credentials and credibility on this topic? Do you find the authors equally credible?
2. What types of evidence are used in each essay?
3. Look for **logical fallacies** in each essay. Are there overgeneralizations? Either-or thinking? Jumping to conclusions? Did you notice flaws in the logic while you were reading the first time? Did your position on the debate influence your attention to logic?
4. Why does Leibowitz begin with the description of Amelia Schoville's hands? Is this an effective opening? Why does he return to this image in the last paragraph?
5. Why do Dwyer, Neufeld, and Scheck begin with an **anecdote** rather than data or a preview of their argument?

WRITING STEP BY STEP

Write an essay in which you argue for the best way to punish people convicted of murder: the death penalty, life imprisonment, or some other type of punishment.

A. Before you start writing, make a list of reasons why you support one form of punishment over the other. Next to each reason, try to write a response or a counterargument.
B. Identify at least four major reasons to support your choice. These reasons will be your claims. If necessary, do additional research to develop arguments for other forms of punishment.
C. Start your essay by stating your position and giving a brief overview of your main claims.
D. Take up each of your claims one by one. Explain each reason in a separate paragraph, and provide evidence to demonstrate the validity of that reason.
E. Be sure to consider any objections to your claims that readers might have. If you're not sure what these objections would be, ask classmates, friends, family members, or instructors for their ideas. Refute the main objections by minimizing their importance, questioning their factuality, or offering a different point of view. You can include

your refutations as part of the argument for each claim, or you can put them in a separate paragraph.

F. Close by summarizing your main ideas. You might end with a **rhetorical question,** as Joshua Shenk does (p. 315), or with a short, emphatic sentence, like the one Barbara Huttmann uses (p. 334).

OTHER WRITING IDEAS

1. WRITING ABOUT READING. Using a point-by-point refutation format, respond to the four major arguments that David Leibowitz makes in his essay "Death Penalty Showdown." Begin each point of refutation by summarizing Leibowitz's position, and then present your reply to that position.
2. WRITING ABOUT READING and USING THE INTERNET. "When Justice Lets Us Down" first appeared in *Newsweek* magazine. Write a letter to the editor of that magazine, responding to the article's claims about DNA. Use the Internet to obtain additional information about a DNA defense.
3. COLLABORATIVE WRITING. Write an essay about a controversial **issue** other than capital punishment. Get together with a group of classmates to discuss possible topics and draw up a list of points to argue. You might even set up a formal debate on the topic you select. Then, either separately or as a group, write an essay as a guest column for the Op-Ed page of your local or school newspaper. Op-Ed means "opposite the editorial" page; it's the place where professional columnists, as well as local citizens, present their opinions on the issues of the day. Take some time to study the articles that appear on the Op-Ed page of any newspaper; pay attention to length, tone, and approach.

EDITING SKILLS: SUBJECT-VERB AGREEMENT

Look at the verbs in the following sentences:

Execution represents a proportional, measured response to mankind's most barbarous act.

The process stops cold.

DNA testing reveals the problems in the system.

Racism trumps the truth.

The verbs that we have underlined are in the present tense; they express actions that are happening at the present time or that happen all the time. We also use the present tense to state facts or general truths. You will notice that the verbs end in *s*. That's because the subject of each verb is singular (execution, process, testing, racism). When the subject of a present-tense verb is *he, she,* or *it*—or a noun that could be replaced by *he, she,* or *it*—we put an *s* on the end of the verb.

This ending is an exception. Present-tense verbs with other subjects do not require the *s* ending:

> The anti-death penalty foes try to make it as expensive as possible.
> Sometimes eyewitnesses make mistakes. Snitches tell lies.
> I imagine his mother's hands—now untied, now set free.

Because we don't always put an ending on a present-tense verb, some people forget to add it. That causes an error in subject-verb agreement. And sometimes it is difficult to tell what the subject of a verb really is, as in this sentence:

> The death penalty, like the police and the courts, produces little change in the murder rate.

The subject of the verb produces is death penalty, but the words in between might lead a writer to think that the police and the courts is the subject—and to mistakenly leave the *s* ending off. Now take a look at this example:

> Foes of the death penalty often argue that execution has no deterrent effect.

Can you figure out why there is no *s* on the verb argue? That's because the subject is foes, a word that does not mean *he, she,* or *it*. When trying to figure out the subject-verb agreement, you just have to forget about the words that come between foes and argue.

EXERCISE

In each of the following sentences, underline the subject and then circle the verb that agrees with it. Example:

> A verb in the present tense take/takes an *s* ending when its subject is *he, she,* or *it*—or a noun that means *he, she,* or *it.*

1. An anthropologist study/studies buildings, tools, and other artifacts of ancient cultures.
2. Anthropologists always look/looks for signs of social change.

3. A box of fruit arrive/arrives at the house every month.
4. An adult student who has children find/finds little time for partying.
5. Low scores on the Scholastic Aptitude Test discourage/discourages students from applying to some colleges.
6. It give/gives me great pleasure to introduce tonight's speaker.
7. The first baseman, along with most of his teammates, refuse/refuses to sign autographs after the game.
8. Bonsai trees require/requires careful pruning.
9. Many movies of the past year contain/contains scenes of violence.

Now check over the essay you have just written. Look at all the verbs, especially those in the present tense. Did you use the *s* ending on the appropriate verbs? Edit your writing carefully for subject-verb agreement.

WEB SITES

www.deathpenaltyinfo.org/

Death Penalty Information Center: one of the best sites for anti–death penalty information online.

www.prodeathpenalty.com

This site provides information and resources that support the death penalty.

DEBATE: THE RIGHT TO SAME-SEX MARRIAGE

In 1993 the Hawaii Supreme Court ruled that denying same-sex couples the right to marry violated the state's constitution, and in 1996 a trial judge ruled that the state of Hawaii had failed to show a valid reason for excluding same-sex couples from marriage. These two decisions launched a national debate over the right of gay people to marry.

Eventually, the state of Hawaii amended its constitution to prohibit same-sex marriages. But in 2000, after Vermont's Supreme Court held that the state must extend to same-sex couples the same benefits that married couples receive, the legislature created the status of "civil union," a separate category that includes most of the legal advantages of marriage.

In November 2003, the Massachusetts Supreme Judicial Court ruled that excluding same-sex couples from the benefits of civil marriage violated the state constitution, and in February 2004 the court further held that a "civil union" law would not be sufficient. So on May 17, 2004, Massachusetts became the first state in the United States where same-sex marriage is legal.

In response to these actions, opponents of same-sex marriage introduced a resolution to amend the U.S. Constitution, defining marriage as the union between one man and one woman. On July 14, 2004, the Senate failed to approve the amendment, but a week later the House of Representatives voted to prevent federal courts from ordering states to recognize gay marriages sanctioned by other states.

The following articles touch on many of the issues in this ongoing debate.

Preparing to Read

If you were voting in a referendum to legalize same-sex marriage in your state, how would you vote? What reasons would you give for your decision? Based on what experiences?

For Better or for Worse?

Mary Ann Glendon

Mary Ann Glendon is the Learned Hand Professor of Law at Harvard University. She writes and teaches in the fields of human rights, comparative law, constitutional law, and legal theory. Her most recent book is A World Made New: Eleanor Roosevelt and the Universal Declaration of Human Rights *(2001).*

In 1994, Glendon was appointed by Pope John Paul II to the newly created Pontifical Academy of Social Science. The following article appeared in The Wall Street Journal *on February 25, 2004.*

TERMS TO RECOGNIZE

jurisprudence *(para. 1)* — a system of laws, the course of court decisions
affluent *(para. 2)* — wealthy, prosperous
retroactive *(para. 5)* — extending to a prior time or previous condition
distributive *(para. 5)* — dealing a proper share to each member of a group
havoc *(para. 6)* — destruction, ruin
wrought *(para. 6)* — built, created
homophobes *(para. 6)* — people with a deep-seated fear of homosexuals
flagrant *(para. 8)* — outrageous, glaring

Cambridge, Mass.—President Bush's endorsement of a constitutional amendment to protect the institution of marriage should be welcomed by all Americans who are concerned about equality and preserving democratic decision-making. "After more than two centuries of American jurisprudence and millennia of human experience," he explained, "a few judges and local authorities are presuming to change the most fundamental institution of civilization." 1

Those judges are here in Massachusetts, of course, where the state is cutting back on programs to aid the elderly, the disabled, and children in poor families. Yet a four-judge majority has ruled in favor of special benefits for a group of relatively affluent households, most of which have two earners and are not raising children. What same-sex marriage advocates have tried to present as a civil rights issue is really a bid for special preferences of the type our society gives to married couples for the very good reason that most of them are raising or have raised children. Now, in the wake of the Massachusetts case, local officials in other parts of the nation have begun to issue marriage licenses to homosexual couples in defiance of state law. 2

A common initial reaction to these local measures has been: "Why should I care whether same-sex couples can get married?" "How will that affect me or my family?" "Why not just live and let live?" But as people began to take stock of the implications of granting special treatment to one group of citizens, the need for a federal marriage amendment has become increasingly clear. As President Bush said yesterday, "The voice of the people must be heard." 3

Indeed, the American people should have the opportunity to deliberate the economic and social costs of this radical social experiment. Astonishingly, in the media coverage of this issue, next to nothing has been said 4 about what this new special preference would cost the rest of society in terms of taxes and insurance premiums.

5 The Canadian government, which is considering same-sex marriage legislation, has just realized that retroactive social-security survivor benefits alone would cost its taxpayers hundreds of millions of dollars. There is a real problem of distributive justice here. How can one justify treating same-sex households like married couples when such benefits are denied to all the people in our society who are caring for elderly or disabled relatives whom they cannot claim as family members for tax or insurance purposes? Shouldn't citizens have a chance to vote on whether they want to give homosexual unions, most of which are childless, the same benefits that society gives to married couples, most of whom have raised or are raising children?

6 If these social experiments go forward, moreover, the rights of children will be impaired. Same-sex marriage will constitute a public, official endorsement of the following extraordinary claims made by the Massachusetts judges in the *Goodridge* case: that marriage is mainly an arrangement for the benefit of adults; that children do not need both a mother and a father; and that alternative family forms are just as good as a husband and wife raising kids together. It would be tragic if, just when the country is beginning to take stock of the havoc those erroneous ideas have already wrought in the lives of American children, we should now freeze them into constitutional law. That philosophy of marriage, moreover, is what our children and grandchildren will be taught in school. They will be required to discuss marriage in those terms. Ordinary words like *husband* and *wife* will be replaced by *partner* and *spouse.* In marriage-preparation and sex-education classes, children will have to be taught about homosexual sex. Parents who complain will be branded as homophobes and their children will suffer.

7 Religious freedom, too, is at stake. As much as one may wish to live and let live, the experience in other countries reveals that once these arrangements become law, there will be no live-and-let-live policy for those who differ. Gay-marriage proponents use the language of openness, tolerance and diversity, yet one foreseeable effect of their success will be to usher in an era of intolerance and discrimination the likes of which we have rarely seen before. Every person and every religion that disagrees will be labeled as bigoted and openly discriminated against. The ax will fall most heavily on religious persons and groups that don't go along. Religious institutions will be hit with lawsuits if they refuse to compromise their principles.

8 Finally, there is the flagrant disregard shown by judges and local officials for the rights of citizens to have a say in setting the conditions under which we live, work and raise our children. Many Americans—however they feel about same-sex marriage—are rightly alarmed that local officials are defying state law, and that four judges in one state took it upon themselves to make the kind of decision that our Constitution says belongs to us, the

people, and to our elected representatives. As one State House wag in Massachusetts put it, "We used to have government of the people, by the people and for the people, now we're getting government by four people!"

9 Whether one is for, against or undecided about same-sex marriage, a decision this important ought to be made in the ordinary democratic way—through full public deliberation in the light of day, not by four people behind closed doors. That deliberation can and must be conducted, as President Bush stated, "in a manner worthy of our country—without bitterness or anger."

Adam and Steve—Together at Last

KATHA POLLITT

Essayist and poet Katha Pollitt is known for her provocative analyses of hot-button issues like abortion, affirmative action, school vouchers, and surrogate motherhood. In addition to her biweekly column in The Nation, *Pollitt also writes for* The New Yorker *and* The New York Times. *Her essays and political commentaries have been published in two collections:* Reasonable Creatures: Essays on Women and Feminism *(1994) and* Subject to Debate: Sense and Dissents on Women, Politics, and Culture *(2001). "Adam and Steve—Together at Last" is from the December 15, 2003, issue of* The Nation.

TERMS TO RECOGNIZE

procreation *(para. 2)*	sexual reproduction
marital *(para. 2)*	relating to marriage
infertile *(para. 2)*	unable to reproduce, barren
celibate *(para. 2)*	abstaining from sexual activity
tribunal *(para. 2)*	a court or forum that makes doctrine
impinge *(para. 3)*	invade, impose
egalitarian *(para. 4)*	equal for everyone
polygyny *(para. 4)*	the state or practice of having more than one wife or female mate at a time

Will someone please explain to me how permitting gays and lesbians to marry threatens the institution of marriage? Now that the Massachusetts Supreme Court has declared gay marriage a constitutional right, opponents really have to get their arguments in line. 1

The most popular theory, advanced by David Blankenhorn, Jean Bethke Elshtain, and other social conservatives is that under the tulle and orange blossom, marriage is all about procreation. There's some truth to this as a practical matter—couples often live together and tie the knot only when baby's on the way. But whether or not marriage is the best framework for child-rearing, having children isn't a marital requirement. As many have pointed out, the law permits marriage to the infertile, the elderly, the impotent, and those with no wish to procreate; it allows married couples to use birth control, to get sterilized, to be celibate. There's something creepily authoritarian and insulting about reducing marriage to procreation, as if intimacy mattered less than biological fitness. It's not a view that anyone outside a right-wing think tank, a Catholic marriage tribunal, or an ultra-Orthodox rabbi's court is likely to find persuasive. 2

So scratch procreation. How about: Marriage is the way women domesticate men. This theory, a favorite of right-wing writer George Gilder, has some statistical support—married men are much less likely than singles to kill people, crash the car, take drugs, commit suicide—although it overlooks such husbandly failings as domestic violence, child abuse, infidelity, and abandonment. If a man rapes his wife instead of his date, it probably won't show up on a police blotter, but has civilization moved forward? Of course, this view of marriage as a barbarian-adoption program doesn't explain why women should undertake it—as is obvious from the state of the world, they haven't been too successful at it, anyway. (Maybe men should civilize men—bring on the Fab Five!) Nor does it explain why marriage should be restricted to heterosexual couples. The gay men and lesbians who want to marry don't impinge on the male-improvement project one way or the other. Surely not even Gilder believes that a heterosexual pothead with plans for murder and suicide would be reformed by marrying a lesbian. 3

What about the argument from history? According to this, marriage has been around forever and has stood the test of time. Actually, though, marriage as we understand it—voluntary, monogamous, legally egalitarian, based on love, involving adults only—is a pretty recent phenomenon. For much of human history, polygyny was the rule—read your Old Testament—and in much of Africa and the Muslim world, it still is. Arranged marriages, forced marriages, child marriages, marriages predicated on the subjugation of women—gay marriage is like a fairy tale romance compared with most chapters of the history of wedlock. 4

The trouble with these and other arguments against gay marriage is that they overlook how loose, flexible, individualized, and easily dissolved the bonds of marriage already are. Virtually any man and woman can marry, no matter how ill assorted or little acquainted. An 80-year-old can marry an 18-year-old; a john can marry a prostitute; two terminally ill patients can marry each other from their hospital beds. You can get married by proxy, like medieval royalty, and not see each other in the flesh for years. Whatever may have been the case in the past, what undergirds marriage in most people's minds today is not some sociobiological theory about reproduction or male socialization. Nor is it the enormous bundle of privileges society awards to married people. It's love, commitment, stability. 5

Speaking just for myself, I don't like marriage. I prefer the old-fashioned ideal of monogamous free love, not that it worked out particularly well in my case. As a social mechanism, moreover, marriage seems to me a deeply unfair way of distributing social goods like health insurance and retirement checks, things everyone needs. Why should one's marital status determine how much you pay the doctor, or whether you eat cat food in old age, or whether a child gets a government check if a parent dies? It's outrageous that, for example, a working wife who pays Social Security all her life gets no more back from the system than if she had married a male worker earning the same amount and stayed home. Still, as long as marriage is here, how can it be right to deny it to those who want it? In fact, you would think that, given how many heterosexuals are happy to live in sin, social conservatives would welcome maritally minded gays with open arms. Gays already have the baby—they can adopt in many states, and lesbians can give birth in all of them—so why deprive them of the marital bathwater? 6

At bottom, the objections to gay marriage are based on religious prejudice: The marriage of man and woman is "sacred" and opening it to same-sexers violates its sacral nature. That is why so many people can live with civil unions but draw the line at marriage—spiritual union. In fact, polls show a striking correlation of religiosity, especially evangelical Protestantism, with opposition to gay marriage and with belief in homosexuality as a choice, the famous "gay lifestyle." For these people gay marriage is wrong because it lets gays and lesbians avoid turning themselves into the straights God wants them to be. 7

As a matter of law, however, marriage is not about Adam and Eve versus Adam and Steve. It's not about what God blesses, it's about what the government permits. People may think "marriage" is a word wholly owned by religion, but actually it's wholly owned by the state. No matter how big your church wedding, you still have to get a marriage license from City Hall. And just as divorced people can marry even if the Catholic Church considers it bigamy, and Muslim and Mormon men can marry only one woman even if their holy books tell them they can wed all the girls in Apart- 8 ment 3G, two men or two women should be able to marry, even if religions oppose it and it makes some heterosexuals, raised in those religions, uncomfortable.

Gay marriage—it's not about sex, it's about separation of church and state. 9

Dearly Beloved, Why Are We Debating This?

LEONARD PITTS

Leonard Pitts has been writing professionally since 1976 when, as an 18-year-old college student, he began doing freelance reviews and profiles for SOUL, *a national black tabloid. He joined the* Miami Herald *in 1991 as its pop music critic, and since 1994, has been writing a syndicated column on pop culture, social issues, and family life. His book,* Becoming Dad: Black Men and the Journey to Fatherhood, *was released in 1999. Pitts won the 2004 Pulitzer Prize for commentary. "Dearly Beloved, Why Are We Debating This?" appeared in the* Miami Herald *and newspapers around the country on August 5, 2003.*

TERMS TO RECOGNIZE

fervently *(para. 1)*	intensely, heatedly
precipitous *(para. 2)*	steep, abrupt
mandates *(para. 3)*	commands, requires
literally *(para. 3)*	word for word, without question
primeval *(para. 4)*	ancient, original
galvanized *(para. 6)*	stirred up, roused into action
bedrock *(para. 7)*	foundation, basis
cohabitation *(para. 7)*	practice of living together (but not married)
poignant *(para. 8)*	moving, emotional, upsetting

So what is it you have against gay marriage? I'm not talking to the guy next to you. He doesn't have a problem with it. No, I'm talking to you, who is fervently opposed. 1

The number of folks who agree with you is up sharply since June, when the U.S. Supreme Court struck down anti-sodomy laws in Texas. As recently as May, 49 percent of us supported some form of gay marriage, according to The Gallup Organization. The figure has since dropped to just 40 percent. That's a precipitous decline. So what's the problem? What is it that bothers you about gay people getting married? 2

Don't read me that part in Leviticus where homosexuality is condemned. I mean, that same book of the gospel mandates the death penalty for sassy kids and fortune tellers, by which standard the Osbourne children and Miss Cleo should have been iced a long time ago. I read The Book. I believe The Book. But I also know that it's impossible to take literally every passage in The Book, unless you want to wind up in prison or a mental ward. So don't hide behind the Bible. 3

Let's just be honest here, you and me. Why do you oppose gay marriage, really? It just feels wrong to you, doesn't it? At some visceral level, it just seems to offend something fundamental. Hey, I understand. It's one of the emotional sticking points for us heterosexual types, this primeval "ick" factor where homosexuality is concerned. I won't try to talk you out of it. 4

I will, though, point out that once upon a time, the same gut-level sense of wrong—and for that matter, the same Bible—were used to keep Jews from swimming in the community pool, women from voting and black people from riding at the front of the bus. All those things once felt as profoundly offensive to some people as gay marriage does to you right now. 5

The issue has been vaulted to the forefront in the last few days. Political conservatives have been galvanized by it. President Bush says he wants to "codify" marriage as a heterosexual union. And the Vatican has told Catholic legislators that they must oppose laws giving legal standing to gay unions, unions the church describes as "gravely immoral." Which is funny, given the level of sexual morality the church has demonstrated lately. Anyway, the reasoning seems to be that gay people will damage or cheapen the sanctity of marriage and that this can't be allowed because marriage is the foundation of our society. 6

I agree that marriage—and I mean legal, not common law—is an institution of vital importance. It stabilizes communities, socializes children, helps create wealth. It is, indeed, our civilization's bedrock. But you know something? That bedrock has been crumbling for years, without homosexual help. We don't attach so much importance to marriage anymore, do we? These days, we marry less, we marry later, we divorce more. And cohabitation, whether as a prelude to, or a substitute for, marriage, has gone from novelty to norm. We say we shack up because we don't need a piece of paper to tell us we are in love. I've always suspected it was actually because we fear the loss of freedom. Or because we're scared to bet forever. 7

I'm not trying to beat up cohabitators. A long time ago, I was one. But it strikes me as intriguing, instructive and poignant, that gay couples so determinedly seek what so many of us scorn, are so ready to take the risk many of us refuse, find such value in an institution we have essentially declared valueless. There's something oddly inspiring in their struggle to achieve the social sanction whose importance many of us long ago dismissed. 8

So tell me again why it is you don't want them to have that? I mean, yeah, some people say they are a threat to the sanctity of marriage. But I'm thinking they might just be its salvation. 9

Student Essay Using Argument

Gay Marriage Should Be a Civic Issue

Mace Boshart

Mace Boshart is a senior, majoring in premedical studies, at Eastern Illinois University in Charleston, IL. "Gay Marriage Should Be a Civic Issue" appeared on the Daily Eastern News *opinion page on March 2, 2004.*

Why does civic marriage apply only to heterosexual men and women in the United States? Why do some conservative groups want to ban same-sex marriage by a Constitutional amendment? These are two questions that I have recently asked myself as these issues have been brought up in the news media recently. 1

I am a Roman Catholic Christian, and for religious and personal reasons, I don't approve of the practice of homosexuality. However, I am not prepared to force others to adhere to my religious beliefs. Two homosexual adults who understand the responsibility and commitment associated with a marriage and who wish to be married should be allowed to do so. As long as the partners uphold their civic duty to one another, they should be allowed to receive the same benefits as any heterosexual couple. 2

I have heard the argument that marriage is a religious sacrament and a right bestowed by God and church upon a heterosexual couple; most if not all Christians would agree with this point of view. But why, then, must couples apply for a marriage at a local courthouse? Aren't couples often married by a justice of the peace? It seems that a long time ago marriage and its Christian meaning became part of the body of laws in this country. If we want to consider our county free from persecution, then we must see how the line drawn between church and state has been blurred and how this has led to injustice and unfairness. 3

I agree with the Massachusetts Supreme Court ruling on the unconstitutionality of civil unions. If someone feels gay couples deserve only civil unions, I offer this suggestion: limit all couples to civil unions and allow 4 them to be united only by a justice of the peace. What I'm saying is that religion has no place in defining the nature of a person's relationship when the government is involved. All citizens should be treated the same by the laws of our country.

5 If homosexuals can meet the requirements of citizenship and uphold the civic duties of marriage, then they should be allowed to enjoy the same freedoms granted to heterosexual couples by our legal system. Arguing that the traditional definition of marriage is between a man and a woman only brings religious practice into a civil issue, and that amounts to discrimination based on religious preferences. You don't have to agree with the act of homosexuality to see that people's freedoms are being infringed upon simply because the word *marriage* is being interpreted in only one way. My solution would be either to allow gay couples the civic right to marriage or to change the wording of our marriage laws in order to remove religious bias from our government.

RESPONDING TO READINGS

After reading these essays, would you change your vote in the referendum on same-sex marriage? Explain your response.

GAINING WORD POWER

Make a list of loaded (highly connotative) language used in each article. Which author uses the most emotionally charged words and phrases? What effect does this usage have on you as a reader? Select a few examples and rewrite them to achieve a different, perhaps more neutral effect.

CONSIDERING CONTENT

1. What objections does Mary Ann Glendon have toward same-sex marriage?
2. Which of Glendon's arguments does Katha Pollitt refute? Which of her refutations do you find most persuasive?
3. What is the main point of Leonard Pitts's argument?
4. What is the main point of Mace Boshart's argument? In what way is he in agreement with Pollitt?

CONSIDERING METHOD

1. What support does Glendon offer for her claim that the success of gay marriage will "usher in an era of intolerance and discrimination the likes of which we have rarely seen before"?
2. Glendon says the debate on same-sex marriage should be conducted "without bitterness or anger." Do you detect any bitterness or anger in her article?
3. Pollitt says she doesn't like marriage, that she prefers "monogamous free love." Why does she include this admission? Do you think it harms or helps her arguments?
4. How does Pitts use the Bible to advance his argument? Is it an effective strategy?
5. Describe the tone of each essay. How did the tone affect your response to the arguments?

WRITING STEP BY STEP

Write an essay in which you argue for or against same-sex marriage.

A. Before you start writing, make a list of reasons why you support or oppose same-sex marriage. Next to each reason, try to write a brief response or a counterargument.
B. Identify at least three major reasons to support your choice. These reasons will be your claims.
C. Start your essay by stating your position and giving a brief overview of your main claims.
D. Take up each of your claims one by one. Explain each reason in its own separate paragraph, and provide evidence to demonstrate the validity of that reason. If you want, you can use quotations from the essays you have just read, but be sure to give credit to the authors you are quoting from.
E. Be sure to consider any objections to your claims that readers might have. Consult the notes you made for step A. If you're not sure what these objections would be, ask classmates, friends, family members, or instructors for their ideas. You might also find additional information on Internet sites such as those listed at the end of this chapter. Refute these objections by minimizing their importance, questioning their factuality, or offering a different point of view. You can include your refutations as part of the argument for each claim, or you can put them in a separate paragraph.

F. Close by summarizing your main ideas. Consider using a clincher sentence to drive home your major claim, as Pollitt and Pitts do.

OTHER WRITING IDEAS

1. COLLABORATIVE WRITING. Write the paper outlined in the step-by-step assignment, but argue for or against one of the following **issues.** Feel free to narrow these topics further if you find them too general.
   a. It should (should not) be harder than it is now to get a divorce.
   b. Having a working mother does (does not) harm a child's welfare.
   c. Free speech should (should not) be restricted on the Internet.
   d. Performance standards should (should not) be raised in our high schools.
   e. Date rape is (is not) a serious problem on college campuses.
   f. People should (should not) give money to panhandlers.
   g. Over-the-counter drugs should (should not) be regulated more closely.

   If these topics don't appeal to you, choose a controversial issue that you feel strongly about. Before you begin writing, get together with a group of classmates to brainstorm arguments for and against the topic you have chosen. If possible, interview people whose experience might provide firsthand insight into the topic.
2. USING THE INTERNET. Find out about same-sex marriage, civil unions, and domestic partnerships in other countries. You can find information at **www.religioustolerance.org/hom_mary.htm**. Write a report that would be suitable for a class in sociology or cultural studies, summarizing your research.
3. WRITING ABOUT READING. Choose one of the articles in this section, and write a letter to the author in which you disagree with his or her point of view. Be sure to cite specific points that you dispute, and explain the reasons for your difference of opinion.

EDITING SKILLS: USING QUESTIONS

Notice how Katha Pollitt, Leonard Pitts, and Mace Boshart each pose questions to open their arguments:

> Will someone please explain to me how permitting gays and lesbians to marry threatens the institution of marriage?

> So what is it you have against gay marriage?

> Why does civic marriage apply only to heterosexual men and women in the United States? Why do some conservative groups want to ban same-sex marriage by a Constitutional amendment?

These are **rhetorical questions;** they don't require answers. In fact, the authors intend to answer them for their readers. You can use rhetorical questions for various purposes: to provoke thought, to set a tone, to add emphasis to a point, to assert or deny a claim indirectly, or to establish the direction of a discussion. They will add variety to your writing, but like any special technique, they lose their effect when used too often.

EXERCISE

There are nineteen more examples of rhetorical questions in the essays by Glendon, Pollitt, Pitts, and Boshart. Find nine of them, and explain what purpose each one serves. Do you think these authors use too many questions?

Now, look at the essay you have written. Add at least one rhetorical question that indicates the direction of your thought. If you like the effect, try using a few more.

WEB SITES

http://news.yahoo.com/fc?tmpl=fc&cid=34&in=us&cat=same_sex_marriage

Yahoo! News Full Coverage provides links, updated weekly, to the latest articles and editorials about same-sex marriage and related issues.

http://pewforum.org/docs/index.php?DocID=45

The Resources on the Federal Marriage Amendment page from the Pew Forum on Religious & Public Life provides links to news coverage, editorials, and sites for advocacy groups on both sides of the issue.

Chapter 11

Combining Strategies: Further Readings

This chapter provides you with additional reading selections. Although some of these readings are developed by one controlling strategy, most of them illustrate combinations of various strategies. As you read, use the following questions to analyze how a writer combines strategies:

- What are the purpose and thesis of the essay? Who is the intended audience?
- Which strategy controls or dominates the essay?
- How does this strategy help readers to understand the essay's thesis and purpose?
- What other strategies appear in the essay?
- What do these strategies contribute to the readers' understanding of the essay's thesis and purpose?

Also keep in mind our suggestions for being an active reader: preview the selection, make predictions, pay attention to conventions, mark the text, use the dictionary, make inferences and associations, and summarize your reactions on paper. After reading the selection actively, you can then follow the process you've used in earlier chapters: reflect on the content, analyze the writer's techniques, and write something of your own that relates to the reading. When possible, discuss the readings and your written responses with your classmates.

Do We Fear the Right Things?

DAVID G. MYERS

David G. Myers is a professor of psychology at Michigan's Hope College, where he has repeatedly been voted "outstanding professor" by students. His writings have appeared in more than four dozen journals and magazines, from Science *to* Scientific American, *and in a dozen books. You can enjoy Myers's flair for instruction in his best-selling introduction-to-psychology textbook,* Psychology. *The essay we reprint here appeared in the* Observer, *the journal of the American Psychological Society. The marginal notes identify the major rhetorical strategies.*

Cause-effect: why we fear flying and how it affects our lives (¶s 1–13).

1 "Freedom and fear are at war," President Bush has told us. The terrorists' goal, he says, is "not only to kill and maim and destroy" but to frighten us into inaction. Alas, the terrorists have made progress in their fear war by diverting our anxieties from big risks toward smaller risks. Flying is a case in point.

2 Even before the horror of September 11 and the ensuing crash at Rockaway Beach, forty-four percent of those willing to risk flying told Gallup they felt fearful. "Every time I get off a plane, I view it as a failed suicide attempt," movie director Barry Sonnenfeld has said. After five crashed airliners, and with threats of more terror to come, cancellations understandably left airlines, travel agencies, and holiday hotels flying into the red.

Comparison and contrast: flying risks vs. driving risks (¶s 3–6).

3 Indeed, the terrorists may still be killing us, in ways unnoticed. If we now fly twenty percent less and instead drive half those unflown miles, we will spend two percent more time in motor vehicles. This translates into eight hundred more people dying as passengers and pedestrians. So, in just the next year the terrorists may indirectly kill three times more people on our highways than died on those four fated planes.

4 Ah, but won't we have spared some of those folks fiery plane crashes? Likely not many, especially now with heightened security, hardened cockpit doors, more reactive passengers, and the likelihood that future terrorists will hit us where

we're not looking. National Safety Council data reveal that in the last half of the 1990s Americans were, mile for mile, thirty-seven times more likely to die in a vehicle crash than on a commercial flight. When I fly to New York, the most dangerous part of my journey is the drive to the Grand Rapids airport. (My highway risk may be muted by my not drinking and driving, but I'm still vulnerable to others who do.)

5 Or consider this: From 1990 through 2000 there were 1.4 deaths per ten million passengers on U.S. scheduled airlines. Flying understandably feels dangerous. But we have actually been less likely to crash and die on any flight than, when coin tossing, to flip twenty-two heads in a row.

6 Will yesterday's safety statistics predict the future? Even if not, terrorists could take down fifty more planes with sixty passengers each and—if we kept flying—we'd still have been safer this year in planes than on the road. Flying may be scary, but driving the same distance should be many times scarier.

7 Why do we fear the wrong things? Why do so many smokers (whose habit shortens their lives, on average, by about five years) fret before flying (which, averaged across people, shortens life by one day)? Why do we fear terrorism more than accidents, which kill nearly as many per week in just the United States as did terrorism with its 2,527 worldwide deaths in all of the 1990s? Why do we fear violent crime more than clogged arteries?

Definition: characteristics of fears, developed through comparison and contrast

8 Psychological science has identified four influences on our intuitions about risk. First, we fear what our ancestral history has prepared us to fear. Human emotions were road tested in the Stone Age. Yesterday's risks prepare us to fear snakes, lizards, and spiders, although all three combined now kill only a dozen Americans a year. Flying may be far safer than biking, but our biological past predisposes us to fear confinement and heights, and therefore flying.

Classification and division: types of fear, developed with examples (¶s 8–11)

9 Second, we fear what we cannot control. Skiing, by one estimate, poses one thousand times the health and injury risk of food preservatives.

Yet many people gladly assume the risk of skiing, which they control, but avoid preservatives. Driving we control, flying we do not. "We are loathe to let others do unto us what we happily do to ourselves," noted risk analyst Chauncey Starr.

10 Third, we fear what's immediate. Teens are indifferent to smoking's toxicity because they live more for the present than the distant future. Much of the plane's threat is telescoped into the moments of takeoff and landing, while the dangers of driving are diffused across many moments to come, each trivially dangerous.

11 Fourth, we fear what's most readily available in memory. Horrific images of a DC-10 catapulting across the Sioux City runway, or the Concorde exploding in Paris, or of United Flight 175 slicing into the World Trade Center, form indelible memories. And availability in memory provides our intuitive rule-of-thumb for judging risks. Small wonder that most of us perceive accidents as more lethal than strokes, and homicide as more lethal than diabetes. (In actuality, the Grim Reaper snatches twice as many lives by stroke as by accident and four times as many by diabetes as by homicide.)

12 Vivid, memorable images dominate our fears. We can know that unprovoked great white shark attacks have claimed merely sixty-seven lives worldwide since 1876. Yet after watching *Jaws* and reading vivid accounts of last summer's Atlantic coastal shark attacks, we may feel chills when an underwater object brushes our leg. A thousand massively publicized anthrax victims would similarly rivet our attention more than yet another 20,000+ annual influenza fatalities, or than another 30,000+ lives claimed by guns (via suicide, homicide, and accident).

Examples: vivid, memorable fears

13 As publicized Powerball lottery winners cause us to overestimate the infinitesimal odds of lottery success, so vivid airline casualties cause us to overestimate the infinitesimal odds of a lethal airline ticket. We comprehend Maria Grasso's winning $197 million in a 1999 Powerball lottery. We don't comprehend the 328 million losing tickets

Examples: overestimated and underestimated risks.

enabling her jackpot. We comprehend the 266 passengers and crew on those four fated flights. We don't comprehend the vast numbers of accident-free flights—16 million consecutive fatality-free takeoffs and landings during one stretch of the 1990s. The result: We overvalue lottery tickets, overestimate flight risk, and underestimate the dangers of driving.

14 The moral: It's perfectly normal to fear purposeful violence from those who hate us. But with our emotions now calming a bit, perhaps it's time to check our fears against facts. "It's time to get back to life," said terror-victim widow Lisa Beamer before boarding the same flight her husband had taken on September 11. To be prudent is to be mindful of the realities of how humans die. By so doing, we can take away the terrorists' most omnipresent weapon: exaggerated fear.

Argument: reasons to resist exaggerated fear (¶s 14–15)

15 And when terrorists strike again, remember the odds. If, God forbid, anthrax or truck bombs kill a thousand Americans, we will all recoil in horror. Small comfort, perhaps, but the odds are 284,000 to one that you won't be among them.

CONSIDERING CONTENT AND METHOD

1. What are the four reasons why we fear the wrong things, according to Myers? Can you think of other examples to go along with each reason?
2. Make a list of the examples Myers uses in pursuing his point. Do you find the examples persuasive?
3. Why does this essay include so many numbers? For instance, look at paragraph five, which has five numbers in two sentences. What was your response to the statistics in the essay? Why do you think you responded the way you did? What was Myers expecting about his audience for this essay?

WEB SITE

www.davidmyers.org/fears/

You will find an annotated version of "Do We Fear the Right Things?" on this site.

Salvation

LANGSTON HUGHES

Langston Hughes (1902–1967) wrote often about the experience of African Americans in the United States in his newspaper columns, poetry, plays, and essays. He was a war correspondent during the Spanish Civil War and traveled the world doing jobs like dock worker and restaurant waiter. Hughes was an outstanding figure in the 1920s arts movement called the Harlem Renaissance in New York City. In the autobiographical essay we reprint here, Hughes reminisces about social pressure.

I was saved from sin when I was going on thirteen. But not really saved. It happened like this. There was a big revival at my Auntie Reed's church. Every night for weeks there had been much preaching, singing, praying, and shouting, and some very hardened sinners had been brought to Christ, and the membership of the church had grown by leaps and bounds. Then just before the revival ended, they held a special meeting for children, "to bring the young lambs to the fold." My aunt spoke of it for days ahead. That night I was escorted to the front row and placed on the mourners' bench with all the other young sinners, who had not yet been brought to Jesus. 1

My aunt told me that when you were saved you saw a light, and something happened to you inside! And Jesus came into your life! And God was with you from then on! She said you could see and hear and feel Jesus in your soul. I believed her. I had heard a great many old people say the same thing and it seemed to me they ought to know. So I sat there calmly in the hot crowded church, waiting for Jesus to come to me. 2

The preacher preached a wonderful rhythmical sermon, all moans and shouts and lonely cries and dire pictures of hell, and then he sang a song about the ninety and nine safe in the fold, but one little lamb was left in the cold. Then he said, "Won't you come? Won't you come to Jesus? Young lambs, won't you come?" And he held out his arms to all us young sinners there on the mourners' bench. And the little girls cried. And some of them jumped up and went to Jesus right away. But most of us just sat there. 3

A great many old people came and knelt around us and prayed, old women with jet-black faces and braided hair, old men with work-gnarled hands. And the church sang a song about the lower lights are burning, some poor sinners to be saved. And the whole building rocked with prayer and song. 4

Still I kept waiting to see Jesus. 5

Finally all the young people had gone to the altar and were saved, but one boy and me. He was a rounder's son named Westley. Westley and I were surrounded by sisters and deacons praying. It was very hot in the church, and getting late now. Finally Westley said to me in a whisper: "Goddamn! I'm tired o' sitting here. Let's get up and be saved." So he got up and was saved. 6

Then I was left all alone on the mourners' bench. My aunt came and knelt at my knees and cried, while prayers and songs swirled all around me in the little church. The whole congregation prayed for me alone, in a mighty wail of moans and voices. And I kept waiting serenely for Jesus, waiting, waiting—but he didn't come. I wanted to see him, but nothing happened to me. Nothing! I wanted something to happen to me, but nothing happened. 7

I heard the songs and the minister saying: "Why don't you come? My dear child, why don't you come to Jesus? Jesus is waiting for you. He wants you. Why don't you come? Sister Reed, what is this child's name?" 8

"Langston," my aunt sobbed. 9

"Langston, why don't you come? Why don't you come and be saved? Oh, Lamb of God! Why don't you come?" 10

Now it was really getting late. I began to be ashamed of myself, holding everything up so long. I began to wonder what God thought about Westley, who certainly hadn't seen Jesus either, but who was now sitting proudly on the platform, swinging his knickerbockered legs and grinning down at me, surrounded by deacons and old women on their knees praying. God had not struck Westley dead for taking his name in vain or for lying in the temple. So I decided that maybe to save further trouble, I'd better lie, too, and say that Jesus had come, and get up and be saved. 11

So I got up. 12

Suddenly the whole room broke into a sea of shouting, as they saw me rise. Waves of rejoicing swept the place. Women leaped into the air. My aunt threw her arms around me. The minister took me by the hand and led me to the platform. 13

When things quieted down, in a hushed silence, punctuated by a few ecstatic "Amens," all the new young lambs were blessed in the name of God. Then joyous singing filled the room. 14

That night, for the last time in my life but one—for I was a big boy twelve years old—I cried. I cried, in bed alone, and couldn't stop. I buried my head under the quilts, but my aunt heard me. She woke up and told my uncle I was crying because the Holy Ghost had come into my life, and because I had seen Jesus. But I was really crying because I couldn't bear to tell her that I had lied, that I had deceived everybody in the church, and I hadn't seen 15

Jesus, and that now I didn't believe there was a Jesus any more, since he didn't come to help me.

CONSIDERING CONTENT AND METHOD

1. How does the style of this narrative reflect the child's point of view? How does it reflect the grown author's point of view?
2. Review the opening of Chapter 3 on narration and description. Point out elements of narrative and descriptive strategies in "Salvation."
3. How does this essay follow a cause and effect strategy (Chapter 9)? In other words, what causes and effects are described or suggested in "Salvation"?
4. In reading this essay, do you get the idea that Hughes is criticizing religion? What else might he be criticizing?

WEB SITE

www.nku.edu/~diesmanj/hughes.html

Contains some of Hughes's poetry as well as links to information about Hughes's work as a whole and about the Harlem Renaissance, of which he was an important part.

Marriage in the U.S.: Early, Often, and Informal

CYNTHIA CROSSEN

Born in 1951 in Battle Creek, Michigan, Cynthia Crossen went to Macalester College in St. Paul, Minnesota, and to graduate school at the University of Minnesota. She has been on the staff of the Wall Street Journal *since 1984. She is the author of two nonfiction books:* Tainted Truth: The Manipulation of Fact in America *and* The Rich and How They Got That Way. *She says she was inspired to write the column reprinted here because of the debate over whether same-sex couples should be allowed to marry. It appeared in the* Wall Street Journal *on February 25, 2004.*

In a Connecticut village in the 17th century, an unmarried couple moved in together. One day, while out for a stroll, they ran into the local magistrate. 1

"John Rogers," the magistrate said, "do you persist in calling this woman your wife?" 2

"Yes, I do." 3

"And Mary, do you really wish this old man to be your husband?" 4

"Indeed I do." 5

"Then by the laws of God and this common wealth, I pronounce you man and wife. . . ." 6

Although probably apocryphal, this story reflects the state of marriage in early America: no license, witness, ceremony, often not even a magistrate. Some couples wanted a blessing from church or state, but common-law marriages—men and women behaving as spouses without a formal contract—were both legal and respectable. An 1843 Indiana marriage law stated, "No particular form of ceremony shall be necessary, except that the parties shall declare . . . that they take each other as husband and wife." 7

Early American settlers adapted traditions imported from England to life on the wide-open plains. A man and woman may have committed themselves to lifelong devotion months, even years, before a circuit judge or preacher happened along. Many couples didn't wait. In the early Chesapeake region, roughly a third of brides were pregnant. 8

Females were marriageable at the age of 12, males at 15. But in Hempstead Harbor, N.Y., in 1838, Edward Tappan, 15 years old, married Harriet Allen, who had just celebrated her 11th birthday. In Green Hollow, Maine, in 1828, a Mr. Williams, age 87, wed Polly Candle, 14. 9

After marrying, most couples had a powerful incentive to sustain their connubial enterprise: The labor of both was crucial to their survival. Further cementing the bond, by law a wife's personal property belonged to her husband; if she left him, she took nothing. Even so, some early settlers, both women and men, sought divorces. But death, not strife, ended most marriages. In the 17th century, one of the spouses, usually the woman, was likely to be dead in seven years. 10

By the 19th century, state legislators were realizing that people who married themselves could—and would—also divorce themselves, leaving a trail of destitution. Lawmakers passed a hodgepodge of bills setting the terms of the marriage contract. In most states, common-law marriage was gradually abolished. Before Americans could marry, they had to ask the government's permission. 11

Thirty states prohibited people with physical or mental disabilities—epileptics or the "feebleminded, idiotic, imbecilic or insane"—from marrying. (In many states, women over 45 were exempted from this rule.) Four states disqualified paupers or inmates in public institutions for the indigent. Washington and North Dakota didn't issue marriage licenses to people suffering from advanced tuberculosis. 12

Most states banned interracial marriage; white citizens of Florida could not marry anyone of "one-eighth or more Negro blood." California's white residents couldn't legally marry "a Negro, mulatto, Mongolian or member of the Malay race." Nevada's racial restrictions were all-inclusive—a white man or woman could not marry "a person of the black, brown, yellow or red races." In Mississippi, the penalty for interracial marriage was life in prison. 13

Slaves could not legally marry. How, legislators argued, could property itself enter into a contract? Nonetheless, slaves got married, "till death or distance do you part," as their preachers sometimes said. Mormon polygamy, deemed "inhumane," was finally banned by the Supreme Court in 1879. 14

If marriage law became a thicket of red tape, divorce was a jungle. In South Carolina, divorce was prohibited for any reason. But in most states, a man or woman could petition the legislature for divorce—a long, expensive and often futile exercise. Grounds for divorce were narrow and literal: adultery, drunkenness, desertion. Some states didn't allow the divorced, especially the "guilty" ones, to remarry while their former spouses were alive. 15

State divorce laws were so different—and so often contested—that an exasperated U.S. Supreme Court justice wrote, "If there is one thing the people are entitled to expect from their lawmakers, it's rules that will enable them to tell whether they are married, and if so, to whom." 16

Not surprisingly, couples often fled to states with the most lenient divorce rules. In the 19th century, Indiana was a favorite. Since then, except for a few brief periods in the 20th century, America's divorce rate has 17 steadily marched upward. In 1880, one in 21 marriages ended in divorce; by 1916, one of every nine couples divorced.

18 Today, a state's marriage law contains dozens of technicalities defining who may and may not marry. In Arizona, for example, first cousins can't marry unless both are at least 65 years old. If one is under 65, however, they may marry "upon approval of any superior court judge in the state if proof has been presented to the judge that one of the cousins is unable to reproduce."

CONSIDERING CONTENT AND METHOD

1. How does Crossen capture our interest in her opening paragraphs?
2. What laws or official regulations existed for marriage in early America?
3. What factors kept married couples together far longer in those days than today?
4. Why were laws concerning marriage introduced in the 19th century?
5. Why do you think there were laws in the past banning certain people from marrying? Were these reasonable laws? Have they been repealed today?
6. This is primarily an essay of example and illustration with a brief narrative opening. Could you make a case for it as a persuasive essay?

WEB SITE

http://opinionjournal.com/taste/?id=95001655

Read another entertaining article by Cynthia Crossen about changes in a social practice: "The Spoon: Its Use and Abuse."

Navajo Code Talkers: The Century's Best Kept Secret

JACK HITT

Journalist Jack Hitt analyzes everyday life and contemporary issues for magazines such as Harper's, Mother Jones, *and the* New York Times Magazine *and is a regular contributor to National Public Radio's popular program* This American Life. *In 1990 he received the Livingston Award for Young Journalists for an article he co-authored for* Esquire. *Hitt also has written and edited several nonfiction books, including* Off the Road: A Walk Down the Pilgrim's Road into Spain *(1994) and* Perfect Murder: Five Great Mystery Writers Create the Perfect Crime *(1991). "Navajo Code Talkers," an essay that brings a little-known historical event to life, was published in 1990 as part of a collection titled* American Greats, *edited by Robert Wilson and Stanley Marcus.*

1 During World War II, on the dramatic day when Marines raised the American flag to signal a key and decisive victory at Iwo Jima, the first word of this momentous news crackled over the radio in odd guttural noises and complex intonations. Throughout the war, the Japanese were repeatedly baffled and infuriated by these seemingly inhuman sounds. They conformed to no linguistic system known to the Japanese. The curious sounds were the military's one form of conveying tactics and strategy that the master cryptographers in Tokyo were unable to decipher. This perfect code was the language of the Navajo tribe. Its application in World War II as a clandestine system of communication was one of the twentieth century's best-kept secrets.

2 After a string of cryptographic failures, the military in 1942 was desperate for a way to open clear lines of communication among troops that would not be easily intercepted by the enemy. In the 1940s there was no such thing as a "secure line." All talk had to go out onto the public airwaves. Standard codes were an option, but the cryptographers in Japan could quickly crack them. And there was another problem: the Japanese were proficient at intercepting short-distance communications, on walkie-talkies for example, and then having well-trained English-speaking soldiers either sabotage the message or send out false commands to set up an ambush. That was the situation in 1942 when the Pentagon authorized one of the boldest gambits of the war.

3 The solution was conceived by the son of missionaries to the Navajos, a former Marine named Philip Johnston. His idea: station a native Navajo speaker at every radio. Since Navajo had never been written down or translated into any other language, it was an entirely self-contained human communication system restricted to Navajos alone; it was virtually indecipherable without Navajo help. Without some key or way into a language, translation is virtually impossible. Not long after the bombing of Pearl Harbor, the military dispatched twenty-nine Navajos to Camp Elliott and Camp Pendleton in California to begin a test program. These first recruits had to develop a Navajo alphabet since none existed. And because Navajo lacked technical terms of military artillery, the men coined a number of neologisms specific to their task and their war.

4 According to Chester Nez, one of the original code talkers: "Everything we used in the code was what we lived with on the reservation every day, like the ants, the birds, bears." Thus, the term for a tank was "turtle," a tank destroyer was "tortoise killer." A battleship was "whale." A hand grenade was "potato," and plain old bombs were "eggs." A fighter plane was "hummingbird," and a torpedo plane "swallow." A sniper was "pick 'em off." Pyrotechnic was "fancy fire."

5 It didn't take long for the original twenty-nine recruits to expand to an elite corps of Marines, numbering at its height 425 Navajo Code Talkers, all from the American Southwest. Each Talker was so valuable, he traveled everywhere with a personal bodyguard. In the event of capture, the Talkers had solemnly agreed to commit suicide rather than allow America's most valuable war code to fall into the hands of the enemy. If a captured Navajo did not follow that grim instruction, the bodyguard's instructions were understood: shoot and kill the Code Talker.

6 The language of the Code Talkers, their mission, and every detail of their messaging apparatus was a secret they were all ordered to keep, even from their own families. They did. It wasn't until 1968, when the military felt convinced that the Code Talkers would not be needed for any future wars, that America learned of the incredible contribution a handful of Native Americans made to winning history's biggest war. The Navajo Code Talkers, sending and receiving as many as 800 errorless messages at fast speed during "the fog of battle," are widely credited with giving U.S. troops the decisive edge at Guadalcanal, Tarawa, Saipan, Iwo Jima, and Okinawa.

CONSIDERING CONTENT AND METHOD

1. What problem did the Navajo language solve for the military in World War II? Explain the solution Hitt outlines.

2. What's the definition of a neologism? If you don't know, use the examples in paragraph 4 to create a definition for the term.
3. How did the military ensure the secrecy of the Navajo code? What caused the history of this group of heroes to be revealed?
4. What do you know about Iwo Jima? Why does Hitt begin the essay with a story about that day?

WEB SITE

www.mgm.com/windtalkers/html

The official site for *Windtalkers,* the 2001 movie about the Navajo Code Talkers, contains links to a lot of information about the code talkers and their code.

The Discus Thrower

RICHARD SELZER

Dr. Richard Selzer (b. 1928) has now retired from practice and from his position as Professor of Surgery at the Yale Medical School. Although he did not begin writing until age 40, he says, "For me, the decision to write was not a frivolous thing; it was a passion from the beginning, and I knew I had to do it." He is the author of ten books, among them Letters to a Young Doctor *and* The Exact Location of the Soul. *His essays and stories have been published in such magazines as* Harper's, Esquire, *and* Redbook. *His work presents the dramatic, sometimes agonizing experiences of a practicing surgeon. The narrative reprinted here is enriched by realistic dialogue and exact descriptive details.*

I spy on my patients. Ought not a doctor to observe his patients by any means and from any stance, that he might the more fully assemble evidence? So I stand in the doorways of hospital rooms and gaze. Oh, it is not all that furtive an act. Those in bed need only look up to discover me. But they never do. 1

From the doorway of Room 542 the man in the bed seems deeply tanned. Blue eyes and close-cropped white hair give him the appearance of vigor and good health. But I know that his skin is not brown from the sun. It is rusted, rather, in the last stage of containing the vile repose within. And the blue eyes are frosted, looking inward like the windows of a snowbound cottage. This man is blind. This man is also legless—the right leg missing from midthigh down, the left from just below the knee. It gives him the look of a bonsai, roots and branches pruned into the dwarfed facsimile of a great tree. 2

Propped on pillows, he cups his right thigh in both hands. Now and then he shakes his head as though acknowledging the intensity of his suffering. In all of this he makes no sound. Is he mute as well as blind? 3

The room in which he dwells is empty of all possessions—no get-well cards, small, private caches of food, day-old flowers, slippers, all the usual kickshaws of the sickroom. There is only the bed, a chair, a nightstand, and a tray on wheels that can be swung across his lap for meals. 4

"What time is it?" he asks. 5

"Three o'clock." 6

"Morning or afternoon?" 7

"Afternoon." 8

He is silent. There is nothing else he wants to know. 9

"How are you?" I say. 10

"Who is it?" he asks. 11

"It's the doctor. How do you feel?" 12

He does not answer right away. 13

"Feel?" he says. 14

"I hope you feel better," I say. 15

I press the button at the side of the bed. 16

"Down you go," I say. 17

"Yes, down," he says. 18

He falls back upon the bed awkwardly. His stumps, unweighted by legs and feet, rise in the air, presenting themselves. I unwrap the bandages from the stumps, and begin to cut away the black scabs and the dead, glazed fat with scissors and forceps. A shard of white bone comes loose. I pick it away. I wash the wounds with disinfectant and redress the stumps. All this while, he does not speak. What is he thinking behind those lids that do not blink? Is he remembering a time when he was whole? Does he dream of feet? Of when his body was not a rotting log? 19

He lies solid and inert. In spite of everything, he remains impressive, as though he were a sailor standing athwart a slanting deck. 20

"Anything more I can do for you?" I ask. 21

For a long moment he is silent. 22

"Yes," he says at last and without the least irony. "You can bring me a pair of shoes." 23

In the corridor, the head nurse is waiting for me. 24

"We have to do something about him," she says. "Every morning he orders scrambled eggs for breakfast, and, instead of eating them, he picks up the plate and throws it against the wall." 25

"Throws his plate?" 26

"Nasty. That's what he is. No wonder his family doesn't come to visit. They probably can't stand him any more than we can." 27

She is waiting for me to do something. 28

"Well?" 29

"We'll see," I say. 30

The next morning I am waiting in the corridor when the kitchen delivers his breakfast. I watch the aide place the tray on the stand and swing it across his lap. She presses the button to raise the head of the bed. Then she leaves. 31

In time the man reaches to find the rim of the tray, then on to find the dome of the covered dish. He lifts off the cover and places it on the stand. He fingers across the plate until he probes the eggs. He lifts the plate in both hands, sets it on the palm of his right hand, centers it, balances it. He hefts it up and down slightly, getting the feel of it. Abruptly he draws back his right arm as far as he can. 32

There is the crack of the plate breaking against the wall at the foot of his bed and the small wet sound of the scrambled eggs dropping to the floor. 33

And then he laughs. It is a sound you have never heard. It is something new under the sun. It could cure cancer. 34

Out in the corridor, the eyes of the head nurse narrow. 35

"Laughed, did he?" 36

She writes something down on her clipboard. 37

A second aide arrives, brings a second breakfast tray, puts it on the nightstand, out of his reach. She looks over at me shaking her head and making her mouth go. I see that we are to be accomplices. 38

"I've got to feed you," she says to the man. 39

"Oh, no you don't," the man says. 40

"Oh, yes I do," the aide says, "after the way you just did. Nurse says so." 41

"Get me my shoes," the man says. 42

"Here's oatmeal," the aide says. "Open." And she touches the spoon to his lower lip. 43

"I ordered scrambled eggs," says the man. 44

"That's right," the aide says. 45

I step forward. 46

"Is there anything I can do?" I say. 47

"Who are you?" the man asks. 48

In the evening I go once more to that ward to make my rounds. The head nurse reports to me that Room 542 is deceased. She has discovered this quite by accident, she says. No, there had been no sound. Nothing. It's a blessing, she says. 49

I go into his room, a spy looking for secrets. He is still there in his bed. His face is relaxed, grave, dignified. After a while, I turn to leave. My gaze sweeps the wall at the foot of the bed, and I see the place where it has been repeatedly washed, where the wall looks very clean and very white. 50

CONSIDERING CONTENT AND METHOD

1. Why does the doctor spy on this particular patient?
2. Besides the interesting story line, what other elements make Selzer's narrative so compelling?
3. What is the basis of the conflict between the head nurse and the doctor? How is the conflict developed?
4. Refer to at least three descriptive passages to illustrate how these specific details enrich the narrative.
5. What is the point of the narrative? Is it stated or implied? Was that a good strategy? Why or why not?

WEB SITE

http://endeavor.med.nyu.edu/lit-med/lit-med-db/webdocs/webauthors/selzer83-au-.html

The Literature, Arts, and Medicine database, maintained by the Medical Humanities Department at New York University, contains summaries and commentaries on a number of essays and stories by Richard Selzer.

Glossary

Abstract words: language that refers to ideas, conditions, and qualities that cannot be observed directly through the five senses. Words such as *beauty, love, joy, wealth, cruelty, power,* and *justice* are abstract. In his essay (p. 127), Isaac Asimov explores the abstract term *intelligence,* offering a series of concrete examples and incidents to make the meaning clearer. *Also see* Concrete words.

Active reader: a reader who gets involved with the reading material by surveying the text, making predictions, writing questions and responses in the margins, rereading difficult passages, and spending time afterward summarizing and reflecting.

Allusion: a passing reference to a person, place, or object in history, myth, or literature. Writers use allusions to enrich or illuminate their ideas. Suzanne Britt, for instance, mentions "Never-Never Land" (p. 204), an allusion to the imaginary land in *Peter Pan*. And in the title of his article "Dearly Beloved, Why Are We Debating This?" (p. 357), Leonard Pitts alludes to the opening of a traditional wedding ceremony.

Analogy: a comparison that uses a familiar or concrete item to explain an abstract or unfamiliar concept. For example, a geologist may compare the structure of the earth's crust to the layers of an onion, or a biologist may explain the anatomy of the eye by comparing it to a camera.

Anecdote: a brief story about an amusing or interesting event, usually told to illustrate an idea or support a point. Writers also use anecdotes to begin essays, as do Barbara Huttmann in "A Crime of Compassion" (p. 332) and Elizabeth Berg in "My Heroes" (p. 78).

Antonym: a word that has the opposite meaning of another word. For example, *wet* is an antonym of *dry; coarse* is an antonym of *smooth; cowardly* is an antonym of *brave.*

Argument: a type of writing in which the author tries to influence the reader's thinking on a controversial topic. See the introduction to Chapter 10.

Attribution: a phrase like "according to," "the FDA reports," "Dr. Weber thinks," which acknowledges the source of a fact, opinion, or quotation.

Audience: the readers for whom a piece of writing is intended. Many essays are aimed at a general audience, but a writer can focus on a specific group of readers. For example, Arthur Ashe directs his essay "Send Your Children to the Libraries" (p. 318) to the parents of young African-American males, while Leonard Pitts in "Dearly Beloved, Why Are We Debating This?" (p. 357) is primarily addressing people who oppose same-sex marriage.

Block-by-block pattern: an organizational pattern used in comparison-and-contrast writing. In this method, a writer presents, in a block, all the important points about the first item to be compared and then presents, in another block, the corresponding points about the second item to be compared.

Brainstorming: a method for generating ideas for writing. In brainstorming, a writer jots down a list of as many details and ideas on a topic as possible without stopping to evaluate or organize them.

Causes: the reasons or explanations for why something happens. Causes can be *immediate* or *remote.* See the introduction to Chapter 9.

Chronological order: the arrangement of events according to time—that is, in the sequence in which they happened.

Claim: a positive statement or assertion that requires support. Claims are the backbone of any argument.

Classification: the process of sorting items or ideas into meaningful groups or categories. See the introduction to Chapter 6.

Cliché: a phrase or expression that has lost its originality or force through overuse. To illustrate, novelist and teacher Janet Burroway writes: "Clichés are *the last word* in bad writing, and it's *a crying shame* to see all you *bright young things* spoiling your *deathless prose* with phrases *as old as the hills*. You must *keep your nose to the grindstone,* because the *sweet smell of success* only comes to those who *march to the tune of a different drummer.*"

Coherence: the logical flow of ideas in a piece of writing. A writer achieves coherence by having a clear thesis and by making sure that all the supporting details relate to that thesis. *Also see* Unity.

Colloquial language: conversational words and expressions that are sometimes used in writing to add color and authenticity. Dave Barry (p. 258) and Rick Reilly (p. 273) use colloquial language to good effect in their writing. *Also see* Informal writing.

Combining form: a modified form of an independent word that occurs only in combination with words, prefixes, suffixes, or other combining forms to create compounds or derivatives, as *electro-* in *electromagnet* or *geo-* in *geography*. *Also see* Root.

Comparison and contrast: a pattern of writing in which an author points out the similarities and differences between two or more subjects. See the introduction to Chapter 7.

Conclusion: the sentences and paragraphs that bring an essay to its close. In the conclusion, a writer may restate the thesis, sum up important ideas, emphasize the topic's significance, make a generalization, offer a solution to a problem, or encourage the reader to take some action. Whatever the strategy, a conclusion should end the essay in a firm and definite way.

Concrete words: language that refers to real objects that can be seen, heard, tasted, touched, or smelled. Words such as *tree, desk, car, orange, Chicago, Roseanne,* or *jogging* are concrete. Concrete examples make abstractions easier to understand, as in "Contentment is a well-fed cat asleep in the sun." *Also see* Abstract words.

Connotation and denotation: terms used to describe the different kinds of meaning that words convey. **Denotation** refers to the most specific or direct meaning of a word—the dictionary definition. **Connotations** are the feelings or associations that attach themselves to words. For example, *assertive* and *pushy* share a similar denotation—both mean "strong" or "forceful." But their differing connotations suggest different attitudes: an assertive person is admirable; a pushy person is offensive.

Controlling idea *See* Thesis.

Conventions: customs or generally accepted practices. The conventions of writing an essay require a title, a subject, a thesis, a pattern of organization, transitions, and paragraph breaks.

Definition: a method of explaining a word or term so that the reader understands what the writer means. Writers use a variety of methods for defining words and terms; see the introduction to Chapter 5.

Denotation *See* Connotation and denotation.

Dependent clause: a group of words that contains a subject and verb but does not stand alone as a sentence. For example, *until the game ended* is a dependent clause; its complete meaning depends on being attached to an independent clause: *Few fans stayed until the game ended. Also see* Independent clause.

Derivation: the historical origin and development of a word. For instance, the English word *verbiage* (meaning "too many words") comes from the French word *verbier* meaning "to chatter." *Also see* Root.

Description: writing that uses sensory details to create a word picture for the reader. See the introduction to Chapter 3.

Details: specific pieces of information (*examples, incidents, dates, statistics, descriptions,* and the like) that explain and support the general ideas in a piece of writing.

Development: the techniques and materials that a writer uses to expand and build on a general idea or topic.

Dialogue: speech or conversation recorded in writing. Dialogue, which is commonly found in narrative writing, reveals character and adds life and authenticity to an essay.

Diction: choice of words in writing or speaking.

Division: the process of breaking a large subject into its components or parts. Division is often used in combination with classification. See the introduction to Chapter 6.

Editing: a step in the writing process that focuses on making small-scale changes to correct mechanics and improve clarity and readability.

Effects: the results or outcomes of certain events. Effects can be *immediate* or *long term*. Writers often combine causes and effects in explaining why something happens. See the introduction to Chapter 9.

Ellipsis: three equally spaced dots that signal an omission of words.

Emphasis: the placement of words and ideas in key positions to give them stress and importance. A writer can emphasize a word or idea by putting it at the beginning or end of a paragraph or essay. Emphasis can also be achieved by using repetition and figurative language to call attention to an idea or term.

Essay: a short prose work on a limited topic. Essays can take many forms, but they usually focus on a central theme or thesis and often convey the writer's personal ideas about the topic.

Evidence *See* Supporting material.

Example: a specific case or instance used to illustrate or explain a general concept. See the introduction to Chapter 4.

Fable: a brief narrative that teaches a lesson or truth.

Figurative language: words that create images or convey symbolic meaning beyond the literal level. Richard Selzer, for example, uses figurative language to portray the dramatic, often agonizing experiences of a practicing surgeon: "And the blue eyes are frosted, looking inward like the windows of a snowbound cottage" (see "The Discus Thrower," p. 379).

Figures of speech: deliberate departures from the ordinary, literal use of words in order to provide fresh perceptions and create lasting impressions. *See* Metaphor, Paradox, Personification, *and* Simile.

First person: the use of *I, me, we,* and *us* in speech and writing to express a personal view or present a firsthand report. *Also see* Point of view.

Focus: the narrowing of a topic to a specific aspect or set of features.

Freewriting: a procedure for exploring a topic that involves writing without stopping for a set period of time.

Generalization: a broad assertion or conclusion based on specific observations. The value of a generalization is determined by the number and quality of the specific instances.

Generic nouns: the name of a class of people, such as *doctor, teacher, student, player, citizen, juror, consumer, reader, author,* and so forth. The use of such nouns in the singular to designate a whole class or group causes problems with pronoun selection. For example, a sentence like "Each applicant is responsible for scheduling his own interview" seems to ignore or exclude female applicants. This same point can be expressed without relying on the masculine pronoun (*his*): "Each applicant is responsible for scheduling his or her own interview" or "Applicants are responsible for scheduling their own interviews."

Homophone: a word that sounds the same as another but is different in spelling and meaning. *Knew* and *new* are homophones of each other.

Hyperbole: a conscious, intentional use of exaggeration, as in "I'm so hungry I could eat a horse" or "All the perfumes of Arabia will not sweeten this little hand." This figure of speech is used to heighten effect, or it may be used for humor, as Dave Barry does in his essay "There Are Rules, You Know" (p. 259).

Illustration: the use of examples, or a single long example, to support or explain an idea. See the introduction to Chapter 4.

Images: descriptions that appeal to our senses of sight, smell, sound, touch, or taste. Images add interest and clarify meaning.

Imperative sentence: a sentence that gives a command or a direction. Imperative sentences usually begin with a verb; they are often used in writing about a process: "Snap the knee up to the chest as close as possible"; "Leave enough space after the complimentary close to sign your name"; "Don't forget to proofread your final copy."

Independent clause: a group of words that contains a subject and verb and can stand alone as a sentence.

Inference: a conclusion drawn by a reader from the hints and suggestions provided by the writer. Writers sometimes express ideas indirectly rather than stating them outright; readers must use their own experience and knowledge to read between the lines and make inferences to gather the full meaning of a selection.

Informal writing: the familiar, everyday level of usage, which includes contractions and perhaps slang but requires standard grammar and punctuation.

Interrupter: a word or phrase that interrupts the normal flow of a sentence without changing the basic meaning. Interrupters are usually set off from the rest of the sentence with commas: "Magnum Oil Company, *our best client,* canceled its account." "Being lucky, *it seems to me,* is better than being smart."

Introduction: the beginning or opening of an essay, which usually presents the topic, arouses interest, and prepares the reader for the development of the thesis.

Irony: the use of verbal clues to express the opposite of what is stated. Writers use irony to expose unpleasant truths or to poke fun at human weakness.

Issue: an important point or problem for discussion or debate such as the issue of global warming or the issue of same-sex marriage.

Jargon: the specialized or technical language of a trade, profession, or similar group. To readers outside the group, jargon is confusing and meaningless.

Journalistic style: the kind of writing found in newspapers and popular magazines. It normally employs informal diction with relatively simple sentences and unusually short paragraphs.

Logical order: arrangement of points and ideas according to some reasonable principle or scheme (e.g., from least important to most important).

Main idea *See* Thesis.

Metaphor: a figure of speech in which a word or phrase that ordinarily refers to one thing is applied to something else, thus making an implied comparison. For example, Mark Twain writes of "the language of this water" and says the river "turned to blood" ("Two Views of the Mississippi," p. 197). Similarly, Judith Ortiz Cofer refers to Mamá's room as "the heart of the house" (p. 61), and Leonard Pitts says that marriage is "our civilization's bedrock" (p. 358).

A *dead metaphor* is an implied comparison that has become so familiar that we accept it as literal: the arm of a chair, dog tired, or time is running out.

Modes *See* Patterns of development.

Narration: writing that recounts an event or series of interrelated events; presentation of a story in order to illustrate an idea or make a point. See the introduction to Chapter 3.

Onomatopoeia: the use of words that suggest or echo the sounds they are describing—*hiss, plop, buzz, whir,* or *sizzle,* for example.

Order: the sequence in which the information or ideas in an essay are presented. *Also see* Chronological order *and* Logical order.

Paradox: a seeming contradiction that may nonetheless be true. For example, "Less is more" or "The simplest writing is usually the hardest to do."

Paragraph: a series of two or more related sentences. Paragraphs are units of meaning; they signal a division or shift in thought. In newspapers and magazines, frequent paragraph divisions are used to break up the narrow columns of print and make articles easier to read.

Parallelism: the presentation of two or more equally important ideas in similar grammatical form. In his essay "Send Your Children to the Libraries"

(p. 318), Arthur Ashe emphasizes his thesis by using parallel structure: "I have become convinced that we blacks spend *too much time on the playing fields* and *too little time in the libraries.*" He also uses parallelism to make other points forceful and memorable: "Somehow, parents must instill *a desire for learning* alongside *the desire to be Walt Frazier*"; "*While we are about 35 percent of* major league baseball, *we are less than 2 percent of* the engineers. *While we are about 40 percent of* the National Football League, *we are less than 11 percent of* construction workers such as carpenters and bricklayers."

Patterns of development: strategies for presenting and expanding ideas in writing. Some of these patterns relate to basic ways of thinking (classification, cause and effect, argument), whereas others reflect the most common means for presenting material (narration, comparison-contrast, process) or developing ideas (example and illustration, definition, description) in writing.

Person *See* Point of view.

Personification: a figure of speech in which an inanimate object or an abstract concept is given human qualities. For example, "Hunger sat shivering on the road"; "Flowers danced on the lawn." In "Two Views of the Mississippi" (p. 197), Mark Twain refers to the "river's face" and describes the river as a subtle and dangerous enemy.

Persuasion: writing that attempts to move readers to action or to influence them to agree with a position or belief.

Point-by-point pattern: an organizational pattern used in comparison-and-contrast writing. In this method (also called the *alternating method*), the writer moves back and forth between the two subjects, focusing on particular features of each in turn: the first point or feature of subject *A* is followed by the first point or feature of subject *B* and so on.

Point of view: the angle or perspective from which a story or topic is presented. Personal essays often take a first-person (or *I*) point of view and sometimes address the reader as *you* (second person). The more formal third person (*he, she, it, one, they*) is used to create distance and suggest objectivity.

Prefix: a syllable or syllables used at the beginning of a word to change or add to the meaning. For example, prefixes change *mature* to ***im**mature* and

***pre**mature;* and *form* can be expanded to ***in**form, **re**form, **per**form, **de**form, **trans**form, **uni**form,* and ***misin**form. Also see* Root *and* Suffix.

Previewing: the first step in active reading in which the reader prepares to read by looking over the text, and making preliminary judgments and predictions about what to expect.

Prewriting: the process that writers use to prepare for the actual writing stage by gathering information, considering audience and purpose, developing a provisional thesis, and mapping out a tentative plan.

Problem-solution: a strategy for analyzing and writing about a topic by identifying a problem within the topic and offering a solution or solutions to it.

Process writing: a pattern in which the author explains the step-by-step procedure for doing something. See the introduction to Chapter 8.

Proper noun: a noun that names a single particular person, place, or historical event, and is written with a capital letter: *Carlo, Warsaw, Mexico, Garfield,* the *Holocaust.*

Purpose: the writer's reasons for writing; what the writer wants to accomplish in an essay.

Refutation: in argumentation, the process of acknowledging and responding to opposing views. See the introduction to Chapter 10.

Revising: the stage in the writing process during which the author makes changes in focus, organization, development, and style to make the writing more effective.

Rhetorical question: a question that a writer or speaker asks to emphasize or introduce a point and usually goes on to answer. The authors of the essays in the same-sex marriage debate (Chapter 10) use a number of rhetorical questions.

Root: the stem or base of a word; the element that carries the primary meaning of a word. The Latin word *videre,* meaning "to see," is the root of such English words as *video, vista, vision, visionary,* and *revision. Also see* Derivation.

Satire: writing that uses wit and irony to attack and expose human folly, weakness, and stupidity. Dave Barry (p. 258) and Suzanne Britt (p. 204) use satire to question human behavior and criticize contemporary values.

Sentence *See* Independent clause.

Sexist language: words and phrases that stereotype or ignore members of either sex. For example, the sentence "A doctor must finish his residency before he can begin to practice" suggests that only men are doctors. Writing in the plural will avoid this exclusion: "Doctors must finish their residencies before they can begin to practice." Terms like *mailman, stewardess, manpower,* and *mothering* are also sexist; try to use gender-neutral terms instead: *mail carrier, flight attendant, workforce, parenting.*

Simile: a figure of speech in which two essentially unlike things are compared, usually in a phrase introduced by *like* or *as.* For example, in "More Room" (p. 60) Judith Ortiz Cofer says the house is "*like* a chambered nautilus" and that "it rested on its perch *like* a great blue bird, more *like* a nesting hen. . . ."

Slang: the informal language of a given group or locale, often characterized by racy, colorful expressions and short-lived usage.

Standard English: the language written or spoken by most educated people.

Structure: the general plan, framework, or pattern of a piece of writing.

Style: individuality of expression, achieved in writing through the selection and arrangement of words, sentences, and punctuation.

Subject: what a piece of writing is about.

Subordination: the process of expressing less important ideas in dependent clauses and combining them with independent clauses. For example, the independent statement "Tim heard a noise" can be subordinated and combined with "Tim began to run" by using the subordinator *when:* "Tim began to run *when he heard a noise.*"

Suffix: a syllable or syllables added to the end of a word to change or affect the meaning. For example, suffixes change *love* to *loved, lover, lovable, loveless, loving,* and *lovely. See also* Prefix *and* Root.

Supporting material: facts, figures, details, examples, reasoning, expert testimony, personal experiences, and the like, which are used to develop and explain the general ideas in a piece of writing.

Symbol: a concrete or material object that suggests or represents an abstract idea, quality, or concept. The lion is a symbol of courage; a voyage or journey can symbolize life; water suggests spirituality; dryness stands for the absence of spirituality. In Richard Selzer's "The Discus Thrower" (p. 379), the stumps of the patient's amputated legs can be seen as symbols of human helplessness and immobility.

Synonym: a word that means the same or nearly the same as another word. *Sad* is a synonym of *unhappy. Also see* Antonym.

Thesis: the main point or proposition that a writer develops and supports in an essay. The thesis is often stated early, normally in the first paragraph, to give the reader a clear indication of the essay's main idea.

Third person: the point of view in which a writer uses *he, she, it, one,* and *they* to give the reader a less-limited and more seemingly objective account than a first-person view would provide. *Also see* Point of view.

Title: the heading a writer gives to an article or essay. The title usually catches the reader's attention and indicates what the selection is about.

Tone: the attitude that a writer conveys toward the subject matter. Tone can be serious or humorous, critical or sympathetic, affectionate or hostile, sarcastic or soothing, passionate or detached—or any of numerous other attitudes.

Topic sentence: the sentence in which the main idea of a paragraph is stated. Writers often state the topic sentence first and develop the rest of the paragraph in support of this main idea. Sometimes a writer will build up to the topic sentence and place it at the end of a paragraph.

Transitions: words and expressions, such as *for example, on the other hand, next,* or *to illustrate,* that help the reader to see the connections between points and ideas.

Unity: the fitting together of all elements in a piece of writing; sticking to the point. *Also see* Coherence.

Usage: the way in which a word or phrase is normally spoken or written.

Voice: the expression of a writer's personality in his or her writing; an author's distinctive style or manner of writing.

Wordiness: the use of roundabout expressions and unnecessary words, such as "majoring in the field of journalism" instead of "majoring in journalism"; or "these socks, which are made of wool" instead of "these wool socks"; or "in this day and age" instead of "today"; or "at this point in time" instead of "now."

Writing process: the series of steps that most writers follow in producing a piece of writing. The five major stages in the writing process are finding a subject (prewriting), focusing on a main idea and mapping out an approach (planning), preparing a rough draft (writing), reworking and improving the draft (revising), and polishing style and correcting errors (editing).

Credits

Michael Agger, "Silent Responsibility" from *The New Yorker,* October 20, 2003: p. 106. Reprinted by permission of the author.

Arthur Ashe, "Send Your Children to the Libraries" from *The New York Times,* Feb. 6, 1977. Copyright © 1977 by The New York Times Co. Reprinted with permission.

Isaac Asimov, "What Is Intelligence Anyway?" Published by permission of The Estate of Isaac Asimov c/o Ralph M. Vicinanza, Ltd.

Russell Baker, "Learning to Write" from *Growing Up* by Russell Baker. Reprinted by permission of The McGraw-Hill Companies.

Dave Barry, "There Are Rules, You Know" in *Miami Herald,* January 30, 2000. Reprinted by permission.

Elizabeth Berg, "My Heroes" from *Parents Magazine,* 1992 [International Creative Management]. Reprinted by permission of International Creative Management, Inc. Copyright © 1992 by Elizabeth Berg.

Suzanne Britt, "Neat People vs. Sloppy People" from *Show and Tell* by Suzanne Britt. Copyright © 1983 by Suzanne Britt. Reprinted by permission of the author.

Bill Bryson, "The War on Drugs" from *I'm a Stranger Here Myself* (New York: Random House, 1998), pp. 151–153. Copyright © 1995 by Bill Bryson. Used by permission. Also published in *Notes from a Big Country* by Bill Bryson. Copyright © 1998 by Bill Bryson. Reprinted by permission of Doubleday Canada.

Paul Chance, "I"m Ok; You're a Bit Odd" from *Psychology Today,* July/August 1988. Reprinted with Permission from Psychology Today Magazine, Copyright © 1988 Sussex Publishers, Inc.

Wayson Choy, "I'm a Banana and Proud of It" from *The Globe and Mail,* copyright © 1997 Wayson Choy. First published in Canada by *The Globe and Mail.* Reprinted by permission of the author.

Judith Ortiz Cofer, "More Room" from *Silent Dancing: A Partial Remembrance of a Puero Rican Childhood* by Judith Ortiz Cofer. Copyright © 1990 by Judith Ortiz Cofer. Reprinted by permission of Arte Publico Press/University of Houston.

Greg Critser, "Supersize Me," from *Fatland* by Greg Critser. Copyright © 2003 by Greg Critser. Reprinted by permission of Houghton Mifflin Company. All right reserved.

Cynthia Crossen, "Couples in the U.S. Used to Marry Early, Often and Informally" from *The Wall Street Journal,* February 2004, p. B1. *Wall Street Journal.* Central Edition [Staff Produced Copy Only] by Crossen, Cynthia. Copyright 2004 by Dow Jones & Co Inc. Reproduced with permission of Dow Jones & Co Inc. in the format Textbook via Copyright Clearance Center.

Jim Dwyer, Peter Neufeld, and Barry Scheck. "When Justice Lets Us Down" from *Newsweek,* February 14, 2000, p. 59. Adapted from *Actual Innocence* by Barry Scheck, Peter Neufeld, and Jim Dwyer, copyright © 2000 by Barry Scheck, Peter Neufeld, and Jim Dwyer. Used by permission of Doubleday, a division of Random House, Inc.

David Elkind, "Types of Stress for Young People". From *All Grown Up and No Place to Go: Teenagers in Crisis,* Revised Ed. by David Elkind. Copyright © 1998 by David Elkind. Reprinted by permission of Perseus Books PLC, a member of Perseus Books, L.L.C.

Carol Fleischman, "Shopping Can Be a Challenge with Curious Onlookers, Puzzled Clerks, and a Guide Dog," from *Buffalo News,* February, 24, 1998. Reprinted by permission of the author, with gratitude to the Seeing Eye for 75 years of dedicated service. www.seeingeye.org.

Mary Ann Glendon, "For Better or for Worse?" from *The Wall Street Journal,* February 25, 2004, p. A-14. *Wall Street Journal.* Central Edition [Staff Produced Copy Only] by Mary Ann Glendon. Copyright 2004 by Dow Jones & Co Inc. Reproduced with permission of Dow Jones & Co Inc. in the format Textbook via Copyright Clearance Center.

Bob Greene, "Handled with Care". Reprinted with the permission of Scribner, an imprint of Simon & Schuster Adult Publishing Group, from *American Beat* by Bob Greene. Copyright © 1983 by Jone Deadline Enterprises, Inc.

Jack Hitt, "Navajo Code Talkers: The Century's Best Kept Secret". From *American Greats* by Robert Wilson. Copyright © 1999 by Robert Wilson. Reprinted by permission of PublicAffairs, a member of Perseus Books, L.L.C.

Langston Hughes, "Salvation" from *The Big Sea* by Langston Hughes. Copyright © 1940 by Langston Hughes. Copyright renewed 1968 by Arna Bontemps and George Houston Bass. Reprinted by permission of Hill and Wang, a division of Farrar, Straus and Giroux, LLC.

Barbara Huttmann, "A Crime of Compassion" from *Newsweek,* August 8,1983. Copyright © 1983 by Barbara Huttmann. Reprinted by permission of the author.

Tim Jones, "The Working Poor" from *Chicago Tribune,* April 25th, 2004. Chicago Tribune Company.

Garrison Keilor, "How to Write a Letter" from *We Are Still Married* (Viking Penguin, 1987), originally titled, "How to Write a Personal Letter," pp. 137–140. Copyright © 1987 by International Paper Company. Reprinted by permission of Garrison Keillor/Prairie Home Productions, LLC.

Stephen King, "Why We Crave Horror Movies". Reprinted with permission. © Stephen King. All rights reserved. Originally appeared in *Playboy* (1982).

Juleyka Lantigua, "The Latino Show" from *The Progressive,* February, 2003, pp. 32–33. Reprinted by permission of the author.

David Leibowitz, "Death Penalty Showdown" from *The Arizona Republic,* May 23, 1999. Copyright © 1999 by Phoenix Newspapers, Inc. Reprinted by permission.

Daniel Meier, "One Man's Kids" from *The New York Times Magazine* (November 1, 1987). Copyright © 1987 by The New York Times Co. Reprinted with permission.

David G. Myers, "Do We Fear the Right Things?" from *The American Psychological Society Observer,* December 2001 (vol. 14, no. 10), pp. 3, 31. Reprinted with permission.

Gloria Naylor, "What Does 'Nigger' Mean?" from *The New York Times,* 2/20/86. Reprinted by permission of Sll/Sterling Lord Literistic, Inc. Copyright 1986 by Gloria Naylor.

Emily Nelson, "Making Fake Flakes Isn't So Easy for TV: Send in the Spuds" from *The Wall Street Journal,* June 6, 2003, p. A-1. Wall Street Journal. Central Edition [Staff Produced Copy Only] by Emily Nelson. Copyright 2003 by Dow Jones Co Inc. Reproduced with permission of Dow Jones & Co Inc. in the format Textbook via Copyright Clearance Center.

Brendan O'Shaughnessy, "It a Whole New Ballgame for a Veteran Coach" from the *Chicago Tribune,* December 1, 2002, sect. 2, p. 3. Chicago Tribune Company.

Leonard Pitts, "Dearly Beloved, Why Are We Debating This?" from *Chicago Tribune,* August 5, 2003: sect. 1, p. 13. Chicago Tribune Company.

Katha Pollitt, "Adam and Steve—Together at Last" from *The Nation,* December 15, 2002, p. 9. Reprinted by permission of the author. Originally printed in *The Nation,* December 15, 2002.

William Recktenwald, "A Guard's First Night on the Job" in *St. Louis Globe Democrat,* 11/13/1978. Reprinted by permission of the author.

Rick Reilly, "The Biggest Play of His Life" from *Sports Illustrated,* May 8, 2000, p. 114. Reprinted courtesy of *Sports Illustrated:* "The Biggest Play of His Life" by Rick Reilly, May 8, 2000. Copyright © 2000. Time Inc. All rights reserved.

Mike Royko, "Jackie's Debut: A Unique Day". October 25, 1972, copyright © 1972 by Mike Royko, from *Slats Grobnik and Some Other Friends* by Mike Royko. Used by permission of Dutton, a division of Penguin Group (USA) Inc.

Kathy Seal, "The Trouble with Talent" from *Lear's,* July 1993. Copyright © 1993 by Kathy Seal. Reprinted by permission of the author.

Richard Selzer, "The Discus Thrower" from *Confessions of a Knife,* by Richard Selzer. Copyright © 1979 by David Goldman and Janet Selzer, Trustees. Reprinted by permission of Georges Borchardt, Inc., Literary Agency.

Joshua Wolf Shenk, "Ignoring the Solution" in *U.S. News & World Report,* 12/20/96. Copyright 1996 U.S. News & World Report, L.P. Reprinted with permission.

Brent Staples, " 'Walk On By': A Black Man Ponders His Power to Alter Public Space" from *Ms.,* September 1986. Copyright © 1986 by Brent Staples. Reprinted by permission of the author.

Evan Thomas, "Rain of Fire" Part IV of "The Day That Changed America" from *Newsweek,* December 31, 2001, © 2001 Newsweek, Inc. All rights reserved.

Judith Viorst, "Friends, Good Friends—and Such Good Friends". Copyright © 1977 by Judith Viorst. Originally appeared in *Redbook.* Reprinted by permission of Lescher & Lescher, Ltd. All rights reserved.

Dictionary entry for "unique" from *Webster's New World Dictionary,* 3rd Edition. Reprinted with permission of Wiley Publishing Inc., a subsidiary of John Wiley & Sons, Inc.

Jade Snow Wong, "The Fifth Chinese Daughter" from *The Immigrant Experience* by Thomas C. Wheeler, copyright © 1971 by Doubleday, a division of Random House, Inc. Used by permission of Doubleday, a division of Random House, Inc.

Index

F

G

H

I

J

K

L

Notes

Notes

Notes

Notes

Notes

Notes

Notes

Notes

Notes

Notes

Notes

Notes

Notes

Notes